D1717141

Mathematical Methods with Applications

Mathematical Methods with Applications

M. Rahman
Dalhousie University, Canada

M. Rahman
Dalhousie University, Canada

Published by

WIT Press
Ashurst Lodge, Ashurst, Southampton, SO40 7AA, UK
Tel: 44 (0) 238 029 3223; Fax: 44 (0) 238 029 2853
E-Mail: witpress@witpress.com
http://www.witpress.com

For USA, Canada and Mexico

Computational Mechanics Inc
25 Bridge Street, Billerica, MA 01821, USA
Tel: 978 667 5841; Fax: 978 667 7582
E-Mail: info@compmech.com
US site: http://www.compmech.com

British Library Cataloguing-in-Publication Data

A Catalogue record for this book is available
from the British Library

ISBN: 1-85312-847-3

Library of Congress Catalog Card Number: 99-67842

Printed in Great Britain by Bookcraft Ltd., Bath.

Contents

About the Author

Professor Matiur Rahman is a Faculty member of Dalhousie University in Halifax, Canada where he is a Professor of Applied Mathematics and Fluid Mechanics, and he has gained extensive experience in teaching mathematical methods in many universities world-wide, including Jorhat Engineering College in Assam, Imperial College in UK, Windsor and Ottawa. Originally from the foothills of the Himalayas in the Province of Assam, in India, he has gone on to become a leading scientist in his field. He received his PhD (1973) from Windsor University, Canada and his DSc (Eng) (1992) from London University, UK. His other academic achievements include a BSc (Hons) (1962) from Cotton College, India; an MSc (1964) from Gauhati University, India; a DIC (1969) from Imperial College, UK and a MPhil (1969) from London University, UK. In 1990 he was elected a Fellow of the Institute of Mathematics and its Applications (FIMA), and in 1992 the American Biographical Institute selected him as "Man of the Year".

Professor Rahman has published 12 textbooks and monographs and over 200 research papers. His graduate textbook *Water Waves: Relating Modern Theory to Advanced Engineering Applications* brought considerable accolades from the scientific community, was reviewed in the SIAM Reviews, and received an award from the Natural Sciences and Engineering Research Council of Canada (NSERC).

Professor Rahman's main research interests are in the areas of waves and hydrodynamic loading, fluid-structure interaction, diffraction and scattering of ocean waves, hydrodynamics, boundary element method, natural convection flows with diffusion and reaction, stability of tubular chemical reactors, temperature stratification in large bodies of water, and non-linear ocean waves.

The author has been associated with Wessex Institute of Technology (WIT) since 1980 and is now a member of its Governing Board. He is currently Managing Editor of the international series *Advances in Fluid Mechanics*, published by WIT Press, twenty-six volumes of which have been published since it was created in 1993. The main aim of the Series is to exchange new research findings and ideas on modern fluid mechanics between scientists and engineers.

Preface

In this computer age, classical mathematics may sometimes appear irrelevant. However, use of computer solutions without real understanding of the underlying mathematics may easily lead to serious errors, and therefore a solid understanding of the relevant mathematics is absolutely necessary.

This book has developed from a two semester course on engineering mathematics taught at DalTech, Dalhousie University (formerly the Technical University of Nova Scotia), since 1980, to undergraduate and graduate students in engineering, mathematics and computer science. Whilst the book is directed towards this audience, senior undergraduate and graduate students in physical sciences will also find many sections have considerable relevance. Its purpose is to present a clear and well-organized description of the mathematical methods required for physical problems, and their solution techniques. It will provide the reader with fundamental concepts, underlying principles, a wide range of applications and various methods of solutions of the differential equations. Many practical problems have been illustrated displaying a wide variety of solution techniques. This book is intended for applied scientists and engineers, and so the rigorous proofs of the theorems have been avoided.

The central topic of this volume is differential equations applied to physical problems. The solution techniques of differential equations are highlighted with many practical examples. In this book emphasis is placed on ordinary differential equations, operator methods, Fourier series, the convolution integral, periodic signals, energy and power spectra, the Frobenius method, Fourier and Laplace Transforms, Hankel and Z-Transforms, Green's function method, similarity techniques, method of characteristics, separation of variables method, Bessel functions and Legendre polynomials. These topics are covered with extensive use of practical examples.

As ordinary differential equations are so useful in many applied fields, such as mechanical systems, electrical circuits, chemical kinetics etc., Chapter 1 is mainly devoted to these equations with various classical solution techniques, and contains many practical illustrations. The operator method has been considered to show how the solutions of practical problems can be obtained with greater ease. Chapters 2 and 3 cover the material of the Fourier series and integrals and Laplace transforms with their applications. Special attention is given to signal analysis, which contains the concepts of energy and power spectral density and the convolution integral and theorem. Their practical usefulness is demonstrated with the help of electric circuit problems. The graphical convolution has been considered in some detail to show how the solution of a very complex problem can be obtained when an analytical solution is not readily available. Chapter 4 deals with the Frobenius type of series solutions. Chapter 5 contains the partial differential equations and their applications to practical problems. The various solution techniques, such as the method of separation of variables, the method of characteristics, D'Alembert's solution of wave equations, Laplace transforms, Fourier transforms and similarity techniques for solving nonlinear boundary value problems are demonstrated. In Chapter 6, Laplace's equations in cylindrical and spherical forms are reduced to Bessel and Legendre equations respectively, by using the method of separation of variables. The Frobenius type of series solutions of these equations is obtained and special attention is given to Bessel functions and Legendre polynomials. Many practical examples are considered in this chapter. Chapter 7 is primarily concerned with the practical solutions of some applied science and engineering problems. Chapter 8 deals with Green's functions and Chapter 9 covers integral transforms. A CD accompanies the book, which includes exercises, selected answers and an Appendix containing short tables of Z-transforms, Fourier, Hankel and Laplace transforms.

Matiur Rahman
Halifax, Canada, 2000.

Acknowledgements

Financial support provided by the Natural Sciences and Engineering Research Council of Canada is gratefully acknowledged. I am very grateful to Mr. Lance Sucharov for his interest in publishing this manuscript, and to all the staff of WIT Press. Thanks are extended to Mr. Isaac Mulolani for having checked the work. Tarjin Rahman deserves special thanks for constructing the manuscript from its original state. I am thankful to Ms. Rhonda L. Sutherland for drafting all the figures of the book for publication. Finally, the author expresses gratitude to his family for their patience and encouragement during the long preparation of this manuscript.

While it has been a joy to write this book over several years, the fruits of this labor will hopefully be in the enjoyment and benefits realized by the reader.

Chapter 1

Ordinary differential equations

The theory of ordinary differential equations plays a very important role in solving many practical problems. The beauty of differential equations lies in their richness and variety. There is always a large class of equations which exhibits a new behaviour or illustrate some counterintuitive notion. This chapter is a synopsis of exact methods for solving ordinary differential equations.

The primary purpose of this chapter is to refresh the reader with those concepts that were introduced at the undergraduate level. Although this chapter is self contained in the sense that it begins with the most elementary aspects of the subject, it would also be useful to someone who has already had some prerequisite knowledge and experience in solving elementary differential equations. In this chapter we highlight applications rather than theory. We state theorems without proving them and stress methods for obtaining analytical solutions to equations. We start our presentation with the first order ordinary differential equations before we discuss the second and higher order differential equations.

1.1 Classification of first order differential equations

A first order ordinary differential equation can be written in its simplest form

$$\frac{dy}{dx} = f(x, y). \tag{1.1}$$

Analytical solution of this equation is not always possible. However, there are some types of this equation which can be solved very easily. They are:

(a) Separable equations

(b) Homogeneous equations

(c) Exact equations

(d) Linear equations

The solution techniques of these first order ordinary differential equations come from John Bernoulli of Bale (1667–1748), the most inspiring teacher of his time and to his student, Leonard Euler (1707–1783), also of Bale.

1.1.1 Separable equations

First order separable equations can be written in the form

$$f(x)dx + g(y)dy = 0 \tag{1.2}$$

and the solution of this equation is simply

$$\int f(x)dx + \int g(y)dy = c. \tag{1.3}$$

1.1.2 Homogeneous equations

A first order differential equation in the form

$$M(x,y)dx = N(x,y)dy \tag{1.4}$$

can be said to be homogeneous if $M(x,y)$ and $N(x,y)$ are of the same total degree in variables x and y, and which can be written in the form

$$\frac{dy}{dx} = f(y/x) \quad or \quad \frac{dx}{dy} = g(x/y). \tag{1.5}$$

1.1.3 Exact equations

First order exact equations can be written in the form

$$M(x,y(x)) + N(x,y(x))\frac{dy}{dx} = \frac{d}{dx}f(x,y(x)) = 0 \tag{1.6}$$

and the solution of this equation is simply $f(x,y(x)) = c$. A necessary and sufficient condition for exactness is that $\frac{\partial M}{\partial y} = \frac{\partial N}{\partial x}$.

1.1.4 Linear equations

An equation which is in the form

$$\frac{dy}{dx} + P(x)y = Q(x) \tag{1.7}$$

where $P(x)$ and $Q(x)$ are functions of x only can be defined as the linear first order differential equation. The solution can be obtained upon multiplying through by the integrating factor $If = e^{\int P(x)dx}$ and then integrating with respect to x. The solution becomes then

$$y(x) = e^{-\int P(x)dx}\{\int Q(x)e^{\int P(x)dx}dx + c\}. \tag{1.8}$$

Example 1: Exact equation

To check that the equation $y' = \frac{x^2-y}{x+y^2}$ is exact, we identify that $M = y - x^2$ and $N = x + y^2$ and observe that $\frac{\partial M}{\partial y} = \frac{\partial N}{\partial x} = 1$. The solution of the exact equation is $y^3 + 3xy - x^3 = c$.

Example 2: Separable equation

Separable equations are exact because they have the form $M(x) + N(y)y'(x) = 0$. Thus $\frac{\partial M}{\partial y} = \frac{\partial N}{\partial x} = 0$.

Example 3: Integrating factor

The equation $(1+xy+y^2)+(1+xy+x^2)y'(x) = 0$ is not exact because of the fact that $\frac{\partial M}{\partial y} \neq \frac{\partial N}{\partial x}$. However, it becomes exact upon multiplying through by the integrating factor $If = e^{xy}$. Once this integrating factor has been guessed, it is easy to rewrite the equation as $\frac{d}{dx}\{(x+y)e^{xy}\} = 0$ and to obtain the solution $(x+y)e^{xy} = c$.

Example 4: Homogeneous equation

The equation $y' = \frac{x^2+y^2}{2xy}$ is homogeneous and so the variables can be separated upon substituting $y = v(x)x$. The solution can be obtained as $x^2 - y^2 = cx$.

Example 5: Linear equation

The equation $y' = y + 3e^x$ is a linear equation. The solution can be obtained upon multiplying through by the integrating factor $If = e^{-x}$. The solution is then obtained as $y = (3x + c)e^x$.

1.2 First order nonlinear differential equations

Although most nonlinear differential equations are too difficult to solve in close form, it is important to be able to recognize those equations which are soluble and to know the appropriate techniques for obtaining a solution. For first order equations the usual procedure is to make a substitution which converts equations into one that is either linear, or exact. We shall discuss two special types in the following:

1.2.1 Bernoulli's equation

The Bernoulli equations have the form

$$y' + P(x)y = Q(x)y^n \tag{1.9}$$

where $P(x)$ and $Q(x)$ are arbitrary functions of x and n is any number. This equation has two elementary cases: when $n = 0$ the equation is linear and when $n = 1$, the equation is separable. Dividing this equation through by y^n suggests the substitution $z(x) = \{y(x)\}^{1-n}$. The new differential equation for $z(x)$,

$$z' + (1-n)Pz = (1-n)Q \tag{1.10}$$

is soluble because it is linear in $z(x)$.

Example 6: Bernoulli's equation

The differential equation $y' = \frac{x}{x^2y^2+y^5}$ is not a Bernoulli equation in y. However, exchanging the dependent and independent variables gives $\frac{d}{dy}x(y) = xy^2 + y^5/x$ which is a Bernoulli equation in $x(n=-1)$. The solution is $x(y) = \pm\{ce^{2y^3/3} - \frac{1}{2}y^3 - \frac{3}{4}\}^{1/2}$.

1.2.2 Riccati equation

Riccati equations are quadratic in $y(x)$:

$$y' = P(x)y^2 + Q(x)y + R(x). \tag{1.11}$$

There are two elementary cases: when $P(x) = 0$, the equation is linear and when $R(x) = 0$ the equation is a Bernoulli equation. Unfortunately, apart from these special cases, there is no general technique for obtaining a solution. Nevertheless, many Riccati equations can be solved. For these equations the procedure is to guess just one solution $y(x) = u(x)$, no matter how trivial, and then to use this solution to reduce the Riccati equation to a linear equation. Specifically, one seeks a general solution of the form $y(x) = u(x) + \frac{1}{z(x)}$, where $z(x)$ is an unknown function. The resulting equation for $z(x)$ reduces to

$$-\frac{dz}{dx} = P(1+2uz) + qz \tag{1.12}$$

which is a linear differential equation in $z(x)$. This equation is solvable.

Example 7: Riccati equation

It is not hard to see that a solution of $y' = y^2 - xy + 1$ is $u(x) = x$. This is not the general solution which must contain an arbitrary integration constant and which is much too difficult to guess; it is merely a solution. Now let $y = x + \frac{1}{z(x)}$. The equation for $z(x)$ is $-\frac{dz}{dx} = 1 + xz$ the solution of which can be obtained easily as $z(x) = e^{-x^2/2}\{c - \int_0^x e^{t^2/2}dt\}$. So the general solution of the Riccati equation is $y(x) = x + \frac{e^{x^2/2}}{c - \int_0^x e^{t^2/2}dt}$.

1.3 Singular solutions of differential equations

We know that the envelope of the curves represented by the complete primitive yields a singular solution. Hence to find the singular solution, the general method is to eliminate the parameter c between $f(x,y,c)=0$, the equation of the family of curves, and $\frac{\partial f}{\partial c}=0$. As for example, if $f(x,y,c)=0$ is $y-cx-a/c=0$, then $\frac{\partial f}{\partial c}=0$ yields $-x+\frac{a}{c^2}=0$ or $c=\pm\sqrt{a/x}$. Substituting this value of c in the complete primitive gives $y=\pm 2\sqrt{ax}$ or $y^2=4ax$ which is exactly the discriminant of $c^2x-cy+a=0$. This method is equivalent to finding the locus of intersection of two curves $f(x,y,c)=0$ and $f(x,y,c+h)=0$ where h is a very small quantity. The result is called the c-discriminant of $f(x,y,c)=0$.

In the treatise of advanced differential calculus, it is found that the c-discriminant may contain the following solution curves: (a) the envelope (b) the node-locus squared and (c) the cusp-locus cubed. The envelope is a singular solution, but the node and cusp-loci are generally not solutions at all.

Example 8

Find the singular solutions of the differential equations:

$$2(\frac{dy}{dx})^2-2x^2(\frac{dy}{dx})+3xy=0.$$

Solution

Substituting $p=\frac{dy}{dx}$, the equation can be written as $2p^2-2x^2p+3xy=0$. Differentiating with respect to x yields $4p\frac{dp}{dx}-xp-2x^2\frac{dp}{dx}+3y=0$ and then multiplying throughout by x we have $4px\frac{dp}{dx}-x^2p-2x^3\frac{dp}{dx}+3xy=0$. Now eliminating y between these two equations yields $4px\frac{dp}{dx}-2x^3\frac{dp}{dx}+x^2p-2p^2=0$. After factorization, it gives

$$x^2-2p=0 \quad \text{or} \quad p=2x\frac{dp}{dx}.$$

The second equation can be written as $\frac{dx}{x}=2\frac{dp}{p}$ which integrates to $cx=p^2$ or $dy=\pm\sqrt{cx}dx$. Integrating again $3y=\pm 2c^{1/2}x^{3/2}-2c$ or $(3y+2c)^2=4cx^3$ which is a family of semi-cubical parabolas with their cusps on the axis of y.

The c-discriminant is $(3y-x^3)^2=9y^2$, i.e., $x^3(6y-x^3)=0$. The cusp-locus appears cubed, and the other factor represents the envelope. It can be easily verified that $y=x^3/6$ is a solution of the differential equation, while $x=0$ (giving $p\to\infty$) is not.

By substitution of p from $x^2-2p=0$ into the differential equation, we obtain $y=\frac{x^3}{6}$ which is the same as before and represents an envelope. This illustrates the method of obtaining singular solutions.

Remark

Next, we shall consider how to obtain the singular solutions of a differential equation directly from the equation itself, without getting the complete primitive.

Consider the differential equation

$$f(x, y, p) = a_0 p^n + a_1 p^{n-1} + \cdots + a_n = 0$$

where $a_n^{\prime} s$ are functions of x and y, and which gives n values of p for a given pair of x and y, corresponding to n curves through any point. Two of these n curves have the same tangent at all points on the locus given by eliminating p from $f(x, y, p) = 0$ and $\frac{\partial f}{\partial p}(x, y, p) = 0$.

This is the condition of existence of two repeated roots in an algebraic equation. We are thus led to the p-discriminant. Like the c-discriminant, the p-discriminant may also contain the following solution curves:

(a) the envelope

(b) the tac-locus squared

(c) the cusp-locus cubed

Only the envelope is a singular solution and not the tac-locus or the cusp-locus.

Example 9

Find the singular solutions of the differential equation

$$x^2 (\frac{dy}{dx})^2 - y(\frac{dy}{dx}) + 1 = 0.$$

Solution

Substituting $p = \frac{dy}{dx}$ in the given equation yields $x^2 p^2 - yp + 1 = 0$. Differentiating with respect to p, we have $p = \frac{y}{2x^2}$ and hence eliminating p from these equations, we obtain $y^2 = 4x^2$ which is the required envelope. This result can be obtained at once using the discriminant of p which is zero for equal roots in p, i.e., $y^2 - 4x^2 = 0$.

1.4 Orthogonal trajectories

The solution of the first order ordinary differential equation contains a single arbitrary constant. Let us consider the first order differential equation in the following form

$$y' = f(x, y). \tag{1.13}$$

The solution of which can be written in the functional form

$$F(x, y, c) = 0, \tag{1.14}$$

which represents a curve in the xy-plane for the constant c. For several numerical values of c we can generate many curves, the totality of these curves is called a parameter family of curves with c called the parameter of the family.

In many engineering applications, it is of considerable interest to determine a family of curves which are mutually perpendicular to a given family of curves. The curves of these two families are said to be mutually orthogonal, and the curves of the family to be obtained are called **orthogonal trajectories** of the given curves.

As for examples, the meridians on the earth's surface are the orthogonal trajectories of the parallels. In fluid mechanics the streamlines and the equipotential lines are orthogonal trajectories of each other.

Mathematically, the orthogonal trajectories of the solution of a first order differential equation $y' = f(x, y)$ are the solutions of the equation $y' = -1/f(x, y)$. That means both the solution sets intersect at right angles.

Example 10

The path of a particle of fluid is called a **streamline** and the orthogonal trajectories of the streamlines are called **equipotential lines.** Suppose the streamlines are given by $xy = c$. Find and graph the equipotential lines.

Solution

The streamlines are given by $xy = c$ and so $xy' + y = 0$. Therefore, $y' = -\dfrac{y}{x}$ is the differential equation of the streamlines. Thus, the differential equation for the orthogonal trajectories of the streamlines is $y' = \frac{x}{y}$ or $xdx - ydy = 0$. Upon integration we obtain $x^2 - y^2 = k^2$ which represents a rectangular hyperbola. Thus the equipotential lines are obtained from this equation for various values of k. Both the **streamlines** and the **equipotential lines** are plotted in Fig. 1.1

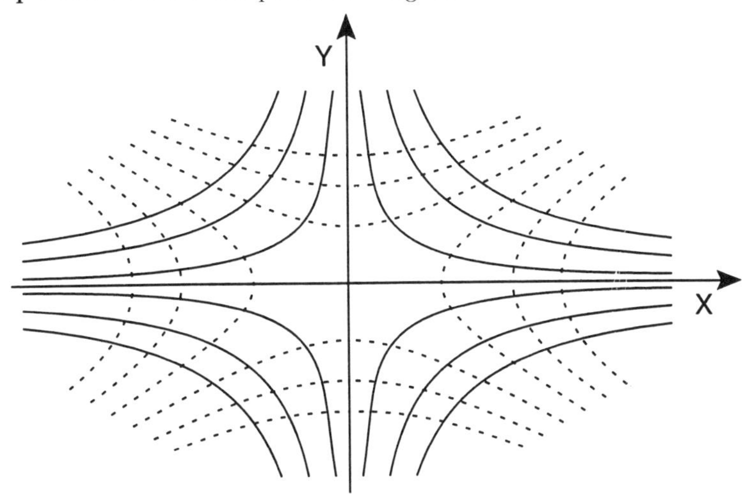

Figure 1.1: Orthogonal trajectories

1.5 Higher order linear differential equations

In this section, we shall concentrate on linear differential equations of order higher than one. The equations to be discussed are of the form

$$a_0(x)y^{(n)} + a_1(x)y^{(n-1)} + \cdots + a_{n-1}(x)y' + a_n(x)y = f(x) \tag{1.15}$$

which contains the nth order derivative of the dependent variable y with respect to the independent variable x. This equation is known as the nth order linear ordinary differential equation. Here $f(x)$ is a function of x and the a's are called the coefficients which may be functions of x or pure constants. Because of the presence of $f(x)$ on the right hand side of (1.15), it is said to be a nonhomogeneous equation and $f(x)$ is called its nonhomogeneous term. If $f(x) = 0$, (1.15) reduces to

$$a_0(x)y^{(n)} + a_1(x)y^{(n-1)} + \cdots + a_{n-1}(x)y' + a_n(x)y = 0 \tag{1.16}$$

and in this case (1.16) is said to be a homogeneous equation. In our work, we shall usually assume that the leading coefficient $a_0(x)$ is never zero. In this chapter, we shall discuss the various methods for solving the second and higher order differential equations with constant and variable coefficients.

The equation of the form

$$y'' + P(x)y' + Q(x)y = R(x) \tag{1.17}$$

where $P(x), Q(x)$ and $R(x)$ are functions of x, is known as the linear second order differential equation. If $R(x) \neq 0$, (1.17) is called a non-homogeneous equation and when $R(x) = 0$, it is said to be a homogeneous equation. In this section, we shall deal with homogeneous second order differential equations of the form

$$y'' + P(x)y' + Q(x)y = 0. \tag{1.18}$$

In the following theorem, we state an extremely important property of every homogeneous linear equation:

Theorem 1

If y_1 and y_2 are any two particular solutions of the linear homogeneous equation (1.18), then for all values of the constants c_1 and c_2

$$y = c_1y_1 + c_2y_2 \tag{1.19}$$

is also a solution of that equation.

This theorem can be extended to the nth order linear homogeneous equation such that if $y_1, y_2, \cdots, y_n$ are n particular solutions of a homogeneous linear differential equation, then for all values of the constants $c_1, c_2, \cdots, c_n$

$$y = c_1y_1 + c_2y_2 + \cdots + c_ny_n$$

is also a solution of that equation. An expression of this form is called a linear combination of the variables $y_1, y_2, \cdots, y_n$ because each term in the expression is linear.

Theorem 2

If y_1 and y_2 are two solutions of the homogeneous equation

$$y'' + P(x)y' + Q(x)y = 0$$

for which the Wronskian of the solutions y_1 and y_2 defined by

$$W(y_1, y_2) = \begin{vmatrix} y_1 & y_2 \\ y_1' & y_2' \end{vmatrix} = y_1y_2' - y_1'y_2$$

is not equal to zero, then there exist two constants c_1 and c_2 such that

$$y_3 = c_1y_1 + c_2y_2$$

is any solution of the homogeneous equation where y_1 and y_2 are two independent solutions. But, when $W(y_1, y_2) = 0$, the two solutions y_1 and y_2 are not independent but related to each other.

Note:

The Wronskian of the n solutions $y_1, y_2, \cdots, y_n$ of the nth order linear differential equation

$$y^{(n)} + P_1(x)y^{(n-1)} + \cdots + P_{n-1}y' + P_ny = 0$$

is defined by the following determinant:

$$W(y_1, y_2, \cdots, y_n) = \begin{vmatrix} y_1 & y_2 & \cdots & y_n \\ y_1' & y_2' & \cdots & y_n' \\ \cdots & \cdots & \cdots & \cdots \\ y_1^{(n-1)} & y_2^{(n-1)} & \cdots & y_n^{(n-1)} \end{vmatrix}. \tag{1.20}$$

In honour of the Polish mathematician Hoene Wronsky (1778–1853), a determinant of this kind is usually referred to as an nth order Wronskian of the solution functions $y_1, y_2, \cdots, y_n$.

Results of Theorem 2 can be extended to the nth order linear differential eqn (1.16), the solution of which can be obtained as the linear combination of n independent solutions as

$$y = c_1y_1 + c_2y_2 + \cdots + c_ny_n.$$

This solution is usually known as the complete solution or the general solution of (1.16).

To find the second solution by Abel's identity

If a solution of the homogeneous second order linear ordinary differential equation is known, then the second independent solution may be obtained from Abel's identity.

We know from Abel's identity, $W(y_1, y_2) = y_1y_2' - y_1'y_2 = K_{12}e^{-\int P dx}$ where y_1 is given and y_2 is unknown.

Then $\frac{y_1y_2'-y_1'y_2}{y_1^2} = \frac{K_{12}e^{-\int Pdx}}{y_1^2}$ or $\frac{d}{dx}(\frac{y_2}{y_1}) = \frac{K_{12}e^{-\int Pdx}}{y_1^2}$ Integrating with respect to x (choose $K_{12} = 1$) $\frac{y_2}{y_1} = \int \frac{e^{-\int Pdx}}{y_1^2}dx$ and therefore, $y_2 = y_1 \int \frac{e^{-\int Pdx}}{y_1^2}dx$. Here y_1 and y_2 are two linearly independent solutions, and knowing y_1, the second solution y_2 can be obtained from the above formula.

Example 11

Show that the two solutions y_1 and $y_1 \int \frac{e^{-\int Pdx}}{y_1^2}dx$ of $y'' + P(x)y' + Q(x)y = 0$ have a non-vanishing Wronskian.

Proof

We know the Wronskian W can be written as a determinant

$$\begin{aligned} W &= \begin{vmatrix} y_1 & y_1 \int \frac{e^{-\int Pdx}}{y_1^2}dx \\ y_1' & \frac{d}{dx}\{y_1 \int \frac{e^{-\int Pdx}}{y_1^2}dx\} \end{vmatrix} \\ &= y_1y_1' \int \frac{e^{-\int Pdx}}{y_1^2}dx + e^{-\int Pdx} - y_1y_1' \int \frac{e^{-\int Pdx}}{y_1^2}dx \\ &= e^{-\int Pdx} \neq 0. \end{aligned}$$

Hence this is the required proof.

1.6 The solution of the nonhomogeneous equations

In the previous sections, we have discussed how to solve the homogeneous second and higher order equations with constant coefficients. In this section, we shall discuss the method for constructing a complete solution of a nonhomogeneous equation

$$a_0(x)y^{(n)} + a_1(x)y^{(n-1)} + \cdots + a_{n-1}(x)y' + a_n(x)y = f(x). \tag{1.21}$$

To obtain a complete solution of (1.21) we state the following theorem.

Theorem 3

If y_p is any specific solution of (1.21) and $y_c = c_1y_1 + c_2y_2 + \cdots + c_ny_n$ is a complete solution of the corresponding homogeneous equation, then a complete solution of the nonhomogeneous equation is

$$y = y_c + y_p.$$

The method to find a complete solution (sometimes called a general solution) can be summarized as follows:

(a) Find n linearly independent particular solutions of the related homogeneous equation which have a nonvanishing Wronskian. Then obtain a complementary function by linear combination of these solutions as $y_c = c_1y_1 + c_2y_2 + \cdots + c_ny_n$.

(b) Find one particular solution y_p of the nonhomogeneous equation itself.

(c) A complete solution of the given nonhomogeneous equation is then given by

$$\begin{aligned} y &= y_c + y_p \equiv \text{complementary function} + \text{particular integral} \\ &= c_1y_1 + c_2y_2 + \cdots + c_ny_n + y_p. \end{aligned} \tag{1.22}$$

In the following sections, we shall illustrate with examples, different methods to obtain a complete solution of the nonhomogeneous differential equation.

The following sections describe the solution techniques of a nonhomogeneous linear differential equation by the methods of

(a) the variation of parameters

(b) the method of differential operators.

Remark

The differential equation of order greater than 2 has the form

$$a_0y^{(n)} + a_1y^{(n-1)} + \cdots + a_{n-1}y' + a_ny = 0 \quad n > 2, a_0 \neq 0. \tag{1.23}$$

The solution of this equation parallels the second order case in all significant details. Let us consider a trial solution $y = Ae^{mx}$. Equation (1.23) becomes an auxiliary equation

$$a_0m^n + a_1m^{n-1} + \cdots + a_{n-1}m + a_n = 0 \tag{1.24}$$

which can be obtained in a specific problem simply by replacing each derivative by the corresponding power of m. The degree of this algebraic equation will be the same as the order of the differential equation (1.21). We shall demonstrate the solution technique by specific problems.

1.7 The method of variation of parameters

It is ordinarily impossible to find a particular integral of a nonhomogeneous linear differential equation by inspection or by trial and error. In this section, however, we shall discuss a method known as the method of variation of parameters whenever a complementary function of a differential equation is known. This method, which was first discovered by J. L. Lagrange (1736–1813), appears to be the most powerful method above all others, including the method of undetermined coefficients and the method of operators. In fact, if one can master this method thoroughly, he can do well without the method of undetermined coefficients and the method of operators.

We will discuss the method of variation of parameters for the second order linear differential equation, and then we generalize the method for higher order equations.

Consider the second order nonhomogeneous differential equation in the following form:

$$y'' + P(x)y' + Q(x)y = R(x). \tag{1.25}$$

Suppose y_1 and y_2 are the two particular solutions for the homogeneous equation $y'' + P(x)y' + Q(x)y = 0$. Then the complementary function can be written as

$$y = Ay_1 + By_2 \tag{1.26}$$

where A and B are arbitrary constants.

Assume that A and B in (1.26) are not constant, but instead are functions of x, denoted by $A(x)$ and $B(x)$ respectively. Then consider that

$$y = A(x)y_1 + B(x)y_2 \tag{1.27}$$

is the complete solution of (1.25). Since the method assumes that the quantities A and B vary, the method is generally known as the method of variation of parameter.

It is obvious that, since we have two unknowns to be determined, we usually need two conditions. One of these conditions arises from the fact that the assumed solution must satisfy the differential equation and the second condition is at our disposal.

Thus, differentiating (1.27) once,

$$y' = A(x)y_1' + B(x)y_2' + (A'(x)y_1 + B'(x)y_2). \tag{1.28}$$

We realize that further differentiation would introduce second derivatives of the unknown variables. We choose a condition which simplifies (1.28)

$$A'(x)y_1 + B'(x)y_2 = 0. \tag{1.29}$$

Then (1.28) reduces to

$$y' = A(x)y_1' + B(x)y_2'.$$

Differentiating again

$$y'' = A(x)y_1'' + B(x)y_2'' + A'(x)y_1' + B'(x)y_2'. \tag{1.30}$$

Substituting the values of y, y' and y'' into (1.25), we obtain

$$(Ay_1'' + By_2'' + A'y_1' + B'y_2') + P(Ay_1' + By_2') + Q(Ay_1 + By_2) = R(x).$$

Simplifying the above yields

$$A(y_1'' + Py_1' + Qy_1) + B(y_2'' + Py_2' + Qy_2) + A'y_1' + B'y_2' = R(x). \tag{1.31}$$

But, y_1 and y_2 are the solutions of the homogeneous equation, and therefore,

$$\begin{aligned} y_1'' + Py_1' + Qy_1 &= 0 \\ y_2'' + Py_2' + Qy_2 &= 0 \end{aligned}$$

and hence (1.30) reduces to

$$A'y_1' + B'y_2' = R(x). \tag{1.32}$$

Now we obtain two equations with two unknowns A' and B' as follows:

$$\begin{aligned} A'y_1 + B'y_2 &= 0 \\ A'y_1' + B'y_2' &= R(x). \end{aligned}$$

Using Cramer's rule we obtain the solutions for A' and B' as

$$A' = \frac{-y_2R}{y_1y_2' - y_1'y_2} = \frac{-y_2R}{W(y_1, y_2)}$$

and

$$B' = \frac{y_1R}{y_1y_2' - y_1'y_2} = \frac{y_1R}{W(y_1, y_2)}.$$

Integrating these two equations, we obtain

$$\begin{aligned} A(x) &= -\int \frac{y_2R}{W}dx + c_1 \\ B(x) &= +\int \frac{y_1R}{W}dx + c_2, \end{aligned} \tag{1.33}$$

where c_1 and c_2 are two arbitrary constants.
Hence the complete solution is given by

$$\begin{aligned} y &= A(x)y_1 + B(x)y_2 \\ &= c_1y_1 + c_2y_2 + (-y_1\int \frac{y_2R}{W}dx) + (y_2\int \frac{y_1R}{W}dx). \end{aligned} \tag{1.34}$$

The first two terms on the right hand side of (1.33) are the complementary function and the third and fourth terms are the particular integral.

Example 12

Find a complete solution of the differential equation $y'' + y = \sec x$ by the method of variation of parameters.

Solution

A complementary function of the given equation is

$$y = A\cos x + B\sin x.$$

Assume that $y = A(x)\cos x + B(x)\sin x$.
Then by the method of variation of parameters, we have the following two equations to be solved for A and B.

$$\begin{aligned} A'\cos x + B'\sin x &= 0 \\ -A'\sin x + B'\cos x &= \sec x. \end{aligned}$$

Solving for A' and B' by Cramer's rule:

$$\begin{aligned} A' &= -\frac{\sin x}{\cos x} \\ B' &= 1. \end{aligned}$$

Then, integrating with respect to x, $A = \ell n(\cos x) + c_1$ and $B = x + c_2$ where c_1 and c_2 are two arbitrary constants.

Therefore, a complete solution is,

$$\begin{aligned} y &= (c_1 + \ell n(\cos x))\cos x + (c_2 + x)\sin x \\ &= \underbrace{c_1\cos x + c_2\sin x}_{I} + \underbrace{\cos x\,\ell n(\cos x) + x\sin x}_{II} \end{aligned}$$

where part I is a complementary function and part II is a particular integral.

1.8 The method of differential operator

We have seen in the previous section how to solve nonhomogeneous linear differential equations with variable and constant coefficients by the method of variation of parameters. The first method can handle almost all the nonhomogeneous functions occurring on the right hand side of the differential equation. In this section, we shall discuss the so-called operator method for solving nonhomogeneous linear equations with constant coefficients. However, the second method can be applied only when the right hand side contains functions of exponential, polynomial and trigonometric types. Although this method is not as powerful as the method of variation of parameters, it has some advantages. Using this method, a problem can be solved very quickly in a less tedious manner. It is worth mentioning here that the method of operators yields the incentive of the discovery of the Laplace transform method to be discussed later.

Consider the nth order nonhomogeneous linear differential equation with constant coefficients in the following manner

$$a_0\frac{d^n y}{dx^n} + a_1\frac{d^{n-1}y}{dx^{n-1}} + \cdots + a_{n-1}\frac{dy}{dx} + a_n y = f(x) \tag{1.35}$$

where $a_0, a_1 \cdots a_n$ are all constants.
Now defining the differential operators as $D = \frac{d}{dx}$, $D^2 = \frac{d^2}{dx^2}$, $\cdots$ $D^n = \frac{d^n}{dx^n}$, we can very compactly write (1.34) in the form

$$\phi(D)y = f(x) \tag{1.36}$$

where $\phi(D)$ is a linear polynomial operator in D and is given by

$$\phi(D) = a_0 D^n + a_1 D^{n-1} + \cdots + a_{n-1} D + a_n. \tag{1.37}$$

We know that the operations performed in an ordinary algebra are based upon three laws:

(a) The distributive law
$a(u + v) = au + av.$

(b) The commutative law
$ab = ba.$

(c) The index law
$a^m.a^n = a^{m+n}.$

The differential operator D satisfies (a) and (c) because $D(u + v) = Du + Dv$ and $D^m x D^n u = D^{m+n} u$ where m and n are positive integers. It is to be noted here that for negative powers, the index law ceases to hold. With regard to (b), we see that $Db \neq bD$ where b is any quantity. Also, $D(cu) = cDu$ is true if c is a constant but not if c is a variable. Thus D satisfies the fundamental laws of algebra except that it does not satisfy the commutative law (b) as explained above.

Now, treating eqn (1.35) as an algebraic equation, we can solve for y

$$y = \frac{1}{\phi(D)} f(x) \tag{1.38}$$

where $\frac{1}{\phi(D)}$ represents an operator to be performed on $f(x)$, and can be recognized later as an inverse operator.

For simplicity, let us consider the equation

$$Dy = f(x) \quad , \quad \text{where } \phi(D) = D.$$

Then the particular integral can be obtained as

$$y_p = \int f(x)dx. \tag{1.39}$$

But, by the definition of (1.37)

$$y_p = \frac{1}{D} f(x). \tag{1.40}$$

Since (1.38) and (1.39) are the same, it is natural to make the definition $\frac{1}{D} = \int$, $\frac{1}{D^2} = \iint$ $\cdots$ $\frac{1}{D^n} = \iint \cdots \int$ where the last one is n-fold integration. Thus, operators such as $\frac{1}{D}$, $\frac{1}{D^2}$, etc., are called the Inverse Operators.

We can illustrate the inverse operator concept by considering another simple example

$$(D-p)y = f(x). \tag{1.41}$$

Solving algebraically

$$y_p = \frac{1}{D-p}f(x) \tag{1.42}$$

where $\frac{1}{D-p}$ is operating upon $f(x)$. Then, using the integrating factor method, the solution of (1.40) yields as

$$y_p = e^{px}\int e^{-px}f(x)dx. \tag{1.43}$$

Thus, comparing (1.41) and (1.42), we have

$$\frac{1}{D-p}f(x) = e^{px}\int e^{-px}f(x)dx. \tag{1.44}$$

By observation as in (1.43) the explicit result is:

$$\begin{aligned}
\frac{1}{D-p}f(x) &= \frac{1}{D-p}e^{px}.e^{-px}f(x) \\
&= e^{px}\frac{1}{D+p-p}e^{-px}f(x) \\
&= e^{px}\frac{1}{D}e^{-px}f(x) \\
&= e^{px}\int e^{-px}f(x)dx.
\end{aligned}$$

We have obviously used a theorem to obtain this result. In the following, we shall establish three very important theorems.

Theorem 4

Show that $\phi(D)e^{ax} = e^{ax}\phi(a)$.

Proof:

Since $De^{ax} = ae^{ax}$, $D^2e^{ax} = a^2e^{ax}$ and $D^ne^{ax} = a^ne^{ax}$, hence

$$\begin{aligned}
\phi(D)e^{ax} &= (a_0D^n + a_1D^{n-1} + \cdots + a_{n-1}D + a_n)e^{ax} \\
&= (a_0a^n + a_1a^{n-1} + \cdots + a_{n-1}a + a_n)e^{ax} \\
&= \phi(a)e^{ax}
\end{aligned}$$

as asserted.

Corollary 1

It can be easily shown that

$$\frac{1}{\phi(D)}e^{ax} = \frac{e^{ax}}{\phi(a)} \quad \text{if} \quad \phi(a) \neq 0.$$

This is in essence a particular integral of the equation

$$\phi(D)y = e^{ax}.$$

Note:

If $\phi(a) = 0$, then the following theorem must be used.

Theorem 5

Show that $\phi(D)(e^{ax}V) = e^{ax}\phi(D+a)V$, where V is a function of x.

Proof

We know

$$\begin{aligned}
D(e^{ax}V) &= (De^{ax})V + e^{ax}DV = e^{ax}(D+a)V \\
D^2(e^{ax}V) &= D[e^{ax}(D+a)V] \\
&= e^{ax}D(D+a)V + ae^{ax}(D+a)V \\
&= e^{ax}(D+a)^2V \\
D^3(e^{ax}V) &= D[e^{ax}(D+a)^2V] \\
&= e^{ax}D(D+a)^2V + ae^{ax}(D+a)^2V \\
&= e^{ax}(D+a)^3V \\
&\vdots \\
D^n(e^{ax}V) &= e^{ax}(D+a)^nV.
\end{aligned}$$

Therefore

$$\begin{aligned}
\phi(D)(e^{ax}V) &= [a_0D^n + a_1D^{n-1} + \cdots + a_{n-1}D + a_n](e^{ax}V) \\
&= e^{ax}[a_0(D+a)^n + a_1(D+a)^{n-1} + \cdots \\
&\quad + a_{n-1}(D+a) + a_n]V \\
&= e^{ax}\phi(D+a)V.
\end{aligned}$$

Hence this is the required proof.

Corollary 2

A particular integral of the equation $\phi(D)y = e^{ax}V$ is then

$$y_p = \frac{1}{\phi(D)}(e^{ax}V) = e^{ax}\frac{1}{\phi(D+a)}V.$$

Theorem 6

Show that

$$\phi(D^2)\left\{\begin{array}{c}\cos ax\\ \sin ax\end{array}\right\} = \phi(-a^2)\left\{\begin{array}{c}\cos ax\\ \sin ax\end{array}\right\}$$

where

$$\phi(D^2) = a_0D^{2n} + a_1D^{2n-2} + \cdots + a_{n-1}D^2 + a_n.$$

Proof

It can be proven as follows:

$$\begin{aligned}
D^2\left\{\begin{array}{c}\cos ax\\ \sin ax\end{array}\right\} &= (-a^2)\left\{\begin{array}{c}\cos ax\\ \sin ax\end{array}\right\}\\
D^4\left\{\begin{array}{c}\cos ax\\ \sin ax\end{array}\right\} &= (-a^2)D^2\left\{\begin{array}{c}\cos ax\\ \sin ax\end{array}\right\} = (-a^2)^2\left\{\begin{array}{c}\cos ax\\ \sin ax\end{array}\right\}\\
&\vdots\\
D^{2n}\left\{\begin{array}{c}\cos ax\\ \sin ax\end{array}\right\} &= (-a^2)^n\left\{\begin{array}{c}\cos ax\\ \sin ax\end{array}\right\}.
\end{aligned}$$

Thus

$$\begin{aligned}
\phi(D^2)\left\{\begin{array}{c}\cos ax\\ \sin ax\end{array}\right\} &= (a_0D^{2n} + a_1D^{2n-2} + \cdots + a_{n-1}D^2 + a_n)\left\{\begin{array}{c}\cos ax\\ \sin ax\end{array}\right\}\\
&= \{(a_0(-a^2)^n + a_1(-a^2)^{n-1} + \cdots + a_n\}\left\{\begin{array}{c}\cos ax\\ \sin ax\end{array}\right\}\\
&= \phi(-a^2)\left\{\begin{array}{c}\cos ax\\ \sin ax\end{array}\right\}.
\end{aligned}$$

Corollary 3

A particular integral of the equation $\phi(D^2)y = \left\{\begin{array}{c}\cos ax\\ \sin ax\end{array}\right\}$ is then given by

$$\begin{aligned}
y_p &= \frac{1}{\phi(D^2)}\left\{\begin{array}{c}\cos ax\\ \sin ax\end{array}\right\}\\
&= \frac{1}{\phi(-a^2)}\left\{\begin{array}{c}\cos ax\\ \sin ax\end{array}\right\}.
\end{aligned}$$

Note:

The method of operators is very powerful with the differential equation with constant coefficients and with $f(x)$ of the type k, x^n, e^{ax}, $\cos ax$, $\sin ax$, or any combination with these functions.

Example 13

Use the method of operators to find a complete solution of the equation $(D-a)^a y = a^x$, where a is a positive integer.

Solution

The auxiliary equation is $(m-a)^a = 0$ which has $m = a, a, \cdots, a$ (a equal roots). Therefore, the complementary function is

$$y_c = (a_0 + a_1 x + a_2 x^2 + \cdots + a_{a-1} x^{a-1}) e^{ax}.$$

By the operator method, a particular integral is (by theorem 4, corollary 1)

$$\begin{aligned} y_p &= \frac{1}{(D-a)^a}.a^x \\ &= \frac{1}{(D-a)^a}.e^{x \ell n a} \\ &= \frac{a^x}{(\ell n a - a)^a}. \end{aligned}$$

Therefore, a complete solution is

$$y = (a_0 + a_1 x + \cdots + a_{a-1} x^{a-1}) e^{ax} + \frac{a^x}{(\ell n a - a)^a}.$$

1.9 Euler-Cauchy differential equations

We have discussed in the previous sections the solution techniques of non-homogeneous linear differential equations with constant coefficients. However, there is a certain class of linear differential equations with variable coefficients which can be solved by transforming them into linear differential equations with constant coefficients. This is the so-called Euler-Cauchy differential equation and has the following form:

$$a_0 x^n y^{(n)} + a_1 x^{n-1} y^{(n-1)} + \cdots + a_{n-1} x y' + a_n y = f(x)$$

or

$$[a_0 (xD)^n + a_1 (xD)^{n-1} + \cdots + a_{n-1}(xD) + a_n] y = f(x) \tag{1.45}$$

in which $a_0, a_1, \cdots, a_n$ are constants and the coefficient of each derivative is a constant multiple of corresponding power of the dependent variable.

We shall demonstrate the method of solving this type of differential equation with some examples. We shall see that, by the change of the independent variable defined by $x = e^z$ or $z = \ell n x, \quad x \neq 0$, the above equation (1.44) can easily be transformed into a linear equation with constant coefficients.

Note

- The transformation $x = e^z$ will transfer (1.44) into the following linear differential equation with constant coefficients

$$\begin{aligned}[a_0\theta(\theta-1)(\theta-2)\cdots(\theta-n+1) \quad + \quad & \cdots + a_{n-2}\theta(\theta-1) \\ & + a_{n-1}\theta + a_n]y = f(e^z)\end{aligned}$$

where

$$\begin{aligned} xD &= \theta \\ x^2D^2 &= \theta(\theta-1) \\ x^3D^3 &= \theta(\theta-1)(\theta-2) \\ &\vdots \\ x^nD^n &= \theta(\theta-1)(\theta-2)\cdots(\theta-n+1). \end{aligned}$$

- The equation

$$[a_0(a+bx)^nD^n + a_1(a+bx)^{n-1}D^{n-1} + \cdots + a_{n-1}(a+bx)D + a_n]y = f(x)$$

can be reduced to a linear equation with constant coefficient by putting

$$a + bx = e^z \quad , \quad z = \ell n(a+bx)$$

giving

$$D = \frac{d}{dx} = \frac{dz}{dx}\frac{d}{dz} = \frac{b}{a+bx}\frac{d}{dz} = \frac{b}{a+bx}\theta.$$

1.10 Applications to practical problems

We shall demonstrate in this section an application of linear differential equations with constant coefficients in areas of chemical reactors. Because of its usefulness to engineers and scientists, we shall devote an entire chapter (Chapter 7) to the comprehensive applications in engineering problems.

Applications to chemical reactors

Consider the tubular chemical flow reactor as shown in Fig. 1.2 below. For simplicity, let us assume a first order chemical reaction $A \to B$ is occurring in the reaction-chamber, $0 \le x \le \ell$.

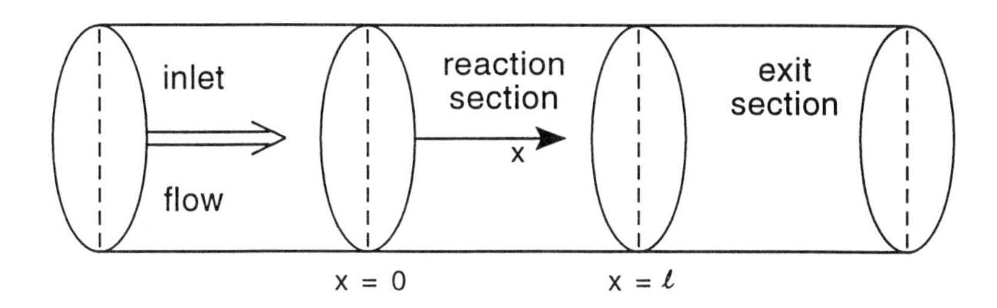

Figure 1.2: Tubular chemical reactor

The simplest type of model in this kind of problem is to assume that there is a flat velocity profile, no radial dispersion and no axial dispersion and the flow is turbulent. With steady flow, the governing equation for this system can be written as (see Crank's *Mathematics of Diffusion,* Oxford University Press, Oxford 1967).

$$u\frac{dC}{dx} = -kC \tag{1.46}$$

where u is the uniform velocity of the fluid, C is the concentration of the reactant, k is the rate of reaction which is a constant and x is the axial coordinate. This simple differential equation can be solved using the boundary condition that, at $x = 0$, the reactant concentration is the feed concentration given by c_0. Thus, $c(0) = c_0$ Solving this equation subject to the boundary condition at $x = 0$ yields $c = c_0 e^{-\frac{kx}{u}}$ which is an exponentially decaying solution with respect to the reactor length x. However, at the exit of the reactor, at $x = \ell$, it becomes $\frac{c}{c_0} = e^{-\frac{k\ell}{u}}$ which represents that this amount of reactant is still left to be converted by the chemical reaction.

Next, to see the importance of axial dispersion, the problem will be solved again taking into consideration that the axial dispersion takes place in the system. The general mass balance now reduces to

$$u\frac{dc}{dx} = \alpha\frac{d^2c}{dx^2} - kc \tag{1.47}$$

where α is the diffusion coefficient which is a measure of turbulent diffusivity.

This is a second order differential equation, and two integration constants result, requiring two boundary conditions. The exact details of the mathematics of the boundary conditions may be found in Rahman (1969). For the simple case of no dispersion in the inlet section $(x < 0)$, a flux balance at $x = 0$ is $uc_0 = [uc - \alpha\frac{dc}{dx}]_{x=0+}$

where c_0 is again the feed concentration. Thus, at $x = 0$ $c_0 = [c - \frac{\alpha}{u}\frac{dc}{dx}]_{x=0+}$. A similar treatment at $x = \ell$ leads to $\frac{dc}{dx}]_{x=\ell} = 0$. The differential equation can be written as

$$\alpha\frac{d^2c}{dx^2} - u\frac{dc}{dx} - kc = 0. \tag{1.48}$$

The auxiliary equation is given by $\alpha m^2 - um - k = 0$ and therefore, the two roots are $m_1, m_2 = \frac{u}{2\alpha}[1 \pm a]$ where $a = \sqrt{1 + \frac{4\alpha k}{u^2}}$. Then, the solution is

$$c = Ae^{\frac{u}{2\alpha}(1+a)x} + Be^{\frac{u}{2\alpha}(1-a)x} \tag{1.49}$$

where A and B are two integration constants.

Differentiating the above solution with respect to x yields

$$\frac{dc}{dx} = \frac{u}{2\alpha}(1+a)Ae^{\frac{u}{2\alpha}(1+a)x} + \frac{u}{2\alpha}(1-a)Be^{\frac{u}{2\alpha}(1-a)x}. \tag{1.50}$$

Using the boundary conditions at $x = 0$ and $x = \ell$, we obtain

$$c_0 = A + B - \frac{1+a}{2}A - \frac{1-a}{2}B$$

or $(1-a)A + (1+a)B = 2c_0$ and

$$(1+a)e^{\frac{ua\ell}{2\alpha}}A + (1-a)e^{-\frac{ua\ell}{2\alpha}}B = 0.$$

Using Cramer's rule, we obtain

$$A = \frac{\begin{vmatrix} 2c_0 & (1+a) \\ 0 & (1-a)e^{-\frac{ua\ell}{2\alpha}} \end{vmatrix}}{\begin{vmatrix} 1-a & 1+a \\ (1+a)e^{\frac{ua\ell}{2\alpha}} & (1-a)e^{-\frac{ua\ell}{2\alpha}} \end{vmatrix}} = \frac{2c_0(1-a)e^{-\frac{ua\ell}{2\alpha}}}{(1-a)^2e^{-\frac{ua\ell}{2\alpha}} - (1+a)^2e^{\frac{ua\ell}{2\alpha}}}$$

and

$$B = \frac{-2c_0(1+a)e^{+\frac{ua\ell}{2\alpha}}}{(1-a)^2e^{-\frac{ua\ell}{2\alpha}} - (1+a)^2e^{\frac{ua\ell}{2\alpha}}}. \tag{1.51}$$

Thus, the solution is

$$\frac{c}{c_0} = \frac{2(1-a)e^{-\frac{u}{2\alpha}(x+ax-a\ell)} - 2(1+a)e^{\frac{u}{2\alpha}(x-ax+a\ell)}}{(1-a)^2e^{-\frac{ua\ell}{2\alpha}} - (1+a)^2e^{\frac{ua\ell}{2\alpha}}}. \tag{1.52}$$

Therefore, the reactant concentration at $x = \ell$ is given by

$$\frac{c}{c_0} = \frac{4a\,exp(\frac{u\ell}{2\alpha})}{(1+a)^2exp(\frac{ua\ell}{2\alpha}) - (1-a)^2exp(-\frac{ua\ell}{2\alpha})}. \tag{1.53}$$

To see the effect of axial dispersions, we expand eqn (1.52) in series for small values of $(\frac{\alpha}{u\ell})$ or small deviations from the plug flow, with the result

$$\frac{c_p}{c_0} = 1 + (\frac{k\ell}{u})(\frac{\alpha}{u\ell}) \tag{1.54}$$

where c_p = exit reactant concentration for same reactor but in plug flow.

Chapter 2

Fourier series and Fourier transform

2.1 Introduction

In the eighteenth century, Jean B.J. Fourier (1768–1830), a French mathematician and physicist, came across a heat conduction problem while researching in heat transfer involving Laplace's equation in two dimensions. He conceived that a periodic function can be expressed as the sum of cosine and sine functions such that they form an orthogonal set of functions if properly defined in some interval. The development of Fourier series is the creation of Jean Fourier in the early eighteenth century. The knowledge of Fourier series is essential in the study of signal analysis. The practical applications of these methods often furnish valuable insights into the signal design aspects of communication systems. Not only that the Fourier series and transforms have enormous applications in many other branches of applied science and technology.

The most useful method of signal representation for any given situation depends upon the type of signal being considered. In electrical engineering, we generally use two kinds of signals, namely, power signals and energy signals. The definition of the energy signal, E, is that the energy dissipated by the time signal $f(t)$ during a time interval $-p < t < p$ where p is a constant parameter is

$$E = \lim_{p\to\infty} \int_{-p}^{p} |f(t)|^2 dt = \int_{-\infty}^{\infty} |f(t)|^2 dt < \infty. \tag{2.1}$$

This signal $f(t)$ has the finite energy and is called the energy signal.

Also, we define the power signal $f(t)$ during the time interval $-p < t < p$ as

$$P_{av} = \lim_{p\to\infty} \frac{1}{2p} \int_{-p}^{p} |f(t)|^2 dt < \infty \tag{2.2}$$

such that $0 < P_{av} < \infty$. This signal $f(t)$ has the finite average power and is called a power signal. For example, a Gaussian pulse $f(t) = e^{-t^2}$ defined in $-\infty < t < \infty$ is an energy signal whereas a function $f(t) = e^{i2\pi t}$ defined over the same interval is a power signal.

These functions can be periodic or nonperiodic. A periodic function can be expressed as the Fourier series and a nonperiodic function can be expressed as a Fourier integral. We will examine these notions one after another. In the following we will define a periodic function.

2.2 Definition of a periodic function

A function f is periodic if, and only if, there exists a positive parameter $2p$ such that for every t in the domain of f, $f(t+2p) = f(t)$. The parameter $2p$ is called a period of f.

Thus with this definition we can show that if f is periodic, then

$$f(t) = f(t+2p) = f(t+4p) = f(t+6p) = \cdots = f(t+2np) = \cdots$$

where $n = 1, 2, 3, \cdots$ and hence $2p, 4p, 6p, 8p, \cdots, 2np$ are also periods of f. Here $2p$ is the smallest of all the periods and is usually defined as the fundamental period of f. As for example, the fundamental period of the function $f(t) = e^{in\pi t/p}$ defined on $-\infty < t < \infty$ where $p > 0$, and n is a positive integer is $2p/n$. It is to be noted that some periodic functions do not have any fundamental period. A typical periodic function is depicted below.

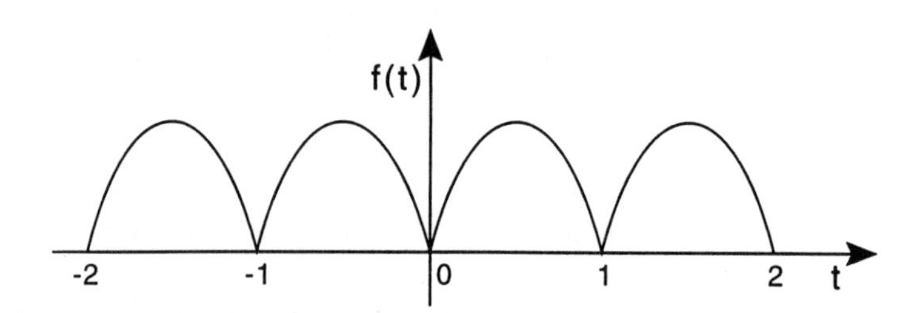

Figure 2.1: A typical periodic function

2.3 Fourier series and Fourier coefficients

Let $f(t)$ satisfy the following conditions:

(a) $f(t)$ is defined in the interval $d < t < d + 2p$.

(b) $f(t)$ and $f'(t)$ are sectionally continuous in the interval $d < t < d + 2p$.

(c) $f(t + 2p) = f(t)$, i.e., $f(t)$ is periodic with period $2p$.

(d) $f(t)$ is absolutely integrable, i.e., $\int_d^{d+2p} |f(t)|dt < \infty$. A weak form will be $\int_d^{d+2p} |f(t)|^2 dt < \infty$.

Then at every point of continuity, $f(t)$ can be expressed as an infinite series of cosine and sine, and is given by

$$f(t) = \frac{a_0}{2} + \sum_{n=1}^{\infty}\{a_n \cos\frac{n\pi t}{p} + b_n \sin\frac{n\pi t}{p}\} \tag{2.3}$$

where

$$a_n = \frac{1}{p}\int_d^{d+2p} f(t)\cos\frac{n\pi t}{p}dt \qquad n = 0, 1, 2, 3, \cdots \tag{2.4}$$

$$b_n = \frac{1}{p}\int_d^{d+2p} f(t)\sin\frac{n\pi t}{p}dt \qquad n = 1, 2, 3, \cdots \tag{2.5}$$

Here (2.3) is known as Fourier series and a_n and b_n are called Fourier coefficients.

At a point of discontinuity, the left hand side of (2.3) should be modified to yield

$$f(t) = \frac{1}{2}[f(t+0) + f(t-0)]. \tag{2.6}$$

The above conditions (a), (b), (c) and (d) together are known as the DIRICHLET conditions and are sufficient (but not necessary) conditions for convergence of Fourier series.

In general if $f(t)$ is defined in the interval $d < t < d + 2p$, $f(t)$ and $f'(t)$ are sectionally continuous in that interval, then $f(t)$ can be expressed as an infinite series (generalized Fourier series) of a set of orthogonal functions $\{\phi_n(t)\}$, $n = 0, 1, 2\ldots$

$$f(t) = \sum_{n=0}^{\infty} f_n\phi_n(t)$$

where f_n is the coefficient of the generalized Fourier series and $\{\phi_n\}$ satisfies the orthogonal property

$$\int_d^{d+2p} \phi_n(t)\phi_n(t)dt = \begin{cases} 0 & m \neq n \\ \lambda_n & m = n \end{cases}$$

where λ_n is a constant which is different from zero.

Thus any periodic function with fundamental period $2p$ can be expressed as in (2.3) with the assumption that the infinite series converges. To see the mathematical development of how the Fourier coefficients a_n and b_n could be obtained from the formulae (2.4) and (2.5), we follow the following steps:

To find a_n, we multiply both sides of (2.3) by $\cos\frac{n\pi t}{p}$, $m = 0, 1, 2, \cdots$ and integrate with respect to t from $t = d$ to $t = d + 2p$ assuming that the series (2.3) can be term by term integrable.

Because of the orthogonality property, namely,

$$\int_d^{d+2p} \cos\frac{m\pi t}{p}\cos\frac{n\pi t}{p}dt = \begin{cases} 0 & m \neq n \\ p & m = n > 0 \end{cases}$$

$$\int_d^{d+2p} \cos\frac{m\pi t}{p}\sin\frac{n\pi t}{p}dt = 0 \text{ for all integers } m, n$$

only one term from the infinite numbers of terms will contribute to the integral which corresponds to $m = n$ and accordingly we have

$$\int_d^{d+2p} f(t) \cos \frac{n\pi t}{p} dt = a_n \int_d^{d+2p} \cos^2 \frac{n\pi t}{p} dt = a_n p.$$

Therefore

$$a_n = \frac{1}{p} \int_d^{d+2p} f(t) \cos \frac{n\pi t}{p} dt \qquad n = 1, 2, 3, \cdots$$

If we multiply (2.3) by $\cos 0t = 1$ and proceed in the same way, we find that (2.4) is also valid for $n = 0$. The notation $\frac{a_0}{2}$ for the first term in (2.3) is chosen to make (2.4) valid for all n.

Similarly, if we multiply (2.3) by $\sin \frac{m\pi t}{p}$, $m = 1, 2, 3, \cdots$ and integrate with respect to t from $t = d$ to $t = d + 2p$ and use the above orthogonality property together with the following

$$\int_d^{d+2p} \sin \frac{m\pi t}{p} \sin \frac{n\pi t}{p} dt = \begin{cases} 0 & \text{for } m \neq n \\ p & \text{for } m = n \end{cases}$$

we find that

$$\int_d^{d+2p} f(t) \sin \frac{n\pi t}{p} dt = b_n \int_d^{d+2p} \sin^2 \frac{n\pi t}{p} dt = b_n p.$$

Therefore

$$b_n = \frac{1}{p} \int_d^{d+2p} f(t) \sin \frac{n\pi t}{p} dt \qquad n = 1, 2, 3, \cdots$$

It is noted here that corresponding to $n = 0$, $\sin \frac{n\pi t}{p} = 0$, and that is why b_0 coefficient does not exist in the series. The Fourier series of a function $f(t)$ is defined whenever all the integrals in (2.4) and (2.5) have meaning.

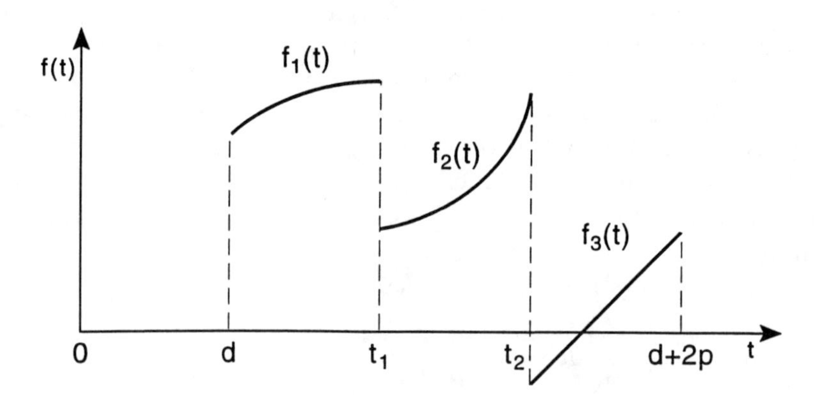

Figure 2.2: A piecewise continuous periodic function

This is certainly the case if $f(t)$ is continuous on the interval $d \leq t \leq d+2p$. However, the integrals also have meaning when $f(t)$ has jump discontinuities, as shown in Fig. 2.2. Thus the mathematical definition is:

$$f(t) = \begin{cases} f_1(t) & d < t < t_1 \\ f_2(t) & t_1 < t < t_2 \\ f_3(t) & t_2 < t < d + 2p. \end{cases}$$

This type of function is defined as the piecewise continuous function. Hence it is necessary to write the Fourier coefficients as

$$\begin{aligned} a_n &= \frac{1}{p}\int_d^{d+2p} f(t)\cos\frac{n\pi t}{p}dt \\ &= \frac{1}{p}\int_d^{t_1} f_1(t)\frac{n\pi t}{p}dt + \frac{1}{p}\int_{t_1}^{t_2} f_2(t)\cos\frac{n\pi t}{p}dt \\ &\quad +\frac{1}{p}\int_{t_1}^{d+2p} f_3(t)\cos\frac{n\pi t}{p}dt \end{aligned} \tag{2.7}$$

$$\begin{aligned} b_n &= \frac{1}{p}\int_d^{d+2p} f(t)\sin\frac{n\pi t}{p}dt \\ &= \frac{1}{p}\int_d^{t_1} f_1(t)\sin\frac{n\pi t}{p}dt + \frac{1}{p}\int_{t_1}^{t_2} f_2(t)\sin\frac{n\pi t}{p}dt \\ &\quad +\frac{1}{p}\int_{t_2}^{d+2p} f_3(t)\sin\frac{n\pi t}{p}dt. \end{aligned} \tag{2.8}$$

At the jump point it is not obvious what value to assign to f. We shall always use the average of left and right limits given by (2.6).

Example 1

A sinusoidal voltage $E_0 \sin\omega t$ is passed through a half-wave rectifier which clips the negative portion of the wave. Find the Fourier series of the resulting function:

$$f(t) = \begin{cases} 0 & -\pi/\omega < t < 0 \\ E_0 \sin\omega t & 0 < t < \pi/\omega. \end{cases}$$

Solution

In this case $2p = \frac{2\pi}{\omega}$, $p = \frac{\pi}{\omega}$ and $d = -\frac{\pi}{\omega}$. Thus the Fourier series of $f(t)$ exists in the form

$$f(t) = \frac{a_0}{2} + \sum_{n=1}^{\infty}\{a_n \cos n\omega t + b_n \sin n\omega t\}.$$

The coefficients a_n and b_n are evaluated from the formulas stated earlier.

$$\begin{aligned} a_n &= \frac{\omega}{\pi}\int_{-\pi/\omega}^{\pi/\omega} f(t)\cos n\omega t dt \quad n = 0, 1, 2, \cdots \\ b_n &= \frac{\omega}{\pi}\int_{-\pi/\omega}^{\pi/\omega} f(t)\sin n\omega t dt \quad n = 1, 2, 3\cdots \end{aligned}$$

Therefore

$$a_n = \begin{cases} \frac{2E_0}{\pi(1-n^2)} & n = 0, 2, 4, 6, \cdots \\ 0 & n = 3, 5, 7, \cdots \end{cases}$$

and $b_n = 0$ for $n \neq 0$. But for $n = 1$, $a_1 = 0$ and $b_1 = E_0/2$.

Thus the Fourier series is given by

$$f(t) = \frac{E_0}{\pi} + \frac{E_0}{2}\sin\omega t - \frac{2E_0}{\pi}\sum_{n=2,4,6,\cdots}^{\infty}\frac{\cos n\omega t}{n^2-1}.$$

2.4 Complex form of Fourier series

A useful form of Fourier Series is the complex exponential form. The form is obtained by substituting the exponential equivalent of the cosine and sine terms into the original form of Fourier series:

$$\begin{aligned} f(t) &= \frac{a_0}{2} + \sum_{n=1}^{\infty}\{a_n\cos\frac{n\pi t}{p} + b_n\sin\frac{n\pi t}{p}\} \\ &= \frac{a_0}{2} + \sum_{n=1}^{\infty}\{a_n(\frac{e^{\frac{ni\pi t}{p}} + e^{-\frac{ni\pi t}{p}}}{2}) + b_n(\frac{e^{\frac{ni\pi t}{p}} - e^{-\frac{ni\pi t}{p}}}{2i})\}. \end{aligned} \tag{2.9}$$

Collecting the coefficients of $e^{\frac{ni\pi t}{p}}$ and $e^{-\frac{ni\pi t}{p}}$, we obtain

$$f(t) = \frac{a_0}{2} + \sum_{n=1}^{\infty}(\frac{a_n - ib_n}{2})e^{\frac{ni\pi t}{p}} + \sum_{n=1}^{\infty}(\frac{a_n + ib_n}{2})e^{-\frac{ni\pi t}{p}}. \tag{2.10}$$

We can combine these expressions into one series if we define

$$c_0 = \frac{a_0}{2} \quad , \quad c_n = \frac{a_n - ib_n}{2} \quad , \quad c_{-n} = \frac{a_n + ib_n}{2} = c_n^*$$

such that the last series can be written in more symmetric form

$$f(t) = \sum_{n=-\infty}^{\infty} c_n e^{\frac{ni\pi t}{p}}. \tag{2.11}$$

Here it is noted that $c_{-n} = c_n^*$ is the complex conjugate of c_n. This series is known as the complex form of the Fourier series and represents a real quantity. The coefficients c_0, c_n and c_{-n} can be calculated directly from their definitions. Thus,

$$\begin{aligned} c_0 &= \frac{a_0}{2} = \frac{1}{2p}\int_d^{d+2p} f(t)dt \\ c_n &= \frac{1}{2}(a_n - ib_n) = \frac{1}{2p}\int_d^{d+2p} f(t)\{\cos\frac{n\pi t}{p} - i\sin\frac{n\pi t}{p}\}dt \\ &= \frac{1}{2p}\int_d^{d+2p} f(t)e^{-\frac{ni\pi t}{p}}dt \\ c_{-n} &= \frac{1}{2}(a_n + ib_n) = \frac{1}{2p}\int_d^{d+2p} f(t)e^{\frac{ni\pi t}{p}}dt. \end{aligned}$$

Obviously, the coefficient c_n can be calculated from the single formula

$$c_n = \frac{1}{2p}\int_d^{d+2p} f(t)e^{-\frac{ni\pi t}{p}}\,dt \quad , \quad n = 0, \pm 1, \pm 2, \cdots \tag{2.12}$$

Another important trigonometric form of the Fourier series is the following. For instance, the original Fourier series

$$f(t) = \frac{a_0}{2} + \sum_{n=1}^{\infty}\{a_n \cos\frac{n\pi t}{p} + b_n \sin\frac{n\pi t}{p}\}$$

can be put in the form

$$f(t) = A_0 + \sum_{n=1}^{\infty} A_n \cos(\frac{n\pi t}{p} - \gamma_n) \tag{2.13}$$

$$= A_0 + \sum_{n=1}^{\infty} A_n \sin(\frac{n\pi t}{p} + \delta_n) \tag{2.14}$$

provided we define

$$\begin{aligned} a_n &= A_n \cos\gamma_n = A_n \sin\delta_n \\ b_n &= A_n \sin\gamma_n = A_n \cos\delta_n \\ \frac{a_0}{2} &= A_0 \end{aligned}$$

such that

$$A_n = \sqrt{a_n^2 + b_n^2} \tag{2.15}$$

$$\gamma_n = \tan^{-1}\frac{b_n}{a_n} = \frac{\pi}{2} - \delta_n. \tag{2.16}$$

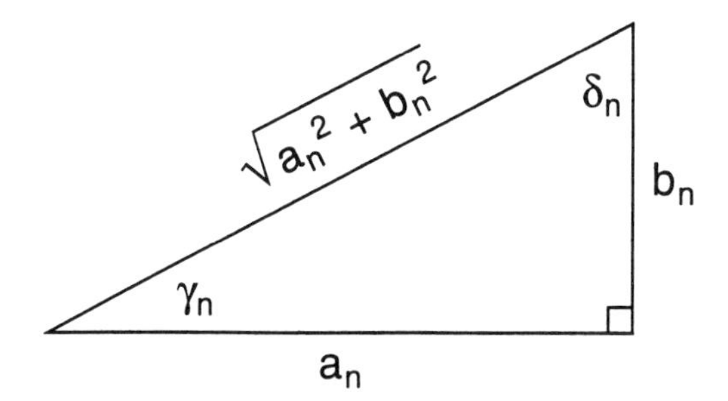

Figure 2.3: The right angled triangle defining phase angles γ_n and δ_n.

Example 2

Find the complex form of the Fourier series of the periodic function whose definition in one period is

$$f(t) = e^t \qquad 0 \le t \le 1.$$

Solution

In this case $2p = 1$, $p = \frac{1}{2}$ and $d = 0$.

The complex form of Fourier series is $f(t) = \sum_{n=-\infty}^{\infty} c_n e^{2ni\pi t}$.

The complex Fourier coefficients are

$$c_n = \int_0^1 e^t e^{-2ni\pi t} dt = \frac{e-1}{1+4n^2\pi^2}(1+2ni\pi).$$

Hence the complex Fourier expansion is

$$f(t) = \sum_{n=-\infty}^{\infty} \frac{(e-1)(1+2ni\pi)}{1+4n^2\pi^2} e^{2ni\pi t}.$$

To find the trigonometric series we have to follow the steps as in the previous example.

2.5 Half-range Fourier sine and cosine series

In many of the applications of Fourier series it may be noticed that $f(t)$ possesses certain symmetry properties. The given function may be symmetric in the origin or it may be symmetric about the y-axis. In that situation, evaluations of the Fourier coefficients are extremely simple. When the function is symmetric in the origin, i.e., when it is an odd function, the Fourier expansion contains only sine terms; on the other hand, when the function is symmetric about the y-axis, i.e., when it is an even function, the Fourier expansion contains only cosine terms. This can be illustrated below.

Suppose $f(t)$ is periodic with period $2p$ and is defined in $(-p, p)$. Then the Fourier series can be written as

$$f(t) = \frac{a_0}{2} + \sum_{n=1}^{\infty}(a_n \cos\frac{n\pi t}{p} + b_n \sin\frac{n\pi t}{p})$$

where

$$a_n = \frac{1}{p}\int_{-p}^{p} f(t)\cos\frac{n\pi t}{p} dt \tag{2.17}$$

$$b_n = \frac{1}{p}\int_{-p}^{p} f(t)\sin\frac{n\pi t}{p} dt. \tag{2.18}$$

Let us consider first that $f(t)$ is an even function. Then by definition, $f(-t) = f(t)$ for all t's in the domain of f. Geometrically this means the graph of $f(t)$ is symmetric about the y-axis.

The Fourier coefficient

$$a_n = \frac{1}{p}\int_{-p}^{p} f(t)\cos\frac{n\pi t}{p} dt = \frac{2}{p}\int_0^p f(t)\cos\frac{n\pi t}{p} dt$$

$$b_n = \frac{1}{p}\int_{-p}^{p} f(t)\sin\frac{n\pi t}{p} dt = 0.$$

Thus, when $f(t)$ is an even periodic function with period $2p$ defined in $(-p, p)$, then the Fourier series expansion can be written as

$$f(t) = \frac{a_0}{2} + \sum_{n=1}^{\infty} a_n \cos\frac{n\pi t}{p} \tag{2.19}$$

where

$$a_n = \frac{2}{p}\int_0^p f(t)\cos\frac{n\pi t}{p}dt. \tag{2.20}$$

Let us consider that $f(t)$ is an odd function. Then by definition, $f(-t) = -f(t)$ for all t's, in the domain of f. Geometrically this means the graph of $f(t)$ is symmetric in the origin. Thus,

$$\begin{aligned} a_n &= \frac{1}{p}\int_0^p f(t)\cos\frac{n\pi t}{p}dt = 0 \\ b_n &= \frac{1}{p}\int_{-p}^p f(t)\sin\frac{n\pi t}{p}dt = \frac{2}{p}\int_0^p f(t)\sin\frac{n\pi t}{p}dt. \end{aligned}$$

Therefore, when $f(t)$ is an odd periodic function with period $2p$ defined in $(-p, p)$, then the Fourier series expansion can be written as

$$f(t) = \sum_{n=1}^{\infty} b_n \sin\frac{n\pi t}{p} \tag{2.21}$$

where

$$b_n = \frac{2}{p}\int_0^p f(t)\sin\frac{n\pi t}{p}dt. \tag{2.22}$$

Here it is noted that this observation of even and odd functions of a periodic function $f(t)$ reduces the labour of calculating the Fourier coefficients by half.

From the integrals of (2.20) and (2.22), it may be observed as if the given periodic function with period $2p$ is defined in the interval $[0, p]$. In practice, in many of the applications of Fourier series a function $f(t)$ is defined on the interval $0 < t < p$. We may then represent this function either as a series consisting of only cosine terms (2.19), or as one consisting of only sine terms (2.21). This is accomplished by making either an even or an odd periodic extension of the given function, respectively.

The Fourier series expansions of (2.19) and (2.21) are respectively called the half-range Fourier cosine and sine series. The concept of even and odd extension of a periodic function is demonstrated below with examples.

Example 3

(a) Expand $f(t) = t^2$ by the half-range Fourier sine series defined in $0 < t < 1$, and then show that

$$1 - \frac{1}{3^3} + \frac{1}{5^3} - \frac{1}{7^3} + \cdots = \frac{\pi^3}{32}.$$

(b) Integrating the result, show that

$$1+\frac{1}{3^4}+\frac{1}{5^4}+\frac{1}{7^4}+\cdots=\frac{\pi^4}{96}.$$

Solution

(a) The half-range sine series can be obtained as

$$\begin{aligned} f(t) &= \sum_{n=1}^{\infty} b_n \sin n\pi t \\ &= -\frac{2}{\pi}\sum_{n=2,4,\cdots}^{\infty}\frac{\sin n\pi t}{n}+\sum_{n=1,3,5,\cdots}^{\infty}\left(\frac{2}{n\pi}-\frac{8}{n^3\pi^3}\right)\sin n\pi t. \end{aligned}$$

Now putting $t=\frac{1}{2}$ in the above expression and after a little reduction we obtain

$$1-\frac{1}{3^3}+\frac{1}{5^3}-\frac{1}{7^3}+\cdots=\frac{\pi^3}{32}.$$

[Note that $\tan^{-1}x = x-\frac{x^3}{3}+\frac{x^5}{5}-\frac{x^7}{7}+\cdots$, and putting $x=1$ we obtain $\frac{\pi}{4}=1-\frac{1}{3}+\frac{1}{5}-\frac{1}{7}+\cdots$]

(b) We have

$$t^2 = \sum_{n=1}^{\infty} b_n \sin n\pi t$$

and using the relation $\sum_{n=1,3,\cdots}^{\infty}\frac{1}{n^2}=\frac{\pi^2}{8}$ and after a little reduction it can be written as $\sum_{n=1,3,5,\cdots}^{\infty}\frac{1}{n^4}=\frac{\pi^4}{96}$.

2.6 Parseval's theorem

Parseval's theorem has enormous applications in the field of science and technology. This theorem concerns with expressing a function $f(t)$ in terms of summation of a set of complete orthogonal functions $\{\phi_n(t)\}$ defined in a certain domain (see section 2.3). Its mathematical development is illustrated with the following three cases.

2.6.1 For generalized Fourier series

If a function $f(t)$ can be expressed in terms of summation of a complete orthogonal set of functions $\{\phi_n(t)\}$, $n=0,1,2,\cdots$, then the generalized Fourier expansion is given by

$$\begin{aligned} f(t) &= \sum_{n=0}^{\infty} f_n\phi_n(t) \\ f_n &= \frac{1}{\lambda_n}\int_{-p}^{p} f(t)\phi_n^*(t)dt \end{aligned}$$

where $\lambda = \int_{-p}^{p} |\phi(t)|^2 dt$.

Multiplying the first equation by the complex conjugate $f_n^*(t)$ throughtout and integrating with respect to t between the limits $-p$ and p, we obtain

$$\begin{aligned}\int_{-p}^{p} |f(t)|^2 dt &= \sum_{n=0}^{\infty} f_n \int_{-p}^{p} f^*(t)\phi_n(t)dt \\ &= \sum_{n=0}^{\infty} f_n(\lambda_n f_n^*) = \sum_{n=0}^{\infty} \lambda_n |f_n|^2.\end{aligned}$$

This is known as Parseval's theorem.

2.6.2 For trigonometric Fourier series

Assuming that the Fourier series of $f(t)$ converges uniformly in $(-p, p)$ then Parseval's theorem can be written as

$$\frac{1}{p}\int_{-p}^{p} \{f(t)\}^2 dt = \frac{a_0^2}{2} + \sum_{n=1}^{\infty} (a_n^2 + b_n^2)$$

where the integral is assumed to exist. To see this result, we use the following steps.

The Fourier expansion of $f(t)$ is

$$f(t) = \frac{a_0}{2} + \sum_{n=1}^{\infty} \{a_n \cos\frac{n\pi t}{p} + b_n \sin\frac{n\pi t}{p}\} \tag{2.23}$$

where

$$\begin{aligned}a_n &= \frac{1}{p}\int_{-p}^{p} f(t)\cos\frac{n\pi t}{p} dt \\ b_n &= \frac{1}{p}\int_{-p}^{p} f(t)\sin\frac{n\pi t}{p} dt.\end{aligned}$$

After multiplying both sides of Fourier expansion by $f(t)$, we get,

$$[f(t)]^2 = \frac{a_0}{2} f(t) + \sum_{n=1}^{\infty} [a_n f(t)\cos\frac{n\pi t}{p} + b_n f(t)\sin\frac{n\pi t}{p}].$$

Integrating now, term by term, with respect to t from $-p$ to p,

$$\begin{aligned}\frac{1}{p}\int_{-p}^{p} [f(t)]^2 dt &= \frac{a_0}{2}\frac{1}{p}\int_{-p}^{p} f(t)dt + \sum_{n=1}^{\infty} [a_n\{\frac{1}{p}\int_{-p}^{p} f(t)\cos\frac{n\pi t}{p} dt\} \\ &\quad + b_n\{\frac{1}{p}\int_{-p}^{p} f(t)\sin\frac{n\pi t}{p} dt\}] \\ &= \frac{a_0^2}{2} + \sum_{n=1}^{\infty} (a_n^2 + b_n^2).\end{aligned} \tag{2.24}$$

2.6.3 For complex Fourier series

We know the complex form of Fourier series is given by

$$\begin{aligned} f(t) &= \sum_{n=-\infty}^{\infty} c_n e^{\frac{n\pi i t}{p}} \\ c_n &= \frac{1}{2p}\int_{-p}^{p} f(t) e^{-\frac{n\pi i t}{p}} dt. \end{aligned}$$

Multiplying the first equation by the complex conjugate $f^*(t)$ throughout and integrating between the limits $-p$ and p and then dividing the result by $2p$, we obtain

$$\begin{aligned} \frac{1}{2p}\int_{-p}^{p} |f(t)|^2 dt &= \sum_{n=-\infty}^{\infty} c_n \{\frac{1}{2p}\int_{-p}^{p} f^*(t) e^{\frac{n\pi i t}{p}} dt\} \\ &= \sum_{n=-\infty}^{\infty} c_n c_n^* = \sum_{n=-\infty}^{\infty} |c_n|^2 \end{aligned}$$

which is known as Parseval's theorem.

2.6.4 Application

This theorem can be used now to determine the electric power delivered by an electric current, $I(t)$, flowing under a voltage, $E(t)$, through a resistor of resistance, R.

From the theory of electricity, it is known that this power is given by $P = EI$. Since $E = RI$, the power is given by $P = RI^2$.

In most applications the current $I(t)$ is a periodic function and it is important to know the average power delivered per cycle.

Thus, the average power is

$$P_{av} = \frac{1}{2p}\int_{-p}^{p} RI^2(t) dt = \frac{R}{2p}\int_{-p}^{p} I^2(t) dt. \tag{2.25}$$

Suppose that $I(t)$ is given as a Fourier series

$$I(t) = \frac{a_0}{2} + \sum_{n=1}^{\infty}[a_n \cos\frac{n\pi t}{p} + b_n \sin\frac{n\pi t}{p}].$$

Then using Parseval's theorem, we have

$$P_{av} = \frac{R}{2}[\frac{a_0^2}{2} + \sum_{n=1}^{\infty}(a_n^2 + b_n^2)] = R[\frac{a_0^2}{4} + \frac{1}{2}\sum_{n=1}^{\infty}(a_n^2 + b_n^2)]. \tag{2.26}$$

But mean square of the current $I(t)$ is given by

$$I_{av} = \frac{1}{2p}\int_{-p}^{p} I^2(t) dt = \frac{a_0^2}{4} + \frac{1}{2}\sum_{n=1}^{\infty}(a_n^2 + b_n^2).$$

such that $P_{av} = RI_{av}$. Therefore, the root mean-square of the current denoted by I_{rms} is simply

$$I_{rms} = \sqrt{\frac{a_0^2}{4} + \frac{1}{2}\sum_{n=1}^{\infty}(a_n^2 + b_n^2)}. \tag{2.27}$$

2.7 Gibbs' phenomenon

In this section we shall demonstrate the Gibbs' phenomenon by considering a very simple periodic function.

Consider the following periodic function whose definition of one period is

$$f(t) = \begin{cases} 1 & 0 < t < 1 \\ 0 & 1 < t < 2. \end{cases}$$

The function can be represented as

$$f(t) = \frac{1}{2} + \frac{2}{\pi} \sum_{n=1}^{\infty} \frac{\sin(2n-1)\pi t}{(2n-1)}. \tag{2.28}$$

This is an infinite series which represents the given periodic function. To see how well this infinite series represents the function, we will graph both sides. To do this, let us truncate the series after N terms. Let the sum of these first N terms of an infinite series be denoted by S_N. In the present case,

$$S_N = \frac{1}{2} + \frac{2}{\pi} \sum_{n=1}^{N} \frac{\sin(2n-1)\pi t}{2n-1}. \tag{2.29}$$

The overshoot at $t = 1 - 0$ and the undershoot at $t = 1 + 0$ are characteristics of Fourier (and other) series at points of discontinuity and are known as the Gibbs' Phenomenon,[1] after the American mathematical physicist Josiah Willard Gibbs (1839–1903) who first explained the phenomenon. It persists even though a large number of terms are considered in the partial sums, S_N.[2] This overshoot and undershoot both generally amount to about 18 percent of the distance between the functional values at a discontinuity.

2.8 Development of Fourier integral and transform

We have already seen that a periodic function can adequately be represented by a Fourier series expansion satisfying the Dirichlet conditions. In many problems of physical interests, the impressed force on the applied voltage is nonperiodic rather than periodic. In that situation, Fourier series expansion cannot represent a nonperiodic function. The nonperiodic function may be obtained when the period of a periodic function goes to infinity. Under this limiting condition, a Fourier series approaches to Fourier integral. For rigorous mathematical treatment of the subject, the reader is directed to I.N. Sneedon's *Fourier Transforms* (New York, McGraw-Hill, 1951).

---

[1]Ref. After Josiah Willard Gibbs (1839–1903) an American mathematical physicist who pointed out this in 1899 (*Nature* **59**, p. 606), and in 1906 Maxime Bocher (1867–1918) gave a mathematical explanation (*Annals of Math*, **2**(7), p.81).

[2]Ref. See also David Shelnpsky, "Derivation of Gibbs Phenomenon," *American Mathematical Monthly*, **87**(3) (March 1980), pp. 210-212.

The following problem will illustrate how a periodic function becomes a nonperiodic function when the period goes to infinity.

For simplicity, consider the following example whose definition in one period is

$$f(t) = e^{-|t|} \qquad -p < t < p. \tag{2.30}$$

This periodic function can be represented in Fourier series

$$f(t) = \frac{a_0}{2} + \sum_{n=1}^{\infty}(a_n \cos\frac{n\pi t}{p} + b_n \sin\frac{n\pi t}{p}) \tag{2.31}$$

where

$$\begin{aligned} a_n &= \frac{1}{p}\int_{-p}^{p} f(t)\cos\frac{n\pi t}{p}dt \\ b_n &= \frac{1}{p}\int_{-p}^{p} f(t)\sin\frac{n\pi t}{p}dt. \end{aligned} \tag{2.32}$$

Since the given function (2.30) is an even periodic function, we would have $b_n = 0$ for all n. Hence, we have

$$f(t) = \frac{a_0}{2} + \sum_{n=1}^{\infty} a_n \cos\frac{n\pi t}{p} \tag{2.33}$$

where

$$a_n = \frac{2}{p}\int_{0}^{p} f(t)\cos\frac{n\pi t}{p}dt.$$

For our convenience, let us denote (2.30) as follows:

$$f_p(t) = e^{-|t|} \qquad -p < t < p$$

where the subscript p denotes the function is periodic.

The graphical representation of this function is shown in Fig. 2.4(a) and when $p \to \infty$, this periodic function becomes a nonperiodic function and is shown in Fig. 2.4(b). This then leads us to define the function $f(t)$ as $f(t) = \lim_{p\to\infty} f_p(t)$. Now $f(t)$ function is no longer periodic. Further we assume that $f(t)$ is absolutely integrable on the real line, i.e., that the improper integral, $\int_{-\infty}^{\infty} |f(t)|dt$ be finite. From our sample problem, it is obvious that $\int_{-\infty}^{\infty} |e^{-|t|}|dt = 2\int_{0}^{\infty} e^{-t}dt = 2\lim_{p\to\infty}\int_{0}^{p} e^{-t}dt = 2$ which is finite.

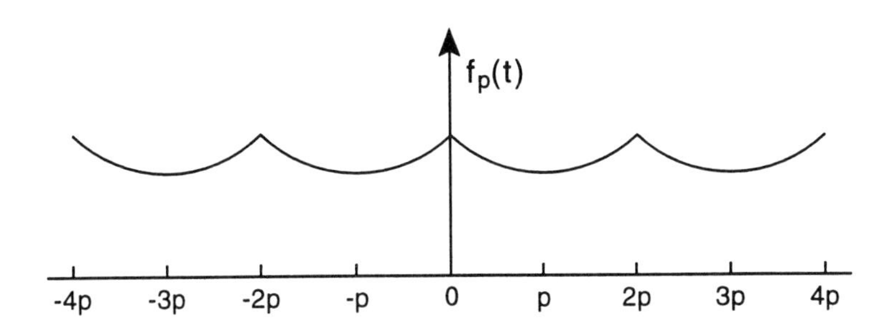

Figure 2.4(a): Sketch of periodic behaviour of function $f(t)$

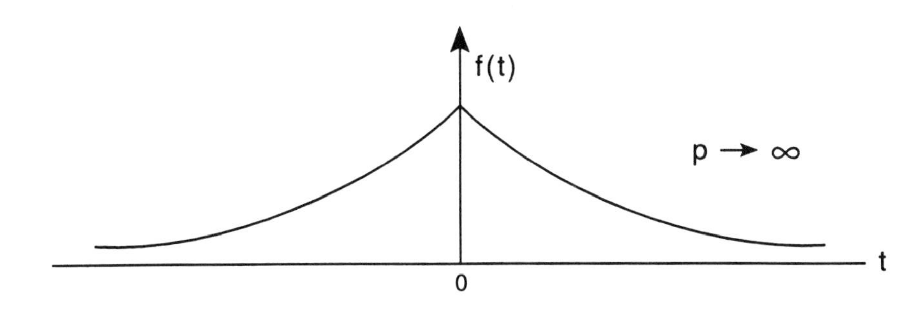

Figure 2.4(b): Sketch of nonperiodic behaviour of function $f(t)$

Now to establish the Fourier integral as the limit of a Fourier series, we begin with the complex exponential form of a Fourier series,

$$f_p(t) = \sum_{n=-\infty}^{\infty} c_n e^{\frac{n\pi i t}{p}} \tag{2.34}$$

where

$$\begin{aligned} c_n &= \frac{1}{2p}\int_{-p}^{p} f_p(t) e^{-\frac{n\pi i t}{p}} dt \\ &= \frac{1}{2p}\int_{-p}^{p} f_p(\tau) e^{-\frac{n\pi i \tau}{p}} d\tau. \end{aligned} \tag{2.35}$$

Substitution of the second expression of c_n into (2.33), yields,

$$\begin{aligned} f_p(t) &= \sum_{n=-\infty}^{\infty} [\frac{1}{2p}\int_{-p}^{p} f_p(\tau) e^{-\frac{n\pi i \tau}{p}} d\tau] e^{\frac{n\pi i t}{p}} \\ &= \sum_{n=-\infty}^{\infty} [\frac{1}{2\pi}\int_{-p}^{p} f_p(\tau) e^{-\frac{n\pi i \tau}{p}} d\tau] e^{\frac{n\pi i t}{p}} (\frac{\pi}{p}). \end{aligned} \tag{2.36}$$

Define the frequency of the general term by

$$\sigma_n = \frac{n\pi}{p} \tag{2.37}$$

and difference in frequency between successive terms by

$$\Delta\sigma = \sigma_{n+1} - \sigma_n = \frac{(n+1)\pi}{p} - \frac{n\pi}{p} = \frac{\pi}{p}.$$

Then (2.35) reduces to

$$f_p(t) = \sum_{n=-\infty}^{\infty} [\frac{1}{2\pi} e^{i\sigma_n t} \int_{-p}^{p} f_p(\tau) e^{-i\sigma_n \tau} d\tau] \Delta\sigma. \tag{2.38}$$

We know that as $p \to \infty$, $\Delta\sigma \to 0$. Thus as $p \to \infty$, eqn (2.38) can be written as an integral,

$$f(t) = \int_{-\infty}^{\infty} [\frac{1}{2\pi} e^{i\sigma t} \int_{-\infty}^{\infty} f(\tau) e^{-i\sigma\tau} d\tau] d\sigma \tag{2.39}$$

which is known as the complex form of Fourier integral.

Equation (2.39) is actually a valid representation of the nonperiodic limit function $f(t)$, provided that (a) in every finite interval $f(t)$ satisfies the Dirichlet conditions; and (b) the improper integral $\int_{-\infty}^{\infty} |f(t)| dt$ exists.

2.8.1 Complex Fourier transform

Referring to eqn (2.39), we are now in a position to define the complex Fourier transform pair.

Define

$$g(\sigma) = \mathcal{F}\{f(t)\} = \int_{-\infty}^{\infty} e^{-i\sigma t} f(t)dt \tag{2.40}$$

then

$$\begin{aligned} f(t) &= \mathcal{F}^{-1}\{g(\sigma)\} \\ &= \frac{1}{2\pi}\int_{-\infty}^{\infty} g(\sigma)e^{i\sigma t}d\sigma. \end{aligned} \tag{2.41}$$

Relations (2.40) and (2.41) are together called Fourier transform pairs. Here $g(\sigma)$ is in the frequency domain and $f(t)$ is in the time domain.

2.8.2 Trigonometric representation of Fourier integral

We next obtain the real trigonometric representation of the Fourier integral. Equation (2.39) can be rewritten as

$$\begin{aligned} f(t) &= \frac{1}{2\pi}\int_{-\infty}^{\infty}\int_{-\infty}^{\infty} f(\tau)e^{-i\sigma(\tau-t)}d\tau d\sigma \\ &= \frac{1}{2\pi}\int_{-\infty}^{\infty}\int_{-\infty}^{\infty} f(\tau)[\cos\sigma(\tau-t) - i\sin\sigma(\tau-t)]d\tau d\sigma \\ &= \frac{1}{2\pi}\int_{-\infty}^{-\infty}\int_{-\infty}^{\infty} f(\tau)\cos\sigma(\tau-t)d\tau d\sigma \\ &\quad -\frac{i}{2\pi}\int_{-\infty}^{\infty}\int_{-\infty}^{\infty} f(\tau)\sin\sigma(\tau-t)d\tau d\sigma. \end{aligned} \tag{2.42}$$

The fact that $\sin\sigma(\tau-t)$ is an odd function of σ, and $\cos\sigma(\tau-t)$ an even function of σ, we have $\int_{-\infty}^{\infty}\sin\sigma(\tau-t)d\sigma = 0$ and $\int_{-\infty}^{\infty}\cos\sigma(\tau-t)d\sigma = 2\int_0^{\infty}\cos\sigma(\tau-t)d\sigma$.

Thus we have

$$f(t) = \frac{1}{\pi}\int_0^{\infty}\int_{-\infty}^{\infty} f(\tau)\cos\sigma(\tau-t)d\tau d\sigma.$$

Expanding $\cos\sigma(\tau-t)$ we can write the integral as follows:

$$f(t) = \frac{1}{\pi}\int_0^{\infty} a(\sigma)\cos\sigma t d\sigma + \frac{1}{\pi}\int_0^{\infty} b(\sigma)\sin\sigma t d\sigma \tag{2.43}$$

where

$$\begin{aligned} a(\sigma) &= \int_{-\infty}^{\infty} f(t)\cos\sigma t dt \\ b(\sigma) &= \int_{-\infty}^{\infty} f(t)\sin\sigma t dt. \end{aligned}$$

Now if $f(t)$ is an even function then $f(-t) = f(t)$ and, hence, we have simply $a(\sigma) = 2\int_0^\infty f(t)\cos\sigma t dt$ and $b(\sigma) = 0$.

Thus we obtain

$$f(t) = \frac{2}{\pi}\int_0^\infty\int_0^\infty f(\tau)\cos\sigma\tau\cos\sigma t d\tau d\sigma. \tag{2.44}$$

This is called the Fourier cosine integral of f and is analogous to the half range cosine expansion of an even periodic function.

Now if $f(t)$ is an odd function, then $f(-t) = -f(t)$ and, hence, we have simply $a(\sigma) = 0$ and $b(\sigma) = 2\int_0^\infty f(t)\sin\sigma t dt$. Thus we obtain

$$f(t) = \frac{2}{\pi}\int_0^\infty\int_0^\infty f(\tau)\sin\sigma\tau\sin\sigma t d\tau d\sigma. \tag{2.45}$$

This is called the Fourier sine integral of f and is analogous to half range sine expansion of an odd periodic function.

2.8.3 Fourier cosine transform

Now referring to (2.44), we can define the Fourier cosine transform pair as follows.

Define

$$a(\sigma) = \mathcal{F}\{f(t)\} = 2\int_0^\infty f(t)\cos\sigma t dt \tag{2.46}$$

then

$$f(t) = \mathcal{F}^{-1}\{g(\sigma)\} = \frac{1}{\pi}\int_0^\infty g(\sigma)\cos\sigma t d\sigma. \tag{2.47}$$

Relations (2.46) and (2.47) are called Fourier cosine transform pairs and this transform exists when $f(t)$ is an even function.

2.8.4 Fourier sine transform

Now referring to (2.45), we can define the Fourier sine transform pair as follows.

Define

$$b(\sigma) = \mathcal{F}\{f(t)\} = \int_0^\infty f(t)\sin\sigma t dt \tag{2.48}$$

then

$$f(t) = \mathcal{F}^{-1}\{g(\sigma)\} = \frac{2}{\pi}\int_0^\infty g(\sigma)\sin\sigma t d\sigma. \tag{2.49}$$

These two relations are called Fourier sine transform pairs which exist only when $f(t)$ is an odd function.

2.8.5 Finite Fourier cosine and sine transforms

Finite Fourier cosine and sine transforms can be deduced from the half range Fourier cosine and sine series of $f(t)$, defined in $0 < t < p$. We know the half range Fourier cosine series is

$$f(t) = \frac{a_0}{2} + \sum_{n=1}^{\infty} a_n \cos \frac{n\pi t}{p}$$

where

$$a_n = \frac{2}{p} \int_0^p f(t) \cos \frac{n\pi t}{p} dt.$$

If we define

$$F_c(n) = \int_0^p f(t) \cos \frac{n\pi t}{p} dt \tag{2.50}$$

then $a_n = \frac{2}{p} F_c(n)$ and $\frac{a_0}{2} = \frac{1}{p} F_c(0)$ and hence,

$$f(t) = \frac{1}{p} F_c(0) + \frac{2}{p} \sum_{n=1}^{\infty} F_c(n) \cos \frac{n\pi t}{p}. \tag{2.51}$$

Here $F_c(n)$ is called the finite Fourier cosine transform and $f(t)$ is the inverse and n is an integer.

Similarly, half range Fourier sine series is

$$f(t) = \sum_{n=1}^{\infty} b_n \sin \frac{n\pi t}{p}$$

where

$$b_n = \frac{2}{p} \int_0^p f(t) \sin \frac{n\pi t}{p} dt.$$

If we define

$$F_s(n) = \int_0^p f(t) \sin \frac{n\pi t}{p} dt \tag{2.52}$$

then $b_n = \frac{2}{p} F_s(n)$ and hence,

$$f(t) = \frac{2}{p} \sum_{n=1}^{\infty} F_s(n) \sin \frac{n\pi t}{p}. \tag{2.53}$$

Here $F_s(n)$ is called the Finite Fourier sine transform, and $f(t)$ is its inverse.

Example 4

Find the Fourier integral representation of the following non-periodic function.

$$f(t) = \begin{cases} e^{at} & t \leq 0 \\ & & a > 0 \\ e^{-at} & t \geq 0. \end{cases}$$

Solution

To obtain the Fourier integral representation of the above function, we first obtain the Fourier transform and then take the inverse of the transform. Thus,

$$\begin{aligned}
\mathcal{F}\{f(t)\} &= g(\sigma) = \int_{-\infty}^{\infty} f(t)e^{-i\sigma t}dt \\
&= \int_{-\infty}^{0} e^{(a-i\sigma)t}dt + \int_{0}^{\infty} e^{-(a+i\sigma)t}dt \\
&= \frac{1}{a-i\sigma} + \frac{1}{a+i\sigma} = \frac{2a}{a^2+\sigma^2}.
\end{aligned}$$

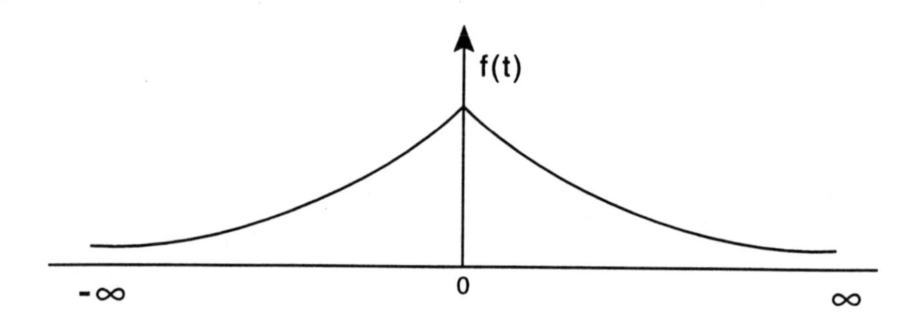

Figure 2.5: Graphical representation of $f(t)$

Now taking the inverse,

$$\begin{aligned}
f(t) &= \mathcal{F}^{-1}[g(\sigma)] = \frac{1}{2\pi}\int_{-\infty}^{\infty} g(\sigma)e^{i\sigma t}d\sigma \\
f(t) &= \frac{1}{2\pi}\int_{-\infty}^{\infty} \frac{2a}{a^2+\sigma^2}e^{i\sigma t}d\sigma \\
&= \frac{a}{\pi}\int_{-\infty}^{\infty} \frac{e^{i\sigma t}}{a^2+\sigma^2}d\sigma \\
&= \frac{a}{\pi}\int_{-\infty}^{\infty} \frac{\cos\sigma t + i\sin\sigma t}{a^2+\sigma^2}d\sigma \\
&= \frac{2a}{\pi}\int_{0}^{\infty} \frac{\cos\sigma t}{a^2+\sigma^2}d\sigma
\end{aligned}$$

since $\sin\sigma t$ is an odd function.

Another short cut approach

From the graph shown in Fig. 2.5, it is obvious that the given function is an even nonperiodic function. Thus, its Fourier integral representation can be obtained from the Fourier cosine integral representation (2.44). Thus, we have

$$\begin{aligned}
f(t) &= \frac{2}{\pi}\int_0^\infty\int_0^\infty f(\tau)\cos\sigma\tau\cos\sigma t d\tau d\sigma \\
&= \frac{2}{\pi}\int_0^\infty [\int_0^\infty e^{-a\tau}\cos\sigma\tau d\tau]\cos\sigma t d\sigma \\
&= \frac{2}{\pi}\int_0^\infty [\frac{e^{-a\tau}(-a\cos\sigma\tau+\sigma\sin\sigma\tau)}{a^2+\sigma^2}]_0^\infty \cos\sigma t d\sigma \\
&= \frac{2a}{\pi}\int_0^\infty \frac{\cos\sigma t}{a^2+\sigma^2}d\sigma.
\end{aligned}$$

Example 5

Find the Fourier integral representation of the following nonperiodic function.

$$f(t) = \begin{cases} \sin t & t^2 \le \pi^2 \\ 0 & t^2 \ge \pi^2. \end{cases}$$

Solution

To obtain the Fourier integral representation of this function, we first obtain the Fourier transform and then obtain the inverse of the transform. The graphical representation is depicted in Fig. 2.6.

$$\begin{aligned}
\mathcal{F}\{f(t)\} &= \int_{-\infty}^\infty f(t)e^{-i\sigma t}dt \\
&= \int_{-\pi}^\pi \sin te^{-i\sigma t}dt \\
&= [\frac{\{(-i\sigma)\sin t-\cos t\}}{1+(-i\sigma)^2}e^{-i\sigma t}]_{-\pi}^\pi \\
&= \frac{-\cos\pi e^{-i\sigma\pi}+\cos\pi e^{i\sigma\pi}}{1-\sigma^2} \\
&= -\frac{e^{i\sigma\pi}-e^{-i\sigma\pi}}{1-\sigma^2} = \frac{-2i\sin\sigma\pi}{1-\sigma^2}.
\end{aligned}$$

Thus

$$\begin{aligned}
f(t) &= \mathcal{F}^{-1}[\frac{-2i\sin\sigma\pi}{1-\sigma^2}] \\
&= \frac{1}{2\pi}\int_{-\infty}^\infty [\frac{-2i\sin\sigma\pi}{1-\sigma^2}]e^{i\sigma t}d\sigma \\
&= -\frac{i}{\pi}\int_{-\infty}^\infty \frac{\sin\sigma\pi[\cos\sigma t+i\sin\sigma t]}{1-\sigma^2}d\sigma.
\end{aligned}$$

Therefore,

$$f(t) = \frac{2}{\pi}\int_0^{\infty} \frac{\sin\sigma\pi \sin\sigma t}{1-\sigma^2} d\sigma.$$

because $\int_{-\infty}^{\infty} \sin\sigma\pi \cos\sigma t d\sigma = 0$ since $\sin\sigma\pi\cos\sigma t$ is an odd function.

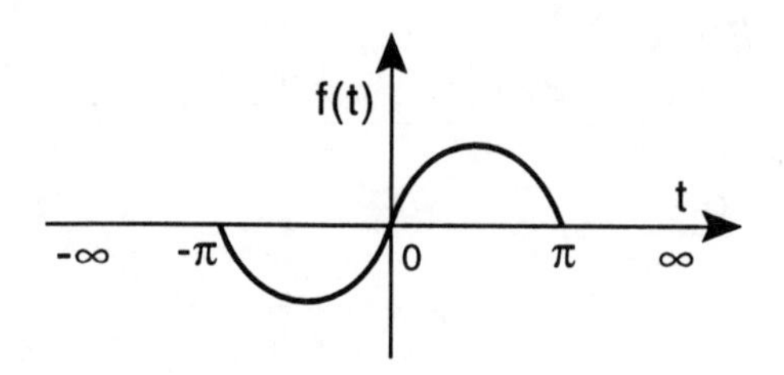

Figure 2.6: Graphical representation of $f(t)$

Short cut approach

Realizing that the given function $f(t)$ is an odd function, Fourier representation can be obtained from the formula (2.45) which is the Fourier sine integral representation. Thus, we have

$$\begin{aligned} f(t) &= \frac{2}{\pi}\int_{\sigma=0}^{\infty}\int_{\tau=0}^{\infty} f(\tau)\sin\sigma\tau \sin\sigma t d\tau d\sigma \\ &= \frac{2}{\pi}\int_0^{\infty}[\int_0^{\pi} \sin\tau \sin\sigma\tau d\tau] \sin\sigma t d\sigma \\ &= \frac{1}{\pi}\int_0^{\infty}[\int_0^{\pi}\{\cos(1-\sigma)\tau - \cos(1+\sigma)\tau\} d\tau] \sin\sigma t d\sigma. \end{aligned}$$

Therefore

$$f(t) = \frac{2}{\pi}\int_0^{\infty} \frac{\sin\sigma\pi \sin\sigma t}{1-\sigma^2} d\sigma$$

which is the same as before.

Example 6

Find the Fourier integral representation of the following nonperiodic function.

$$f(t) = \begin{cases} 0 & -\infty < t < -1 \\ -1 & -1 < t < 0 \\ 1 & 0 < t < 1 \\ 0 & 1 < t < \infty. \end{cases}$$

Solution

This function is graphed in Fig. 2.7.

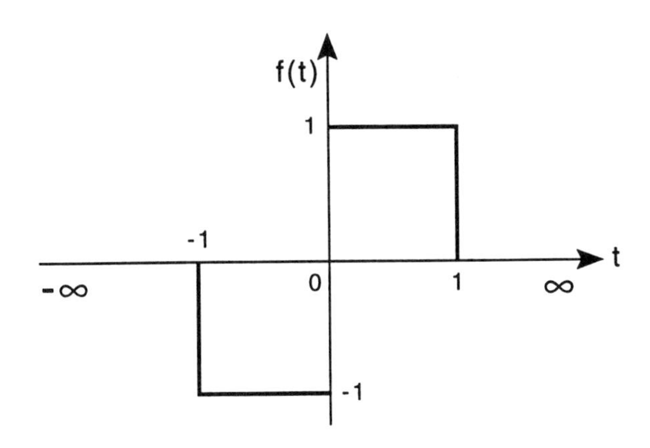

Figure 2.7: Graphical representation of $f(t)$.

From the graph it is obvious that the given function is an odd function, and hence its Fourier integral representation can be obtained from the Fourier sine integral representation (2.45). Thus, we have

$$\begin{aligned} f(t) &= \frac{2}{\pi}\int_0^\infty \int_0^\infty f(\tau)\sin\sigma\tau \sin\sigma t d\tau d\sigma \\ &= \frac{2}{\pi}\int_0^\infty [\int_0^1 \sin\sigma\tau d\tau]\sin\sigma t d\sigma \\ &= \frac{2}{\pi}\int_0^\infty \frac{(1-\cos\sigma)}{\sigma}\sin\sigma t d\sigma. \end{aligned}$$

Suppose we consider only frequencies below σ_0 to approximate the given function. Then,

$$\begin{aligned} f(t) &= \frac{2}{\pi}\int_0^{\sigma_0} [\frac{\sin\sigma t}{\sigma} - \frac{\cos\sigma\sin\sigma t}{\sigma}]d\sigma \\ &= \frac{2}{\pi}\int_0^{\sigma_0} \frac{\sin\sigma t}{\sigma}d\sigma - \frac{1}{\pi}\int_{-0}^{\sigma_0} \frac{2\cos\sigma\sin\sigma t}{\sigma}d\sigma. \end{aligned}$$

Now

$$\frac{1}{\pi}\int_0^{\sigma_0} \frac{2\cos\sigma\sin\sigma t}{\sigma}d\sigma = \frac{1}{\pi}\int_0^{\sigma_0} [\frac{\sin(1+t)\sigma}{\sigma} + \frac{\sin(t-1)\sigma}{\sigma}]d\sigma.$$

This result is obtained using the following trigonometric identity

$$2\cos B \sin A = \sin(A+B) + \sin(A-B)$$

and similarly,

$$\int_0^{\sigma_0} \frac{\sin \sigma t}{\sigma} d\sigma = \int_0^{\sigma_0 t} \frac{\sin \omega}{\omega} d\omega.$$

But by definition

$$S_i(x) = \int_0^x \frac{\sin u}{u} du$$

is known as the sine integral function of x, and is tabulated in many handbooks.

Thus,

$$\begin{aligned} f(t) &= \frac{2}{\pi} \int_0^{\sigma_0} \frac{(1-\cos\sigma)\sin\sigma t}{\sigma} d\sigma \\ &= \frac{2}{\pi} S_i(\sigma_0 t) - \frac{1}{\pi} S_i[\sigma_0(1+t)] - \frac{1}{\pi} S_i[\sigma_0(t-1)]. \end{aligned}$$

Note that the cosine integral function of x is defined as $C_i(x) = \int_x^\infty \frac{\cos u}{u} du$.

Example 7

Find the finite Fourier cosine and sine transforms of the function $f(t) = 2t, 0 < t < 4$.

Solution

(a) Finite Fourier cosine transform:
Here $p = 4$ and therefore,

$$F_c(n) = \int_0^p f(t) \cos \frac{n\pi t}{p} dt$$

where

$$\begin{aligned} F_c(n) &= \begin{cases} \frac{-64}{n^2\pi^2} & n = 1, 3, 5, \cdots \\ 0 & n = 2, 4, 6, \cdots \end{cases} \\ F_c(0) &= \int_0^4 (2t) dt = [t^2]_0^4 = 16. \end{aligned}$$

(b) Finite Fourier sine transform:

$$\begin{aligned} F_s(n) &= \int_0^p f(t) \sin \frac{n\pi t}{p} dt \\ &= \frac{32}{n\pi} (-1)^{n+1}. \end{aligned}$$

2.9 Relationship of Fourier and Laplace transforms

In using applications of the Fourier integral the function to be represented is identically zero before some instant, usually $t = 0$. In that situation the general Fourier transform pair given by (2.40) and (2.41) reduce to,

$$g(\sigma) = \mathcal{F}\{f(t)\} = \int_0^\infty e^{-i\sigma t} f(t)dt \tag{2.54}$$

where

$$f(t) = \mathcal{F}^{-1}\{g(\sigma)\} = \frac{1}{2\pi}\int_{-\infty}^{\infty} g(\sigma)e^{i\sigma t}d\sigma. \tag{2.55}$$

Now consider the following function,

$$F(t) = \begin{cases} e^{-at}f(t) & t > 0 \\ & \qquad a > 0 \\ 0 & t < 0 \end{cases} \tag{2.56}$$

where $f(t)$ is the function in which we are interested. Then taking the Fourier transform of $F(t)$, we obtain,

$$G(\sigma) = \mathcal{F}\{F(t)\} = \int_{-\infty}^{\infty} e^{-i\sigma t}F(t)dt = \int_0^\infty e^{-i\sigma t}[e^{-at}f(t)]dt. \tag{2.57}$$

Therefore,

$$G(\sigma) = \int_0^\infty e^{-(a+i\sigma)t}f(t)dt \tag{2.58}$$

whose inverse transform is

$$F(t) = e^{-at}f(t) = \frac{1}{2\pi}\int_{-\infty}^{\infty} G(\sigma)e^{i\sigma t}d\sigma.$$

Therefore,

$$f(t) = \frac{1}{2\pi}\int_{-\infty}^{\infty} G(\sigma)e^{(a+i\sigma)t}d\sigma. \tag{2.59}$$

Now if we redefine (2.58) to emphasize the fact that the left hand side is really a function of $a + i\sigma$, then

$$g(a+i\sigma) = G(\sigma) = \int_0^\infty e^{-(a+i\sigma)t}f(t)dt.$$

Then the equation of the transform pair is rewritten as

$$g(a+i\sigma) = \int_0^\infty e^{-(a+i\sigma)t}f(t)dt \tag{2.60}$$

where

$$f(t) = \frac{1}{2\pi} \int_{-\infty}^{\infty} e^{(a+i\sigma)t} g(a+i\sigma) d\sigma. \tag{2.61}$$

Now make the substitution $a + i\sigma = s$ such that $d\sigma = \frac{ds}{i}$ and when

$$\begin{array}{ll} \sigma \to -\infty & , \quad s \to a - i\infty \\ \sigma \to \infty & , \quad s \to a + i\infty. \end{array}$$

Then (2.60) and (2.61) may be written as

$$g(s) = \int_0^{\infty} f(t) e^{-st} dt \tag{2.62}$$

$$f(t) = \frac{1}{2\pi i} \int_{a-i\infty}^{a+i\infty} g(s) e^{st} ds. \tag{2.63}$$

These equations for $f(t)$ and $g(s)$ represent a Laplace transform pair. The function $g(s)$ is known as the Laplace transform of $f(t)$. The integral for $f(t)$ is known as the complex inversion integral. It is obvious that in general the parameter s which is called the Laplace parameter is a complex number. We shall continue our discussion on Laplace transform in the next chapter. Before we conclude this chapter, we will discuss in the next section the notion of convolution in the Fourier sense and then establish the relationship between Fourier and Laplace convolution.

2.9.1 Convolution integral and convolution theorem

In this section we define and establish the convolution integral and convolution theorem, which is of considerable theoretical as well as practical interest.

Mathematically the convolution of two functions $f(t)$ and $g(t)$, where $-\infty < t < \infty$, is defined as

$$\begin{aligned} y(t) &= \int_{-\infty}^{\infty} f(\tau) g(t-\tau) d\tau \\ &= \int_{-\infty}^{\infty} f(t-\tau) g(\tau) d\tau \\ &= f(t) * g(t). \end{aligned} \tag{2.64}$$

This integral is usually known as the convolution integral and is denoted by $f * g$. Here τ is a dummy variable.

Theorem

If $y(t)$ is the response obtained by convolving the function $f(t)$ with $g(t)$, then

$$\mathcal{F}\{y(t)\} = \mathcal{F}\{f * g\} = \mathcal{F}\{f(t)\}\mathcal{F}\{g(t)\}.$$

In words we can say that the Fourier transform of convolution of $f(t)$ and $g(t)$ is the product of the Fourier transforms $f(t)$ and $g(t)$.

Proof

We have by definition,

$$\begin{aligned}
\mathcal{F}\{y(t)\} &= \mathcal{F}\{f * g\} = \int_{-\infty}^{\infty} y(t)e^{-i\sigma t}dt \\
&= \int_{-\infty}^{\infty}[\int_{-\infty}^{\infty} f(\tau)g(t-\tau)d\tau]e^{-i\sigma t}dt \\
&= \int_{-\infty}^{\infty} f(\tau)[\int_{-\infty}^{\infty} g(t-\tau)e^{-i\sigma t}dt]d\tau.
\end{aligned}$$

Now put $t - \tau = w$ and $dt = dw$ in the inner integral

$$\begin{aligned}
&= \int_{-\infty}^{\infty} f(\tau)[\int_{-\infty}^{\infty} g(w)e^{-i\sigma(\tau+w)}dw]d\tau \\
&= \{\int_{-\infty}^{\infty} f(\tau)e^{-i\sigma\tau}d\tau\}\{\int_{-\infty}^{\infty} g(w)e^{-i\sigma w}dw\} \\
&= \{\int_{-\infty}^{\infty} f(t)e^{-i\sigma t}dt\}\{\int_{-\infty}^{\infty} g(t)e^{-i\sigma t}dt\} \\
&= \mathcal{F}\{f(t)\}\mathcal{F}\{g(t)\}.
\end{aligned}$$

as asserted.

2.9.2 Convolution integrals from Fourier to Laplace

We shall now deduce the convolution integral in Laplace sense. Suppose

$$\begin{aligned}
f(t) &= 0 \quad t < 0 \\
g(t) &= 0 \quad t < 0
\end{aligned} \tag{2.65}$$

and $f(t)$ and $g(t)$ are defined in $0 < t < \infty$. Then by the Fourier convolution we have,

$$\begin{aligned}
y(t) &= f(t) * g(t) = \int_{-\infty}^{\infty} f(\tau)g(t-\tau)d\tau \\
&= \int_{-\infty}^{0} f(\tau)g(t-\tau)d\tau + \int_{0}^{\infty} f(\tau)g(t-\tau)d\tau.
\end{aligned}$$

The first integral is zero because

$$f(\tau) = 0 \quad \tau < 0.$$

Therefore,

$$\begin{aligned}
y(t) &= \int_{0}^{\infty} f(\tau)g(t-\tau)d\tau \\
&= \int_{0}^{t} f(\tau)g(t-\tau)d\tau + \int_{t}^{\infty} f(\tau)g(t-\tau)d\tau.
\end{aligned}$$

We know

$$\begin{aligned}
g(t-\tau) &= 0 \quad \text{for} \quad t-\tau < 0, \text{i.e., } \tau > t \\
&\neq 0 \quad \text{for} \quad \tau < t
\end{aligned}$$

Thus it is obvious that the second integral contributes nothing. Therefore, the convolution integral in Laplace sense is given by

$$\begin{aligned} y(t) &= \int_0^t f(\tau)g(t-\tau)d\tau \\ &= \int_0^t f(t-\tau)g(\tau)d\tau \\ &= f(t) * g(t). \end{aligned} \tag{2.66}$$

This is called convolution or the Faltung integral and is frequently denoted by $f(t) * g(t)$. This integral has many important applications in the field of electrical engineering. We will exploit this integral and see its usefulness in electrical network problems when we take up this matter in the next section.

Example 8

A linear system has the impulse response given by $f(t)$ and is subjected to the rectangular pulse given by $g(t)$ as defined below.

$$\begin{aligned} f(t) &= \begin{cases} 0 & -\infty < t < 0 \\ ae^{-at} & 0 < t < \infty. \end{cases} \quad a > 0 \\ g(t) &= \begin{cases} 0 & -\infty < t < -1 \\ 1 & -1 < t < 1 \\ 0 & 1 < t < \infty. \end{cases} \end{aligned}$$

Find the output time function convolving $f(t)$ with $g(t)$ from the convolution integral

$$y(t) = \int_{-\infty}^{\infty} f(t-\tau)g(\tau)d\tau$$

using Fourier transforms.

Solution

We know

$$\begin{aligned} y(t) &= \int_{-\infty}^{\infty} f(t-\tau)g(\tau)d\tau \\ &= f(t) * g(t) \\ \mathcal{F}\{y(t)\} &= \mathcal{F}\{f(t)\}\mathcal{F}\{g(t)\}. \end{aligned}$$

Then

$$\begin{aligned} \mathcal{F}\{f(t)\} &= \int_{-\infty}^{\infty} f(t)e^{-i\sigma t}dt = \int_0^{\infty} ae^{-at}e^{-i\sigma t}dt \\ \mathcal{F}\{f(t)\} &= a\int_0^{\infty} e^{-(a+i\sigma)t}dt = \frac{1}{a+i\sigma} = \frac{a(a-i\sigma)}{a^2+\sigma^2} \\ \mathcal{F}\{g(t)\} &= \int_{-\infty}^{\infty} g(t)e^{-i\sigma t}dt = \int_0^1 (1)e^{-i\sigma t}dt = \frac{2\sin\sigma}{\sigma}. \end{aligned}$$

Therefore,

$$\begin{aligned}\mathcal{F}\{y(t)\} &= \mathcal{F}\{f\}\mathcal{F}\{g\}\\ &= \frac{a(a-i\sigma)}{a^2+\sigma^2}\cdot\frac{2\sin\sigma}{\sigma}\\ &= \frac{2a\sin\sigma}{\sigma(a^2+\sigma^2)}(a-i\sigma).\end{aligned}$$

Taking the Fourier inverse,

$$\begin{aligned}y(t) &= \frac{1}{2\pi}\int_{-\infty}^{\infty}\frac{2a\sin\sigma}{\sigma(a^2+\sigma^2)}(a-i\sigma)e^{i\sigma t}d\sigma\\ &= \frac{a}{\pi}\int_{-\infty}^{\infty}\frac{\sin\sigma}{\sigma(a^2+\sigma^2)}(a-i\sigma)(\cos\sigma t+i\sin\sigma t)d\sigma\\ &= \frac{a}{\pi}\int_{-\infty}^{\infty}\frac{\sin\sigma(a\cos\sigma t+\sigma\sin\sigma t)}{\sigma(a^2+\sigma^2)}d\sigma\\ &\quad+\frac{ia}{\pi}\int_{-\infty}^{\infty}\frac{\sin\sigma(a\sin\sigma t-\sigma\cos\sigma t)}{\sigma(a^2+\sigma^2)}d\sigma.\end{aligned}$$

Because the integrand in the second integral is an odd function, therefore, the value of the integral is zero.

Thus,

$$\begin{aligned}y(t) &= \frac{a}{\pi}\int_{-\infty}^{\infty}\frac{\sin\sigma(a\cos\sigma t+\sigma\sin\sigma t)}{\sigma(a^2+\sigma^2)}d\sigma\\ &= \frac{2a}{\pi}\int_{0}^{\infty}\frac{\sin\sigma(a\cos\sigma t+\sigma\sin\sigma t)}{\sigma(a^2+\sigma^2)}d\sigma\end{aligned}$$

which is the output time function.

2.10 Applications of Fourier transforms

The Fourier transform pair is rewritten as follows:

$$\begin{aligned}\mathcal{F}\{f(t)\} &= F(\sigma)=\int_{-\infty}^{\infty}f(t)e^{-i\sigma t}dt\\ f(t) &= \mathcal{F}^{-1}\{F(\sigma)\}=\frac{1}{2\pi}\int_{-\infty}^{\infty}e^{i\sigma t}F(\sigma)d\sigma.\end{aligned}$$

The second equation represents $f(t)$ as a continuous sum of exponential functions with frequencies lying in the interval $(-\infty,\infty)$. The relative amplitude of the components at any frequency σ is proportional to $F(\sigma)$. If the signal represents a voltage, $F(\sigma)$ has the dimensions of voltage multiplied by the time. Because frequency has the dimension of inverse time, we can consider $F(\sigma)$ as a voltage-density spectrum or, more generally speaking, it is known as the spectral-density function of $f(t)$.

In contrast to this, a periodic wave form has all its amplitude components at discrete frequencies. At each of these discrete frequencies there is some definite contribution; to either side there is none. It follows then, that to portray the amplitude components of a periodic wave form of spectral-density graph requires a representation with area equal to the respective amplitude components yet occupying zero frequency width. We recognize that this can be done formally by representing each amplitude component of the periodic function by an impulse function. The area (weight) of the impulse is equal to the amplitude component and the position of the impulse is determined by the particular discrete frequency.

Remark

A signal of finite energy can be described by a continuous spectral density function. This spectral density function is found by taking the Fourier transform of the signal. A periodic signal of finite average energy can be described either by a set of lines on a spectral graph or by a set of impulse functions on a spectral density graph. Each impulse on the latter graph has an area corresponding to the height of each line, respectively, on the former graph.

Sufficient conditions for the existence of the Fourier transforms are:

- $f(t)$ has a finite number of maxima and minima in any finite time interval;
- $f(t)$ has only a finite number of finite discontinuities in any finite time interval;
- $f(t)$ is absolutely integrable; that is, $\int_{-\infty}^{\infty} |f(t)|dt$ is less than infinity.

A somewhat weaker condition is $\int_{-\infty}^{\infty} |f(t)|^2 dt$ which is less than infinity. Because this condition corresponds to the distribution of an energy signal, we can state that the Fourier transform can be used to uniquely represent any energy signal.

Example 9

Find the Fourier transform (spectral density) of a gate function having unit amplitude, width τ, and centred at the origin.

Solution

The gate function is defined by

$$f(t) = rect(t/\tau) = \begin{cases} 1 & |t| < \tau/2 \\ 0 & |t| > \tau/2 \end{cases}$$

where τ describes the width. A graphical representation of this function is depicted below.

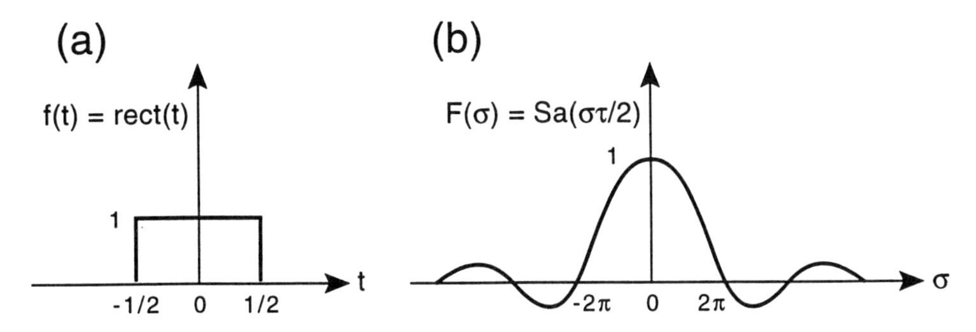

Figure 2.8: (a) The gate function and (b) its Fourier transform

The Fourier transform of the gate function is

$$2\ 2\ \mathcal{F}\{f(t)\} = F(\sigma) = \int_{-\infty}^{\infty} rect(t/\tau)e^{-i\sigma t}dt = \int_{-\tau/2}^{\tau/2} e^{-i\sigma t}dt$$

$$= \tau\frac{\sin(\sigma\tau/2)}{\sigma\tau/2} = \tau Sa(\sigma\tau/2).$$

where a function $Sa(x)$ is defined as $Sa(x) = \frac{\sin x}{x}$. Thus we have $\mathcal{F}\{rect(t/\tau)\} = \tau Sa(\sigma\tau/2)$.

It is helpful to note that because $\lim_{x\to 0} S_a(x) = 1$, the amplitude coefficient of this function is equal to the area beneath the gate function as shown in Fig. 2.8(b).

2.11 Parseval's theorem for energy signals

We know that the energy, E, delivered to a 1-ohm resistor is given by

$$E = \int_{-\infty}^{\infty} |f(t)|^2 dt = \int_{-\infty}^{\infty} f(t)f^*(t)dt.$$

We would like to express the energy in terms of the frequency components of $f(t)$. Replacing $f^*(t)$ by the inverse Fourier transform, we obtain

$$E = \int_{-\infty}^{\infty} f(t)[\frac{1}{2\pi}\int_{-\infty}^{\infty} F^*(\sigma)e^{-i\sigma t}d\sigma]dt.$$

Interchanging the order of integration on t and σ, we obtain

$$E = \frac{1}{2\pi}\int_{-\infty}^{\infty} F(\sigma)F^*(\sigma)d\sigma.$$

Thus Parseval's theorem can be written as

$$E = \int_{-\infty}^{\infty} |f(t)|^2 dt = \frac{1}{2\pi}\int_{-\infty}^{\infty} |F(\sigma)|^2 d\sigma.$$

From this relation we can find the energy of a signal either in the time domain or the frequency domain.

2.12 Heaviside unit step function and Dirac delta function

We define the Heaviside unit step function as follows:

$$u(t) = \begin{cases} 1 & t > 0 \\ 0 & t < 0. \end{cases} \tag{2.67}$$

This definition implies that the value of the function is always unity when the argument is greater than zero and its value is zero when the argument is less than zero. The value of the function at the origin is not defined but can be assigned as the average of unity and zero. An impulse function which is singular at a certain point can be called the Dirac delta function, $\delta(t)$. This function has enormous applications in many physical problems including signal analysis. This function can be mathematically defined as

$$\delta(t) = \begin{cases} \infty & t = 0 \\ 0 & t \neq 0 \end{cases} \tag{2.68}$$

which has the property exhibited by the following integral

$$\int_a^b f(t)\delta(t - t_0)dt = \begin{cases} f(t_0) & a < t_0 < b \\ 0 & \text{elsewhere} \end{cases} \tag{2.69}$$

for any $f(t)$ continuous at $t = t_0$, t_0 is finite.

The operations indicated involving the impulse function all arose formally from its integral definition. We know that the impulse function is not really a true function in mathematical sense. The impulse function has, however, been justified mathematically using a theory of **generalized functions** as described by Lighthill *Fourier Analysis and Generalized Function*, Cambridge University Press, 1958. With this approach, the impulse function is defined as the limit of a sequence of regular well-behaved functions which have the required property that the area remains constant (unity) as the width is reduced. Finally the limit of this sequence is taken to define the impulse function as the width is reduced toward zero.

The defining sequence of pulses is not unique and many pulse shapes can be chosen. In fact, the shape of the particular pulse is relatively unimportant as long as the sequence satisfies the conditions that (a) the sequence formed describes a function which becomes infinitely high and infinitesimally narrow in such a way that (b) the enclosed area is a constant (unity). For example, the following sequences all satisfy these conditions (see the following figures):

(a) Rectangular pulse: $\delta(t) = \lim_{\tau \to 0} \frac{1}{\tau}[u(t + \tau/2) - u(t - \tau/2)]$.

(b) Triangular pulse: $\delta(t) = \lim_{\tau \to 0} \frac{1}{\tau}[1 - \frac{|t|}{\tau}]$, $|t| < \tau$.

(c) Two-sided exponential: $\delta(t) = \lim_{\tau \to 0} \frac{1}{\tau} e^{-|2t|/\tau}$.

(d) Gaussian pulse: $\delta(t) = \lim_{\tau \to 0} \frac{1}{\tau} e^{-\pi(\frac{t}{\tau})^2}$.

(e) $Sa(t)$ function: $\delta(t) = \lim_{\tau \to 0} \frac{1}{\tau} Sa(\frac{\pi t}{\tau})$.

(f) $Sa^2(t)$ function: $\delta(t) = \lim_{\tau \to 0} \frac{1}{\tau} Sa^2(\frac{\pi t}{\tau})$.

Note that $\frac{1}{\pi} \int_{-\infty}^{\infty} Sa(x)dx = 1$. In the following figure, we have depicted graphically these definitions of the Dirac delta function.

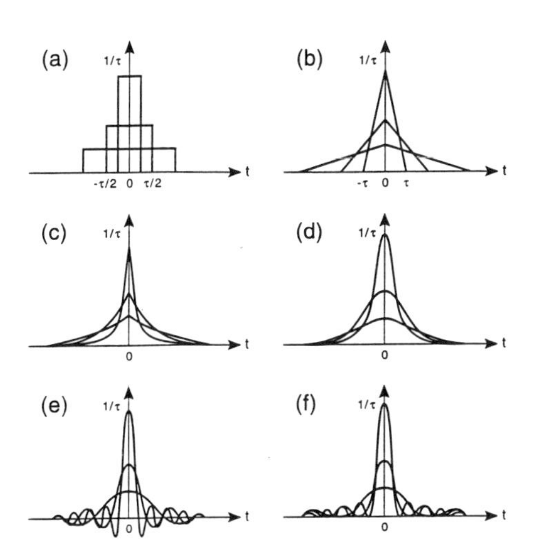

Figure 2.9: Function sequence definition of the impulse function: (a) Rectangular pulse; (b) triangular pulse; (c) two-sided exponential; (d) Gaussian pulse; (e) $Sa(t)$ function; (f) $Sa^2(t)$ function.

2.13 Some Fourier transforms involving impulse functions

The procedure for finding the Fourier transform of signals of finite energy is straight forward. In contrast, this is not always true for signals of infinite energy. Of particular interest are those cases that involve the unit impulse function.

2.13.1 The impulse function

Using the integral properties of the impulse function, the Fourier transform of a unit impulse, $\delta(t)$, is

$$\mathcal{F}\{\delta(t)\} = \int_{-\infty}^{\infty} \delta(t - t_0) e^{-i\sigma t} dt = e^{-i\sigma 0} = 1. \tag{2.70}$$

If the impulse is time shifted, we have

$$\mathcal{F}\{\delta(t-t_0)\} = \int_{-\infty}^{\infty} \delta(t-t_0)e^{-i\sigma t}dt = e^{-i\sigma t_0}. \tag{2.71}$$

From these two results it is evident that an impulse function has a uniform magnitude spectrum over the entire frequency interval ($-\infty < \sigma < \infty$). This type of spectrum is called "white" in analogy to white light. The phase spectrum of the time shifted impulse is linear with a slope that is proportional to the time-shift.

2.13.2 The eternal complex exponential

We could expect that the spectral density of $e^{\pm i\sigma t}$ will be concentrated at $\sigma = \pm\sigma_0$. That this is the case is demonstrated below.

$$\mathcal{F}^{-1}\{\delta(\sigma \mp \sigma_0)\} = \frac{1}{2\pi}\int_{-\infty}^{\infty} \delta(\sigma \mp \sigma_0)e^{i\sigma t}d\sigma = \frac{1}{2\pi}e^{\pm i\sigma_0 t}. \tag{2.72}$$

Taking the Fourier transform of both sides of the above equation, we obtain

$$\mathcal{F}\mathcal{F}^{-1}\{\delta(\sigma \mp \sigma_0)\} = \frac{1}{2\pi}\mathcal{F}\{e^{\pm i\sigma_0 t}\}. \tag{2.73}$$

or by interchanging the sides, we obtain

$$\mathcal{F}\{e^{\pm i\sigma_0 t}\} = 2\pi\delta(\sigma \mp \sigma_0). \tag{2.74}$$

This result should hardly surprise us because the complex exponential describes a phasor whose angular rate is σ_0. The spectral description of such a phasor is a line at this angular rate (which is greater than $\sigma = 0$ if rotating in a positive direction, less than $\sigma = 0$ if in a negative direction). The spectral density description is an impulse at these angular rates and the factor of 2π arises because we are using radian frequency.

Now if $\sigma_0 = 0$ then $\mathcal{F}\{1\} = 2\pi\delta(\sigma)$. It follows that any signal with a non zero average value over the infinite interval $-\infty < t < \infty$ has an impulse in its spectral density function at $\sigma = 0$.

2.13.3 Eternal sinusoidal signals

The sinusoidal functions $\cos\sigma_0 t$ and $\sin\sigma_0 t$ can be written in terms of the complex exponentials using Euler's formulas. Their Fourier transforms can be found directly from the eqn (2.74)

$$\begin{aligned}
\mathcal{F}\{\cos\sigma_0 t\} &= \mathcal{F}\{\frac{e^{i\sigma_0 t} + e^{-i\sigma_0 t}}{2}\} \\
&= \pi\{\delta(\sigma - \sigma_0) + \delta(\sigma + \sigma_0)\}, \qquad (2.75) \\
\mathcal{F}\{\sin\sigma_0 t\} &= \mathcal{F}\{\frac{e^{i\sigma_0 t} - e^{-i\sigma_0 t}}{2i}\} \\
&= \frac{\pi}{i}\{\delta(\sigma - \sigma_0) - \delta(\sigma + \sigma_0)\}. \qquad (2.76)
\end{aligned}$$

Note that it is convenient to multiply the eqn (2.76) by the factor i so that the result is real valued for the graph.

2.13.4 The signum function and the unit step function

The signum function, $sgn(t)$, is that function which changes sign when its argument is zero:

$$sgn(t) = \frac{|t|}{t} = \begin{cases} 1 & t > 0 \\ 0 & t = 0 \\ -1 & t < 0. \end{cases} \tag{2.77}$$

The signum function has an average value of zero and is piecewise continuous, but not absolutely integrable. In order to make it absolutely integrable we multiply $sgn(t)$ by $e^{-a|t|}$ and then take the limit as $a \to 0$.

$$\mathcal{F}\{sgn(t)\} = \mathcal{F}\{\lim_{a\to 0}[e^{-a|t|}sgn(t)]\}.$$

Interchanging the operations of taking the limit and integrating,

$$\begin{aligned} \mathcal{F}\{sgn(t)\} &= \lim_{a\to 0}\{\int_{-\infty}^{\infty} e^{-a|t|}sgn(t)e^{-i\sigma t}dt\} \\ &= \lim_{a\to 0}\{-\int_{-\infty}^{0} e^{(a-i\sigma)t}dt + \int_{0}^{\infty} e^{-(a+i\sigma)t}dt\}. \end{aligned} \tag{2.78}$$

Proceeding, we obtain

$$\mathcal{F}\{sgn(t)\} = \lim_{a\to 0}\{\frac{-2i\sigma}{a^2+\sigma^2}\} = \frac{2}{i\sigma}. \tag{2.79}$$

The unit step function, expressed in terms of its average value and the signum function, is

$$u(t) = \frac{1}{2} + \frac{1}{2}sgn(t). \tag{2.80}$$

The Fourier transform is

$$\mathcal{F}\{u(t)\} = \frac{1}{2}\mathcal{F}\{1\} + \frac{1}{2}\mathcal{F}\{sgn(t)\} \tag{2.81}$$

which becomes

$$\mathcal{F}\{u(t)\} = \pi\delta(\sigma) + \frac{1}{i\sigma}. \tag{2.82}$$

Therefore, the spectral density function of the unit step function contains an impulse at $\sigma = 0$ corresponding to the average value of $\frac{1}{2}$ in the step function. It also has all the high frequency components of the signum function reduced by one half.

Remark

Note that eqn (2.78) is the integral of a complex function taken symmetrically about the origin. Taking the integral of a complex function in a nonsymmetrical manner may lead to errors. For example, if one attempted to evaluate the Fourier transform of the unit step directly,

$$\lim_{a\to 0}\int_{0}^{\infty} e^{-at}e^{-i\sigma t}dt = \frac{1}{i\sigma}$$

which is not in agreement with the correct result in eqn (2.82).

2.13.5 Periodic functions

The Fourier series of a periodic function $f_p(t)$ of period $2p$ can be expressed as

$$f_p(t) = \sum_{n=-\infty}^{\infty} c_n e^{\frac{in\pi t}{p}}.$$

Taking the Fourier transform, we find

$$\mathcal{F}\{f_p(t)\} = \mathcal{F} \sum_{n=-\infty}^{\infty} c_n e^{\frac{in\pi t}{p}}.$$

If we assume the operations of summation and integration can be interchanged,

$$\mathcal{F}\{f_p(t)\} = \sum_{n=-\infty}^{\infty} c_n \mathcal{F}\{e^{\frac{in\pi t}{p}}\}.$$

Using eqn (2.74), we obtain

$$\mathcal{F}\{f_p(t)\} = 2\pi \sum_{n=-\infty}^{\infty} c_n \delta(\sigma - \frac{n\pi}{p}). \tag{2.83}$$

Thus the Fourier transform (spectral density) of a periodic function consists of a set of impulses located at the harmonic frequencies of the function. The area (weight) of each impulse is 2π times the value of its corresponding coefficient in the exponential Fourier series. This result permits us to handle both periodic and nonperiodic functions in one unified treatment. A listing of some selected Fourier transform pairs appears in Table 2.1.

2.14 Properties of the Fourier transform

The Fourier transform is a method of expressing a given function of time in terms of a continuous set of exponential components of frequency. The resulting spectal density function gives the relative weighting of each frequency component. The Fourier transform relationship between time and frequency is assumed to be unique; that is, $\mathcal{F}^{-1}\mathcal{F}[f(t)] = f(t)$ for all t. The purpose of this section is to introduce some properties of the Fourier transform in a general form.

2.14.1 Linearity (Superposition)

The Fourier transform is a linear operation based on the properties of integration and therefore superposition applies. Thus for any arbitrary constants a_1 and a_2,

$$\mathcal{F}\{a_1 f_1(t) + a_2 f_2(t)\} = a_1 F_1(\sigma) + a_2 F(\sigma). \tag{2.84}$$

This follows from the integral definition of the Fourier transform. Although the proof is trivial, the consequences of this property to the study of linear systems are of major importance.

2.14.2 Complex conjugate

For any complex valued function, we have

$$\mathcal{F}\{f^*(t)\} = F^*(-\sigma). \tag{2.85}$$

Proof

$$\begin{aligned}\mathcal{F}\{f^*(t)\} &= \int_{-\infty}^{\infty} f^*(t)e^{-i\sigma t}dt \\ &= [\int_{-\infty}^{\infty} f(t)e^{i\sigma t}dt]^* \\ &= F^*(-\sigma).\end{aligned}$$

An important consequence of this property is that if the function is real valued, then $f^*(t) = f(t)$ and $F^*(-\sigma) = F(\sigma)$.

2.14.3 Duality

A duality exists between the time domain and the frequency domain. This is exhibited in the Fourier transform pairs and can be stated explicitly in the following manner.

If $\mathcal{F}\{f(t)\} = F(\sigma)$, then $\mathcal{F}\{F(t)\} = 2\pi f(-\sigma)$. A proof of this property involves an interchange in t and σ in the Fourier transform integrals. Its use is shown in the following example.

Example 10

Given that $\mathcal{F}\{rect(t)\} = Sa(\frac{\sigma}{2})$, determine $\mathcal{F}\{Sa(\frac{t}{2})\}$.

Solution

Let $F(\sigma) = Sa(\frac{\sigma}{2})$, then $F(t) = Sa(\frac{t}{2})$. Using the above duality property $\mathcal{F}\{F(t)\} = 2\pi f(-\sigma)$, we obtain

$$\mathcal{F}\{Sa(\frac{t}{2})\} = 2\pi rect(-\sigma) = 2\pi rect(\sigma).$$

2.14.4 Coordinate scaling

The expansion or compression of a time waveform affects the spectral density of the waveform. For a real valued scaling constant β and any pulse signal $f(t)$,

$$\mathcal{F}\{f(\beta t)\} = \frac{1}{|\beta|}F(\frac{\sigma}{\beta}). \tag{2.86}$$

Proof

By integral definition of a Fourier transform

$$\mathcal{F}\{f(\beta t)\} = \int_{-\infty}^{\infty} f(\beta t)e^{-i\sigma t}dt.$$

Table 2.1: Fourier transforms of some elementary functions

	$f(t)$	$\mathcal{F}\{f(t)\} = F(\sigma) = \int_{-\infty}^{\infty} f(t)e^{-i\sigma t}dt$
1.	$e^{-at}u(t) \quad a > 0$	$\frac{1}{a+i\sigma}$
2.	$e^{-a\vert t\vert} \quad a > 0$	$\frac{2a}{a^2+\sigma^2}$
3.	$te^{-at}u(t) \quad a > 0$	$\frac{1}{(a+i\sigma)^2}$
4.	$e^{-\frac{t^2}{2\omega^2}}$	$\omega\sqrt{2\pi}e^{-\frac{\sigma^2\omega^2}{2}}$
5.	$sgn(t) = \begin{cases} 1 & t > 0 \\ -1 & t < 0 \end{cases}$	$\frac{2}{i\sigma}$
6.	$\frac{i}{\pi t}$	$sgn(t)$
7.	$u(t) = \begin{cases} 1 & t > 0 \\ 0 & t < 0 \end{cases}$	$\pi\delta(\sigma) + \frac{1}{i\sigma}$
8.	$\delta(t) = \begin{cases} \infty & t = 0 \\ 0 & t \neq 0 \end{cases}$	1
9.	1	$2\pi\delta(\sigma)$
10.	$e^{\pm i\sigma_0 t}$	$2\pi\delta(\sigma \mp \sigma_0)$
11.	$\cos\sigma_0 t$	$\pi\{\delta(\sigma-\sigma_0)+\delta(\sigma+\sigma_0)\}$
12.	$\sin\sigma_0 t$	$-i\pi\{\delta(\sigma-\sigma_0)-\delta(\sigma+\sigma_0)\}$
13.	$rect(\frac{t}{\tau})$	$\tau Sa(\frac{\sigma\tau}{2})$
14.	$Sa(\frac{t}{2})$	$2\pi rect(\sigma)$
15.	$Sa(t)$	$\pi rect(\frac{\sigma}{2})$

We consider positive and negative values of β separately. For β positive, and changing the variable of integration to $x = \beta t$, we have

$$\begin{aligned}\mathcal{F}\{f(\beta t)\} &= \int_{-\infty}^{\infty} f(x)e^{-i\sigma x/\beta}dx/\beta \\ &= \frac{1}{\beta}F(\frac{\sigma}{\beta}) \quad \text{for } \beta > 0.\end{aligned}$$

When β is negative, the limits on the integral are reversed when the variable of integration is changed so that

$$\mathcal{F}\{f(\beta t)\} = -\frac{1}{\beta}F(\frac{\sigma}{\beta}).$$

These two cases can be combined into more compact form

$$\mathcal{F}\{f(\beta t)\} = \frac{1}{|\beta|}F(\frac{\sigma}{\beta}).$$

2.14.5 Time shifting

Another geometric operation of the time origin, causing the signal to be delayed (or advanced) in time by some time t_0 is the corresponding effect on signal spectral density which is given below:

$$\mathcal{F}\{f(t-t_0)\} = F(\sigma)e^{-i\sigma t_0}. \tag{2.87}$$

Proof

$$\mathcal{F}\{f(t-t_0)\} = \int_{-\infty}^{\infty} f(t-t_0)e^{-i\sigma t}dt.$$

Changing the variable of integration, let $x = t - t_0$,

$$\begin{aligned}
\mathcal{F}\{f(t-t_0)\} &= \int_{-\infty}^{\infty} f(x)e^{-i\sigma(x+t_0)}dx \\
&= e^{-i\sigma t_0}\int_{-\infty}^{\infty} f(x)e^{-i\sigma x}dx \\
&= e^{-i\sigma t_0}F(\sigma).
\end{aligned}$$

Thus if a signal $f(t)$ is delayed in time by t_0, its magnitude spectral density remains unchanged and a negative phase $(-\sigma t_0)$ is added to each frequency component.

2.14.6 Frequency shifting (Modulation)

The dual of the delay property is the frequency translation property,

$$\mathcal{F}\{f(t)e^{i\sigma_0 t}\} = F(\sigma - \sigma_0). \tag{2.88}$$

Proof

$$\begin{aligned}
\mathcal{F}\{f(t)e^{i\sigma_0 t}\} &= \int_{-\infty}^{\infty} f(t)e^{i\sigma_0 t}e^{-i\sigma t}dt \\
&= \int_{-\infty}^{\infty} e^{-i(\sigma-\sigma_0)t}dt \\
&= F(\sigma - \sigma_0).
\end{aligned}$$

Therefore multiplying a time function by $e^{i\sigma_0 t}$ causes its spectral density to be translated in frequency by σ_0 rad/sec.

2.14.7 Differentiation and integration

If $\frac{df}{dt}$ is absolutely integrable, then

$$\mathcal{F}\{\frac{df}{dt}\} = i\sigma F(\sigma). \tag{2.89}$$

Proof

Using the inverse Fourier transform, we have

$$\begin{aligned} f(t) &= \frac{1}{2\pi}\int_{-\infty}^{\infty} F(\sigma)e^{i\sigma t}d\sigma \\ \frac{df}{dt} &= \frac{1}{2\pi}\int_{-\infty}^{\infty} (i\sigma)F(\sigma)e^{i\sigma t}d\sigma. \end{aligned}$$

Taking the Fourier transform of both sides,

$$\mathcal{F}\{\frac{d}{dt}f(t)\} = i\sigma F(\sigma).$$

Therefore time differentiation enhances the high frequency components of a signal. The corresponding integration property is

$$\mathcal{F}\{\int_{-\infty}^{t} f(\tau)d\tau\} = \frac{1}{i\sigma}F(\sigma) + \pi F(0)\delta(\sigma) \tag{2.90}$$

where

$$F(0) = \int_{-\infty}^{\infty} f(t)dt. \tag{2.91}$$

Proof

Now integrating the inverse Fourier transform,

$$\begin{aligned} \int_{-\infty}^{t} f(\tau)d\tau &= \frac{1}{2\pi}\int_{-\infty}^{t}\{\int_{-\infty}^{\infty} F(\sigma)e^{i\sigma\tau}d\sigma\}d\tau \\ &= \frac{1}{2\pi}\int_{-\infty}^{\infty} F(\sigma)\{\int_{-\infty}^{t} e^{i\sigma\tau}d\tau\}d\sigma \\ &= \frac{1}{2\pi}\int_{-\infty}^{\infty} F(\sigma)\{\frac{e^{i\sigma t}}{i\sigma} - \lim_{\tau\to\infty}\frac{e^{-i\sigma\tau}}{i\sigma}\}d\sigma. \end{aligned}$$

Now using the known values of

$$\lim_{\tau\to\infty}\frac{\sin\sigma\tau}{\sigma\pi} = \delta(\sigma), \quad \lim_{\tau\to\infty}\frac{\cos\sigma\tau}{\sigma\pi} = 0$$

we obtain

$$\begin{aligned} \mathcal{F}\{\int_{-\infty}^{t} f(\tau)d\tau\} &= \frac{F(\sigma)}{i\sigma} + \mathcal{F}\{\frac{F(0)}{2}\} \\ &= \frac{F(\sigma)}{i\sigma} + \pi F(0)\delta(\sigma) \end{aligned}$$

where $F(0) = \int_{-\infty}^{\infty} f(t)dt$. Integration in time suppresses the high frequency components of a signal. This conclusion agrees with the time domain view point that integration smooths out the time fluctuation in a signal.

2.14.8 Time convolution

One method of characterizing a system is by its frequency transfer function, a second method is by its impulse response. We now wish to relate these methods using the principle of convolution.

For the test signal $f(t) = \delta(t-\tau)$, the system impulse response is defined as

$$\mathcal{D}\{\delta(t-\tau)\} = h(t-\tau) \tag{2.92}$$

where τ is the delay or "age" variable. The input signal $f(t)$ may be expressed in terms of the impulse function by

$$\begin{aligned} f(t) &= \int_{-\infty}^{\infty} f(\tau)\delta(\tau - t)d\tau \\ &= \int_{-\infty}^{\infty} f(\tau)\delta(t-\tau)d\tau. \end{aligned} \tag{2.93}$$

Now using the operator $\mathcal{D}$ in the above equation, we have

$$\mathcal{D}\{f(t)\} = \int_{-\infty}^{\infty} f(\tau)\mathcal{D}\{\delta(t-\tau)\}d\tau$$

such that

$$y(t) = \int_{-\infty}^{\infty} f(\tau)h(t-\tau)d\tau = f(t) * h(t). \tag{2.94}$$

This result is known as the Convolution integral. Now taking the Fourier transform of this convolution integral, we obtain

$$\mathcal{F}\{y(t)\} = \mathcal{F}\{f(t) * h(t)\} = \mathcal{F}\{f(t)\}\mathcal{F}\{h(t)\} = F(\sigma)H(\sigma).$$

Thus convolution in the time domain corresponds to multiplication in the frequency domain.

2.14.9 Frequency convolution

A dual to the preceding property is the following:

If $\mathcal{F}\{f_1(t)\} = F_1(\sigma)$, and $\mathcal{F}\{f_2(t)\} = F_2(\sigma)$; then

$$\mathcal{F}\{f_1(t)f_2(t)\} = \frac{1}{2\pi}\{F_1(\sigma) * F_2(\sigma)\}. \tag{2.95}$$

where

$$F_1(\sigma) * F_2(\sigma) = \int_{-\infty}^{\infty} F(u)F(\sigma - u)du.$$

Proof

We know that

$$f_1(t) = \frac{1}{2\pi}\int_{-\infty}^{\infty} F_1(\sigma)e^{i\sigma t}d\sigma$$
$$f_2(t) = \frac{1}{2\pi}\int_{-\infty}^{\infty} F_2(\sigma)e^{i\sigma t}d\sigma.$$

Multiply the first equation by $f_2(t)$, and we have

$$f_1(t)f_2(t) = \frac{1}{2\pi}\int_{-\infty}^{\infty} F_1(u)f_2(t)e^{iut}du.$$

Table 2.2: Fourier transforms of some mathematical operations

	Operation	$f(t)$	$\mathcal{F}\{f(t)\} = F(\sigma)$
1.	Linearity	$a_1f_1(t) + a_2f_2(t)$	$a_1F_1(\sigma) + a_2F_2(\sigma)$
2.	Complex conjugate	$f^*(t)$	$F^*(-\sigma)$
3.	Scaling	$f(\beta t)$	$\frac{1}{\lvert\beta\rvert}F(\frac{\sigma}{\beta})$
4.	Delay	$f(t - t_0)$	$e^{-i\sigma t_0}F(\sigma)$
5.	Frequency translation	$e^{i\sigma_0 t}f(t)$	$F(\sigma - \sigma_0)$
6.	Amplitude modulation	$f(t)\cos\sigma_0 t$	$\frac{1}{2}\{F(\sigma+\sigma_0) + F(\sigma - \sigma_0)\}$
7.	Time convolution	$\int_{-\infty}^{\infty} f_1(\tau)f_2(t-\tau)d\tau$	$F_1(\sigma)F_2(\sigma)$
8.	Frequency convolution	$f_1(t)f_2(t)$	$\frac{1}{2\pi}\int_{-\infty}^{\infty} F_1(u)F_2(\sigma - u)du$
9.	Duality-time frequency	$F(t)$	$2\pi f(-\sigma)$
10.	Time-differentiation	$\frac{d}{dt}f(t)$	$i\sigma F(\sigma)$
11.	Time-integration	$\int_{-\infty}^{t} f(\tau)d\tau$	$\frac{1}{i\sigma}F(\sigma) \quad +\pi F(0)\delta(\sigma)$ where $F(0) \quad = \int_{-\infty}^{\infty} f(t)dt$

Then taking the Fourier transforms

$$\begin{aligned}\mathcal{F}\{f_1(t)f_2(t)\} &= \int_{-\infty}^{\infty}\{\frac{1}{2\pi}\int_{-\infty}^{\infty} F_1(u)f_2(t)e^{iut}du\}e^{-i\sigma t}dt \\ &= \frac{1}{2\pi}\int_{-\infty}^{\infty} F_1(u)\{\int_{-\infty}^{\infty} f_2(t)e^{-i(\sigma-u)t}dt\}du \\ &= \frac{1}{2\pi}\int_{-\infty}^{\infty} F_1(u)F_2(\sigma - u)du \\ &= \frac{1}{2\pi}\{F_1(\sigma) * F_2(\sigma)\}.\end{aligned}$$

Thus the multiplication of two functions in the time domain is equivalent to the convolution of their spectral densities in the frequency domain.

Example 11

Find $f(t) * h(t)$ for the $f(t) = A\sin\pi t u(t)$ and $h(t) = \delta(t) - \delta(t-2)$ as shown in the following figure by using three methods: (a) Direct integration, (b) Fourier transform method and (c) Laplace transform method.

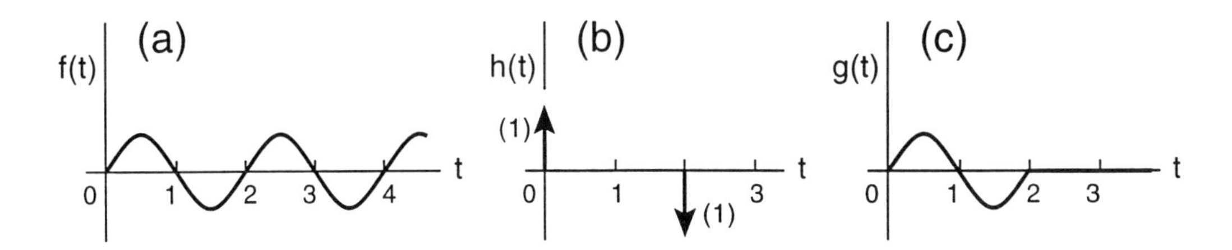

Figure 2.10: The sketch of $f(t)$ and $h(t)$

Solution

(a) By direct integration:

$$\begin{aligned} y(t) &= f(t) * h(t) \\ &= \int_{-\infty}^{\infty} \{A \sin \pi\tau u(\tau)\}\{\delta(t-\tau) - \delta(t-\tau-2)\}d\tau \\ &= A \sin \pi t u(t) - A \sin \pi(t-2)u(t-2) \\ &= \begin{cases} 0 & t < 0 \\ A \sin \pi t & 0 < t < 2 \\ 0 & t > 2. \end{cases} \end{aligned}$$

(b) Fourier transform method: We know

$$\mathcal{F}\{y(t)\} = \mathcal{F}\{f(t)\}\mathcal{F}\{h(t)\}$$

and

$$\mathcal{F}\{h(t)\} = \mathcal{F}\{\delta(t) - \delta(t-2)\} = 1 - e^{-i2\sigma}.$$

Thus

$$\begin{aligned} \mathcal{F}\{y(t)\} &= \mathcal{F}\{f(t)\}[1 - e^{-i2\sigma}] \\ &= \mathcal{F}\{f(t)\} - e^{-i2\sigma}\mathcal{F}\{f(t)\}. \end{aligned}$$

Hence taking the inverse Fourier transform

$$y(t) = A \sin \pi t u(t) - A \sin \pi(t-2)u(t-2)$$

which is the same as the previous result.

(c) By Laplace transform method (see Chapter 3):

$$\begin{aligned}\mathcal{L}\{f(t)\} &= \mathcal{L}\{A\sin\pi t\} = \frac{A\pi}{s^2+\pi^2} \\ \mathcal{L}\{h(t)\} &= \mathcal{L}\{\delta(t)-\delta(t-2)\} = 1-e^{-2s}.\end{aligned}$$

Thus we have

$$\begin{aligned}\mathcal{L}\{y(t)\} &= [\mathcal{L}f(t)][\mathcal{L}h(t)] \\ &= [\frac{A\pi}{s^2+\pi^2}][1-e^{-2s}] \\ &= \frac{A\pi}{s^2+\pi^2} - \frac{A\pi e^{-2s}}{s^2+\pi^2}.\end{aligned}$$

Now taking the inverse Laplace transform

$$y(t) = A\sin\pi t u(t) - a\sin\pi(t-2)u(t-2)$$

which is the same as the previuos two results.

2.15 The frequency transfer function

A fundamental property of a linear time-invariant system is that the input, $f(t)$, and the output, $y(t)$, are related by linear differential equations with constant coefficients. A typical system could be described by

$$\phi(D)\{y(t)\} = \psi(D)\{f(t)\} \tag{2.96}$$

where

$$\begin{aligned}\phi(D) &= a_0D^n + a_1D^{n-1} + \cdots + a_{n-1}D + a_n \\ \psi(D) &= b_0D^n + b_1D^{n-1} + \cdots + b_{n-1}D + b_n\end{aligned}$$

in which the a's and b's are constants. Now we use the input signal

$$f(t) = e^{i\sigma t} \tag{2.97}$$

to test the system. A particular solution can be written as

$$y(t) = H(\sigma)e^{i\sigma t}. \tag{2.98}$$

Using eqn (2.97) and eqn (2.98) in eqn (2.96), we obtain

$$H(\sigma) = \frac{\psi(i\sigma)}{\phi(i\sigma)}. \tag{2.99}$$

This important ratio is called the frequency transfer function of the system. Note that the right-hand side of eqn (2.99) depends only on the system.

Physically, eqn (2.99) tells us that a way to test a linear time-invariant system is to apply a sinusoid of known amplitude, frequency, and phase to the input of the system. The output will be another sinusoid at the same frequency but the amplitude and phase will, in general, differ from that of the input. Taking the ratio of these two complex coefficients gives the value (in amplitude and phase) of the system transfer function at that frequency.

Example 12

Determine the frequency transfer function of the system shown in Fig. 2.11

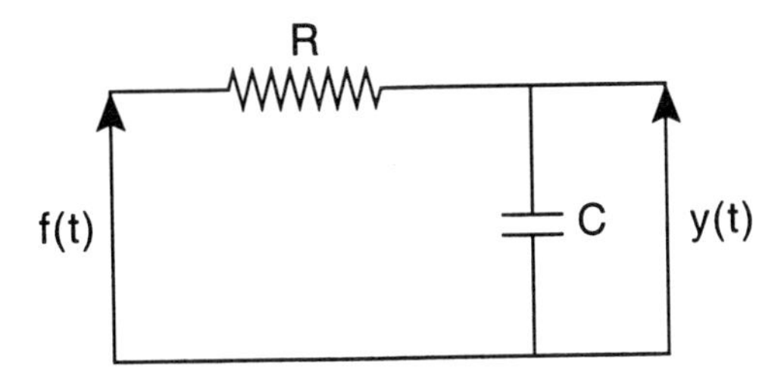

Figure 2.11: The RC low pass filter

Solution

The differential equation describing this system is found easily by using the Kirchoff's law:

$$RC\frac{d}{dt}\{y(t)\} + y(t) = f(t).$$

For the calculation of the system frequency transfer function we let $f(t) = e^{i\sigma t}$ and $y(t) = H(\sigma)e^{i\sigma t}$ so that substitution of the particular solution in the differential equation above becomes

$$RC(i\sigma)H(\sigma)e^{i\sigma t} + H(\sigma)e^{i\sigma t} = e^{i\sigma t}.$$

The system frequency transfer function is then

$$H(\sigma) = \frac{1}{i\sigma RC + 1}.$$

Chapter 3

Laplace transforms

3.1 Introduction

In the previous chapter we saw that the evolution of the Laplace transform was a special case of the general Fourier transform. In this chapter we develop the theory of Laplace transforms and some related concepts. It will be seen later that the method of Laplace transforms is a powerful technique for solving both ordinary and partial differential equations.

In the preceding chapters we learned how to solve linear differential equations with constant coefficients subject to given boundary or initial conditions. During the 19th century, scientists and engineers were encouraged with using the operator method to solve differential equations. The English electrical engineer Oliver Heaviside (1850–1925) was the first to develop the operator method. In this method, operators were treated as algebraic symbols and the resulting equations were manipulated according to the rules of algebra. Remarkably, the method led to correct answers. These successes encouraged scientists and engineers to use the method even more. Some thoughtful mathematicians reasoned that there ought to be some way of placing the procedures on rigorous mathematical foundations. Research towards this goal led to the development of the method of Laplace transforms. Although Pierre Simon de Laplace, a French mathematician, initiated this method in the 18th century, the British electrical engineer Oliver Heaviside developed it and mathematicians put it in its present form.

3.2 Definition of Laplace transform

Let $f(t)$ be a function of t specified for $t > 0$. The Laplace transform of $f(t)$ denoted by $\mathcal{L}\{f(t)\}$ is defined by

$$\mathcal{L}\{f(t)\} = \int_0^\infty e^{-st} f(t) dt = F(s) \tag{3.1}$$

where the parameter s is, in general, a complex quantity. The real part of s is always greater than zero which guarantees the convergence of the integral provided $f(t)$ is a "well-behaved" function. The Laplace transform of $f(t)$ is usually denoted by the symbol $F(s)$.

The inverse of Laplace transforms is given by the complex inversion integral

$$f(t) = \frac{1}{2\pi i} \int_{a-i\infty}^{a+i\infty} e^{st} F(s) ds \tag{3.2}$$

where the path of integration is taken along a Bromwich contour. We shall learn about this contour when we take up the theory of residues.

For the Laplace transform of $f(t)$ defined by eqn (3.1) to exist and for its inverse, defined by (3.2), to exist, it is sufficient that the following two conditions be satisfied.

(a) The function $f(t)$ must be piecewise continuous in the interval $0 \leq t < \infty$. That is, every interval of the form $0 \leq t_1 \leq t \leq t_2$ can be divided into a finite number of subintervals such that $f(t)$ is continuous in the interior of each subinterval and approaches finite limits as t approaches either end-point of the interval.

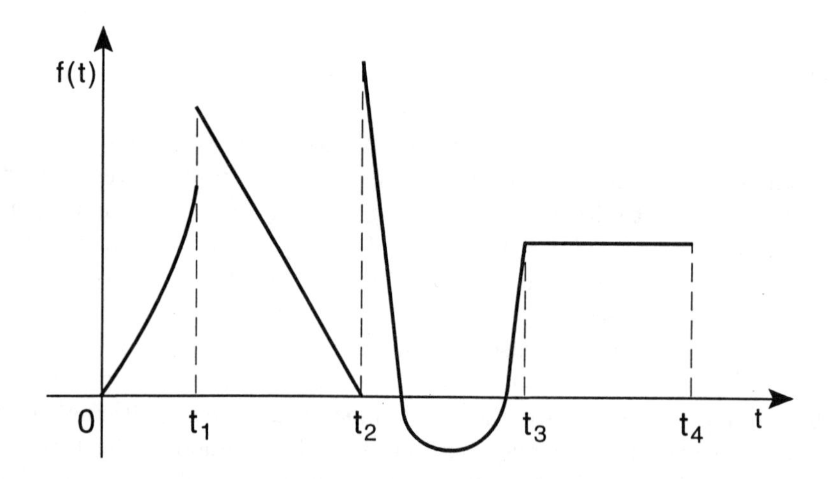

Figure 3.1. Piecewise continuous function $f(t)$

(b) The function $f(t)$ must be of exponential order. That is, there is a constant λ with the property that

$$e^{-\lambda t}|f(t)| \text{remains bounded as } t \to \infty$$

i.e., there are constants λ, M and T such that

$$e^{-\lambda t}|f(t)| < M \quad \text{for} \quad t > T.$$

When these two conditions are satisfied, then a Laplace transform of $f(t)$ exists and the transform can be determined from the definition (3.1). Items (a) and (b) together define the "well-behaved" function referred to earlier.

In connection with condition (b) we need the concept of the abscissa of convergence, since for $t > 0, e^{-\lambda t}$ is a monotonically-decreasing function of λ. It is clear that if

$$e^{-\lambda t}|f(t)| < M \quad t > T$$

then, for all $\lambda_1 > \lambda$,

$$e^{-\lambda_1 t}|f(t)| < M \quad t > T.$$

Thus the λ required by condition (b) is not unique. Thus we define the Abscissa of convergence of $f(t)$ as the greatest lower bound λ_0 of the set of all λ's which can be used in condition (b). For example, if $f(t) = e^{3t}$, then

$$\mathcal{L}\{f(t)\} = \int_0^\infty e^{-\lambda t} f(t)dt = \int_0^\infty e^{-(\lambda-3)t}dt.$$

For the convergence of this integral λ must be greater than 3 and the greatest lower bound of λ must be $\lambda_0 = 3$. The integral may or may not exist at the greatest lower bound; in this particular example, the integral does not exist at $\lambda_0 = 3$. In a Laplace transform, we usually use the s as a parameter rather than λ.

Thus we can summarize: if $f(t)$ is piecewise regular in $0 \le t < \infty$ and of exponential order, then for any value of s which is greater than the abscissa of convergence of $f(t)$, the integral $\int_0^\infty e^{-st} f(t)dt$ converges absolutely.

3.3 Laplace transform properties

In the following section, we are going to illustrate three theorems which will be useful in obtaining the Laplace transforms of some functions.

Theorem 3.1

If $f_1(t)$ and $f_2(t)$ are piecewise regular functions of exponential order and c_1 and c_2 are any two constants, then

$$\mathcal{L}\{c_1 f_1(t) \pm c_2 f_2(t)\} = c_1 \mathcal{L}\{f_1(t)\} \pm c_2 \mathcal{L}\{f_2(t)\}.$$

Proof

By the definition of the Laplace transform, we know

$$\begin{aligned}
\mathcal{L}\{c_1 f_1(t) \pm c_2 f_2(t)\} &= \int_0^\infty \{c_1 f_1(t) \pm c_2 f_2(t)\} e^{-st} dt \\
&= c_1 \int_0^\infty f_1(t) e^{-st} dt \pm c_2 \int_0^\infty f_2(t) e^{-st} dt \\
&= c_1 \mathcal{L}\{f_1(t)\} \pm c_2 \mathcal{L}\{f_2(t)\}.
\end{aligned}$$

This is the required proof.

This result may be extended to any finite number of terms and may be extended to an infinite number if the sum is convergent. This theorem expresses the fact that the Laplace transform operator is a linear operator.

Theorem 3.2

If $f(t)$ is a continuous function of exponential order on $(0, \infty)$ whose derivative $f'(t)$ is also of exponential order and at least piecewise regular on $(0, \infty)$, then the Laplace transform of $f'(t)$ is given by the formula

$$\mathcal{L}\{f'(t)\} = s\mathcal{L}\{f(t)\} - f(0)$$

provided s is greater than the abscissa of convergence of $f(t)$.

Proof

Using the definition of the Laplace transform and integrating by parts, we obtain

$$\begin{aligned} \mathcal{L}\{f'(t)\} &= \int_0^\infty e^{-st} f'(t) dt \\ &= [\{e^{-st} f(t)\}]_{0+}^\infty + s \int_0^\infty e^{-st} f(t) dt \\ &= s\mathcal{L}\{f(t)\} - f(0_+). \end{aligned}$$

This is the required proof.

This theorem can be extended to the nth derivative of the given function as follows: If $f(t)$ and $f'(t)$ are continuous for $0 \leq t < \infty$ and of exponential order as $t \to \infty$, while $f'(t)$ is piecewise continuous for $0 \leq t < \infty$, then

$$\begin{aligned} \mathcal{L}\{f''(t)\} &= \mathcal{L}\{[f'(t)]'\} \\ &= s\mathcal{L}\{f'(t)\} - f'(0); \qquad \text{using Theorem 7.2} \\ &= s[s\mathcal{L}\{f(t)\} - f(0)] - f'(0); \qquad \text{using Theorem 7.2 again.} \end{aligned}$$

Therefore,

$$\mathcal{L}\{f''(t)\} = s^2\mathcal{L}\{f(t)\} - sf(0) - f'(0).$$

Similarly, if $f(t)$ and its first $(n-1)$ derivatives are continuous for $0 \leq t < \infty$ and of exponential order as $t \to \infty$, while $f^{(n)}(t)$ is piecewise continuous for $0 \leq t < \infty$, then

$$\begin{aligned} \mathcal{L}\{f^{(n)}(t)\} &= s^n\mathcal{L}\{f(t)\} - s^{n-1}f(0) - s^{n-2}f'(0) \\ &\quad - \cdots - sf^{(n-2)}(0) - f^{(n-1)}(0). \end{aligned}$$

In many physical situations, we find that the physics of the problem is governed by an integrodifferential equation, in addition to a differential equation. For example, the series-electrical L-C-R circuit can be described either by a differential equation or by an integrodifferential equation: $L\frac{dI}{dt} + IR + \frac{Q}{C} = E$ or $L\frac{dI}{dt} + IR + \frac{1}{C}\int Idt = E$.

We saw in Theorem 3.2 how to obtain the Laplace transform of a derivative of a function. To solve an integrodifferential equation by the Laplace transform, we need a formula. Theorem 3.3 provides the formula for solving an integrodifferential equation.

Theorem 3.3

If $f(t)$ is piecewise regular for $0 \le t < \infty$ and of exponential order, then the Laplace transform of $\int_{t_0}^{t} f(t)dt$ is given by the formula

$$\mathcal{L}\{\int_{t_0}^{t} f(t)dt\} = \frac{1}{s}\mathcal{L}\{f(t)\} + \frac{1}{s}\int_{t_0}^{0} f(t)dt \quad t \ge 0.$$

Proof

Since $f(t)$ is piecewise regular for $0 \le t < \infty$ and of exponential order as $t \to \infty$, and since $t \ge 0$, the function $g(t) = \int_{t_0}^{t} f(\lambda)d\lambda$ is continuous for $0 \le t < \infty$ and is also of exponential order. Then $g'(t) = f(t)$, and hence using Theorem 3.2, we have $\mathcal{L}\{g(t)\} = \frac{1}{s}\mathcal{L}\{f(t)\} + \frac{1}{s}g(0)$ which implies $\mathcal{L}\{\int_{t_0}^{t} f(\lambda)d\lambda\} = \frac{1}{s}\mathcal{L}\{f(t)\} + \frac{1}{s}\int_{t_0}^{0} f(t)dt$. This is the required proof.

Extending this theorem to repeated integrals we obtain:

$$\begin{aligned}
\mathcal{L}\{\int_{t_0}^{t}\int_{t_0}^{t} f(t)dtdt\} &= \mathcal{L}\{\int_{t_0}^{t}[\int_{t_0}^{t} f(t)dt]dt \\
&= \frac{1}{s}\mathcal{L}\{\int_{t_0}^{t} f(t)dt\} + \frac{1}{s}\int_{t_0}^{0}\int_{t_0}^{t} f(t)dtdt \\
&= \frac{1}{s^2}\mathcal{L}\{f(t)\} + \frac{1}{s^2}\int_{t_0}^{0} f(t)dt + \frac{1}{s}\int_{t_0}^{0}\int_{t_0}^{t} f(t)dtdt.
\end{aligned}$$

Example 1

Find the Laplace transform of the following integrodifferential equation

$$y'(t) + ay(t) + b\int_0^t y(\lambda)d\lambda = 0; \qquad y(0) = 2.$$

Solution

Take the Laplace transform of the given equation:

$$\mathcal{L}[y' + ay + b\int_0^t y(\lambda)d\lambda] = \mathcal{L}(0).$$

Then, using Theorems 3.2 and 3.3, we obtain $s\mathcal{L}(y) - y(0) + a\mathcal{L}\{y\} + \frac{b}{s}\mathcal{L}(y) = 0$. Therefore, $Y(s) = \mathcal{L}\{y\} = \frac{2s}{s^2+as+b}$.

3.4 Laplace transforms of special functions

In this section we will use some formulae for finding the transforms of special functions. Some of the more elementary functions are the following: e^{-at}, e^{at}, $\sinh at$, $\cosh at$, $\cos bt$, $\sin bt$, t^n; and the unit step function:

$$u(t) = \begin{cases} 0 & t < 0 \\ 1 & t > 0. \end{cases}$$

The unit step function is shown in Fig. 3.2.

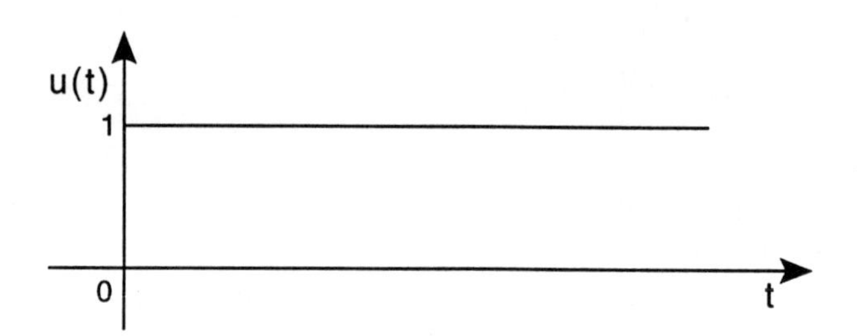

Figure 3.2. The unit step function $u(t)$

The transforms of these functions are the following:

Formula 1: $\mathcal{L}\{e^{-at}\} = \frac{1}{s+a}, \quad s > -a.$

Formula 2: $\mathcal{L}\{e^{at}\} = \frac{1}{s-a}, \quad s > a.$

Formula 3: $\mathcal{L}\{\sinh at\} = \frac{a}{s^2-a^2}, \quad s > a.$

Formula 4: $\mathcal{L}\{\cosh at\} = \frac{s}{s^2-a^2}, \quad s > a.$

Formula 5: $\mathcal{L}\{\cos bt\} = \frac{s}{s^2+b^2}, \quad s > 0.$

Formula 6: $\mathcal{L}\{\sin bt\} = \frac{b}{s^2+b^2}, \quad s > 0.$

Formula 7: $\mathcal{L}\{u(t)\} = \frac{1}{s}, \quad s > 0.$

Formula 8: $\mathcal{L}\{t^n\} = \begin{cases} \frac{\Gamma(n+1)}{s^{n+1}} & , \quad n > -1 \\ \frac{n!}{s^{n+1}} & , \quad n \text{ positive integer }, s > 0. \end{cases}$

Proofs of all these formulas are trivial and will not be given except Formula 8.

Proof of Formula 8

By the definition of Laplace transform

$$\mathcal{L}\{t^n\} = \int_0^\infty e^{-st} t^n dt$$

and substituting $st = z$ such that $dt = \frac{1}{s}dz$, we obtain $\mathcal{L}\{t^n\} = \frac{1}{s^{n+1}}\Gamma(n+1)$ where $\Gamma(n+1)$ is called the Gamma function and is defined as

$$\Gamma(n+1) = \int_0^\infty e^{-z} z^n dz \quad , \quad n+1 > 0. \tag{3.3}$$

Performing this integration by parts, we see that

$$\Gamma(n+1) = [-e^{-z}z^n]_0^\infty + n\int_0^\infty e^{-z}z^{n-1}dz.$$

Therefore,

$$\Gamma(n+1) = n\Gamma(n). \tag{3.4}$$

This is known as the recurrence relation of a Gamma function. If n is an integer, then $\Gamma(n+1) = (n)(n-1)(n-2)\cdots 2.1\Gamma(1)$ where $\Gamma(1) = 1$, and hence

$$\Gamma(n+1) = n! \tag{3.5}$$

Therefore,

$$\mathcal{L}\{t^n\} = \begin{cases} \frac{\Gamma(n+1)}{s^{n+1}} & n+1 > 0 \\ \frac{n!}{s^{n+1}} & n \text{ a positive integer.} \end{cases}$$

From the definition of the Gamma function, it can be seen that

$$\Gamma(0) = \int_0^\infty e^{-z}z^{-1}dz$$

approaches to ∞ at the lower limit. Therefore, $\Gamma(0)$ does not exist and $\Gamma(1) = 0! = 1$ by definition. Thus

$$\begin{aligned} \Gamma(n) &= \frac{\Gamma(n+1)}{n} \\ \Gamma(-1) &= -\infty \\ \Gamma(-2) &= \infty \\ \Gamma(-3) &= -\infty \end{aligned}$$

and so on. It is interesting to see the numerical value of $\Gamma(\frac{1}{2})$, which is defined as

$$\Gamma(\frac{1}{2}) = \int_0^\infty e^{-z}z^{-\frac{1}{2}}dz.$$

Put $z = x^2$ such that $dz = 2xdx$. Therefore,

$$\Gamma(\frac{1}{2}) = 2\int_0^\infty e^{-x^2}dx. \tag{3.6}$$

Also,

$$\Gamma(\frac{1}{2}) = 2\int_0^\infty e^{-y^2}dy. \tag{3.7}$$

Multiplying (3.6) and (3.7):

$$[\Gamma(\frac{1}{2})]^2 = 4\int_0^\infty\int_0^\infty e^{-(x^2+y^2)}dxdy.$$

Substitute $x = r\cos\theta$ and $y = r\sin\theta$, and under this transformation, $dxdy \to rd\theta dr$.

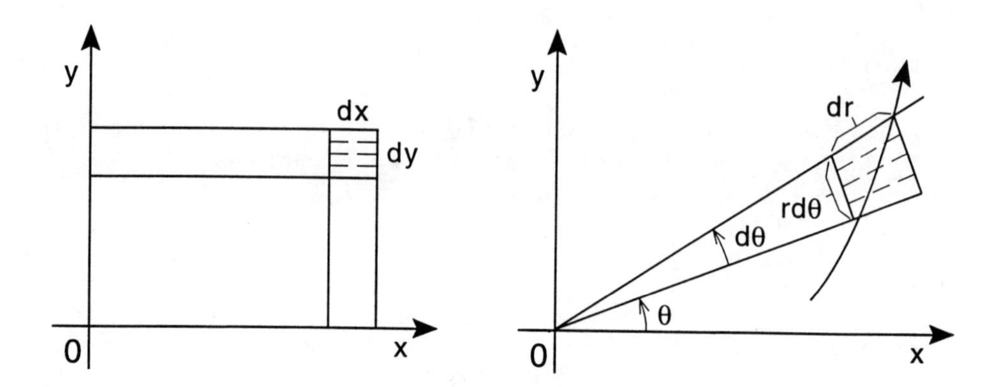

Figure 3.3. Relationship between the elementary areas in Cartesian and polar coordinate systems

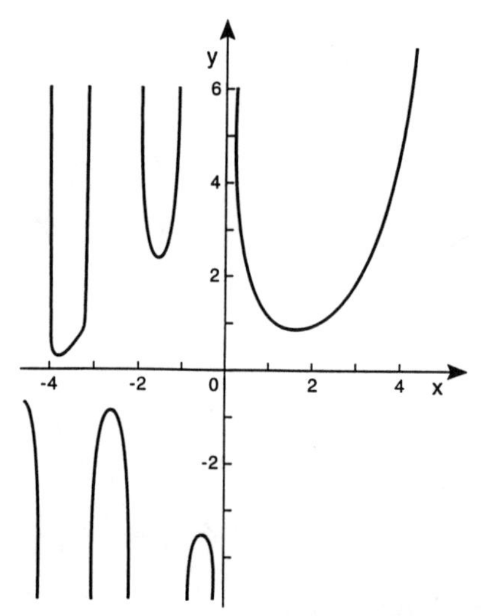

Figure 3.4. Plot of $\Gamma(n)$ against n

Hence

$$[\Gamma(\frac{1}{2})]^2 = 4\int_{r=0}^{\infty}\int_{\theta=0}^{\pi/2} e^{-r^2} r dr d\theta = \pi.$$

Therefore,

$$\Gamma(\frac{1}{2}) = \sqrt{\pi}. \tag{3.8}$$

We are now in a position to plot the Gamma function.

In the following, we state and prove three important theorems of Laplace transforms.

Theorem 3.4: (Scale change)

If $\mathcal{L}\{f(t)\} = F(s)$ when $s > a$, then for $b > 0$,

$$\mathcal{L}\{f(bt)\} = \frac{1}{b}F(\frac{s}{b}) \quad , \quad \frac{s}{b} > a.$$

Proof

By definition

$$\mathcal{L}\{f(bt)\} = \int_0^\infty e^{-st} f(bt)dt.$$

Substituting $bt = \tau$ so that $dt = \frac{1}{b}d\tau$ it can be easily shown that $\mathcal{L}\{f(bt)\} = \frac{1}{b}F(\frac{s}{b})$ for $\frac{s}{b} > a$. This is the required proof.

Theorem 3.5: (First shifting theorem)

If $f(t)$ is piecewise regular for $0 \le t < \infty$ and of exponential order and if $\mathcal{L}\{f(t)\} = F(s) = \int_0^\infty e^{-st} f(t)dt$ then $\mathcal{L}\{e^{-at}f(t)\} = F(s+a)$.

Proof

By definition, $\mathcal{L}\{e^{-at}f(t)\} = \int_0^\infty \{e^{-at}f(t)\}dt = \int_0^\infty e^{-(s+a)t}f(t)dt = F(s+a)$

Similarly, we can prove that

$$\mathcal{L}\{e^{at}f(t)\} = F(s-a).$$

Hence the following two results can be obtained:

Corollary 1

If $\mathcal{L}\{e^{-at}f(t)\} = F(s+a)$ then $f(t) = e^{at}\mathcal{L}^{-1}\{F(s+a)\}$.

Corollary 2

If $\mathcal{L}\{e^{at}f(t)\} = F(s-a)$ then $f(t) = e^{-at}\mathcal{L}^{-1}\{F(s-a)\}$.

By means of this theorem the following formulas can be established:

Formula 9: $\mathcal{L}\{e^{-at}\cos bt\} = \frac{s+a}{(s+a)^2+b^2}.$

Formula 10: $\mathcal{L}\{e^{at}\cos bt\} = \frac{s-a}{(s-a)^2+b^2}.$

Formula 11: $\mathcal{L}\{e^{-at}\sin bt\} = \frac{b}{(s+a)^2+b^2}.$

Formula 12: $\mathcal{L}\{e^{at}\sin bt\} = \frac{b}{(s-a)^2+b^2}.$

Formula 13: $\mathcal{L}\{e^{-at}t^n\} = \begin{cases} \frac{\Gamma(n+1)}{(s+a)^{n+1}} & n+1>0 \\ \frac{n!}{(s+a)^{n+1}} & n \text{ is a positive number.} \end{cases}$

Formula 14: $\mathcal{L}\{e^{at}t^n\} = \begin{cases} \frac{\Gamma(n+1)}{(s-a)^{n+1}} & \\ \frac{n!}{(s-a)^{n+1}} & n \text{ is a positive number.} \end{cases}$

Theorem 3.6: (Second shifting theorem)

If $f(t)$ is piecewise continuous for $0 \leq t < \infty$ and of exponential order, then

$$\mathcal{L}\{f(t-a)u(t-a)\} = e^{-as}\mathcal{L}\{f(t)\}$$

where $u(t-a)$ is a unit step function (shown by the following graph) shifted from the origin by a quantity a.

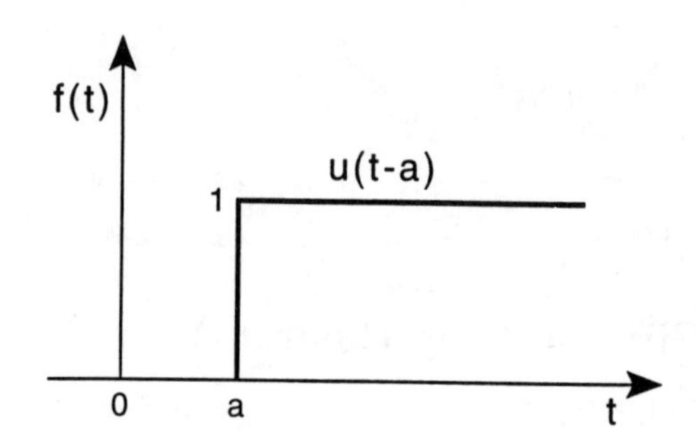

Figure 3.5: Graph of unit step function $u(t-a)$

Proof

By definition,

$$\mathcal{L}\{f(t-a)u(t-a)\} = \int_0^{\infty} e^{-st}\{f(t-a)u(t-a)\}dt.$$

Therefore letting $t-a=\tau$, the result can be obtained easily as $\mathcal{L}\{f(t-a)u(t-a)\} = e^{-as}\mathcal{L}\{f(t)\}$. This is the required proof.

Theorem 3.7: (Third shifting theorem)

$$\mathcal{L}\{f(t)u(t-a)\} = e^{-as}\mathcal{L}\{f(t+a)\}.$$

Proof

By definition,

$$\mathcal{L}\{f(t)u(t-a)\} = \int_0^{\infty} e^{-st}\{f(t)u(t-a)\}dt.$$

Therefore substituting $t=\tau+a$ the result can easily be obtained as $\mathcal{L}\{f(t)u(t-a)\} = e^{-as}\mathcal{L}\{f(t+a)\}$. This is the required proof.

Corollary 3

If $\mathcal{L}\{f(t-a)u(t-a)\} = e^{-as}\mathcal{L}\{f(t)\} = e^{-as}F(s)$ then $\mathcal{L}^{-1}\{e^{-as}F(s)\} = f(t-a)u(t-a)$.

Corollary 4

If $\mathcal{L}\{f(t)u(t-a)\} = e^{-as}\mathcal{L}\{f(t+a)\}$ then $\mathcal{L}^{-1}\{e^{-as}\mathcal{L}\{f(t+a)\}\} = f(t)u(t-a)$.

Definition of a filter function

By Fig. 3.6, a filter function may be defined as

$$f(t) = u(t-a) - u(t-b).$$

Example 2

Find the solution of the equation

$$y' + 3y + 2\int_0^t y(\lambda)d\lambda = f(t)$$

for which $y(0) = 0$, if $f(t)$ is the function whose graph is shown in Fig. 3.7.

Solution

$f(t)$ can be represented in terms of a unit step function:

$$\begin{aligned} f(t) &= [u(t) - u(t-1)] - [u(t-1) - u(t-2] \\ &= u(t) - 2u(t-1) + u(t-2) \\ \mathcal{L}\{f(t)\} &= \mathcal{L}\{u(t)\} - 2\mathcal{L}\{u(t-1)\} + \mathcal{L}\{u(t-2)\} \\ &= \frac{1}{s} - \frac{2e^{-s}}{s} + \frac{e^{-2s}}{s}. \end{aligned}$$

Taking the Laplace transform of the given differential equation, we obtain

$$\begin{aligned} \mathcal{L}\{y'\} + 3\mathcal{L}\{y\} + 2\mathcal{L}\{\int_0^t y(\lambda)d\lambda\} &= \mathcal{L}\{f(t)\} \\ s\mathcal{L}\{y\} + 3\mathcal{L}\{y\} + \frac{2}{s}\mathcal{L}\{y\} &= \frac{1 - 2e^{-s} + e^{-2s}}{s} \\ \mathcal{L}\{y\} &= \frac{1 - 2e^{-s} + e^{-2s}}{s^2 + 3s + 2} \\ &= \frac{1 - 2e^{-s} + e^{-2s}}{(s+2)(s+1)}. \end{aligned}$$

The Laplace inverse can easily be found by using Formula 1 and Corollary 4 of Theorem 3.6,

$$\begin{aligned} y(t) &= (e^{-t} - e^{-2t}) - 2(e^{-(t-1)} - e^{-2(t-1)})u(t-1) \\ &\quad + (e^{-(t-2)} - e^{-2(t-2)})u(t-2). \end{aligned}$$

This is the required result.

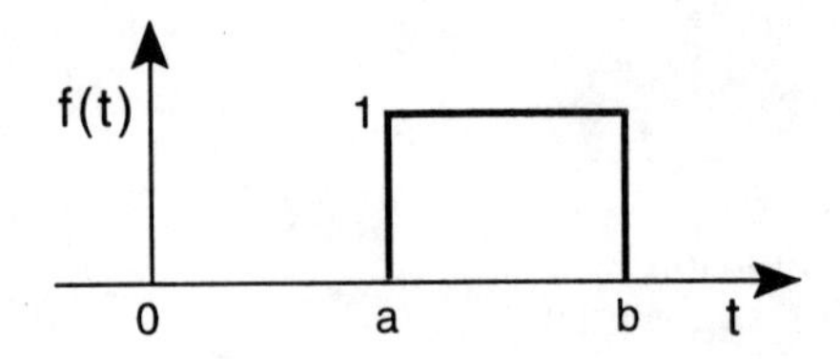

Figure 3.6: Graph of a filter function

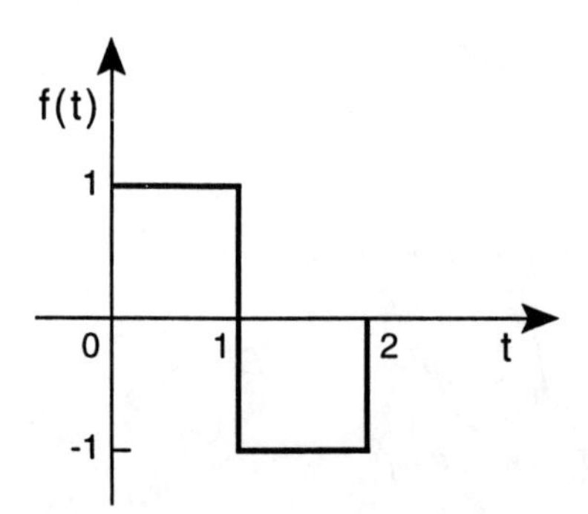

Figure 3.7: A square wave

Remark

As a rule, to obtain $\mathcal{L}^{-1}\{\frac{e^{-2s}}{s+2}\}$, first find $\mathcal{L}^{-1}(\frac{1}{s+2})$, which is e^{-2t}. Then, because the term e^{-2s} is a part of the transform, the t has to be changed to $(t-2)$ in e^{-2t}. The whole term must then be multiplied by the unit step function $u(t-2)$. Thus

$$\mathcal{L}^{-1}(\frac{e^{-2s}}{s+2}) = e^{-2(t-2)}u(t-2).$$

Similarly, to evaluate $\mathcal{L}^{-1}\{\frac{se^{-\pi s}}{s^2+25}\}$, we first find $\mathcal{L}^{-1}(\frac{s}{s^2+25})$, which is $\mathcal{L}^{-1}(\frac{s}{s^2+25}) = \cos 5t$. Then $\mathcal{L}^{-1}\{\frac{se^{-\pi s}}{s^2+25}\} = \cos 5(t-\pi)u(t-\pi)$.

3.5 Some important theorems

In this section we shall consider some theorems which will be useful in the application of Laplace transforms to practical problems. Before we state these theorems, we shall establish two important properties of the Laplace transform of the function $f(t)$.

For the existence of the Laplace transform of $f(t)$, we require that $f(t)$ be of exponential order which, in turn, means that there exist two constants λ and M such that $e^{-\lambda t}|f(t)| < M$ or $|f(t)| < Me^{\lambda t}$.
Hence

$$|\int_0^\ell e^{-st}f(t)dt| \le \int_0^\ell |f(t)|e^{-st}dt < \int_0^\ell (Me^{\lambda t})e^{-st}dt \le M\int_0^\ell e^{-(s-\lambda)t}dt.$$

Now as $\ell \to \infty$, we obtain $|\mathcal{L}\{f(t)\}| < \frac{M}{s-\lambda}$. From this inequality we can derive two important properties:

Property 1

If $f(t)$ is piecewise regular over $0 \le t < \infty$ and of exponential order, then

$$\lim_{s\to\infty} |\mathcal{L}\{f(t)\}| < \lim_{s\to\infty} \frac{M}{s-\lambda} \to 0.$$

Property 2

If $f(t)$ is piecewise regular over $0 \le t < \infty$ and of exponential order, then

$$\lim_{s\to\infty} |s\mathcal{L}\{f(t)\}| < \lim_{s\to\infty} \frac{sM}{s-\lambda} = M$$

which is bounded.
To illustrate, consider $\mathcal{L}\{\cos bt\} = \frac{s}{s^2+b^2}$. Then

$$\lim_{s\to\infty} |\mathcal{L}\{\cos bt\}| = \lim_{s\to\infty} \frac{s}{s^2+b^2} \to 0, \text{ and } \lim_{s\to\infty} |s\mathcal{L}\{\cos bt\}| = \lim_{s\to\infty} \frac{s^2}{s^2+b^2} \to 1.$$

With these two results we are now in a position to state the following theorems:

Theorem 3.8

If

(i) $f(t)$ is continuous over $0 \le t < \infty$.

(ii) $\lim_{t\to 0+} f(t)$ exists.

(iii) $f'(t)$ is at least piecewise regular over $0 \le t < \infty$.

(iv) $f(t)$ and $f'(t)$ are of exponential order;

then

$$\lim_{s\to\infty} s\mathcal{L}\{f(t)\} = \lim_{t\to 0+} f(t) = f(0^+).$$

Theorem 3.9

If $f(t)$ and $f'(t)$ are piecewise regular over $0 \le t < \infty$ and of exponential order, and $\lim_{t\to\infty} f(t)$ exists, then $\lim_{s\to 0} s\mathcal{L}\{f(t)\} = \lim_{t\to\infty} f(t)$.

Theorem 3.10

If $f(t)$ is piecewise regular over $0 \le t < \infty$ and of exponential order, and if $\mathcal{L}\{f(t)\} = sG(s)$, then $f(t) = \frac{d}{dt}\mathcal{L}^{-1}\{G(s)\}$.

Thus, if a Laplace transform contains the factor s, the inverse of that transform can be found by first suppressing the factor s. Then determine the inverse of the remaining portion of the transform, and finally differentiate that inverse with respect to t.

For example, let $\mathcal{L}\{f(t)\} = \frac{s}{s^2-a^2}$. Then $f(t) = \frac{d}{dt}\mathcal{L}^{-1}\{\frac{1}{s^2-a^2}\} = \frac{d}{dt}\{\frac{1}{a}\sinh at\} = \cosh at$. Since $f(t) = \mathcal{L}^{-1}\{\frac{s}{s^2-a^2}\} = \cosh at$, these two results are identical.

Theorem 3.11

If $f(t)$ is piecewise regular function over $0 \le t < \infty$ and of exponential order, and if $\mathcal{L}\{f(t)\} = \frac{G(s)}{s}$ and if $G(s)$ possesses an inverse, then $f(t) = \int_0^t \mathcal{L}^{-1}\{G(s)\}dt$.

Thus, if a Laplace transform contains the factor $\frac{1}{s}$, the inverse of that transform can be found by suppressing the factor $\frac{1}{s}$. Then determine the inverse of the remaining portion of the transform, and finally integrate that inverse with respect to t from 0 to t.

For example, let $\mathcal{L}\{f(t)\} = \frac{1}{s(s+1)}$. Then $f(t) = \int_0^t \mathcal{L}^{-1}\{\frac{1}{s+1}\}dt = 1 - e^{-t}$. Since $f(t) = \mathcal{L}^{-1}\{\frac{1}{s} - \frac{1}{s+1}\} = 1 - e^{-t}$, they are identical.

Theorem 3.12

If $f(t)$ is piecewise regular over $0 \le t < \infty$ and of exponential order and if $\mathcal{L}\{f(t)\} = F(s)$, then $\mathcal{L}\{tf(t)\} = -\frac{d}{ds}F(s)$.

Proof

By definition, $F(s) = \mathcal{L}\{f(t)\} = \int_0^\infty e^{-st} f(t)dt$. Then differentiating with respect to s,

$$\begin{aligned}\frac{d}{ds}F(s) &= \frac{d}{ds}\int_0^\infty e^{-st} f(t)dt = \int_0^\infty -te^{-st} f(t)dt \\ &= -\int_0^\infty e^{-st}\{tf(t)\}dt = -\mathcal{L}\{tf(t)\}.\end{aligned}$$

Therefore, $\mathcal{L}\{tf(t)\} = -\frac{d}{ds}F(s)$. This is the required proof.

Theorem 3.12 can very easily be extended as follows:

$$\begin{aligned}\mathcal{L}\{t^2 f(t)\} &= (-1)^2 \frac{d^2}{ds^2}F(s) \\ \mathcal{L}\{t^3 f(t)\} &= (-1)^3 \frac{d^3}{ds^3}F(s) \\ \mathcal{L}\{t^n f(t)\} &= (-1)^n \frac{d^n}{ds^n}F(s).\end{aligned}$$

For example,

$$\begin{aligned}\mathcal{L}\{t^2 e^{at}\} &= (-1)^2 \frac{d^2}{ds^2}\mathcal{L}\{e^{at}\} \\ &= \frac{d^2}{ds^2}\frac{1}{s-a} \\ &= \frac{2}{(s-a)^3}.\end{aligned}$$

Corollary 5

If $\mathcal{L}\{f(t)\} = F(s)$, then $f(t) = \mathcal{L}^{-1}\{F(s)\} = -\frac{1}{t}\mathcal{L}^{-1}\{\frac{d}{ds}F(s)\}$.

Note

Corollary 5 is very useful for obtaining the inverse of a transform provided that the derivative of the transform with respect to s is easily found.

For example,

$$\begin{aligned}\mathcal{L}^{-1}\{\ell n(\frac{s^2+1}{s(s+1)})\} &= -\frac{1}{t}\mathcal{L}^{-1}[\frac{d}{ds}\{\ell n(s^2+1) - \ell n s - \ell n(s+1)\}] \\ &= -\frac{1}{t}\mathcal{L}^{-1}[\frac{2s}{s^2+1} - \frac{1}{s} - \frac{1}{s+1}] \\ &= \frac{1+e^t-2\cos t}{t}.\end{aligned}$$

Theorem 3.13

If $f(t)$ is piecewise regular over $0 \le t < \infty$ and of exponential order, and if $\lim_{t\to 0}\{\frac{f(t)}{t}\}$ exists, and if $\mathcal{L}\{f(t)\} = F(s)$, then $\mathcal{L}\{\frac{f(t)}{t}\} = \int_s^\infty F(s)ds$

Proof

By definition, we know $F(s) = \mathcal{L}\{f(t)\} = \int_0^\infty e^{-st} f(t)dt$.

Then

$$\begin{aligned}
\int_s^\infty F(s)ds &= \int_s^\infty \{\int_0^\infty e^{-st} f(t)dt\}ds \\
&= \int_0^\infty f(t)[\int_s^\infty e^{-st}ds]dt \\
&= \int_0^\infty \{\frac{f(t)}{t}\}e^{-st}dt \\
&= \mathcal{L}\{\frac{f(t)}{t}\}
\end{aligned}$$

provided $\lim_{t\to 0}\{\frac{f(t)}{t}\}$ exists. This is the required proof.
For example,

$$\mathcal{L}\{\frac{e^{-3t}\sin 2t}{t}\} = \int_s^\infty \mathcal{L}\{e^{-3t}\sin 2t\}ds = \int_s^\infty \frac{2}{(s+3)^2+4}ds = \cot^{-1}(\frac{s+3}{2}).$$

Corollary 6

If $\mathcal{L}\{f(t)\} = F(s)$, then $f(t) = \mathcal{L}^{-1}\{F(s)\} = t\mathcal{L}^{-1}\{\int_s^\infty F(s)ds\}$.

Note

Corollary 6 is very useful for obtaining the inverse of a transform provided the integral of the transform with respect to s is easily found. For example,

$$\begin{aligned}
\mathcal{L}^{-1}\{\frac{s+2}{(s^2+4s+5)^2}\} &= \mathcal{L}^{-1}[\frac{s+2}{\{(s+2)^2+1\}^2}] \\
&= t\mathcal{L}^{-1}\int_s^\infty \frac{s+2}{\{(s+2)^2+1\}^2}ds \\
&= \frac{t}{2}e^{-2t}\sin t.
\end{aligned}$$

In general, Theorem 3.13 can be extended as follows: If $\mathcal{L}\{f(t)\} = F(s)$, then $\mathcal{L}\{\frac{f(t)}{t}\} = \int_s^\infty \mathcal{L}\{f(t)\}ds = \int_s^\infty F(s)ds$ Similarly

$$\begin{aligned}
\mathcal{L}\{\frac{f(t)}{t^2}\} &= \int_s^\infty\int_s^\infty F(s)dsds \\
\mathcal{L}\{\frac{f(t)}{t^n}\} &= \int_s^\infty\int_s^\infty\int_s^\infty \cdots \int_s^\infty F(s)dsds\cdots ds.
\end{aligned}$$

which is an n-fold integration. Also, Corolary 6 can be extended as follows: If $\mathcal{L}\{f(t)\} = F(s)$, then

$$\begin{aligned} f(t) = \mathcal{L}^{-1}\{F(s)\} &= t\mathcal{L}^{-1}\int_s^\infty F(s)ds \\ &= t^2\mathcal{L}^{-1}\{\int_s^\infty \int_s^\infty F(s)dsds\} \\ &\vdots \\ &= t^n\mathcal{L}^{-1}\{\int_s^\infty \int_s^\infty \int_s^\infty \cdots \int_s^\infty F(s)dsds\cdots ds\}. \end{aligned}$$

which is an n-fold integration. We shall now solve problems demonstrating the use of these theorems and corollaries.

Example 3

Find

$$\mathcal{L}^{-1}\{\frac{1}{s}\tan^{-1}\frac{1}{s}\}.$$

Solution

$$\begin{aligned} \mathcal{L}^{-1}\{\frac{1}{s}\tan^{-1}\frac{1}{s}\} &= \int_0^t \mathcal{L}^{-1}\{\tan^{-1}\frac{1}{s}\}dt \\ &= \int_0^t (-\frac{1}{t})[\mathcal{L}^{-1}\{\frac{d}{ds}\tan^{-1}\frac{1}{s}\}]dt \\ &= \int_0^t (\frac{\sin t}{t})dt. \end{aligned}$$

Example 4

Find

$$\mathcal{L}^{-1}\{\frac{s}{(s^2+4)^2}\}.$$

Solution

$$\begin{aligned} \mathcal{L}^{-1}\{\frac{s}{(s^2+4)^2}\} &= \frac{1}{2}\mathcal{L}^{-1}\{\frac{2s}{(s^2+4)^2}\} \\ &= \frac{1}{2}t\mathcal{L}^{-1}\{\int_s^\infty \frac{2s}{(s^2+4)^2}ds\} \\ &= \frac{t}{4}\sin 2t. \end{aligned}$$

3.6 The unit step function and the Dirac delta function

Two generalized functions which are used frequently in applied problems are the Heaviside unit step function, $u(t)$, and the Dirac delta function, $\delta(t)$. We have already come across the definition of the unit step function and its Laplace transform. In this section we will see some of its properties and its relation to the Dirac delta function. It should be noted that, formerly, many writers called these two functions the unit function and the unit impulse function, respectively.

3.6.1 The unit step function and its transform

The unit step function is defined as (see Fig. 3.5)

$$u(t-a) = \begin{cases} 0 & t < a \\ 1 & t > a. \end{cases}$$

Thus, this function allows us to express any function $f(t)$ which is non-zero in $a \leq t \leq b$ and zero elsewhere.

For example, consider the function shown in Fig. 3.8. The equation of this function is, obviously,

$$g(t) = f(t)[u(t-a) - u(t-b)].$$

The expression $u(t-a) - u(t-b)$ is called the filter function or window function, as we have already seen.

The Laplace transform of the step function is easily calculated as follows:

$$\mathcal{L}\{u(t-a)\} = \int_0^\infty u(t-a)dt = \frac{e^{-as}}{s}.$$

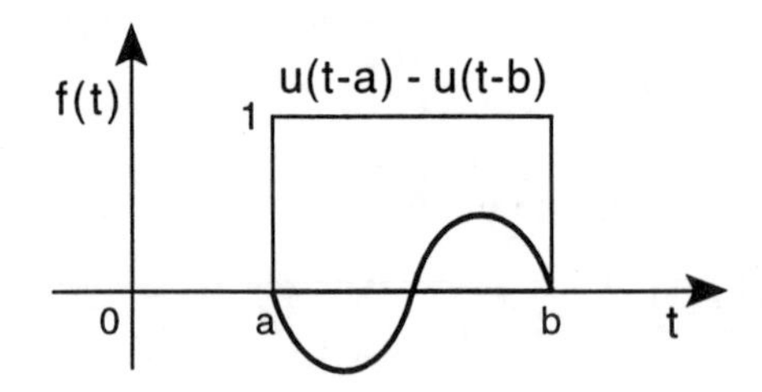

Figure 3.8: An arbitrary pulse

3.6.2 The Dirac delta function

In many physical applications, it is necessary to find the response of a system to an impulsive forcing function. By an impulsive force, we mean a force that has a very large magnitude but acts over a very small period of time. An obvious example is the force encountered when a circuit is subjected to a high voltage for a short period of time, as in a lightning strike. Similarly, the force which results from the collision of two bodies can be considered to be an impulsive force. The precise distribution in time of a very short impulse is much less important than its total value, measured by its integral over time. Moreover, it is much easier to measure this total value than to determine the distribution of a short impulse in space or time. Such a distribution of the impulse is called the Dirac delta function, $\delta(t)$, after the British theoretical physicist P.A.M. Dirac (1902–).

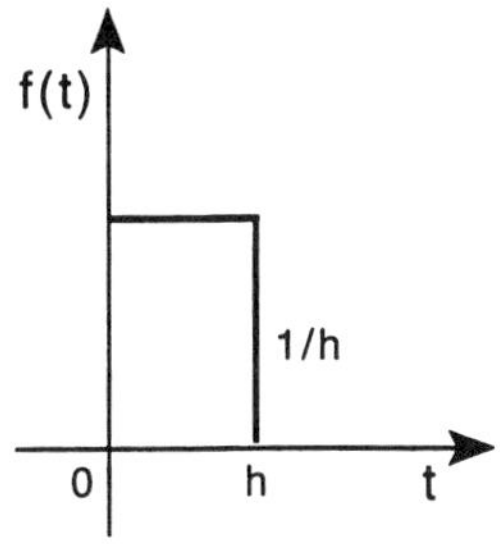

Figure 3.9: A rectangular pulse

It is defined by

$$\delta(t-h) = \begin{cases} 0 & t \neq h \\ \infty & t = h \end{cases} \tag{3.9}$$

$$\int_{-\infty}^{\infty} \delta(t-h)dt = 1 \tag{3.10}$$

Readers who wish to know more about its properties should refer to Dirac, *The Principles of Quantum Mechanics*, Oxford University Press, (1935).

We will arrive at a definition of an impulse function by starting with the definition of a rectangular pulse of height $\frac{1}{h}$ and width h, as shown in Fig. 3.9. Mathematically we write

$$\delta_h(t) = \frac{1}{h}[u(t) - u(t-h)] \quad , \quad h > 0 \tag{3.11}$$

where $u(t)$ and $u(t-h)$ are two unit step functions. Thus $\delta_h(t)$ can be expressed as

$$\delta_h(t) = \begin{cases} \frac{1}{h} & 0 \leq t \leq h \\ 0 & t > h. \end{cases} \tag{3.12}$$

Now in the limiting process, when $t \to 0$, eqn (3.12) reduces to

$$\delta(t) = \lim_{h \to 0} \delta_h(t) = \begin{cases} \infty & t = 0 \\ 0 & t \neq 0. \end{cases} \tag{3.13}$$

The physical implication of (3.12) can be visualized from Fig. 3.9. We allow the pulse to shrink in width and, at the same time, to increase in height, so that its area remains constant. Continuing this process, we shall eventually obtain an extremely narrow, very large amplitude pulse at $t = 0$. If we proceed to the limit, where the width approaches zero and the height approaches infinity (but still with the product width X height $= 1$), we approach the delta function $\delta(t)$ defined in (3.13).

Thus, the strength of the $\delta(t)$ function is given by

$$\int_0^\infty \delta(t)dt = 1. \tag{3.14}$$

This may be written as $\int_0^\infty \delta(t-h)dt = 1$ if the source of the impulse is at $t = h$. An important property of the impulse function is expressed as

$$\int_a^b f(t)\delta(t-h)dt = f(h), \qquad a < h < b. \tag{3.15}$$

Thus, the Dirac delta function can be related to the Heaviside step function as follows:

$$\delta(t) = \lim_{h\to 0} \frac{u(t) - u(t-h)}{h} = \frac{d}{dt}u(t). \tag{3.16}$$

The unit impulse is only the first of infinitely many singular functions. As a generalization of the unit impulse, we define the unit doublet as

$$\begin{aligned} D(t) &= \lim_{h\to 0} \frac{u(t) - 2u(t-h) + u(t-2h)}{h^2} \\ &= \frac{d^2}{dt^2}u(t) = \frac{d}{dt}\delta(t). \end{aligned} \tag{3.17}$$

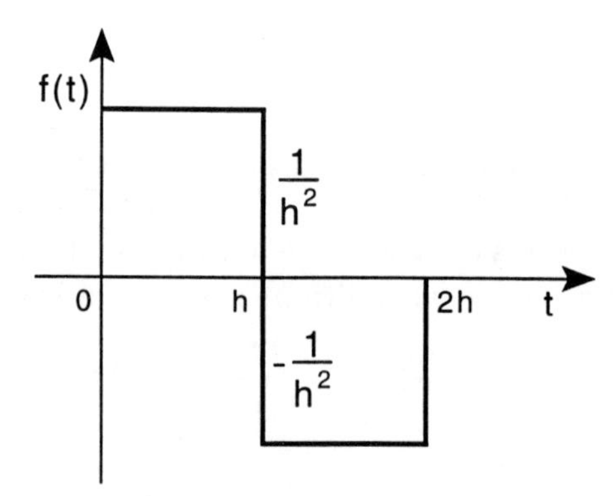

Figure 3.10: Formulation of a doublet

The unit triplet is defined as

$$\begin{aligned} T(t) &= \lim_{h\to 0} \frac{u(t) - 3u(t-h) + 3u(t-2h) - u(t-3h)}{h^3} \\ &= \frac{d^3}{dt^3}u(t) = \frac{d^2}{dt^2}\delta(t) = \frac{d}{dt}D(t). \end{aligned}$$

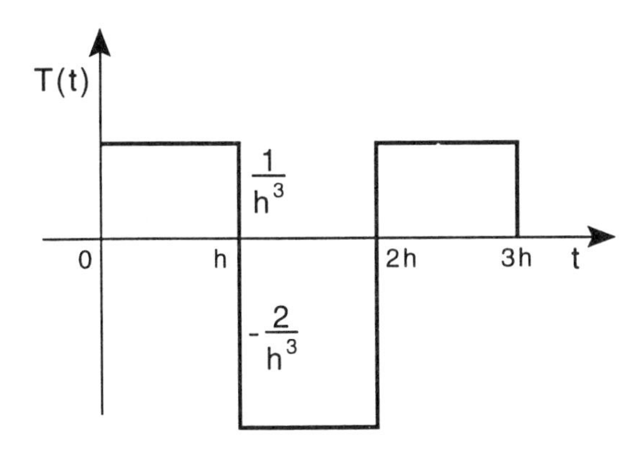

Figure 3.11: Formation of a triplet

The same procedure can be followed to define other step functions.

Whatever its theoretical limitations, the δ function is a very useful device in many applications. All the singularity functions defined above are known as generalized functions or distributions. For fuller discussion the reader is referred to Lighthill (1964).

To determine the Laplace transform of the δ function we start from its definition:

$$\begin{aligned}\delta(t) &= \lim_{h\to 0}\frac{u(t)-u(t-h)}{h}\\ \mathcal{L}\{\delta(t)\} &= \lim_{h\to 0}\mathcal{L}\{\frac{u(t)-u(t-h)}{h}\}\\ &= \lim_{h\to 0}\frac{1-e^{-hs}}{hs}=1.\end{aligned}\tag{3.18}$$

Also,

$$\mathcal{L}\{\delta(t-h)\}=e^{-hs}.\tag{3.19}$$

Similarly, the Laplace transform of the unit doublet $D(t)$ can be obtained as follows:

If $D(t)=\lim_{h\to 0}\frac{\delta(t)-\delta(t-h)}{h}$ then $\mathcal{L}\{D(t)\}=\lim_{h\to 0}\frac{1}{h}[\mathcal{L}\{\delta(t)\}-\mathcal{L}\{\delta(t-h)\}]=\lim_{h\to 0}\frac{1}{h}[1-e^{-hs}]=s$ and if the unit triplet is defined as $T(t)=\lim_{h\to 0}\frac{D(t)-D(t-h)}{h}$ then $\mathcal{L}\{T(t)\}=\lim_{h\to 0}\frac{1}{h}[\mathcal{L}\{D(t)\}-\mathcal{L}\{D(t-h)\}]=\lim_{h\to 0}s(\frac{1-e^{-hs}}{h})=s^2$ and so on.

Example 5

Solve the initial value problem

$$\begin{aligned}y'+2y &= 3u(t-1)-\delta(t-2)\\ y(0) &= 1.\end{aligned}$$

Solution

Taking the Laplace transform of the given equation,

$$\begin{aligned}\mathcal{L}\{y'\} + 2\mathcal{L}\{y\} &= 3\mathcal{L}\{u(t-1)\} - \mathcal{L}\{\delta(t-2)\} \\ (s+2)\mathcal{L}\{y\} &= 1 + 3\frac{e^{-s}}{s} - e^{-2s}.\end{aligned}$$

Thus we obtain

$$\mathcal{L}\{y\} = \frac{1}{s+2} + 3\frac{e^{-s}}{s(s+2)} - \frac{e^{-2s}}{s+2}.$$

Therefore, the inverse is given by

$$y = e^{-2t} + \frac{3}{2}\{1 - e^{-2(t-1)}\}u(t-1) - e^{-2(t-2)}u(t-2).$$

3.6.3 Properties of generalized functions

We have already seen the graphical behaviour of a Dirac delta function. A smooth curve can be drawn, as shown in the following figure:

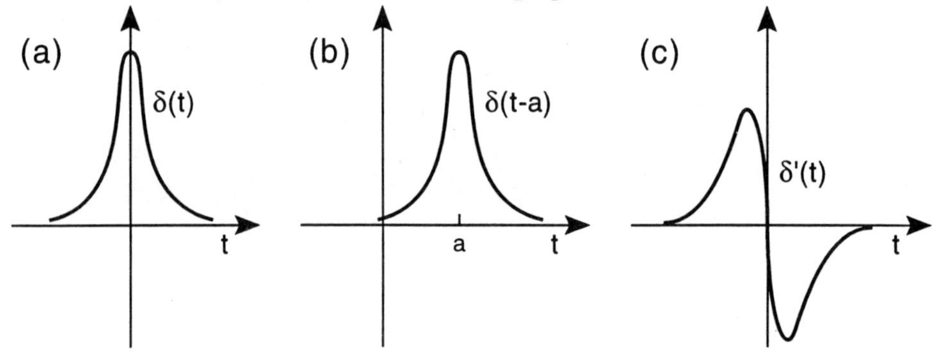

Figure 3.12: Generalized functions: (a) $\delta(t)$; (b) $\delta(t-a)$; (c) $\delta'(t)$

Under appropriate conditions, an ordinary function $f(t)$ can also be multiplied by a generalized function. Thus, we have the following properties:

$$\begin{aligned}\delta(t)f(t) &= f(t)\delta(t) = f(0)\delta(t) \\ \delta(t-a)f(t) &= f(t)\delta(t-a) = f(a)\delta(t-a)\end{aligned}$$

provided $f(t)$ is any continuous function at $t = 0$ and $t = 1$ respectively.

(I) Differentiation rule of generalized functions

Consider $y = f(t)u(t)$ then we have

$$\begin{aligned}y' &= f'(t)u(t) + f(t)u'(t) \\ &= f'(t)u(t) + f(t)\delta(t) \\ &= f'(t)u(t) + f(0)\delta(t).\end{aligned}$$

An expression for $f(t)\delta'(t)$ is deduced from the product rule as follows. We know

$$[f(t)\delta(t)]' = [f(0)\delta(t)]' = f(0)\delta'(t).$$

Also we have

$$\begin{aligned} [f(t)\delta(t)]' &= f'(t)\delta(t) + f(t)\delta'(t) \\ &= f'(0)\delta(t) + f(t)\delta'(t). \end{aligned}$$

Equating these two expressions gives

$$f(0)\delta'(t) = f'(0)\delta(t) + f(t)\delta'(t)$$

or,

$$f(t)\delta'(t) = f(0)\delta'(t) - f'(0)\delta(t)$$

provided f and f' are both continuous at $t = 0$.

In the same way, we may deduce the rules:

$$\begin{aligned} f(t)\delta'(t-a) &= f(a)\delta'(t-a) - f'(a)\delta(t-a) \\ f(t)\delta''(t-a) &= f(a)\delta''(t-a) - 2f'(a)\delta'(t-a) + f''(a)\delta(t-a). \end{aligned}$$

(II) Integration rule of generalized functions

Generalized functions can also be integrated. For instance, if $a < b$ and $a \neq 0$ and $b \neq 0$, then

1. $\int_a^b \delta(t)dt = \int_a^b u'(t)dt = u(b) - u(a) = \begin{cases} 1 & , \quad a < 0 < b \\ 0 & , \quad \text{otherwise.} \end{cases}$
2. $\int_a^b \delta'(t)dt = \delta(b) - \delta(a) = 0.$
3. $\int_a^b \delta''(t)dt = 0.$
4. $\int_a^b g(t)\delta(t)dt = g(0)\int_a^b \delta(t)dt = g(0), \qquad a < 0 < b.$
5. $\int_a^b g(t)\delta'(t)dt = \int_a^b [g(0)\delta'(t) - g'(0)\delta(t)]dt = -g'(0), \qquad a < 0 < b.$
6. $\int_a^b g(t)\delta^{(k)}(t-c)dt = (-1)^k g^k(c), \qquad a < c < b$, provided $g^k(c)$ is continuous at $t = c$.
7. $\delta(-t) = \delta(t)$ is an even function.
8. $\delta'(-t) = -\delta'(t)$ is an odd function.

Example 6

Let $f(t) = e^{3t}u(t)$, then find $f'(t)$.

Solution

Given that $f(t) = e^{3t}u(t)$, then

$$\begin{aligned} f'(t) &= 3e^{3t}u(t) + e^{3t}u'(t) \\ &= 3e^{3t}u(t) + e^{3t}\delta(t) \\ &= 3e^{3t}u(t) + \delta(t). \end{aligned}$$

3.7 The Heaviside expansion theorems to find inverses

We have already seen the usefulness of the partial fraction for obtaining inverse Laplace transforms. In the following section, we extend this technique and systematically write this procedure in the form of those theorems usually known as the Heaviside expansion theorems, after the British electrical engineer Oliver Heaviside.

Theorem 3.14

If

$$\mathcal{L}\{f(t)\} = \frac{p(s)}{q(s)}$$

where $p(s)$ and $q(s)$ are polynomials and the degree of $q(s)$ is greater than the degree of $p(s)$, and $q(s)$ has n distinct zeros $a_1, a_2, \cdots, a_n$, then

$$f(t) = \mathcal{L}^{-1}\{\frac{p(s)}{q(s)}\} = \sum_{k=1}^{n} \frac{p(a_k)}{q'(a_k)} e^{a_k t}.$$

Proof

Since $q(s)$ is a polynomial with n distinct zeros $a_1, a_2, \cdots, a_n$, we can write, according to the method of partial fractions

$$\frac{p(s)}{q(s)} = \frac{A_1}{s-a_1} + \frac{A_2}{s-a_2} + \cdots + \frac{A_k}{s-a_k} + \cdots + \frac{A_n}{s-a_n} \tag{3.20}$$

where $A_1, A_2, \cdots, A_k, \cdots A_n$ are constants. Multiplying both sides of the above equation by $s - a_k$ and letting $s \to a_k$, we find, using L'Hospital's rule,

$$\begin{aligned} A_k &= \lim_{s\to a_k} \frac{(s-a_k)p(s)}{q(s)} \\ &= \lim_{s\to a_k} \frac{p(s)}{(\frac{q(s)}{s-a_k})} \\ &= \frac{p(a_k)}{q'(a_k)} \qquad k = 1, 2, \cdots, n. \end{aligned}$$

Thus (3.20) can be written

$$\begin{aligned}\frac{p(s)}{q(s)} &= \frac{p(a_1)}{q'(a_1)}.\frac{1}{s-a_1} + \frac{p(a_2)}{q'(a_2)}.\frac{1}{s-a_2} + \cdots + \frac{p(a_k)}{q'(a_k)}.\frac{1}{s-a_k} \\ &+ \cdots + \frac{p(a_n)}{q'(a_n)}.\frac{1}{s-a_n}.\end{aligned}$$

Then taking the inverse Laplace transform, we have

$$\begin{aligned}f(t) &= \mathcal{L}^{-1}\{\frac{p(s)}{q(s)}\} \\ &= \frac{p(a_1)}{q'(a_1)}e^{a_1t} + \frac{p(a_2)}{q'(a_2)}e^{a_2t} + \cdots \\ &+ \frac{p(a_k)}{q'(a_k)}e^{a_kt} + \cdots + \frac{p(a_n)}{q'(a_n)}e^{a_nt} \\ &= \sum_{k=1}^{n}\frac{p(a_k)}{q'(a_k)}e^{a_kt}.\end{aligned}$$

This is the required proof.

Theorem 3.15

If

$$\mathcal{L}\{f(t)\} = \frac{p(s)}{q(s)}$$

where $p(s)$ and $q(s)$ are polynomials, and where $q(s) = 0$ has a repeated root b of multiplicity r while the remaining roots, $a_1, a_2, \cdots, a_n$, do not repeat, then

$$\begin{aligned}f(t) &= \mathcal{L}^{-1}\{\frac{p(s)}{q(s)}\} \\ &= [A_1 + A_2t + \frac{A_3t^2}{2!} + \cdots + A_r\frac{t^{r-1}}{(r-1)!}]e^{bt} \\ &+ \sum_{k=1}^{n}\frac{p(a_k)}{q'(a_k)}e^{a_kt}\end{aligned}$$

where

$$A_{r-k} = \lim_{s\to b}\frac{1}{k!}\frac{d^k}{ds^k}[(s-b)^r\frac{p(s)}{q(s)}] \quad k = 0, 1, 2, \cdots, (r-1).$$

Proof

Since $q(s)$ has repeated roots b of multiplicity r and also has n distinct roots, $a_1, a_2, \cdots, a_n$, we can write, according to the method of partial fractions,

$$\begin{aligned} F(s) &= \frac{p(s)}{q(s)} \\ &= \frac{p(s)}{(s-b)^r \Pi_{k=1}^n (s-a_k)} \\ &= \frac{A_1}{s-b} + \frac{A_2}{(s-b)^2} + \cdots + \frac{A_r}{(s-b)^r} + \sum_{k=1}^n \frac{B_k}{s-a_k} \end{aligned} \tag{3.21}$$

where Π represents the product.

From the preceding theorem we have seen that the inverse of $\sum_{k=1}^n \frac{B_k}{s-a_k}$ is equal to

$$\sum_{k=1}^n \frac{p(a_k)}{q'(a_k)} e^{a_k t}.$$

We need only find the inverse corresponding to the repeated roots of $q(s)$. The inverses of (3.21) can be written at once,

$$\begin{aligned} f(t) &= \mathcal{L}^{-1}\{\frac{p(s)}{q(s)}\} \\ &= [A_1 + A_2 t + A_3 \frac{t^2}{2!} + \cdots + A_r \frac{t^{r-1}}{(r-1)!}] e^{bt} \\ &\quad + \sum_{k=1}^n \frac{p(a_k)}{q'(a_k)} e^{a_k t} \end{aligned} \tag{3.22}$$

provided $A_1, A_2, \cdots, A_r$ can be determined.

To find A_r's:

Define

$$h(s) = \sum_{k=1}^n \frac{B_k}{s-a_k}.$$

Multiplying both sides of (3.21) by $(s-b)^r$, we get

$$\begin{aligned} (s-b)^r F(s) &= A_1(s-b)^{r-1} + A_2(s-b)^{r-2} + \cdots + A_{r-2}(s-b)^2 \\ &\quad + A_{r-1}(s-b) + A_r + (s-b)^r h(s). \end{aligned} \tag{3.23}$$

Thus

$$\begin{aligned} A_r &= \lim_{s \to b}\{(s-b)^r F(s)\} \\ &= \lim_{s \to b}\{(s-b)^r \frac{p(s)}{q(s)}\}. \end{aligned}$$

Now differentiating (3.23) with respect to s and evaluating the differential value at $s = b$, we obtain

$$A_{r-1} = \lim_{s \to b} \frac{1}{1!} \frac{d}{ds} \{(s-b)^r \frac{p(s)}{q(s)}\}.$$

Similarly,

$$\begin{aligned} A_{r-2} &= \lim_{s \to b} \frac{1}{2!} \frac{d^2}{ds^2} \{(s-b)^r \frac{p(s)}{q(s)}\} \\ &\vdots \\ A_{r-k} &= \lim_{s \to b} \frac{1}{k!} \frac{d^k}{ds^k} \{(s-b)^r \frac{p(s)}{q(s)}\} \end{aligned} \tag{3.24}$$

where $k = 0, 1, 2, \cdots, r-1$.

For example, consider the double root, $r = 2$, and $k = 0, 1$. Hence, we have

$$A_{2-k} = \lim_{s \to b} \frac{1}{k!} \frac{d^k}{ds^k} \{(s-b)^2 \frac{p(s)}{q(s)}\}.$$

Therefore,

$$\begin{aligned} A_2 &= \lim_{s \to b} \{(s-b)^2 \frac{p(s)}{q(s)}\} \\ A_1 &= \lim_{s \to b} \frac{1}{1!} \frac{d}{ds} \{(s-b)^2 \frac{p(s)}{q(s)}\}. \end{aligned}$$

Theorem 3.16

If

$$f(t) = \mathcal{L}^{-1}\{\frac{p(s)}{q(s)}\},$$

where $p(s)$ and $q(s)$ are polynomials and the degree of $q(s)$ is greater than the degree of $p(s)$, then the terms in $f(t)$ which correspond to an unrepeated complex factor $(s+a)^2 + b^2$ of $q(s)$ are

$$\frac{e^{-at}}{b}(\phi_i \cos bt + \phi_r \sin bt)$$

where ϕ_r and ϕ_i are, respectively, the real and imaginary parts of $\phi(-a+ib)$ and where $\phi(s)$ is the quotient of $p(s)$ and all the factors of $q(s)$ except the factor $(s+a)^2 + b^2$.

Proof

We can write, according to the method of partial fractions,

$$\frac{p(s)}{q(s)} = \frac{\phi(s)}{(s+a)^2+b^2} = \frac{As+B}{(s+a)^2+b^2} + h(s) \tag{3.25}$$

where $h(s)$ = sum of the fractions corresponding to all other factors of $q(s)$. Multiplying both sides of (3.25) by $(s+a)^2+b^2$, we obtain

$$\phi(s) = As + B + [(s+a)^2+b^2]h(s). \tag{3.26}$$

Now substitute $(s+a)^2+b^2 = 0$ where $s = -a \pm ib$ so that s has two complex roots. Then putting $s = -a+ib$ into (3.26) and we obtain $\phi(-a+ib) = A(-a+ib)+B$ or, $\phi_r + i\phi_i = -Aa+B+ibA$. Equating real and imaginary parts, we have $\phi_r = -Aa+B$ and $\phi_i = bA$. Solving for A and B: $A = \frac{\phi_i}{b}$ and $B = \frac{b\phi_r + a\phi_i}{b}$.

Thus,

$$\begin{aligned} \frac{As+B}{(s+a)^2+b^2} &= \frac{1}{b}[\frac{(s+a)\phi_i}{(s+a)^2+b^2} + \frac{b\phi_r}{(s+a)^2+b^2}] \\ \mathcal{L}^{-1}\{\frac{As+B}{(s+a)^2+b^2}\} &= \frac{1}{b}(\phi_i \cos bt + \phi_r \sin bt)e^{-at} \end{aligned}$$

as asserted.

Theorem 3.17

If

$$f(t) = \mathcal{L}^{-1}\{\frac{p(s)}{q(s)}\},$$

where $p(s)$ and $q(s)$ are polynomials and the degree of $q(s)$ is greater than the degree of $p(s)$, then the terms in $f(t)$ which correspond to a repeated complex factor $[(s+a)^2+b^2]^2$ of $q(s)$ are

$$\begin{aligned} &\frac{e^{-at}}{b}\{Cb\cos bt + (D-aC)\sin bt\} \\ &\quad + \frac{Ae^{-at}}{2b} t \sin bt \\ &\quad + \frac{B-aA}{2b^3}e^{-at}[\sin bt - bt\cos bt] \end{aligned}$$

where

$$\begin{aligned} A &= \frac{\phi_i}{b} \\ B &= \frac{1}{b}[b\phi_r + a\phi_i] \\ C &= \frac{1}{2b^3}[\phi_i - b\phi_r'] \\ D &= \frac{1}{2b^3}[b^2\phi_i' - ab\phi_r' + a\phi_i] \end{aligned}$$

in which ϕ_r and ϕ_i are, respectively, the real and imaginary parts of $\phi(-a+ib)$ and where $\phi(s)$ is the quotient of $p(s)$ and all the factors of $q(s)$ except the factor $[(s+a)^2+b^2]$; a prime denotes the differentiation.

Proof

We can write, according to the method of partial fractions,

$$\frac{p(s)}{q(s)} = \frac{\phi(s)}{[(s+a)^2+b^2]^2} = \frac{As+B}{[(s+a)^2+b^2]^2} + \frac{Cs+D}{(s+a)^2+b^2} + h(s) \tag{3.27}$$

where $h(s)$ = sum of the factors corresponding to all other factors of $q(s)$. Multiplying both sides of (3.27) by $[(s+a)^2+b^2]^2$, we obtain

$$\phi(s) = As + B + (Cs+D)[(s+a)^2+b^2] + [(s+a)^2+b^2]^2 h(s). \tag{3.28}$$

Now when $s=-a+ib$, (3.28) becomes

$$\phi(-a+ib) = A(-a+ib) + B$$

or

$$\phi_r + \phi_i = B - aA + ibA.$$

Equating real and imaginary parts, we obtain $\phi_i = bA$ and $\phi_r = B - aA$. Solving for A and B we have $A = \frac{\phi_i}{b}$ and $B = \frac{1}{b}[a\phi_i + b\phi_r]$.

Differentiating (3.28) with respect to s:

$$\begin{aligned}\phi'(s) &= A + C[(s+a)^2+b^2] + (Cs+D)[2(s+a)]\\ &\quad +2[(s+a)^2+b^2]\{2(s+a)\}h(s)\\ &\quad +[(s+a)^2+b^2]^2 h'(s).\end{aligned}$$

Substituting $s=-a+ib$ gives $\phi'_r + i\phi'_i = A + 2ib[D + C(-a+ib)]$ where $\phi'_r = A - 2b^2C$ and $\phi'_i = 2b[D-aC]$.

Solving for C and D, we obtain

$$\begin{aligned}C &= \frac{1}{2b^3}[\phi_i - b\phi'_r]\\ D &= \frac{1}{2b^3}[b^2\phi'_i - ab\phi'_r + a\phi_i].\end{aligned} \tag{3.29}$$

Once we know the values of A, B, C, and D, then the Laplace inverse of (3.27) can be obtained by using formulas and theorems.

Therefore,

$$\begin{aligned}\mathcal{L}^{-1}\{\frac{p(s)}{q(s)}\} &= \mathcal{L}^{-1}\{\frac{As+B}{[(s+a)^2+b^2]^2}\} + \mathcal{L}^{-1}\{\frac{Cs+D}{(s+a)^2+b^2}\}\\ &\quad +\mathcal{L}^{-1}\{h(s)\}\end{aligned}$$

$$\begin{aligned}
\mathcal{L}^{-1}\{\frac{Cs+D}{(s+a)^2+b^2}\} &= \mathcal{L}^{-1}\{\frac{C(s+a)+D-aC}{(s+a)^2+b^2}\} \\
&= e^{-at}[C\cos bt + \frac{D-ac}{b}\sin bt] \\
&= \frac{e^{-at}}{b}[Cb\cos bt + (D-aC)\sin bt] \\
\mathcal{L}^{-1}\{\frac{As+B}{[(s+a)^2+b^2]^2}\} &= \mathcal{L}^{-1}\{\frac{A(s+a)}{[(s+a)^2+b^2]^2} \\
&\quad +\mathcal{L}^{-1}\{\frac{B-aA}{[(s+a)^2+b^2]^2}\}. \qquad (3.30)
\end{aligned}$$

Then,

$$\begin{aligned}
\mathcal{L}^{-1}\{\frac{A(s+a)}{[(s+a)^2+b^2]^2}\} &= e^{-at}\mathcal{L}^{-1}\{\frac{As}{(s^2+b^2)^2}\} \\
&= A\frac{e^{-at}}{2}t\mathcal{L}^{-1}\int_s^\infty \frac{2sds}{(s^2+b^2)^2} \\
&= \frac{A}{2}e^{-at}t\mathcal{L}^{-1}(\frac{1}{s^2+b^2}) \\
&= \frac{At}{2b}e^{-at}\sin bt. \qquad (3.31)
\end{aligned}$$

$$\begin{aligned}
\mathcal{L}^{-1}\{\frac{B-aA}{[(s+a)^2+b^2]^2}\} &= (B-aA)e^{-at}\mathcal{L}^{-1}\{\frac{1}{(s^2+b^2)^2}\} \\
&= (B-aA)\frac{e^{-at}}{2}\mathcal{L}^{-a}\{\frac{2s}{s(s^2+b^2)^2}\} \\
&= (B-aA)\frac{e^{-at}}{2}\int_0^t \mathcal{L}^{-1}\{\frac{2s}{(s^2+b^2)^2}\}dt \\
&= (B-aA)\frac{e^{-at}}{2b}\int_0^t t\sin btdt \\
&= (B-aA)\frac{e^{-at}}{2b^3}[\sin bt - bt\cos bt]. \qquad (3.32)
\end{aligned}$$

The result follows if we combine (3.30), (3.31), and (3.32).

In this theorem we have demonstrated only a way to handle a problem. For a particular problem it is better to work step-by-step.

Example 7

Find

$$\mathcal{L}^{-1}\{\frac{s}{(s+2)^2(s^2+2s+10)}\}.$$

Solution

Here

$$\begin{aligned} p(s) &= s \\ q(s) &= (s+2)^2(s^2+2s+10) \\ &= (s+2)^2\{(s+1)^2+3^2\}. \end{aligned}$$

Thus $q(s) = 0$ yields $s = -2$: two repeated roots and $s = -1 \pm i3$: two complex roots. Now the inverse corresponding to $s = -2$ is $(A_1 + A_2 t)e^{-2t}$ where

$$A_2 = \lim_{s\to -2}\{(s+2)^2\frac{s}{(s+2)^2(s^2+2s+10)}\} = -\frac{1}{5}$$

and

$$\begin{aligned} A_1 &= \lim_{s\to -2}\frac{d}{ds}\{(s+2)^2\frac{s}{(s+2)^2(s^2+2s+10)}\} \\ &= \lim_{s\to -2}[\frac{1}{(s^2+2s+10)} - \frac{2s+2}{(s^2+2s+10)^2}] \\ &= \lim_{s\to -2}\frac{-s^2+10}{(s^2+2s+10)^2} = \frac{3}{50}. \end{aligned}$$

Thus the inverse corresponding to $s = -2$ is

$$(\frac{3}{50} - \frac{1}{5}t)e^{-2t}.$$

To obtain the inverse corresponding to the complex roots $s = -1 \pm i3$ $(a = 1, b = 3)$, we recognize here (Theorem 3.16) that $\phi(s) = \frac{s}{(s+2)^2}$ and so $\phi_r = \frac{13}{50}$ and $\phi_i = -\frac{9}{50}$. Thus the inverse corresponding to the factor $(s^2 + 2s + 10)$ is

$$\frac{e^{-t}}{3}\{\frac{-9\cos 3t + 13\sin 3t}{50}\} = \frac{e^{-t}}{150}\{-9\cos 3t + 13\sin 3t\}.$$

Hence,

$$\begin{aligned} \mathcal{L}^{-1}\frac{s}{(s+2)^2(s^2+2s+10)}\} &= (\frac{3}{50} - \frac{1}{5}t)e^{-2t} \\ &\quad + \frac{e^{-t}}{150}(-9\cos 3t + 13\sin 3t). \end{aligned}$$

3.8 The method of residue to find inverses

In this section, we shall demonstrate the method of residues for finding inverse Laplace transforms without going into the details of the theory. Here we shall only state the necessary theorem and illustrate the applications of that theorem.

Theorem 3.18

If the Laplace transform $F(s)$ is an analytic function of s, except at a finite number of singular points called the poles, each of which lies to the left of the vertical line $Re(s) = a$, and if $sF(s)$ is bounded as s approaches infinity through the half plane $Re(s) \leq a$, then

$$\begin{aligned}\mathcal{L}^{-1}\{F(s)\} &= \frac{1}{2\pi i}\int_{a-i\infty}^{a+i\infty} F(s)e^{st}ds \\ &= \sum \text{Residues of } F(s)e^{st}\text{at all the poles of } F(s) \\ &= R_1 + R_2 + \cdots + R_n\end{aligned}$$

where R_k's are defined as residues and are given by the formula

$$R_k = \lim_{s\to a_k}\{(s-a_k)F(s)e^{st}\} \qquad k = 1, 2, \cdots, n$$

if $a_1, a_2, a_3, \cdots, a_n$ are simple poles of $F(s)$. Now if $s = a_k$ is a multiple pole of $F(s)$ of order $(r+1)$, then

$$R_k = \lim_{s\to a_k}\frac{1}{r!}\frac{d^r}{ds^r}\{(s-a_k)^{r+1}F(s)e^{st}\}.$$

We shall now demonstrate the application of this theory to some problems.

Example 8

Find

$$\mathcal{L}^{-1}\{\frac{s^2 - s + 3}{s^3 + 6s^2 + 11s + 6}\}.$$

Solution

It can easily be shown that

$$\begin{aligned}\lim_{s\to\infty} sF(s) &= \lim_{s\to\infty}\{\frac{s(s^2 - s + 3)}{s^3 + 6s^2 + 11s + 6}\} \\ &= 1,\end{aligned}$$

which is bounded.
Therefore, by Theorem 3.18, we know

$$\begin{aligned}\mathcal{L}^{-1}\{\frac{s^2 - s + 3}{s^3 + 6s^2 + 11s + 6}\} &= \mathcal{L}^{-1}\{F(s)\} \\ &= \sum \text{Residues of } F(s)e^{st} \\ &\quad \text{at poles } s = -1, -2, \text{and } -3.\end{aligned}$$

These are the simple poles.

$$\begin{aligned} R_1 &= \text{Residue at } s = -1 \\ &= \lim_{s\to -1}\{(s+1)\frac{(s^2-s+3)e^{st}}{(s+1)(s+2)(s+3)}\} = \frac{5}{2}e^{-t} \\ R_2 &= \text{Residue at } s = -2 \\ &= \lim_{s\to -2}\{(s+2)\frac{(s^2-s+3)e^{st}}{(s+1)(s+2)(s+3)}\} = -9e^{-2t} \\ R_3 &= \text{Residue at } s = -3 \\ &= \lim_{s\to -3}\{(s+3)\frac{(s^2-s+3)e^{st}}{(s+1)(s+2)(s+3)}\} = \frac{15}{2}e^{-3t}. \end{aligned}$$

Then

$$\begin{aligned} \mathcal{L}^{-1}\{\frac{s^2-s+3}{(s+1)(s+2)(s+3)}\} &= R_1 + R_2 + R_3 \\ &= \frac{5}{2}e^{-t} - 9e^{-2t} + \frac{15}{2}e^{-3t}. \end{aligned}$$

3.9 The Laplace transform of a periodic function

In this section we shall state and prove a theorem concerning the application of Laplace transforms to general periodic functions.

Theorem 3.19

If $f(t)$ is of exponential order and is a periodic function with period $2p$ on $(0,\infty)$ and is piecewise regular over $0 \le t \le 2p$, then

$$\mathcal{L}\{f(t)\} = \frac{\int_0^{2p} f(t)e^{-st}dt}{1-e^{-2ps}} \qquad s > 0.$$

Proof

By definition

$$\begin{aligned} \mathcal{L}\{f(t)\} &= \int_0^\infty e^{-st}f(t)dt \\ &= \int_0^{2p} e^{-st}f(t)dt + \int_{2p}^{4p} e^{-st}f(t)dt \\ &\quad + \int_{4p}^{6p} e^{-st}f(t)dt + \cdots \end{aligned}$$

In the second integral, let $t = x + 2p$; in the third integral, let $t = x + 4p$; and, in general, let $t = x + 2np$. Then,

$$\begin{aligned} \mathcal{L}\{f(t)\} &= \int_0^{2p} e^{-st}f(t)dt + e^{-2ps}\int_0^{2p} e^{-sx}f(x+2p)dx \\ &\quad + e^{-4ps}\int_0^{2p} e^{-sx}f(x+4p)dx + \cdots \end{aligned}$$

However,

$$f(x+2p) = f(x+4p) = \cdots = f(x+2np) = f(x).$$

Therefore,

$$\begin{aligned}\mathcal{L}\{f(t)\} &= \int_0^{2p} e^{-st} f(t)dt + e^{-2ps}\int_0^{2p} e^{-sx} f(x)dx \\ &\quad + e^{-4ps}\int_0^{2p} e^{-sx} f(x)dx + \cdots\end{aligned}$$

Changing the dummy variable x to t, we obtain

$$= (1 + e^{-2ps} + e^{-4ps} + \cdots)\int_0^{2p} e^{-st} f(t)dt.$$

But the sum of the geometric series

$$1 + e^{-2ps} + e^{-4ps} + \cdots = \frac{1}{1-e^{-2ps}}.$$

Therefore,

$$\mathcal{L}\{f(t)\} = \frac{\int_0^{2p} e^{-st} f(t)dt}{1-e^{-2ps}}.$$

Example 9

Find the Laplace transform of the periodic function $f(t)$ shown graphically in Fig. 3.13.

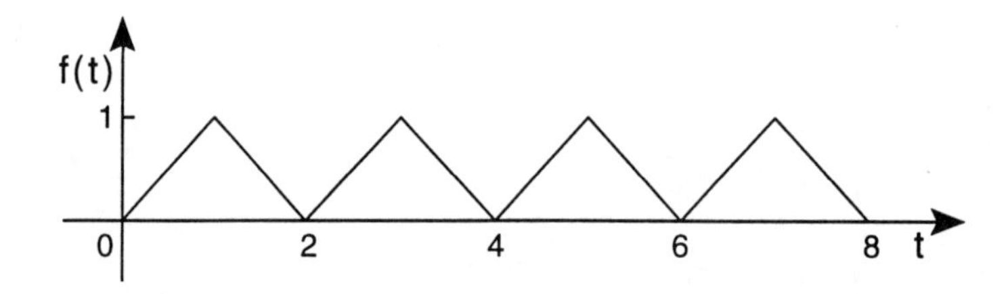

Figure 3.13: A triangular wave

Solution

In this case, the period of the function is 2 and

$$f(t) = \begin{cases} t & 0 < t < 1 \\ 2-t & 1 < t < 2. \end{cases}$$

It follows that,

$$\begin{aligned}\mathcal{L}\{f(t)\} &= \frac{1}{1-e^{-2s}}\int_0^2 f(t)e^{-st}dt \\ &= \frac{1}{1-e^{-2s}}[\int_0^1 te^{-st}dt + \int_1^2 (2-t)e^{-st}dt].\end{aligned}$$

Now,

$$\begin{aligned}\int_0^1 te^{-st}dt &= (-\frac{te^{-st}}{s})|_0^1 + \frac{1}{s}\int_0^1 e^{-st}dt \\ &= \frac{1-e^{-s}-se^{-s}}{s^2} \\ \int_1^2 (2-t)e^{-st}dt &= \{(2-t)\frac{e^{-st}}{-s}\}_1^2 - \frac{1}{s}\int_1^2 e^{-st}dt \\ &= \frac{se^{-s}+e^{-2s}-e^{-s}}{s^2} \\ \mathcal{L}\{f(t)\} &= \frac{1-e^{-s}-se^{-s}+se^{-s}+e^{-2s}-e^{-s}}{s^2(1-e^{-2s})} \\ &= \frac{1-2e^{-s}+e^{-2s}}{s^2(1-e^{-2s})} \\ &= \frac{1}{s^2}\tanh(\frac{s}{2}).\end{aligned}$$

3.10 Convolution

An important property of Laplace transforms exists with the product of transforms. If we know the inverse transforms $f(t)$ of $F(s)$ and $g(t)$ of $G(s)$, then from $f(t)$ and $g(t)$, we can calculate the inverse transform $y(t)$ of the product $Y(s) = F(s)G(s)$. In other words, if two functions, $f(t)$ and $g(t)$, are given, they may be combined to give a new function $y(t)$, which is called the **convolution** of $f(t)$ and $g(t)$, and is often written $f(t) * g(t)$.

The convolution of $f(t) * g(t)$ of $f(t)$ and $g(t)$ is defined by the integral equation

$$\begin{aligned}y(t) &= f(t) * g(t) \\ &= \int_0^t f(\tau)g(t-\tau)d\tau.\end{aligned} \tag{3.33}$$

This integral is usually known as the Convolution Integral.

By substituting $\lambda = t - \tau$ such that $d\lambda = -d\tau$, it can be seen that the convolution integral (3.33) may be written in one of two ways:

$$\begin{aligned}y(t) &= f(t) * g(t) \\ &= \int_0^t f(\tau)g(t-\tau)d\tau \\ &= \int_0^t f(t-\tau)g(\tau)d\tau.\end{aligned} \tag{3.34}$$

This integral is referred to as the "Faltung" (German for "folding") integral in some texts. The graphical interpretation of the physical significance of this integral will be given later. Until then we regard this mathematical combination of the two functions, f and g, as a special product under an integral sign.

We shall now state and prove the following theorem concerning the product of two transforms.

Theorem 3.20: (Convolution theorem)

If $f(t)$ and $g(t)$ have Laplace transforms $F(s)$ and $G(s)$, respectively, then the inverse transform $y(t)$ of the product $Y(s) = F(s)G(s)$ is given by the convolution of $f(t)$ and $g(t)$. Therefore,

$$\begin{aligned}
\mathcal{L}\{y(t)\} &= \mathcal{L}\{f(t) * g(t)\} \\
&= \mathcal{L}\{\int_0^t f(\tau)g(t-\tau)d\tau\} \\
&= \mathcal{L}\{\int_0^t f(t-\tau)g(\tau)d\tau\} \\
&= \mathcal{L}\{f(t)\}\mathcal{L}\{g(t)\} \qquad (3.35)
\end{aligned}$$

which can be written as $Y(s) = F(s)G(s)$.

Alternatively, the transform of the convolution of two functions is equal to the product of their transforms.

Proof

By definition, we know

$$\begin{aligned}
\mathcal{L}\{f(t) * g(t)\} &= \mathcal{L}[\int_0^t f(t-\tau)g(\tau)d\tau] \\
&= \mathcal{L}[\int_0^\infty f(t-\tau)g(\tau)u(t-\tau)d\tau].
\end{aligned}$$

$$\text{Because } u(t-\tau) = \begin{cases} 0 & \tau > t \\ 1 & \tau < t \end{cases}$$

$$\begin{aligned}
&= \int_0^\infty e^{-st}[\int_0^\infty f(t-\tau)g(\tau)u(t-\tau)d\tau]dt \\
&= \int_0^\infty g(\tau)[\int_0^\infty f(t-\tau)u(t-\tau)e^{-st}dt]d\tau.
\end{aligned}$$

Substituting $t - \tau = \lambda$ into the inner integral, we obtain

$$\begin{aligned}
\mathcal{L}\{f(t) * g(t)\} &= \int_0^\infty g(\tau)[\int_{-\tau}^\infty f(\lambda)e^{-s(\tau+\lambda)}d\lambda]d\tau \\
&= \int_0^\infty g(\tau)e^{-s\tau}[\int_0^\infty f(\lambda)e^{-s\lambda}d\lambda]d\tau \\
&= \{\int_0^\infty f(\lambda)e^{-s\lambda}d\lambda\}\{\int_0^\infty g(\tau)e^{-s\tau}d\tau\} \\
&= \mathcal{L}\{f(t)\}\mathcal{L}\{g(t)\}.
\end{aligned}$$

as asserted.

Using the definition, we can show that the convolution $f * g$ has the properties:

$$\begin{aligned} f * g &= g * f \quad \text{(commutative law)} \\ f * (g_1 + g_2) &= f * g_1 + f * g_2 \quad \text{(distributive law)} \\ (f * g) * h &= f * (g * h) \quad \text{(associative law)} \\ f * 0 &= 0 * f = 0. \end{aligned}$$

But $1 * g \neq g$ in general. For example, if $g(t) = t$, then

$$(1 * g) = \int_0^t 1.(t - \tau)d\tau = \frac{t^2}{2}.$$

Another important property is that $(f * f) \geq 0$ may not hold.

In the following examples, we shall illustrate that the convolution theorem is useful for obtaining the inverses of Laplace transforms and for solving differential equations.

Example 10

Applying the convolution theorem, solve the following initial value problem,

$$y'' + y = \sin t \qquad y(0) = 0 \ , \ y'(0) = 0.$$

Solution

Taking the Laplace transform of the given equation, we have

$$\mathcal{L}(y'') + \mathcal{L}(y) = \mathcal{L}(\sin t)$$

which after reduction yields $\mathcal{L}\{y\} = \frac{1}{(s^2+1)^2}$. Then the inverse is given by

$$y = \mathcal{L}^{-1}\{\frac{1}{(s^2+1)^2}\} = \int_0^t \sin\tau \sin(t - \tau)d\tau = \frac{t \sin t}{2}$$

which is the required solution.

3.10.1 Transfer function, impulse response and indicial response

The application of the convolution theorem makes it possible to determine the response of a system to a general excitation. To develop this idea we need the concepts of:

(a) Transfer function (system function)

(b) impulsive response

(c) indicial response (indicial admittance).

Any physical system capable of responding to an excitation $f(t)$ can be described mathematically by a differential equation satisfying the output function $y(t)$. Assume that all initial conditions are zero when a single excitation $f(t)$ begins to act. Then, taking the Laplace transform of the governing differential equation, we obtain a relationship in the form,

$$\mathcal{L}\{y(t)\} = \frac{1}{Z(s)}\mathcal{L}\{f(t)\}. \tag{3.36}$$

In the electrical network problem $f(t)$ is the input voltage, and $y(t)$ is the output current known as the response. Here $\frac{1}{Z(s)}$ is defined as the transfer function of the system. Graphically, this transfer function can be defined as

$$\frac{1}{Z(s)} = \frac{\mathcal{L}\{y(t)\}}{\mathcal{L}\{f(t)\}} = \frac{\mathcal{L}\{\text{output}\}}{\mathcal{L}\{\text{input}\}}.$$

The importance of $Z(s)$ functions is apparent in systems other than the electrical system.

The output of a system to a unit impulse input is called the impulsive response of the system, and is usually denoted by $I(t)$. Importantly, if $I(t)$ is known, then the response to any general input may be obtained from the convolution integral.

One immediate result can be deduced as follows. We know that if $f(t)$ is the general excitation, and $y(t)$ is the response, then

$$\mathcal{L}\{y(t)\} = \frac{1}{Z(s)}\mathcal{L}\{f(t)\}.$$

Let $f(t) = \delta(t) =$ unit impulse, the response of $\delta(t) = I(t)$; hence,

$$\mathcal{L}\{I(t)\} = \frac{1}{Z(s)}\mathcal{L}\{\delta(t)\} = \frac{1}{Z(s)}. \tag{3.37}$$

The transfer function is equal to the transform of the impulsive response.

It is more convenient, from a practical point of view, to deal with the response of a unit step function $u(t)$. This is called the indicial response and is normally denoted by $h(t)$. If the indicial response is known, then the response to a general input can be determined from the Duhamel theorem. This theorem will be discussed later.

In the relationship (3.36), let $f(t) = u(t) =$ unit step function. Then the response of $u(t) = h(t)$, and, hence,

$$\mathcal{L}\{h(t)\} = \frac{1}{Z(s)}\mathcal{L}\{u(t)\} = \frac{1}{sZ(s)}. \tag{3.38}$$

The significance of the impulsive response and the indicial response is that once these two responses are determined (easily accomplished through measuring), then the responses of the system can be determined by the convolution integral and the Duhamel theorem, respectively.

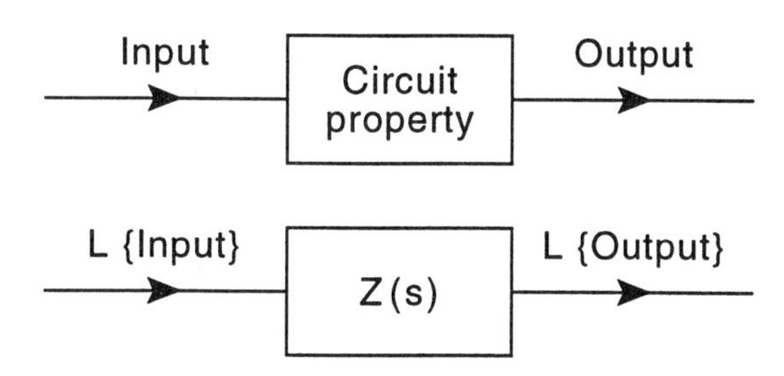

Figure 3.14: Transfer function

3.10.2 Significance of convolution theorem

Given an impulsive response $I(t)$ corresponding to a unit impulse, we obtain from (3.36)

$$\mathcal{L}\{y(t)\} = \frac{1}{Z(s)}\mathcal{L}\{f(t)\} = \mathcal{L}\{I(t)\}\mathcal{L}\{f(t)\}. \tag{3.39}$$

Then by the convolution integral, we have

$$\begin{aligned} y(t) &= \int_0^t I(\tau)f(t-\tau)d\tau \\ &= \int_0^t I(t-\tau)f(\tau)d\tau \\ \text{output} &= \text{(Input)} * \text{(Impulsive Response)}. \end{aligned} \tag{3.40}$$

3.10.3 Significance of Duhamel's theorem

An alternate method of determining the output response of a system is provided by Duhamel's theorem, named after the French mathematician (1797–1872), and used when the indicial response is known. The response to unit step function is normally easier to determine than the unit impulse functions.

From (3.36), we know

$$\mathcal{L}\{y(t)\} = \frac{1}{Z(s)}\mathcal{L}\{f(t)\}.$$

Now rearranging (3.38) into the above equation,

$$\begin{aligned} \mathcal{L}\{y(t)\} &= s\mathcal{L}\{h(t)\}\mathcal{L}\{f(t)\} \\ &= \mathcal{L}\{h(t)\}[\mathcal{L}\{(\frac{df}{dt}) + f(0)\}] \\ &= \mathcal{L}\{h(t)\}\mathcal{L}\{(\frac{df}{dt})\} + f(0)\mathcal{L}\{h(t)\} \end{aligned} \tag{3.41}$$

we have

$$\begin{aligned}\mathcal{L}\{y(t)\} &= [\mathcal{L}\{(\frac{dh}{dt}) + h(0)\}]\mathcal{L}\{f(t)\} \\ &= \mathcal{L}\{(\frac{dh}{dt})\}\mathcal{L}\{f(t)\} + h(0)\mathcal{L}\{f(t)\}. \end{aligned} \tag{3.42}$$

Then applying the convolution theorem to (3.41) and (3.42), we obtain

$$y(t) = f(0)h(t) + \int_0^t \frac{df}{dt}(\tau)h(t-\tau)d\tau \tag{3.43}$$

$$= f(t)h(0) + \int_0^t \frac{dh}{dt}(\tau)f(t-\tau)d\tau. \tag{3.44}$$

Equations (3.43) and (3.44) express the output in terms of the input and the indicial response. These formulas are normally called Duhamel's Theorem.

We shall illustrate the usefulness of the convolution theorem and Duhamel's theorem with a practical example of network theory.

Example 11

Given that the following circuit is initially dead, find the current which flows in a series RC circuit shown in Fig. 3.15 when the voltage applied is $\sin\omega t u(t)$

(a) by finding the system function and using the Laplace transform;

(b) by using the impulsive response method; and

(c) by using the indicial response method.

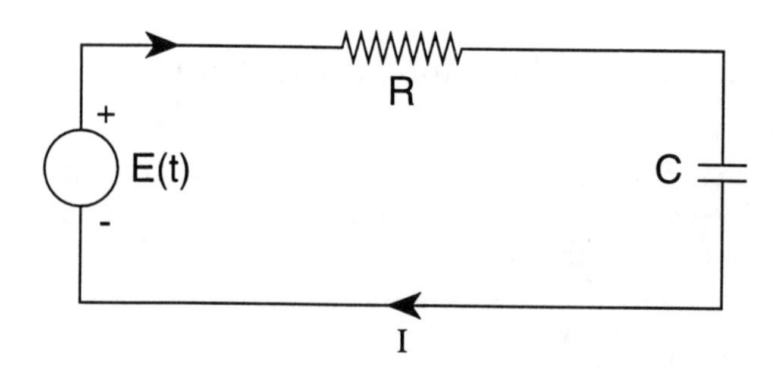

Fig. 3.15: R-C circuit

Solution

The differential equation governing the system can be written as

$$Ry + \frac{1}{C}\int_0^t ydt = f(t) \tag{3.45}$$

where $y(t)$ is the output current and $f(t) = \sin\omega t$ is the input voltage. Then, taking the Laplace transform of the given equation with $y(0) = 0$, we obtain

$$RL\{y\} + \frac{1}{C}\mathcal{L}\{\int_0^t ydt\} = \mathcal{L}\{\sin\omega t\}. \tag{3.46}$$

Therefore we obtain

$$\mathcal{L}\{y\} = (\frac{Cs}{1+Rcs})(\frac{\omega}{s^2+\omega^2}). \tag{3.47}$$

Here the transform function $= \frac{Cs}{1+RCs}$.

(a) By transfer function method

$$y = \mathcal{L}^{-1}\{\frac{(\omega/R)s}{(s+\frac{1}{RC})(s^2+\omega^2)}\}.$$

To use the theory of residues we need to find the poles which are at $s = -\frac{1}{RC}$ and $s = \pm i\omega$.

$$\begin{aligned}
R_1 &= \text{residue at } s = -\frac{1}{RC} \\
&= \lim_{s\to -1/RC}\{(s+\frac{1}{RC})\frac{(\omega/R)se^{st}}{(s+\frac{1}{RC})(s^2+\omega^2)}\} \\
&= -\frac{\omega Ce^{-\frac{t}{RC}}}{1+(RC\omega)^2}. \\
R_2 &= \text{residue at } s = +i\omega \\
&= \lim_{s\to i\omega}\{(s-i\omega)\frac{(\omega/R)se^{st}}{(s+\frac{1}{RC})(s+i\omega)(s-i\omega)}\} \\
&= \frac{\omega}{R}\frac{i\omega e^{i\omega t}}{(i\omega+\frac{1}{RC})2i\omega} = \frac{\omega}{2R}\frac{e^{i\omega t}}{(i\omega+\frac{1}{RC})} \\
R_2 &= \frac{\omega(\frac{1}{RC}-i\omega)e^{+i\omega t}}{2R(\omega^2+\frac{1}{R^2C^2})}. \\
R_3 &= \text{residue at } s = -i\omega \\
&= \frac{\omega(\frac{1}{RC}+i\omega)e^{-i\omega t}}{2R(\omega^2+\frac{1}{R^2C^2})}. \\
R_2+R_3 &= \frac{\omega}{2R(\omega^2+\frac{1}{R^2C^2})}[\frac{1}{RC}(e^{i\omega t}+e^{-i\omega t}) - i\omega[e^{i\omega t}-e^{-i\omega t})] \\
&= \frac{\omega}{2R(\omega^2+\frac{1}{R^2C^2})}[\frac{2}{RC}\cos\omega t + 2\omega\sin\omega t].
\end{aligned}$$

Therefore, the solution is

$$y(t) = \frac{-\omega Ce^{-\frac{t}{RC}}}{1+(RC\omega)^2} + \frac{\omega C}{[1+(RC\omega)^2]}[\cos\omega t + \omega RC\sin\omega t]. \tag{3.48}$$

This is the required response.

(b) By the impulsive response method

$$\mathcal{L}\{y\} = (\frac{\frac{s}{R}}{\frac{1}{RC} + s})(\frac{\omega}{s^2 + \omega^2}).$$

We know $\mathcal{L}\{I(t)\} = (\frac{\frac{s}{R}}{s + \frac{1}{RC}}) = \frac{1}{R}[1 - \frac{1}{RC}\frac{1}{(s + \frac{1}{RC})}]$. Then the inverse is

$$I(t) = \frac{1}{R}[\delta(t) - \frac{1}{RC}e^{-\frac{t}{RC}}].$$

Now using the convolution integral

$$\begin{aligned} y(t) &= \int_0^t I(\tau) f(t-\tau) d\tau \\ &= \int_0^t \frac{1}{R}\{\delta(\tau) - \frac{1}{RC}e^{-\frac{\tau}{RC}}\} \sin\omega(t-\tau) d\tau \\ &= \frac{1}{R}\int_0^t \delta(\tau)\sin\omega(t-\tau)d\tau - \frac{1}{R^2C}\int_0^t e^{-\frac{\tau}{RC}}\sin\omega(t-\tau)d\tau. \end{aligned}$$

$$\begin{aligned} \text{The first integral} &= \frac{1}{R}\int_0^t \delta(\tau)\sin\omega(t-\tau)d\tau \\ &= \frac{1}{R}\sin\omega t. \\ \text{The second integral} &= \frac{1}{R^2C}\int_0^t e^{-\frac{1}{RC}(t-\tau)}\sin\omega\tau d\tau \\ &= \frac{1}{R^2C}e^{-\frac{t}{RC}}\int_0^t e^{\frac{\tau}{RC}}\sin\omega\tau d\tau \\ &= \frac{1}{R^2C}e^{-\frac{t}{RC}}[\frac{e^{\frac{\tau}{RC}}\{\frac{1}{RC}\sin\omega\tau - \omega\cos\omega\tau\}}{(\omega^2 + \frac{1}{R^2C^2})}]_0^t \\ &= \frac{e^{-\frac{t}{RC}}}{R^2C(\omega^2 + \frac{1}{R^2C^2})}[e^{\frac{t}{RC}}\{\frac{1}{RC}\sin\omega t - \omega\cos\omega t\} + \omega]. \end{aligned}$$

Therefore,

$$\begin{aligned} & -\frac{1}{R^2C}\int_0^t e^{-\frac{\tau}{RC}}\sin\omega(t-\tau)d\tau \\ &= \frac{-\omega e^{-\frac{t}{RC}}}{R^2C(\omega^2 + \frac{1}{R^2C^2})} + \frac{[-\sin\omega t + \omega RC\cos\omega t]}{R(1+\omega^2R^2C^2)} \\ &= \frac{-\omega e^{-\frac{t}{RC}}}{1+\omega^2R^2C^2} + \frac{(-\sin\omega t + \omega RC\cos\omega t)}{R(1+\omega^2R^2C^2)}. \end{aligned}$$

Therefore,

$$\begin{aligned} y(t) &= \frac{-\omega C e^{-\frac{t}{RC}}}{1+\omega^2R^2C^2} + \frac{-\sin\omega t + \omega RC\cos\omega t + (1+\omega^2R^2C^2)\sin\omega t}{R(1+\omega^2C^2R^2)} \\ &= \frac{-\omega C e^{-\frac{t}{RC}}}{1+\omega^2R^2C^2} + \frac{\omega C[\cos\omega t + \omega CR\sin\omega t]}{1+\omega^2C^2R^2}. \end{aligned} \tag{3.49}$$

(c) By the indicial response method

We know that $\mathcal{L}\{h(t)\} = (\frac{\frac{s}{R}}{s+\frac{1}{RC}})\mathcal{L}\{u(t)\} = \frac{\frac{1}{R}}{s+\frac{1}{RC}}$.

Therefore the inverse is $h(t) = \frac{1}{R}e^{-\frac{t}{RC}}$.

Then, using Duhamel's formula,

$$\begin{aligned} y(t) &= f(0)h(t) + \int_0^t \frac{df}{d\tau}h(t-\tau)d\tau \\ &= \int_0^t \omega\cos\omega\tau\frac{1}{R}e^{-\frac{1}{RC}(t-\tau)}d\tau \\ &= \frac{\omega}{R}e^{-\frac{t}{RC}}\int_0^t e^{\frac{\tau}{RC}}\cos\omega\tau d\tau \\ &= \frac{\omega}{R}e^{-\frac{t}{RC}}\left[\frac{e^{\frac{\tau}{RC}}\{\frac{1}{RC}\cos\omega\tau + \omega\sin\omega\tau\}}{\omega^2+(\frac{1}{RC})^2}\right]_0^t \\ &= \frac{\frac{-\omega}{R}e^{-\frac{t}{RC}}}{\omega^2+(\frac{1}{RC})^2}\left[e^{\frac{t}{RC}}(\frac{1}{RC}\cos\omega t + \omega\sin\omega t) + \frac{1}{RC}\right]_0^t \\ y(t) &= \frac{-\omega C e^{-\frac{t}{RC}}}{1+\omega^2R^2C^2} + \frac{\omega C[\cos\omega t + \omega RC\sin\omega t]}{1+\omega^2R^2C^2}. \end{aligned} \tag{3.50}$$

Note from (3.48), (3.49) and (3.50) that the responses $y(t)$ produced by $\sin\omega t$ are identical. In this particular problem, Duhamel's formula yields the result very easily, while the other methods seem laborious. In the following section, we shall be concerned with the graphical interpretation of the convolution integral.

3.10.4 Graphical convolution

As indicated earlier, convolution may be carried out graphically. Here we will show how this might be done in an idealized situation. Suppose the voltage applied to an electric circuit of elements $L-R$ in series is given to be a rectangular pulse beginning at $t = 0$, of amplitude E volts and duration a seconds. Initially the system is dead.

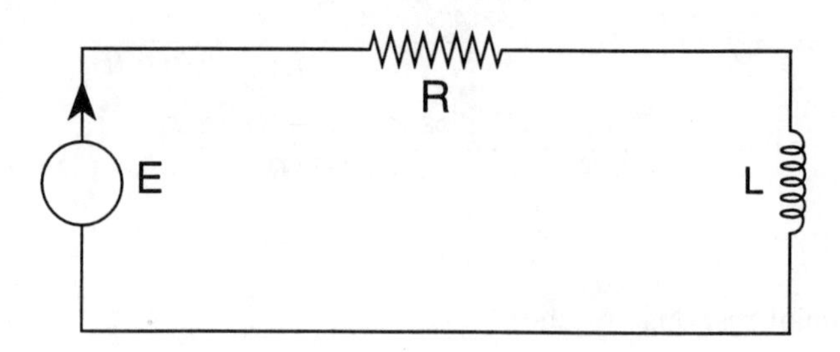

Figure 3.16: L-R series circuits

The differential equation governing the system is written

$$L\frac{dy}{dt} + Ry = f(t) \tag{3.51}$$

where $y(t)$ is the response of the system and $f(t) = E[u(t) - u(t-a)]$ is the input function. Taking the Laplace transform of (3.51), we obtain

$$L\mathcal{L}\{\frac{dy}{dt}\} + R\mathcal{L}\{y\} = \mathcal{L}\{f(t)\}.$$

In view of the zero initial condition (i.e. $y(0) = 0$), we can write $(Ls + R)\mathcal{L}\{y\} = \mathcal{L}\{f(t)\}$ and therefore, $\mathcal{L}\{y\} = \frac{1}{R+Ls}\mathcal{L}\{f(t)\}$.

Therefore, the system function is $\frac{1}{R+Ls}$. This indeed is the transform of the impulsive response. Hence the impulsive response is

$$I(t) = \mathcal{L}^{-1}\{\frac{1}{R+Ls}\} = \frac{1}{L}e^{-\frac{Rt}{L}}. \tag{3.52}$$

By the convolution integral we can determine the response of the system,

$$y(t) = \int_0^t I(t-\tau)f(\tau)d\tau. \tag{3.53}$$

Now to obtain the graphical solution of this integral, we first depict the two graphs representing the applied voltage $f(\tau)$, and then depict the impulsive response $I(\tau)$ (shown in Fig. 3.17 (a) and (b)).

Let us assume that the only information we have are the above graphs of the input applied voltage $f(\tau)$ and of the impulsive response $I(\tau)$ obtained from oscillograms. From the convolution integral (3.53), we know that we need a graph to represent the function $I(t-\tau)$. This can be done as follows. If we shift the graph of $I(\tau)$ t units to the right as in Fig. 3.17 (c), we obtain $I(\tau - t)$ and the reversal of this graph about $\tau = 0$ yields the graph of $I(t-\tau)$ as in Fig. 3.17 (d). The graph of $I(t-\tau)$ is the image of $I(\tau - t)$ when we place a mirror at $\tau = 0$.

In Fig. 3.17 (e), we bring both functions into the same time scale τ. We may choose any value of t in order to find the output at a particular instant. The convolution integral involves the product of $f(\tau)$ times $I(t-\tau)$, integrated with respect to τ between 0 and t; hence, convolution will start when $t=0$, as in Fig. 3.17 (f). After the t_1 instant, (Fig. 3.17 (g)), we superimpose the two graphs from Fig. 3.17 (f). Then integral (3.53), and therefore the output current, may be represented as the area of the graph. It is convenient but not necessary to draw the curves with the scale changed as shown, the shaded area being proportional to the output, and the precise output at time t being given by the area times E/L. By changing the scale, we can produce a very clear picture, and since $f(\tau)$ is either 0 or 1 in the new scale, the product becomes the area under $I(t-\tau)$ which is in common interval between 0 and a.

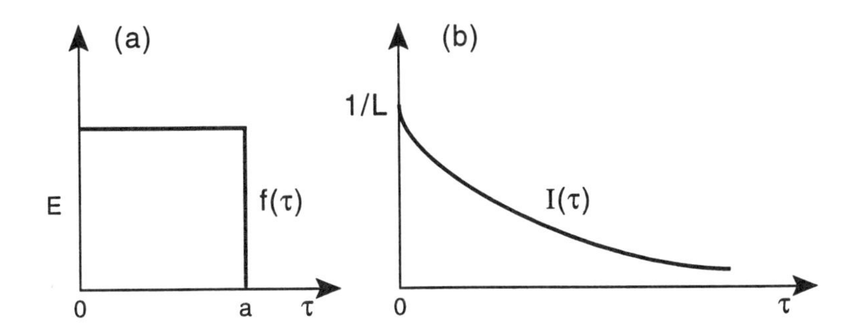

Figure 3.17a and 3.17b: Graphical convolution

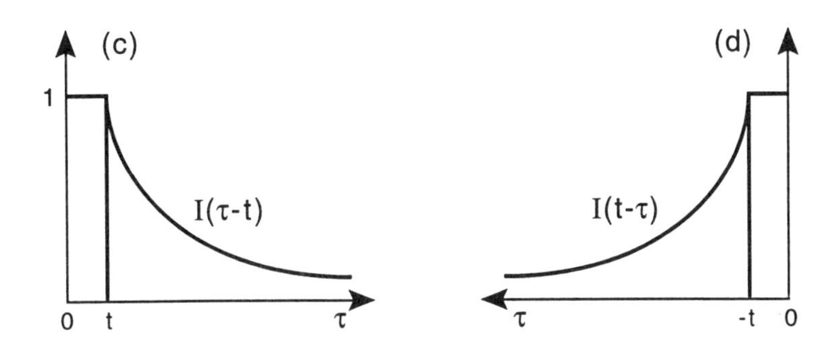

Figure 3.17c and 3.17d: Graphical convolution

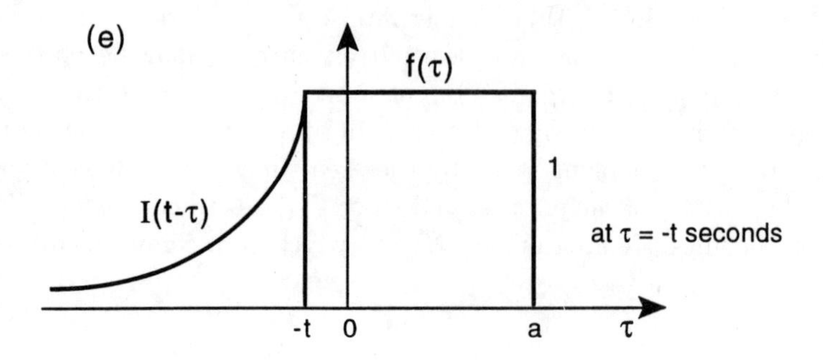

Figure 3.17e: Graphical convolution

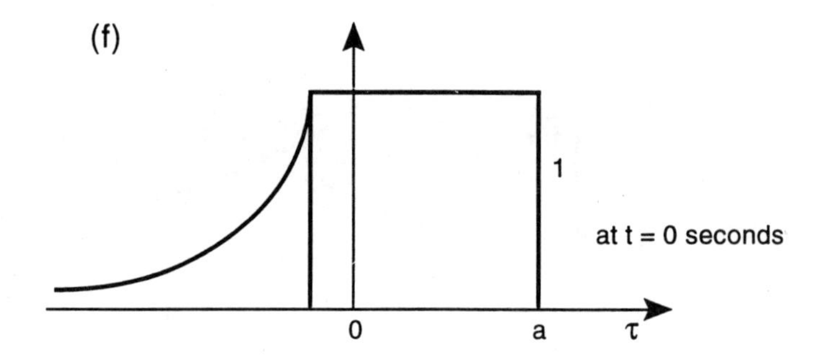

Figure 3.17f: Graphical convolution

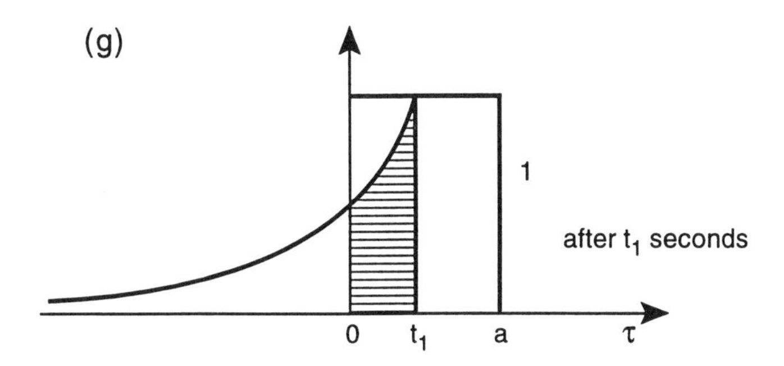

Figure 3.17g: Graphical convolution

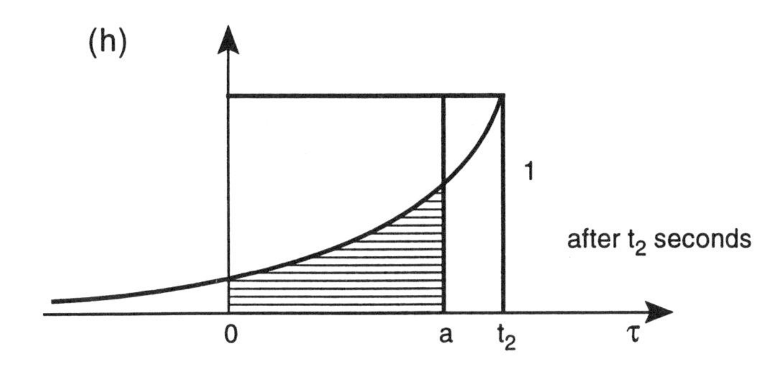

Figure 3.17h: Graphical convolution

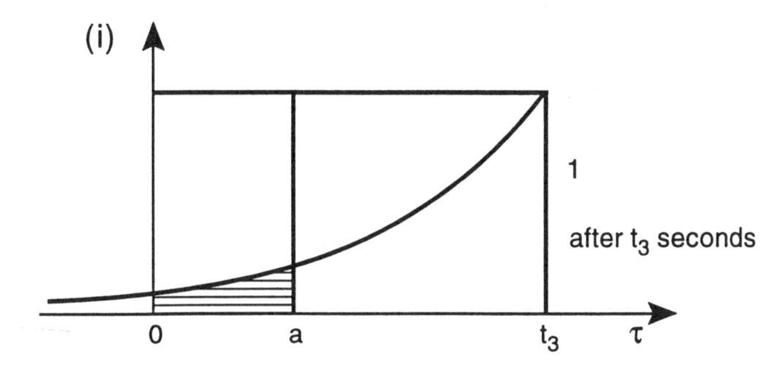

Figure 3.17i: Graphical convolution

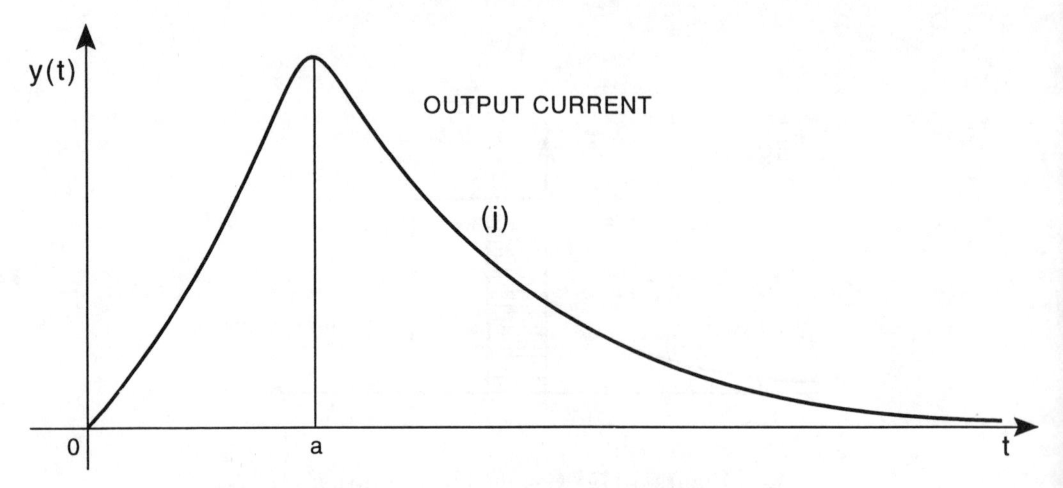

Figure 3.17j: Output current by graphical convolution

The process may be repeated for other instants of time (Fig. 3.17(h) and (i)). In each case, the shaded area gives the response at the chosen instant. From the results for the various instants t_1, t_2, t_3, etc., a graph of the response against time t may be constructed (Fig. 3.17 (j)).

Note that graphical convolution is important in its own right, and does not involve any Laplace transforms. This method is extremely valuable when the evaluation of the integral is necessary, but the use of transforms is either very difficult or impossible.

In the following section, we shall give an example which can be solved by using graphical convolution. The result then can be verified by the Laplace transform method and subsequently by direct integration.

Example 12

A linear system has an impulsive response given by $f(t)$ and is subjected to a pulse given by $g(t)$ as follows:

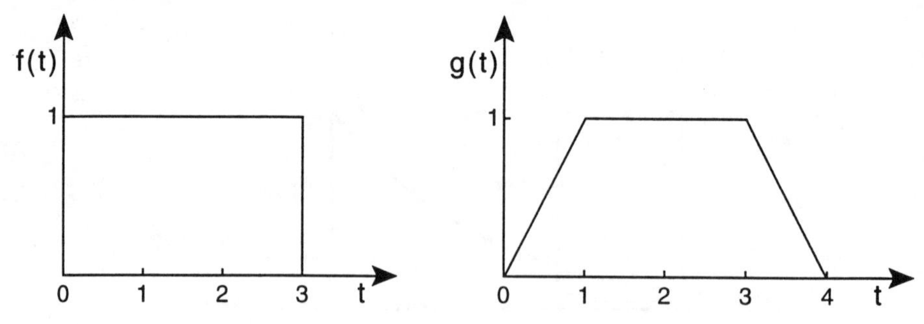

Find the output time function convolving $f(t)$ with $g(t)$ by

(a) graphical method

(b) Laplace transform

(c) direct integration.

Solution

(a) Graphical convolution:

Using the method illustrated above, graphical convolution can be obtained as follows:

$$y(t) = f(t) * g(t) = \int_0^t f(t-\tau)g(\tau)d\tau.$$

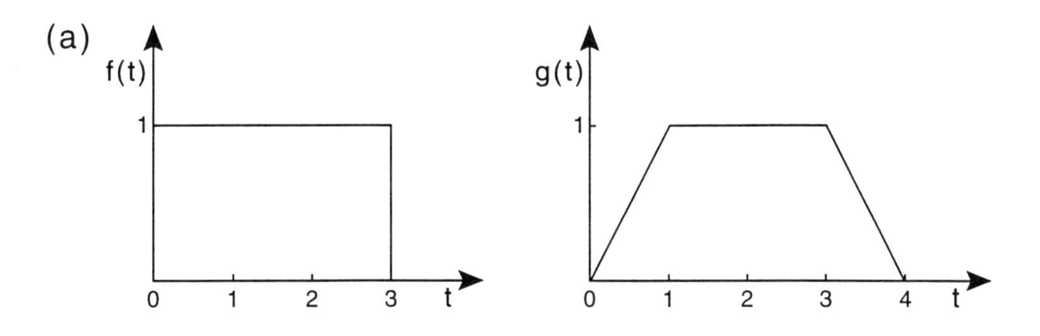

Figure 3.18a: Graph to describe convolution

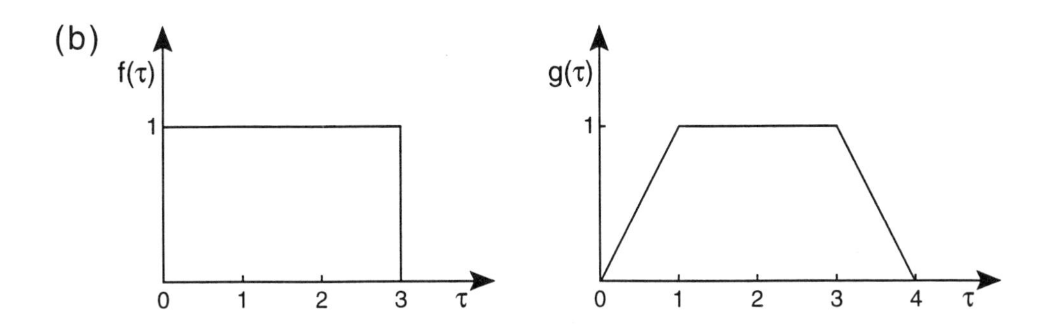

Figure 3.18b: Graph to describe convolution

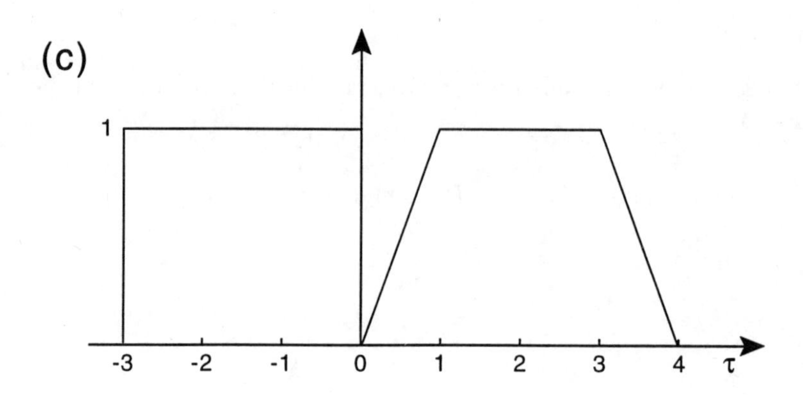

Figure 3.18c: Graph to describe convolution

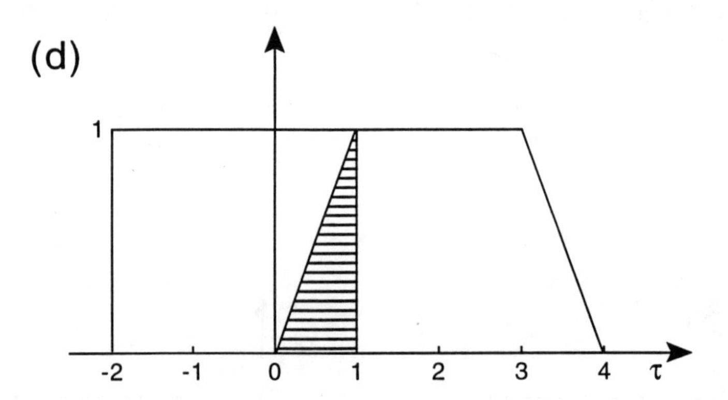

Figure 3.18d: Graph to describe convolution

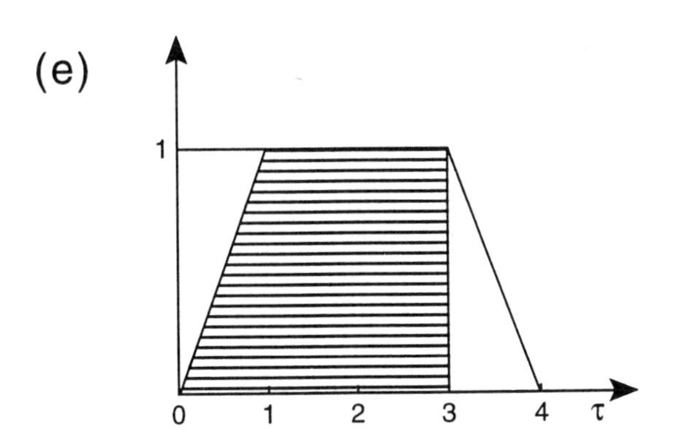

Figure 3.18e: Graph to describe convolution

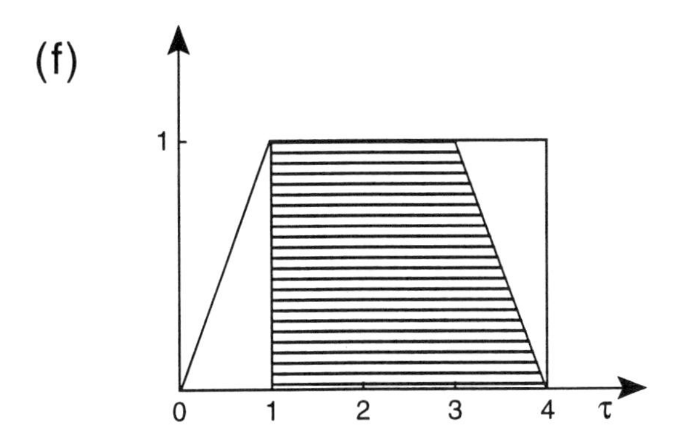

Figure 3.18f: Graph to describe convolution

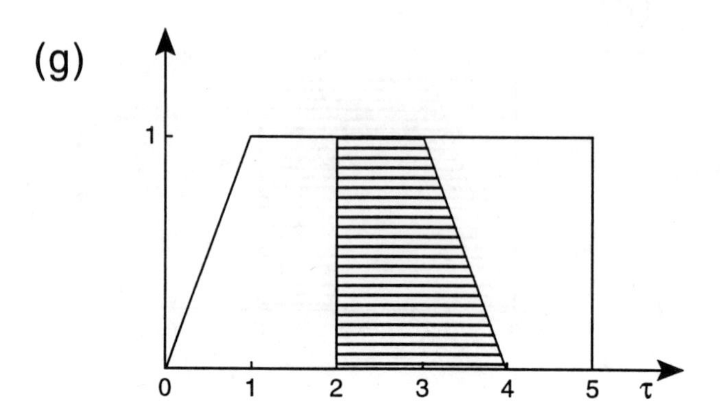

Figure 3.18g: Graph to describe convolution

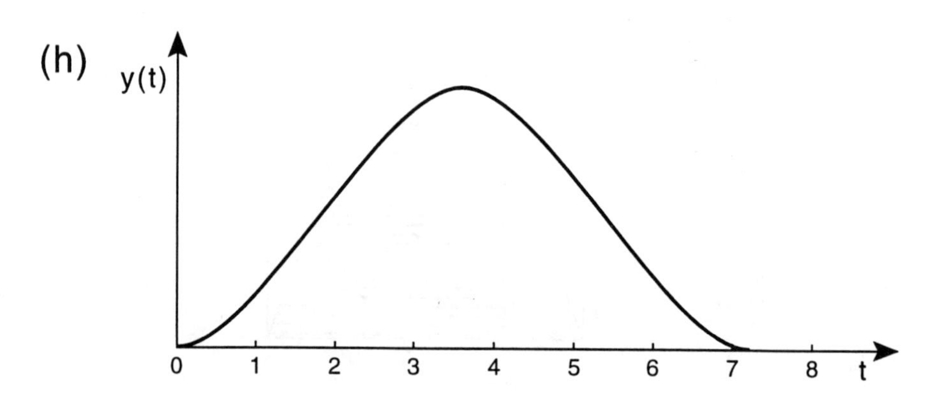

Figure 3.18h: Graph of output time function by convolution

(b) Laplace transform method

Functions $f(t)$ and $g(t)$ may be written,

$$\begin{aligned}
f(t) &= u(t) - u(t-3) \\
g(t) &= t[u(t) - u(t-1)] + u(t-1) - u(t-3) - (t-4)[u(t-3) - u(t-4)] \\
&= tu(t) - (t-1)u(t-1) - (t-3)u(t-3) + (t-4)u(t-4).
\end{aligned}$$

From the convolution integral,

$$\begin{aligned} y(t) = \int^t f(\lambda)g(t-\lambda)d\lambda &= f(t) * g(t) \\ \mathcal{L}\{\int_0^t f(\lambda)g(t-\lambda)d\lambda\} &= \mathcal{L}\{f(t)\}\mathcal{L}\{g(t)\}. \end{aligned}$$

Thus we have

$$\begin{aligned} \mathcal{L}\{f(t)\} &= \frac{1}{s} - \frac{e^{-3s}}{s} \\ \mathcal{L}\{f(t)\} &= \frac{1}{s^2} - \frac{e^{-s}}{s^2} - \frac{e^{-3s}}{s^2} + \frac{e^{-4s}}{s^2} \end{aligned}$$

and, hence,

$$\begin{aligned} \mathcal{L}\{f\}\mathcal{L}\{g\} &= \{\frac{1}{s} - \frac{e^{-3s}}{s}\}\{\frac{1}{s^2} - \frac{e^{-s}}{s^2} - \frac{e^{-3s}}{s^2} + \frac{e^{-4s}}{s^2}\} \\ &= \frac{1}{s^3} - \frac{e^{-s}}{s^3} - \frac{e^{-3s}}{s^3} + \frac{e^{-4s}}{s^3} - \frac{e^{-3s}}{s^3} + \frac{e^{-4s}}{s^3} + \frac{e^{-6s}}{s^3} - \frac{e^{-7s}}{s^3} \\ &= \frac{1}{s^3} - \frac{e^{-s}}{s^3} - \frac{2e^{-3s}}{s^3} + \frac{2e^{-4s}}{s^3} + \frac{e^{-6s}}{s^3} - \frac{e^{-7s}}{s^3}. \end{aligned}$$

Taking the Laplace inverse, we obtain

$$\begin{aligned} y(t) &= \int_0^t f(\lambda)g(t-\lambda)d\lambda = \frac{t^2}{2} - \frac{1}{2}(t-1)^2u(t-1) - (t-3)^2u(t-3) \\ &+(t-4)^2u(t-4) + \frac{1}{2}(t-6)^2u(t-6) - \frac{1}{2}(t-7)^2u(t-7). \end{aligned}$$

(c) Direct integration

$$\begin{aligned} y(t) &= \int_0^t f(t-\tau)g(\tau)d\tau \\ &= \int_0^t [u(t-\tau) - u(t-3-\tau)][\tau u(\tau) - (\tau-1)u(\tau-1) \\ &\quad -(\tau-3)u(\tau-3) + (\tau-4)u(\tau-4)]d\tau \\ &= \int_0^t \tau u(\tau)u(t-\tau)d\tau - \int_0^t (\tau-1)u(t-\tau)u(\tau-1)d\tau \\ &\quad - \int_0^t (\tau-3)u(t-\tau)u(\tau-3)d\tau + \int_0^t (\tau-4)u(t-\tau)u(\tau-4)d\tau \\ &\quad - \int_0^t \tau u(\tau)u(t-3-\tau)d\tau + \int_0^t (\tau-1)u(\tau-1)u(t-3-\tau)d\tau \\ &\quad + \int_0^t (\tau-3)u(\tau-3)u(t-3-\tau)d\tau \\ &\quad - \int_0^t (\tau-4)u(\tau-4)u(t-3-\tau)d\tau \end{aligned}$$

$$= \frac{t^2}{2} - \frac{1}{2}(t-1)^2u(t-1) - (t-3)^2u(t-3) + (t-4)^2u(t-4)$$
$$+\frac{1}{2}(t-6)^2u(t-6) - \frac{1}{2}(t-7)^2u(t-7).$$

Note that in deriving the final results we have taken care of the appropriate Heaviside unit step functions. As for example, the integral

$$\int_0^t (\tau-4)u(\tau-4)u(t-\tau-3)d\tau = \{\int_4^{t-3}(\tau-4)d\tau\}u(t-7).$$

It can easily be seen that these three methods produce identical results.

Chapter 4

Series solution: method of Frobenius

4.1 Introduction

In previous chapters, we learned various solution techniques for ordinary differential equations. However, there are certain types of differential equations of great importance to scientific problems which cannot be solved by an elementary method. The usual procedure in such cases is to obtain a solution in the form of an infinite series of ascending integral powers of x assuming that the series converges under certain domains. The idea of a power series solution stems from the fact that many functions, such as e^x, $\sin x$ and $\cos x$, possess a series expansion of the type

$$a_0 + a_1 x + a_2 x^2 + \cdots$$

which is often called a power series.

In the present chapter, we shall discuss a series solution of the form

$$y = (x - x_0)^\gamma \sum_{n=0}^{\infty} a_n (x - x_0)^n$$

which can be written as

$$y = \sum_{n=0}^{\infty} a_n (x - x_0)^{n+\gamma} \quad , \qquad a_0 \neq 0 \tag{4.1}$$

for the general linear second-order differential equation

$$y'' + P(x)y' + Q(x)y = 0 \tag{4.2}$$

in the neighbourhood of a point $x = x_0$ even if $P(x)$ and $Q(x)$ may not be continuous at that point. The method is usually attributed to the German mathematician F.A. Frobenius (1849–1917), and is known as the method of Frobenius.

We know a complete solution of a differential equation contains a number of arbitrary constants equal to the order of the differential equation. Thus, if we assume an infinite series solution, it can easily be deduced that the number of arbitrary constants is equal to the order of the equation. This can be verified with the help of the following example:

$$y'' + y = 0. \tag{4.3}$$

From the elementary method, we know that a complete solution of (4.3) is

$$y = a\cos x + b\sin x.$$

Now, assume a series solution

$$y = \sum_{n=0}^{\infty} a_n x^n, \qquad a_0 \neq 0 \tag{4.4}$$

in the neighbourhood of the origin.
Then,

$$y'' = \sum_{n=0}^{\infty} a_n n(n-1)x^{n-2}. \tag{4.5}$$

Substituting (4.4) and (4.5) into (4.3) yields

$$\sum_{n=0}^{\infty} a_n n(n-1)x^{n-2} + \sum_{n=0}^{\infty} a_n x^n = 0.$$

Changing n to $n-2$ in the second series gives

$$\sum_{n=0}^{\infty} a_n n(n-1)x^{n-2} + \sum_{n=2}^{\infty} a_{n-2}x^{n-2} = 0. \tag{4.6}$$

Thus, equating the like powers of x, we have the following relations:

$$\begin{aligned} n=0: \quad a_0(0)(-1) &= 0 \\ n=1: \quad a_1(1)(0) &= 0. \end{aligned}$$

Thus, a_0 and a_1 are both non-zero constants. Corresponding to $n \geq 2$, the following recurrence relation results:

$$a_n n(n-1) + a_{n-2} = 0$$

or

$$a_n = -\frac{a_{n-2}}{n(n-1)} \qquad n \geq 2.$$

More explicitly, we have

$$\begin{aligned} a_2 &= -\frac{a_0}{2.1} = -\frac{a_0}{2!} \\ a_3 &= -\frac{a_1}{3.2} = -\frac{a_1}{3!} \\ a_4 &= -\frac{a_2}{4.3} = \frac{a_0}{4!} \\ a_5 &= -\frac{a_3}{5.4} = \frac{a_1}{5!}. \end{aligned}$$

Hence, the solution (4.4) can be written as

$$\begin{aligned} y &= a_0(1 - \frac{x^2}{2!} + \frac{x^4}{4!} - \frac{x^6}{6!} + \cdots) \\ &\quad + a_1(x - \frac{x^3}{3!} + \frac{x^5}{5!} - \frac{x^7}{7!} + \cdots). \end{aligned} \tag{4.7}$$

Solution (4.7) can be written as

$$y = a_0 \cos x + a_1 \sin x, \tag{4.8}$$

where a_0 and a_1 are the two arbitrary constants. This solution is identical to the one obtained by the elementary method.

4.2 Definition of ordinary and singular points

In this section, we shall give the formal definitions of the ordinary point, the regular singular point, and the irregular singular point of a differential equation.

Definition 1: Ordinary point

The point $x = x_0$ is said to be an ordinary point of the differential equation (4.2) if, and only if, both $P(x)$ and $Q(x)$ are analytic at $x = x_0$. By definition, a function is said to be analytic if, and only if, the function can be expanded by Taylor series expansion in the neighbourhood of a point. By using the limit process, if both $\lim_{x \to x_0} P(x)$ and $\lim_{x \to x_0} Q(x)$ exist, then $x = x_0$ is an ordinary point of a differential equation.

Definition 2: Singular point

A point which is not an ordinary point of a differential equation is called a singular point or simply a singularity of the differential equation. There are two kinds of singular points. These are regular singular points and irregular singular points.

Definition 2 (a): Regular singular point

The point $x = x_0$ is said to be a regular singular point of the differential equation (4.2) if, and only if, all three of the following conditions are satisfied:

(i) $x = x_0$ is a singularity of the differential equation

(ii) $P(x)$ is analytic at $x = x_0$ or has a pole of order one at $x = x_0$ such that $(x - x_0)P(x)$ possesses the Taylor series expansion about $x = x_0$

(iii) $Q(x)$ is analytic at $x = x_0$, or has a pole of order one or two at $x = x_0$ such that $(x - x_0)^2 Q(x)$ possesses Taylor series expansion about $x = x_0$.

Note

For example, $f(x) = \frac{1}{x-x_0}$ has a pole of order one at $x = x_0$ and $g(x) = \frac{1}{(x-x_0)^2}$ has a pole of order two at $x = x_0$.

Conditions (ii) and (iii) can be replaced by using the limit processes

$$\lim_{x\to x_0}(x - x_0)P(x) \quad \text{and} \quad \lim_{x\to x_0}(x - x_0)^2 Q(x)$$

where both these limits must exist if $x = x_0$ is a regular singular point of the differential equation.

Definition 2(b): Irregular singular point

Any singular point which is not a regular singular point is called an irregular singular point of the differential equation.

In the following, we shall cite three of the most important theorems. The proofs of these theorems are beyond the scope of this book, but may be found in advanced treatments of the theory of differential equations (see Ince 1957).

Theorem 1

If $x = x_0$ is an ordinary point of the differential equation

$$y'' + P(x)y' + Q(x)y = 0,$$

then there exists a series solution about $x = x_0$ in the form

$$\begin{aligned} y &= a_0 + a_1(x - x_0) + a_2(x - x_0)^2 + \cdots \\ &= \sum_{n=0}^{\infty} a_n(x - x_0)^n. \end{aligned}$$

This series will converge for $0 < |x - x_0| < R$, where R, defined as the radius of convergence, is not less than the distance from x_0 to the nearest singular point of the equation. The series may or may not converge for $|x - x_0| = R$, but definitely diverges for $|x - x_0| > R$.

Theorem 2

If $x = x_0$ is a regular singular point of the differential equation

$$y'' + P(x)y' + Q(x)y = 0,$$

then there exists a series solution about $x = x_0$ in the form

$$\begin{aligned} y &= (x - x_0)^\gamma (a_0 + a_1(x - x_0) + a_2(x - x_0)^2 + \cdots) \\ &= (x - x_0)^\gamma \sum_{n=0}^{\infty} a_n(x - x_0)^n \\ &= \sum_{n=0}^{\infty} a_n(x - x_0)^{n+\gamma}. \end{aligned}$$

This series will converge for $0 < |x - x_0| < R$, where the radius of convergence R is not less than the distance from x_0 to the nearest other singular point of the equation. The series may or may not converge for $|x - x_0| = R$, but definitely diverges for $|x - x_0| > R$.

Theorem 3

If $x = x_0$ is an irregular singular point of the differential equation

$$y'' + P(x)y' + Q(x)y = 0,$$

then there usually does not exist any series solution about $x = x_0$.

Note

The singular points of a differential equation may also be complex, although the point of expansion is real. In that situation, the radius of convergence R is the absolute distance from the point of expansion to the nearest of all other singular points of the differential equation. To illustrate this situation, we consider the following example:

For the differential equation,

$$y'' - \frac{1}{x(x^2+2)}y' + \frac{6}{x^2+2}y = 0$$

where $P(x) = -\frac{1}{x(x^2+2)}$ and $Q(x) = \frac{6}{x^2+2}$.

There is only one real singular point which is at $x = 0$. $x = \pm i\sqrt{2}$ are the two imaginary singular points of the differential equation; all other points are ordinary points. Therefore, a series solution around the ordinary point $x = 1$, say, would have the radius of convergence, $R = 1$, shown clearly in Fig. 4.1, because the distance from the point of expansion $x = 1$ to the nearest singular point $x = 0$ is unity. If, however, the point of expansion is the origin, $x = 0$, which is a singular point itself, then the radius of convergence would be $R = \sqrt{2}$, shown in Fig. 4.1, since in the complex plane the distance from the point of expansion is $x = 0$ to the nearest of all other singular points (namely, $x = i\sqrt{2}$ and $x = -i\sqrt{2}$) is $\sqrt{2}$.

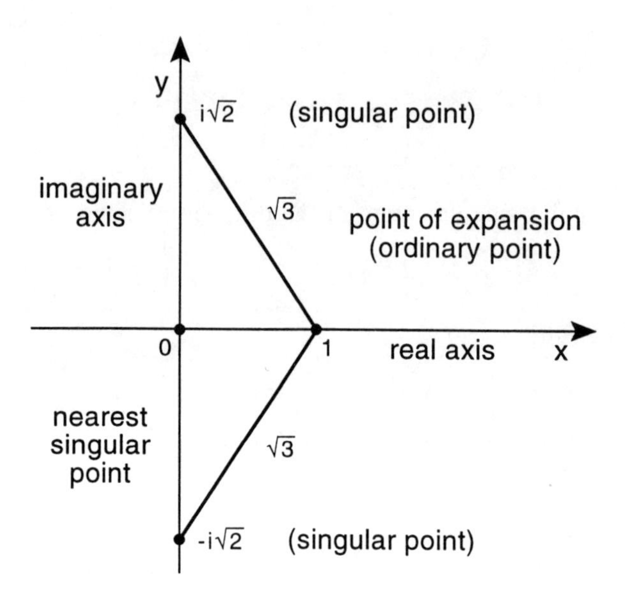

Fig. 4.1: Radius of convergence

In the following, we illustrate some examples:

Example 1

Find the singular points of the following differential equation, and determine whether they are regular or irregular

$$y'' + xy' + y = 0.$$

Solution

Here $P(x) = x$ and $Q(x) = 1$. Since $P(x)$ and $Q(x)$ are both analytic at any finite point, the differential equation has no singular points.

To see that $x = \infty$ is a singular point, we follow the steps below: Substituting $x = \frac{1}{z}$, the above equation can be reduced to

$$z^4 \frac{d^2y}{dz^2} + [2z^3 - z]\frac{dy}{dz} + y = 0$$

or

$$\frac{d^2y}{dz^2} + \frac{2z^2 - 1}{z^3}\frac{dy}{dz} + \frac{1}{z^4}y = 0$$

where $P_1(z) = \frac{2z^2-1}{z^3}$ and $Q_1(z) = \frac{1}{z^4}$. Evidently, $z = 0$, is a singular point. Now, $\lim_{z\to 0} zP_1(z) = \infty$ and $\lim_{z\to 0} z^2 Q_1(z) = \infty$ Therefore, $z = 0$ (i.e., $x = \infty$) is an irregular point of the equation.

Example 2

Find the singular points of the following differential equation, and determine whether they are regular or irregular:

$$(1 - x^2)y'' + y' + y = 0.$$

Solution

Here $P(x) = \frac{1}{1-x^2}$ and $Q(x) = \frac{1}{1-x^2}$. Therefore, $x = \pm 1$ are the two singular points of the differential equation. Now

$$\lim_{x\to 1}(x-1)P(x) = \lim_{x\to 1}\frac{-1}{1+x} = -\frac{1}{2}$$

and

$$\lim_{x\to 1}(x-1)^2 Q(x) = 0.$$

Since both the limits exist, $P(x)$ and $Q(x)$ are analytic at $x = 1$. Thus, $x = 1$ is a regular singular point of the differential equation. Similarly, we can prove that $x = -1$ is a regular singular point of the equation. Using the transformation given in Example 1, we see that $x = \infty$ is an irregular singular point of the differential equation.

Example 3

Find the singular points of the differential equation, and determine whether they are regular or irregular:

$$x^2(1 - x)y'' + (1 - x)y' + y = 0.$$

Solution

Here $P(x) = \frac{1}{x^2}$ and $Q(x) = \frac{1}{x^2(1-x)}$ so that $x = 0$ and $x = 1$ are the two singular points.

Now, $\lim_{x\to 0} xP(x) = \infty$ and $\lim_{x\to 0} x^2 Q(x) = 1$

One of the two limits does not exist. Therefore, $x = 0$ is an irregular singular point of the equation. But,

$$\begin{aligned}\lim_{x\to 1}(x-1)P(x) &= 0\\ \lim_{x\to 1}(x-1)^2 Q(x) &= \lim \frac{1-x}{x^2} = 0.\end{aligned}$$

Here both of the limits exist. Hence $x = 1$ is a regular singular point of the differential equation. $x = \infty$ can be shown to be an irregular singular point.

4.3 Series expansion about an ordinary point

In this section, we shall consider that $x = x_0$ is an ordinary point of the differential equation $y'' + P(x)y' + Q(x)y = 0$. By Theorem 1, a series expansion $x = x_0$ exists in the form

$$y = \sum_{n=0}^{\infty} a_n (x - x_0)^n.$$

We shall demonstrate this with the help of the following example.

Example 4

Find a series solution of the differential equation

$$y'' + x^2 y' + y = 0 \tag{4.9}$$

about the point $x = 0$.

Solution

Here $P(x) = x^2$ and $Q(x) = 1$. It is clear that $x = 0$ is an ordinary point of the equation as $P(x)$ and $Q(x)$ are both analytic at $x = 0$.

We assume the solution

$$y = \sum_{n=0}^{\infty} a_n x^n, \qquad a_0 \neq 0. \tag{4.10}$$

Then, we have

$$\begin{aligned} y' &= \sum_{n=0}^{\infty} a_n n x^{n-1} \\ y'' &= \sum_{n=0}^{\infty} n(n-1) x^{n-2}. \end{aligned}$$

Substituting these values into (4.9), we obtain

$$\sum_{n=0}^{\infty} a_n n(n-1) x^{n-2} + x^2 \sum_{n=0}^{\infty} a_n n x^{n-1} + \sum_{n=0}^{\infty} a_n x^n = 0$$

or

$$\sum_{n=0}^{\infty} a_n n(n-1) x^{n-2} + \sum_{n=0}^{\infty} a_n n x^{n+1} + \sum_{n=0}^{\infty} a_n x^n = 0. \tag{4.11}$$

Then, we shift the index to reduce all exponents on x to the lowest one present. Here, we need to shift the index in the second series in (4.11) from n to $n - 3$ and in the third series in (4.11) from n to $n - 2$. This then gives

$$\sum_{n=0}^{\infty} a_n n(n-1) x^{n-2} + \sum_{n=3}^{\infty} a_{n-3}(n-3) x^{n-2} + \sum_{n=2}^{\infty} a_{n-2} x^{n-2} = 0. \tag{4.12}$$

We are now in a position to equate to zero the coefficients of the like powers of x. For $n = 0$ and $n = 1$, the second and third series have not started yet.

For

$$\begin{aligned} n &= 0: \quad a_0(0)(-1) = 0 \\ n &= 1: \quad a_1(1)(0) = 0. \end{aligned}$$

Thus, a_0 and a_1 are two arbitrary constants different from zero. For

$$n = 2: \quad \text{the second series has not started yet.}$$

Thus

$$n = 2: \quad a_2(2)(1) + a_0 = 0.$$

Therefore,

$$a_2 = -\frac{a_0}{2}.$$

and, in general, for $n \geq 3$, we have

$$a_n n(n-1) + a_{n-2} + a_{n-3}(n-3) = 0.$$

Hence,

$$a_n = -\frac{1}{n(n-1)}a_{n-2} - \frac{n-3}{n(n-1)}a_{n-3}. \tag{4.13}$$

This is known as the recurrence relation. It gives a_n in terms of the preceding $a's$. Therefore, the explicit values of $a_n's$ are the following:

$$\begin{aligned} a_3 &= -\frac{1}{3.2}a_1 = -\frac{1}{3!}a_1 \\ a_4 &= -\frac{1}{4.3}a_2 - \frac{1}{4.3}a_1 = \frac{a_0}{4!} - \frac{2!}{4!}a_1 \\ a_5 &= -\frac{1}{5.4}a_3 - \frac{2}{5.4}a_2 = \frac{3!}{5!}a_0 + \frac{1}{5!}a_1. \end{aligned} \tag{4.14}$$

Substituting these values of $a_n's$ into the series solution (4.10), we have

$$\begin{aligned} y &= a_0(1 - \frac{x^2}{2!} + \frac{x^4}{4!} + \frac{3!}{5!}x^5 - \cdots) \\ &\quad + a_1(x - \frac{x^3}{3!} - \frac{2!}{4!}x^4 + \frac{1}{5!}x^5 - \cdots). \end{aligned} \tag{4.15}$$

This infinite series is convergent for $|x| < \infty$ because the radius of convergence R, the distance from the point of expansion $x = 0$ to the nearest singularity which is at $x = \infty$, is the infinity.

Example 5

Find two linearly independent series solutions of the differential equation

$$y'' + xy' + y = 0$$

about the point $x = 1$.

Solution

Here, $P(x) = x$ and $Q(x) = 1$. They are analytic at $x = 1$. Therefore, $x = 1$ is an ordinary point of the differential equation. We assume the solution

$$y = \sum_{n=0}^{\infty} a_n (x-1)^n, \qquad a_0 \neq 0. \tag{4.16}$$

Then

$$y' = \sum_{n=0}^{\infty} a_n (n)(x-1)^{n-1} \tag{4.17}$$

$$y'' = \sum_{n=0}^{\infty} a_n (n)(n-1)(x-1)^{n-2}. \tag{4.18}$$

Substituting these values into the equation, we obtain

$$\sum_{n=0}^{\infty} a_n (n)(n-1)(x-1)^2 + x\sum_{n=0}^{\infty} a_n (n)(x-1)^{n-1} + \sum_{n=0}^{\infty} a_n (x-1)^n = 0$$

or

$$\sum_{n=0}^{\infty} a_n (n)(n-1)(x-1)^{n-2} + \sum_{n=0}^{\infty} a_n (n)(x-1)^n + \sum_{n=0}^{\infty} a_n (n)(x-1)^{n-1} + \sum_{n=0}^{\infty} a_n (x-1)^n = 0$$

or

$$\sum_{n=0}^{\infty} a_n (n)(n-1)(x-1)^{n-2} + \sum_{n=0}^{\infty} a_n (n)(x-1)^{n-1} + \sum_{n=0}^{\infty} a_n (n+1)(x-1)^n = 0. \tag{4.19}$$

Shifting the indexes to reduce all exponents on $(x-1)$ to the lowest one present, we have

$$\sum_{n=0}^{\infty} a_n(n)(n-1)(x-1)^{n-2} + \sum_{n=1}^{\infty} a_{n-1}(n-1)(x-1)^{n-2} + \sum_{n=2}^{\infty} a_{n-2}(n-1)(x-1)^{n-2} = 0. \tag{4.20}$$

Now equate the coefficients of the like powers of $(x-1)$ to zero. Thus, we have

$$\begin{aligned} n &= 0: \quad a_0(0)(-1) = 0, \qquad a_0 \neq 0 \\ n &= 1: \quad a_1(1)(0) + a_0(0) = 0, \qquad a_1 \neq 0. \end{aligned}$$

Therefore, a_0 and a_1 are two arbitrary constants.

$$n \geq 2: \quad a_n(n)(n-1) + a_{n-1}(n-1) + a_{n-2}(n-1) = 0$$

or

$$a_n = -\frac{a_{n-1}}{n} - \frac{a_{n-2}}{n}. \tag{4.21}$$

This is known as a recurrence relation.

$$\begin{aligned} a_2 &= -\frac{a_1}{2} - \frac{a_0}{2} \\ a_3 &= -\frac{a_2}{3} - \frac{a_1}{3} = \frac{1}{3}\left(\frac{a_1}{2} + \frac{a_0}{2}\right) - \frac{a_1}{3} = \frac{a_0}{6} - \frac{a_1}{6} \\ a_4 &= -\frac{a_3}{4} - \frac{a_2}{4} = -\frac{1}{4}\left(\frac{a_0}{6} - \frac{a_1}{6}\right) + \frac{1}{4}\left(\frac{a_1}{2} + \frac{a_0}{2}\right) = \frac{a_0}{12} + \frac{a_1}{6} \\ a_5 &= -\frac{a_4}{5} - \frac{a_3}{5} = -\frac{1}{5}\left(\frac{a_0}{12} + \frac{a_1}{6}\right) - \frac{1}{5}\left(\frac{a_0}{6} - \frac{a_1}{6}\right) = -\frac{a_0}{20}. \end{aligned}$$

Substituting these values of $a_n's$, we have

$$\begin{aligned} y &= a_0\left[1 - \frac{(x-1)^2}{2} + \frac{(x-1)^3}{6} + \frac{(x-1)^4}{12} - \frac{(x-1)^5}{20} - \cdots\right] \\ &\quad + a_1\left[(x-1) - \frac{(x-1)^2}{2} - \frac{(x-1)^3}{6} + \frac{(x-1)^4}{6} - \cdots\right]. \end{aligned} \tag{4.22}$$

The series converges for $|x-1| < \infty$.

4.4 Series expansion about a regular singular point

Using the method of Frobenius we shall demonstrate in this section the series solution about a regular singular point of a differential equation.

We consider that $x = 0$ is a regular singular point of the differential equation, $y'' + P(x)y' + Q(x)y = 0$. Therefore, $x\,P(x)$ and $x^2Q(x)$ possess Taylor Series expansions about $x = 0$ in the following manner:

$$\begin{aligned} x\,P(x) &= \sum_{n=0}^{\infty} p_n x^n \\ x^2Q(x) &= \sum_{n=0}^{\infty} q_n x^n. \end{aligned}$$

Now, we assume the series solution

$$y = \sum_{n=0}^{\infty} a_n x^{n+\gamma}.$$

Multiplying the given equation by x^2 and then substituting these series expressions, we have

$$\begin{aligned} & x^2 \sum_{n=0}^{\infty} a_n(n+\gamma)(n+\gamma-1)x^{n+\gamma-2} \\ & \qquad + x \sum_{m=0}^{\infty}\sum_{n=0}^{\infty} p_m a_n (n+\gamma)x^{n+\gamma-1+m} + \sum_{m=0}^{\infty}\sum_{n=0}^{\infty} q_m a_n x^{n+\gamma} = 0 \end{aligned}$$

or

$$\begin{aligned} & \sum_{n=0}^{\infty} a_n(n+\gamma)(n+\gamma-1)x^{n+\gamma} \\ & \qquad + \sum_{m=0}^{\infty}\sum_{n=0}^{\infty} p_m a_n (n+\gamma)x^{n+m+\gamma} + \sum_{m=0}^{\infty}\sum_{n=0}^{\infty} q_m a_n x^{n+\gamma+m} = 0. \end{aligned}$$

Collecting the coefficient of x^γ, when both $n = 0$ and $m = 0$, we have

$$a_0[\gamma(\gamma-1) + p_0\gamma + q_0] = 0 \tag{4.23}$$

and, since $a_0 \neq 0$, it follows that

$$\gamma(\gamma-1) + p_0\gamma + q_0 = 0. \tag{4.24}$$

The quadratic equation (4.24) is known as the indicial equation of the differential equation. The two roots of this equation are called the exponents of the differential equation at the regular singular point. Depending on the nature of the roots, we have four different cases which are discussed below with examples. These four cases will demonstrate how to obtain two independent series solutions of the second-order differential equation.

Case I. Unequal roots of indicial equation differing by a quantity not an integer

Example 6

Find two linearly independent series solutions of the equation $9x^2y'' + (x+2)y = 0$ about the origin.

Solution

Here $P(x) = 0$ and $Q(x) = \frac{x+2}{9x^2}$ so $x = 0$ is a singular point.

Now $\lim_{x\to 0} x\,P(x) = 0 = p_0$ and $\lim_{x\to 0} x^2 Q(x) = \frac{2}{9} = q_0$

Both the limits exist. Therefore, $x = 0$ is a regular singular point of the differential equation.

The indicial equation is then

$$\gamma(\gamma - 1) + p_0\gamma + q_0 = 0$$

or

$$\gamma(\gamma - 1) + 2/9 = 0.$$

Therefore, $\gamma = 2/3, 1/3$ are the two roots of the indicial equation and they differ by $\frac{1}{3}$ which is not an integer. Thus, two independent solutions corresponding to these two roots are guaranteed.

Now, let us proceed to find these solutions. Assume the solution

$$y = \sum_{n=0}^{\infty} a_n x^{n+\gamma}$$

and

$$y'' = \sum_{n=0}^{\infty} a_n(n+\gamma)(n+\gamma-1)x^{n+\gamma-2}.$$

Substituting these into the given equation, we obtain

$$9x^2 \sum_{n=0}^{\infty} a_n(n+\gamma)(n+\gamma-1)x^{n+\gamma-2} + (x+2)\sum_{n=0}^{\infty} a_n x^{n+\gamma} = 0$$

or

$$\sum_{n=0}^{\infty} a_n[9(n+\gamma)(n+\gamma-1)+2]x^{n+\gamma} + \sum_{n=0}^{\infty} a_n x^{n+\gamma+1} = 0.$$

Equating the coefficients of like powers of x to zero, we have

$$n = 0: \qquad a_0[9\gamma(\gamma-1)+2] = 0, \qquad a_0 \neq 0$$

so

$$9\gamma(\gamma - 1) + 2 = 0.$$

This is known as the indicial equation. Solving, we get $\gamma = \frac{2}{3}, \frac{1}{3}$, two roots which we have obtained already.

$$\text{For } n \geq 1: \qquad a_n = \frac{-a_{n-1}}{9(n+\gamma)(n+\gamma-1)+2}.$$

This is known as a recurrence relation.

Now, corresponding to $\gamma = 1/3$, we have

$$\begin{aligned} a_1 &= \frac{-a_0}{9.\gamma(\gamma+1)+2} = -\frac{a_0}{2.3} \\ a_2 &= \frac{-a_1}{9(2+\frac{1}{3})(1+\frac{1}{3})+2} = \frac{a_0}{2.3.5.6} \\ a_3 &= \frac{-a_2}{9(3+\frac{1}{3})(2+\frac{1}{3})+2} = \frac{-a_0}{2.3.5.6.8.9}. \end{aligned}$$

Thus,

$$y_1 = x^{1/3}[1 - \frac{x}{2.3} + \frac{x^2}{2.3.5.6} - \frac{x^3}{2.3.5.6.8.9} + \cdots]$$

choosing

$$a_0 = 1.$$

Also, corresponding to $\gamma = 2/3$, we have

$$\begin{aligned} a_1 &= \frac{-a_0}{9(1+2/3)(2/3)+2} = -\frac{a_0}{3.4} \\ a_2 &= \frac{-a_1}{9(2+2/3)(1+2/3)+2} = \frac{a_0}{3.4.6.7} \\ a_3 &= \frac{-a_2}{9(3+2/3)(2+2/3)+2} = \frac{-a_0}{3.4.6.7.9.10}. \end{aligned}$$

Thus

$$y_2 = x^{2/3}[1 - \frac{x}{3.4} + \frac{x^2}{3.4.6.7} - \frac{x^3}{3.4.6.7.9.10} + \cdots].$$

Therefore

$$y = A\,y_1 + B\,y_2$$

is a complete solution which contains two arbitrary constants.

Note

If the indicial equation has two unequal roots, γ_1 and γ_2, differing by a quantity not an integer, then we get two independent solutions by substituting these values of γ in the series solution of y.

Case II: Equal roots of indicial equation

When the indicial equation has equal roots, then we get only one independent solution. To get the second independent solution, we have to use some other trick, as is illustrated in the next example.

Example 7

Find two independent series solutions of the equation

$$y'' + \frac{1}{x}y' + y = 0$$

about the origin.

Solution

Here $P(x) = \frac{1}{x}$ and $Q(x) = 1$. Thus, $x = 0$ is a singular point of the equation. Now, $\lim_{x\to 0} xP(x) = 1 = p_0$ and $\lim_{x\to 0} x^2Q(x) = 0 = q_0$. Both the limits exist and so $x = 0$ is a regular singular point of the equation. The indicial equation is then

$$\gamma(\gamma - 1) + p_0\gamma + q_0 = 0$$

or

$$\gamma(\gamma - 1) + \gamma = 0$$

or

$$\gamma^2 = 0.$$

So $\gamma = 0, 0$ are the two equal roots of the indicial equation. We assume the solution

$$y = \sum_{n=0}^{\infty} a_n x^{\gamma+n}, \tag{4.25}$$

where γ is assumed not to be zero. Then, the given differential equation reduces to

$$\sum_{n=0}^{\infty} a_n(n+\gamma)(n+\gamma-1)x^{n+\gamma-2} + \sum_{n=0}^{\infty} a_n(n+\gamma)x^{n+\gamma-2} + \sum_{n=0}^{\infty} a_n x^{n+\gamma} = 0$$

or

$$\sum_{n=0}^{\infty} a_n(n+\gamma)^2 x^{n+\gamma-2} + \sum_{n=2}^{\infty} x^{n+\gamma-2} = 0.$$

In the second series we have shifted the index.

Now, equating the coefficients of like powers of x to zero, we have

$$n = 0: \qquad a_0\gamma^2 = 0, \qquad a_0 \neq 0$$

so

$$\gamma = 0, 0 \qquad \text{two equal roots.}$$

For

$$n = 1: \qquad a_1(1+\gamma)^2 = 0,$$

which shows that $a_1 = 0$. For the time being, we do not substitute $\gamma = 0$. Hence the following recurrence relation results for $n \geq 2$

$$a_n = -\frac{a_{n-2}}{(n+\gamma)^2}.$$

We can write the coefficients as follows

$$\begin{aligned} a_2 &= -\frac{a_0}{(2+\gamma)^2} \\ a_4 &= -\frac{a_2}{(4+\gamma)^2} = \frac{a_0}{(2+\gamma)^2(4+\gamma)^2} \\ a_6 &= -\frac{a_4}{(6+\gamma)^2} = -\frac{a_0}{(2+\gamma)^2(4+\gamma)^2(6+\gamma)^2} \qquad \text{etc.} \end{aligned}$$

and

$$a_1 = a_3 = a_5 = \cdots = 0.$$

Substituting these coefficients into (4.26), we get

$$y = a_0 x^{\gamma}[1 - \frac{x^2}{(2+\gamma)^2} - \frac{x^4}{(2+\gamma)^2(4+\gamma)^2} - \frac{x^6}{(2+\gamma)^2(4+\gamma)^2(6+\gamma)^2} + \cdots] \qquad (4.26)$$

Substituting $\gamma = 0$ into (4.27), and choosing $a_0 = 1$, we have only one solution and it is given by

$$y_1 = 1 - \frac{x^2}{2^2} + \frac{x^4}{2^2.4^2} - \frac{x^6}{2^2.4^2.6^2} + \cdots \qquad (4.27)$$

To obtain the second independent solution, we observe that if we substitute the series (4.27) into the differential equation without letting $\gamma = 0$, we get a single term $a_0\gamma^2 x^{-1}$. Therefore, (4.27) satisfies not the given homogeneous differential equation, but the following nonhomogeneous equation

$$(x\,D^2 + D + x)y = a_0\gamma^2 x^{\gamma-1}. \qquad (4.28)$$

But, when $\gamma = 0$, we get our given homogeneous equation. It can easily be seen that the right-hand side of (4.29) involves the square of γ; its partial derivative with respect to γ, i.e. $2a_0\gamma x^{\gamma-1} + a_0\gamma^2 x^{\gamma-1}\ell n x$, will also vanish when $\gamma = 0$. Therefore, differentiating (4.29) partially with respect to γ, we get

$$\frac{\partial}{\partial\gamma}[x\,D^2 + D + x]y = 2a_0\gamma\, x^{\gamma-1} + a_0\gamma^2\, x^{\gamma-1}\ell n x.$$

As the differential operators are commutative, the above equation becomes

$$(x\,D^2 + D + x)\frac{\partial y}{\partial\gamma} = 2a_0\gamma\, x^{\gamma-1} + a_0\gamma^2\, x^{\gamma-1}\ell n x.$$

The right-hand side vanishes at $\gamma = 0$. So,

$$(\frac{\partial y}{\partial \gamma})_{\gamma=0} = y_2$$

must be a second independent solution of the differential equation. Therefore, differentiating (4.27) (choose $a_0 = 1$),

$$\begin{aligned}
y_2 &= (\frac{\partial y}{\partial \gamma})_{\gamma=0} \\
&= \{x^\gamma \ell n x[1 - \frac{x^2}{(2+\gamma)^2} + \frac{x^4}{(2+\gamma)(4+\gamma)^2} - \cdots] \\
&\quad + x^\gamma[\frac{x^2}{(2+\gamma)^2}\cdot\frac{2}{(2+\gamma)} - \frac{x^4}{(2+\gamma)^2(4+\gamma)^2}(\frac{2}{2+\gamma} + \frac{4}{4+\gamma}) + \cdots]\}_{\gamma=0} \\
&= y_1 \ell n x + \{(\frac{x}{2})^2 - \frac{(x/4)^4}{(2!)^2}(1+\frac{1}{2}) + \frac{(x/2)^6}{(3!)^2}(1+\frac{1}{2}+\frac{1}{3}) - \cdots\} \\
&= y_1 \ell n x - \sum_{k=1}^{\infty}(-1)^k \frac{(x/2)^{2k}}{(k!)^2}(1+\frac{1}{2}+\frac{1}{3}+\cdots+\frac{1}{k}).
\end{aligned} \tag{4.29}$$

Solutions y_1 and y_2 can be recognized as Bessel's functions of the first and second kinds respectively and are given the special symbols $J_0(x)$ and $Y_0(x)$.

A complete solution is then given by

$$y = A\,J_0(x) + B\,Y_0(x).$$

Note

If the indicial equation has two equal roots $\gamma = \alpha$, we get two independent solutions by substituting $\gamma = \infty$ into y and $(\frac{\partial y}{\partial \gamma})$. In addition to this method of finding a second independent solution of the differential equation when the first solution is known, we can follow Abel's identity, which is

$$y_2 = \int \frac{e^{-\int P(x)dx}}{y_1^2} dx.$$

Case III: Difference of roots of indicial equation: an integer which makes a coefficient of a solution infinite

To demonstrate this case, we consider the following example.

Example 8

Find two independent series solutions of the equation

$$x^2 y'' + xy' + (x^2 - 1)y = 0 \tag{4.30}$$

about the origin.

Solution

Equation (4.30) can be recognized as the Bessel equation of order one. Here, $P(x) = \frac{1}{x}$ and $Q(x) = 1 - \frac{1}{x^2}$.

So, $x = 0$ is a singular point of the equation. Also, $\lim_{x\to 0} x\,P(x) = 1 = p_0$ and $\lim_{x\to 0} x^2 Q(x) = -1 = q_0$ and so both limits exist at $x = 0$. Therefore, $x = 0$ is a regular singular point of the given differential equation.

The indicial equation is then

$$\gamma(\gamma - 1) + p_0\gamma + q_0 = 0$$

or

$$\gamma^2 = 1 \tag{4.31}$$

so that $\gamma = \pm 1$.
These two roots differ by an integer. Now, consider a series solution in the form

$$y = \sum_{n=0}^{\infty} a_n x^{n+\gamma}. \tag{4.32}$$

Substituting the values of y, y' and y'' into the given equation, we obtain

$$\sum_{n=0}^{\infty} a_n(n+\gamma)(n+\gamma-1)x^{n+\gamma} + \sum_{n=0}^{\infty} a_n(n+\gamma)x^{n+\gamma} + (x^2-1)\sum_{n=0}^{\infty} a_n x^{n+\gamma} = 0$$

or

$$\sum_{n=0}^{\infty} a_n[(n+\gamma)^2 - 1]x^{n+\gamma} + \sum_{n=0}^{\infty} a_n x^{n+\gamma+2} = 0$$

or

$$\sum_{n=0}^{\infty} a_n[(n+\gamma)^2 - 1]x^{n+\gamma} + \sum_{n=2}^{\infty} a_{n-2} x^{n+\gamma} = 0.$$

Now, equating the coefficients of like powers of x to zero, we have

$$n = 0: \qquad a_0(\gamma^2 - 1) = 0, \qquad a_0 \neq 0$$

so

$$\gamma = \pm 1.$$

These are the two indicial roots obtained earlier.

$$n = 1: \qquad a_1[(1+\gamma)^2 - 1] = 0.$$

It follows that

$$a_1 = 0.$$

$$n \geq 2: \qquad a_n[(n+\gamma)^2 - 1] + a_{n-2} = 0$$

or

$$\begin{aligned} a_n &= -\frac{a_{n-2}}{(n+\gamma)^2 - 1} \\ &= -\frac{a_{n-2}}{(n+\gamma+1)(n+\gamma-1)}. \end{aligned} \tag{4.33}$$

This is known as the recurrence relation.

The coefficients are given by

$$\begin{aligned} a_2 &= -\frac{a_0}{[(2+\gamma)^2 - 1]} \\ a_4 &= -\frac{a_2}{[(4+\gamma)^2 - 1]} = \frac{a_0}{[(2+\gamma)^2 - 1][(4+\gamma)^2 - 1]} \\ a_6 &= \frac{a_4}{[(6+\gamma)^2 - 1]} = \frac{a_0}{[(2+\gamma)^2 - 1][(4+\gamma)^2 - 1][(6+\gamma)^2 - 1]} \end{aligned}$$

and

$$a_1 = a_3 = a_5 = \cdots = 0.$$

Substituting these values of the coefficients into (4.33) we obtain

$$y = a_0 x^\gamma [1 - \frac{x^2}{[(\gamma+2)^2 - 1]} + \frac{x^4}{[(2+\gamma)^2 - 1][(4+\gamma)^2 - 1]} - \cdots]. \tag{4.34}$$

When we substitute $\gamma = 1$,

$$y = a_0 x [1 - \frac{x^2}{8} + \frac{x^4}{8.24} - \cdots]. \tag{4.35}$$

It is obvious that when $\gamma = -1$ the coefficients become infinite because $[(2+\gamma)^2 - 1]$, which approaches zero at $\gamma = -1$, occurs as a factor in the denominator. To overcome this difficulty, we replace a_0 by $b_0(\gamma+1)$ where $b_0 \neq 0$. So, from (4.34), we have

$$\begin{aligned} y &= b_0(\gamma+1)x^\gamma [1 - \frac{x^2}{(\gamma+1)(\gamma+3)} + \frac{x^4}{(\gamma+1)(\gamma+3)^2(\gamma+5)} - \cdots] \\ &= b_0 x^\gamma [(\gamma+1) - \frac{x^2}{\gamma+3} + \frac{x^4}{(\gamma+3)^2(\gamma+5)} - \cdots]. \end{aligned} \tag{4.36}$$

Substituting (4.36) into the given differential equation, we see that

$$[x^2 D^2 + xD + (x^2 - 1)]y = b_0 x^\gamma (\gamma+1)^2(\gamma-1).$$

As in Case II, the presence of the squared factor $(\gamma+1)^2$ shows that $(\frac{\partial y}{\partial \gamma})$ and y satisfy the differential equation when $\gamma = -1$. These two solutions, together with (4.35) when $\gamma = 1$, constitute three solutions of a second-order differential equation. Clearly, one of these three solutions must be related to one of the other two. It can be easily verified that solution (4.35) is related to (4.36)

Then, the two independent solutions are:

$$y_1 = x^{-1}[-\frac{x^2}{2} + \frac{1}{2^2.4^2}x^4 - \frac{x^6}{2^2.4^2.6} + \cdots] \tag{4.37}$$

$$\begin{aligned} y_2 &= (\frac{\partial y}{\partial \gamma})_{\gamma=-1} \\ &= b_0 y_1 \ell n x + b_0 x^{-1}\{1 + \frac{x^2}{2^2} - \frac{1}{2^2.4}(\frac{2}{2} + \frac{1}{4})x^4 \\ &\quad + \frac{1}{2^2.4^2.6}(\frac{2}{2} + \frac{2}{4} + \frac{1}{6})x^6 - \cdots\}. \end{aligned} \tag{4.38}$$

Thus (4.37) and (4.38) are the required solutions, both of which is independent of the other. Therefore, a complete solution is given by $y = Ay_1 + By_2$.

Note

If the indicial equation has two roots γ_1 and γ_2 ($\gamma_2 > \gamma_1$, say) differing by an integer, and if some of the coefficients of solution y become infinite at $\gamma = \gamma_1$, we then modify the solution form of y by replacing a_0 with $b_0(\gamma - \gamma_1)$. Thus the two independent solutions are obtained by letting $\gamma = \gamma_1$ in the modified form of y and $\frac{\partial y}{\partial \gamma}$.

Case IV: Difference of roots of indicial equation: an integer which makes a coefficient of a solution indeterminate

We consider the following example to demonstrate this case.

Example 9

Find two independent series solutions of the equation

$$(1 - x^2)y'' - 2xy' + 2y = 0 \tag{4.39}$$

about the origin.

Solution

Equation (4.39) is Legendre's equation of order one. Proceeding as in the previous cases, we see that $x = 0$ is an ordinary point of the differential equation. Assume

$$y = \sum_{n=0}^{\infty} a_n x^{n+\gamma} \tag{4.40}$$

is a solution of (4.39).

Substituting the values of y, y' and y'' into (4.39), we obtain

$$(1-x^2)\sum_{n=0}^{\infty} a_n(n+\gamma)(n+\gamma-1)x^{n+\gamma-2}$$
$$-2x\sum_{n=0} a_n(n+\gamma)x^{n+\gamma-1}$$
$$+2x\sum_{n=0} a_n x^{n+\gamma} = 0$$

or

$$-\sum_{n=0}^{\infty} a_n[(n+\gamma)(n+\gamma-1)+2(n+\gamma)-2]x^{n+\gamma}$$

$$+\sum_{n=0}^{\infty} a_n(n+\gamma)(n+\gamma-1)x^{n+\gamma-2} = 0$$

$$\sum_{n=0}^{\infty} a_n(n+\gamma)(n+\gamma-1)x^{n+\gamma-2} - \sum_{n=0}^{\infty} a_n(n+\gamma+2)(n+\gamma-1)x^{n+\gamma} = 0$$

or

$$\sum_{n=0}^{\infty} a_n(n+\gamma)(n+\gamma-1)x^{n+\gamma-2} - \sum_{n=2}^{\infty} a_{n-2}(n+\gamma)(n+\gamma-3)x^{n+\gamma-2} = 0. \quad (4.41)$$

Now, comparing the coefficients of the like powers of x to zero:

$$n=0: \qquad a_0\gamma(\gamma-1)=0, \qquad a_0 \neq 0$$

or

$$\begin{aligned} \gamma(\gamma-1) &= 0 && \text{is the Indicial Equation} \\ \gamma=0, \gamma &= 1 && \text{are the two roots} \end{aligned}$$

$$n=1: \qquad a_1(1+\gamma)(\gamma)=0$$

therefore,

$$a_1 = 0 \qquad \text{when} \quad \gamma = 1.$$

But, a_1 is indeterminate when $\gamma = 0$.

$n \geq 2$: the recurrence relation is

$$a_n(n+\gamma)(n+\gamma-1) = a_{n-2}(n+\gamma)(n+\gamma-3)$$

or

$$a_n = \frac{n+\gamma-3}{n+\gamma-1}a_{n-2}.$$

Thus,

$$\begin{aligned} a_2 &= \frac{\gamma-1}{\gamma+1}a_0 \\ a_3 &= \frac{\gamma}{\gamma+2}a_1 \\ a_4 &= \frac{\gamma+1}{\gamma+3}a_2 = \frac{(\gamma-1)}{(\gamma+3)}a_0 \\ a_5 &= \frac{\gamma+2}{\gamma+4}a_1 = \frac{\gamma}{\gamma+4}a_1. \end{aligned}$$

Substituting these coefficients into (4.40), we get

$$\begin{aligned} y &= x^\gamma[a_0\{1+\frac{\gamma-1}{\gamma+1}x^2+\frac{\gamma-1}{\gamma+3}x^4+\cdots\} \\ &\quad +a_1\{x+\frac{\gamma}{\gamma+2}x^3+\frac{\gamma}{\gamma+4}x^5+\cdots\}]. \end{aligned} \tag{4.42}$$

When $\gamma = 0$, we have

$$y = a_0[x^2+\frac{x^4}{3}+\cdots]+a_1x+a_0. \tag{4.43}$$

As this solution contains two arbitrary constants, it may be considered a complete solution of the given equation. If we consider the other root of the indicial equation, i.e. $\gamma = 1$, a particular solution can be found:

$$y = a_1[x^2+\frac{x^4}{3}+\frac{x^6}{5}+\cdots] \tag{4.44}$$

which shows that (4.44) is the constant multiple of the first part of (4.43). Hence, (4.43) is the required complete solution.

Note

If the indicial equation has two roots γ_1 and γ_2 (say $\gamma_2 > \gamma_1$) differing by an integer, and if one of the coefficients of y becomes indeterminate when $\gamma = \gamma_1$, then the complete solution is given by substituting $\gamma = \gamma_1$. The complete solution then contains two arbitrary constants. Substituting $\gamma = \gamma_2$ gives a numerical multiple of one of the series contained in the first solution.

Chapter 5

Partial differential equations

5.1 Introduction

We saw in the previous chapters how an ordinary differential equation plays an important role in the solution of physical problems arising in the field of engineering and physics. We know ordinary differential equations involve derivatives of one or more dependent variables with respect to a single independent variable. By using ordinary differential equations to solve applied problems, we are in effect greatly simplifying the mathematical model governing the physical situation. In practice, the solution function of a physical problem depends upon the space variables in addition to time t. For example, the temperature in a given body depends upon the point of measurement, which may be defined by Cartesian coordinates x, y, z and the time of measurement t. In such cases, a realistic approach must take into consideration the fact that the dependent variable depends not only on t but also on one or more space variables. Whenever more than a single independent variable must be taken into consideration, the formulation of such problems leads to Partial Differential Equations rather than ordinary differential equations. In this chapter we shall discuss such equations as they commonly arise in applied mathematics and engineering physics. We shall in detail discuss the derivation of some important partial differential equations governing the physical system, and then describe the solution techniques of these equations and illustrate their applications to specific problems.

In the next section, we will develop the following important partial differential equations in Cartesian coordinates.

5.1.1 The wave equation in three dimensions

$$\frac{\partial^2 \eta}{\partial x^2} + \frac{\partial^2 \eta}{\partial y^2} + \frac{\partial^2 \eta}{\partial z^2} = \frac{1}{c^2}\frac{\partial^2 \eta}{\partial t^2}. \tag{5.1}$$

5.1.2 The heat conduction equation in three dimensions

$$\frac{\partial^2 \theta}{\partial x^2} + \frac{\partial^2 \theta}{\partial y^2} + \frac{\partial^2 \theta}{\partial z^2} = \frac{1}{\alpha}\frac{\partial \theta}{\partial t}. \tag{5.2}$$

5.1.3 Laplace's equation in three dimensions

$$\frac{\partial^2 u}{\partial x^2} + \frac{\partial^2 u}{\partial y^2} + \frac{\partial^2 u}{\partial z^2} = 0. \tag{5.3}$$

5.1.4 Telegraph equations in one dimension

$$LC\frac{\partial^2 E}{\partial t^2} + (RC + GL)\frac{\partial E}{\partial t} + RGE = \frac{\partial^2 E}{\partial x^2} \tag{5.4}$$

$$LC\frac{\partial^2 I}{\partial t^2} + (RC + GL)\frac{\partial I}{\partial t} + RGI = \frac{\partial^2 I}{\partial x^2} \tag{5.5}$$

in which c^2, α, L, C, R, and G are physical constants, and η, θ, u, E and I are dependent variables.

5.2 Mathematical formulation of equations

Corresponding to partial differential equations (5.1) to (5.5) stated in the previous section, there are four important types of physical problems which we will consider in this section:

I Problems involving vibrations or oscillations.

II Problems involving heat conduction or diffusion.

III Problems involving Laplace's equation.

IV Problems involving transmission lines.

I. Problems Involving Vibrations or Oscillations

One of the first physical problems to be studied through the partial differential equations is the problem of a vibrating string.

Suppose that such a string is tightly stretched between two fixed points $x = 0$ and $x = \ell$ (Fig. 5.1(a)). At time $t = 0$ the string is pulled up in the middle (Fig. 5.1 (b)), to a distance d. Then the string is released. The problem is to find the motion described by the vibration of the string.

Let us suppose that at some instant t, the string has the shape as shown in Fig. 5.1 (c) $y(x, t)$. Consider the elementary length Δx [Fig. 5.2] of the string.

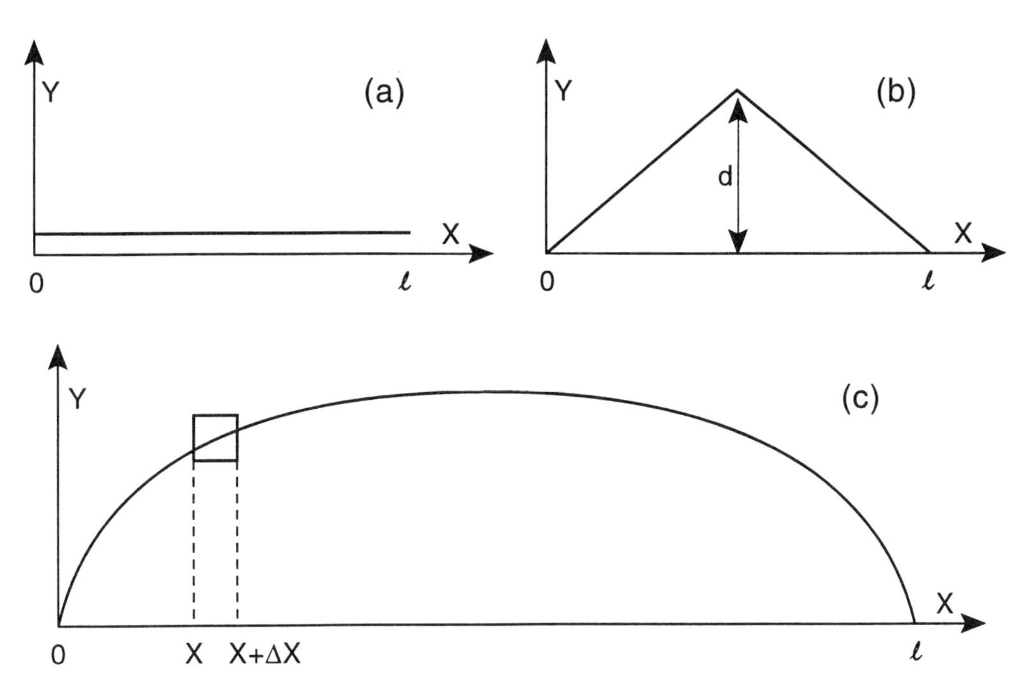

Figure 5.1 A typical element of a vibrating string

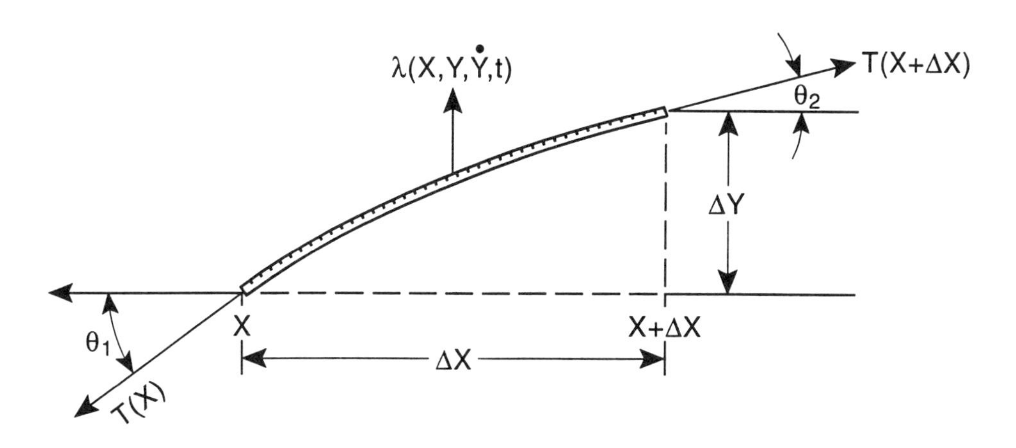

Figure 5.2 Elementary length of the string. Displacement of the string at $x = y(x,t)$ Displacement of the string at $x + \Delta x = y(x + \Delta x, t)$

The forces acting on the element are the tensions as shown in Fig. 5.2. Besides the elastic forces inherent in the system, the string may be acted upon by a known distributed transverse load $\lambda(x, y, \dot{y}, t)$, per unit length, which is a function of x, y, t and the transverse velocity of $\dot{y}$. Gravitational and frictional forces may be taken into account in this term. The weight of the string per unit length is assumed to be a known function $\omega(x)$.

Assumptions in formulating the mathematical model:

(a) It is assumed that the motion takes place entirely in one plane and in this plane each particle moves at right angles to the equilibrium position of the string.

(b) The deflection of the string during the motion is so small that the resulting change of the string has no effect on the tension T.

(c) The string is perfectly flexible, i.e., it can transmit force only in the direction of its length.

(d) The slope of the deflection curve of the string is at all points and at all times so small that with satisfactory accuracy we can approximate $\sin\theta \simeq \tan\theta$, where θ is the angle of inclination.

Now elementary mass of the string is

$$\begin{aligned} g\,\Delta m &= \omega(x)\Delta x \\ \Delta m &= \frac{\omega(x)}{g}\Delta x \end{aligned}$$

where g is the acceleration due to gravity. Now referring to Fig. 5.2 we have the vertical component of tension at x,

$$-T\sin\theta_1$$

which can be written

$$\begin{aligned} -T\sin\theta_1 &= -T\sin\theta|_x \\ &= -T\tan\theta|_x. \end{aligned}$$

The vertical component of the tension at $x + \Delta x$ is

$$T\sin\theta_2$$

which can be written as

$$\begin{aligned} T\sin\theta_2 &= T\sin\theta|_{x+\Delta x} \\ &= T\tan\theta|_{x+\Delta x}. \end{aligned}$$

The vertical acceleration produced on the elementary mass Δm by these forces and the $\lambda(x, y, \dot{y}, t)\Delta x$ is approximately equal to $\frac{\partial^2 y(x,t)}{\partial t^2}$. Here we have used partial time derivative because y is now a function of x and t.

Now applying Newton's law, which is (mass) $\times$ (acceleration) = external forces acting on the string, we have,

$$\begin{aligned}[\frac{\omega(x)\Delta x}{g}]\frac{\partial^2 y}{\partial t^2} &= T\tan\theta|_{x+\Delta x} - T\tan\theta|_x + \lambda(x,y,\dot{y},t)\Delta x \\ \frac{\omega(x)}{g}\frac{\partial^2 y}{\partial t^2} &= T[\frac{\tan\theta|_{x+\Delta x} - \tan\theta|_x}{\Delta x}] + \lambda(x,y,\dot{y},t).\end{aligned}$$

Now when $\Delta x \to 0$

$$\frac{\omega(x)}{g}\frac{\partial^2 y}{\partial t^2} = T\frac{\partial}{\partial x}(\tan\theta) + \lambda(x,y,\dot{y},t).$$

But $\tan\theta = \frac{\partial y}{\partial x}$. Therefore,

$$\frac{\partial^2 y}{\partial t^2} = \frac{gT}{\omega(x)}\frac{\partial^2 y}{\partial x^2} + \frac{g\,\lambda(x,y,\dot{y},t)}{\omega(x)}. \tag{5.6}$$

In most practical applications, $\omega(x)$ is a constant and $\lambda(x,y,\dot{y},t) = 0$. Thus, eqn (5.6) reduces to

$$\frac{\partial^2 y}{\partial t^2} = c^2\frac{\partial^2 y}{\partial x^2} \tag{5.7}$$

which is known as the one-dimensional wave equation. Here,

$$c^2 = \frac{gT}{\omega}$$

the dimension of which is

$$\begin{aligned} c^2 &= \frac{(\text{force}) \times (\text{acceleration})}{(\text{weight})/(\text{length})} \\ &= \frac{(ML/T^2)(L/T^2)}{(ML/T^2)\frac{1}{L}} \\ &= (\frac{L}{T})^2 \\ &= (\text{velocity})^2. \end{aligned}$$

Thus c has the dimension of velocity.

It is now possible to generalize the vibrating string problem to vibrating membrane to describe wave motion in two dimensions. The partial differential equation describing this behavior may be obtained by considering Fig. 5.3.

Suppose the membrane is stretched across a closed circle C in the xy plane and that when it vibrates, each particle moves in a direction perpendicular to the xy plane.

Let at some instant the elementary membrane occupy the position $ABCD$. Let α and β be the angles of deflection made by the tangents with the xy plane.

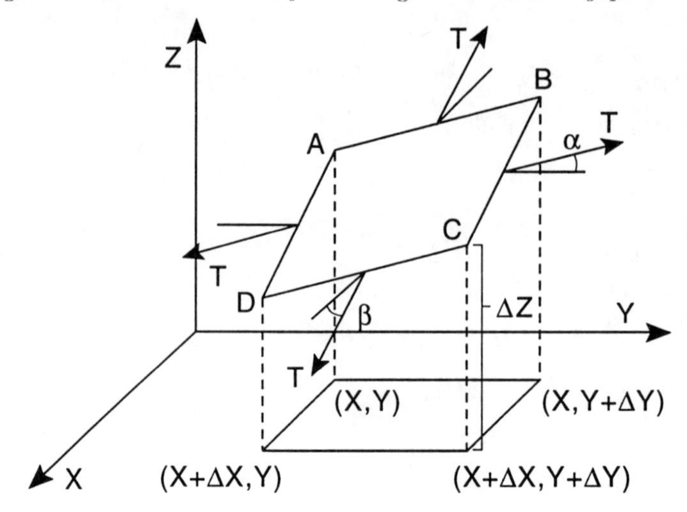

Figure 5.3. Vibration of an elementary membrane

The force acting on this elementary membrane is the tension T per unit length as shown in Fig. 5.3. We further assume that there is a known distributed load $\lambda(x, y, z, \dot{z}, t)$ per unit area which is a function of x, y, z, t and vertical velocity $\dot{z}$. Consider that $\omega(x, y)$ is the weight of the membrane per unit area.

The elementary mass of the membrane is,

$$\Delta m = \frac{\omega(x, y)\Delta x \Delta y}{g}.$$

The vertical acceleration of the membrane is,

$$\frac{\partial^2 z}{\partial t^2}.$$

Net vertical tensions due to tensions perpendicular to BC and AD

$$= (\Delta x T \tan\alpha)|_{(x+\frac{\Delta x}{2}, y+\Delta y)} - (\Delta x T \tan\alpha)|_{(x+\frac{\Delta x}{2}, y)}.$$

Net vertical tensions due to tensions perpendicular to DC and AB

$$= (\Delta y T \tan\beta)|_{(x+\Delta x, y+\frac{\Delta y}{2})} - (\Delta y T \tan\beta)|_{(x, y+\frac{\Delta y}{2})}.$$

The subscript $(x + \frac{\Delta x}{2}, y + \Delta y)$ means the tension T is evaluated at centre of BC. Similarly, the subscript $(x + \Delta x, y + \frac{\Delta y}{2})$ means the tension T is evaluated at the centre of DC. Hence, by Newton's law,

$$\begin{aligned}
(\text{mass}) \times (\text{acceleration}) &= \text{external forces} \\
\left[\frac{\omega(x, y)\Delta x \Delta y}{g}\right]\frac{\partial^2 z}{\partial t^2} &= (\Delta x T \tan\alpha)|_{(x+\frac{\Delta x}{2}, y+\Delta y)} - (\Delta x T \tan\alpha)|_{(x+\frac{\Delta x}{2}, y)} \\
&\quad + (\Delta y T \tan\beta)|_{(x+\Delta x, y+\frac{\Delta y}{2})} - (\Delta y T \tan\beta)|_{(x, y+\frac{\Delta y}{2})} \\
&\quad + \lambda(x, y, z, \dot{z}, t)\Delta x \Delta y.
\end{aligned}$$

But

$$\tan\alpha = \frac{\partial z}{\partial y} \quad \text{and} \quad \tan\beta = \frac{\partial z}{\partial x}.$$

Introducing these relations and dividing throughout by $\Delta x \Delta y$,

$$\begin{aligned}
[\frac{\omega(x,y)}{g}]\frac{\partial^2 z}{\partial t^2} &= T[\frac{(\frac{\partial z}{\partial x})|_{(x+\Delta x, y+\frac{\Delta y}{2})} - (\frac{\partial z}{\partial x})|_{(x, y+\frac{\Delta y}{2})}}{\Delta x} \\
&+T[\frac{(\frac{\partial z}{\partial y})|_{(x+\frac{\Delta x}{2}, y+\Delta y)} - (\frac{\partial z}{\partial y})|_{(x+\frac{\Delta x}{2}, y)}}{\Delta y}] \\
&+\lambda(x,y,z,\dot{z},t)
\end{aligned} \tag{5.8}$$

where

$$\begin{aligned}
\Delta x &\to 0 \\
\Delta y &\to 0.
\end{aligned}$$

Then eqn (5.8) can be written as

$$\frac{\partial^2 z}{\partial t^2} = \frac{gT}{\omega(x,y)}[\frac{\partial^2 z}{\partial x^2} + \frac{\partial^2 z}{\partial y^2}] + \frac{g\,\lambda(x,y,z,\dot{z},t)}{\omega(x,y)}. \tag{5.9}$$

If

$$\begin{aligned}
\omega(x,y) &= \text{constant} \\
\lambda(x,y,z,\dot{z},t) &= 0,
\end{aligned}$$

then eqn (5.9) reduces to

$$\frac{\partial^2 z}{\partial t^2} = c^2[\frac{\partial^2 z}{\partial x^2} + \frac{\partial^2 z}{\partial y^2}] \tag{5.10}$$

which is known as the two-dimensional wave equation where c has the dimension of velocity, because

$$c^2 = \frac{gT}{\omega} = \frac{(LT^{-2})(MLT^{-2}L^{-1})}{(MLT^{-2})L^{-2}} = (\frac{L}{T})^2.$$

Similarly extending to the three-dimensional case the wave equation may be written as

$$\frac{\partial^2 \eta}{\partial t^2} = c^2[\frac{\partial^2 \eta}{\partial x^2} + \frac{\partial^2 \eta}{\partial y^2} + \frac{\partial^2 \eta}{\partial z^2}]. \tag{5.11}$$

II. Problems involving heat conduction or diffusion

The second type of problem leading to partial differential equations is encountered in the study of flow of heat in a thermally conducting medium. To visualize the physical situation of this type of system, let us suppose that a thin metal bar is immersed in boiling water at a temperature of 100^oC. Then it is removed and the ends of the bar are kept in ice at 0^oC. We may assume that the surfaces of the bar are insulated. Now the problem is to find the temperature distribution at certain points in the bar at any time. This problem is obviously a one-dimensional heat conduction problem. The problem can be generalized to study the temperature behaviour in a three-dimensional thermally conducting body. In the following we shall develop the equations governing this phenomenon.

Using experimental facts, we make the following assumptions:

(i) Heat flows in the direction of decreasing temperature.

(ii) Fourier's law says that the rate at which heat flows through an area is proportional to the temperature gradient in degrees per unit distance, in the direction perpendicular to the area.

(iii) The quantity of heat gained or lost by a body when the temperature changes is proportional to the mass of the body and to the temperature change.

The mathematical theory of heat conduction is therefore based primarily on the hypothesis stated above. According to the second assumption,

$$\dot{q} \propto \frac{\partial T}{\partial x}$$

or,

$$\dot{q} = -k\frac{\partial T}{\partial x} \tag{5.12}$$

where $\dot{q}$ is the rate of heat flux per unit area, T is the temperature and k is the thermal conductivity of the body. The minus sign is necessary because the element gains heat.

According to the third assumption, we have

$$\Delta H \propto (\Delta m)(\Delta T)$$

or,

$$\Delta H = c(\Delta m)(\Delta T) \tag{5.13}$$

where ΔH is the quantity of heat stored in the element, Δm is the elementary mass of the element, ΔT is the temperature change and the proportionality constant c is known as the specific heat of the material which is assumed constant.

Referring to the elementary volume of the conducting material in Fig. 5.4,

$$\begin{aligned} \text{mass of the element} &= \Delta m \\ &= \rho \Delta x \Delta y \Delta z \end{aligned}$$

where ρ is the density of the body (mass per unit volume), $\Delta x, \Delta y$ and Δz are respectively the elementary lengths along x, y and z directions.

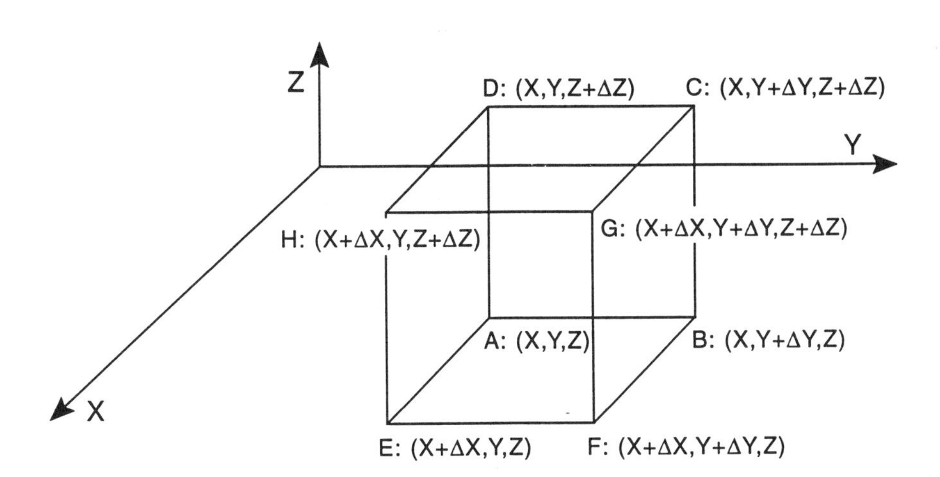

Figure 5.4. A three-dimensional and elementary volume of a heat conducting body

Therefore, eqn (5.13) may be written as

$$\Delta H = c(\rho \Delta x \Delta y \Delta z)(\Delta T). \tag{5.14}$$

The rate at which the heat being stored in the element is therefore

$$\frac{\Delta H}{\Delta t} = c(\rho \Delta x \Delta y \Delta z)(\frac{\Delta T}{\Delta t}). \tag{5.15}$$

Let us evaluate the rate of flow of heat into the element through the six faces of the elementary volume.

The rate at which heat flows into the element through $ABCD$ face is

$$-k\, \Delta y \Delta z \frac{\partial T}{\partial x}|_{(x,y+\frac{\Delta y}{2},z+\frac{\Delta z}{2})}.$$

where an average figure of the temperature gradient $\frac{\partial T}{\partial x}$ is considered at the centre of $(x, y + \frac{\Delta y}{2}, z + \frac{\Delta z}{2})$ of the face $ABCD$. Similarly, the element gains heat through the face $EFGH$ at the rate

$$k\, \Delta y \Delta z \frac{\partial T}{\partial x}|_{(x+\Delta x,y+\frac{\Delta y}{2},z+\frac{\Delta z}{2})}.$$

Therefore, the net rate at which the element gains heat because of the heat flow in the x-direction is

$$-k\,\Delta y\Delta z\frac{\partial T}{\partial x}|_{(x,y+\frac{\Delta y}{2},z+\frac{\Delta z}{2})} + k\,\Delta y\Delta z\frac{\partial T}{\partial x}|_{(x+\Delta x,y+\frac{\Delta y}{2},z+\frac{\Delta z}{2})}. \tag{5.16}$$

Similarly, the rate at which the element gains heat because of heat flow in the direction of y and z, respectively, are

$$-k\,\Delta x\Delta z\frac{\partial T}{\partial y}|_{(x+\frac{\Delta x}{2},y,z+\frac{\Delta z}{2})} + k\,\Delta x\Delta z\frac{\partial T}{\partial y}|_{(x+\frac{\Delta x}{2},y+\Delta y,z+\frac{\Delta z}{2})} \tag{5.17}$$

and,

$$-k\,\Delta x\Delta y\frac{\partial T}{\partial z}|_{(x+\frac{\Delta x}{2},y+\frac{\Delta y}{2},z)} + k\,\Delta x\Delta y\frac{\partial T}{\partial z}|_{(x+\frac{\Delta x}{2},y+\frac{\Delta y}{2},z+\Delta z)}. \tag{5.18}$$

Also heat may be generated within the element by means of electrical devices and chemical reaction.

Suppose the rate at which heat is generated by these agents within the element is

$$\lambda(x,y,z,t)\Delta x\Delta y\Delta z. \tag{5.19}$$

We know the rate at which heat is stored in the element = the rate at which the heat is transferred into the element + the rate at which heat is produced in the element. Hence,

$$\begin{aligned}
&c(\rho\Delta x\Delta y\Delta z)(\frac{\Delta T}{\Delta z})\\
&= [k\,\Delta y\Delta z\frac{\partial T}{\partial x}|_{(x+\Delta x,y+\frac{\Delta y}{2},z+\frac{\Delta z}{2})} - k\,\Delta y\Delta z\frac{\partial T}{\partial x}|_{(x,y+\frac{\Delta y}{2},z+\frac{\Delta z}{2})}\\
&\quad +k\,\Delta x\Delta z\frac{\partial T}{\partial y}|_{(x+\frac{\Delta x}{2},y+\Delta y,z+\frac{\Delta z}{2})} - k\,\Delta x\Delta z\frac{\partial T}{\partial y}|_{(x+\frac{\Delta x}{2},y,z+\frac{\Delta z}{2})}\\
&\quad +k\,\Delta x\Delta y\frac{\partial T}{\partial z}|_{(x+\frac{\Delta x}{2},y+\frac{\Delta y}{2},z+\Delta z)} - k\,\Delta x\Delta y\frac{\partial T}{\partial z}|_{(x+\frac{\Delta x}{2},y+\frac{\Delta y}{2},z)}\\
&\quad +\lambda(x,y,z,t)\Delta x\Delta y\Delta z.
\end{aligned} \tag{5.20}$$

Dividing throughout by $c\rho\Delta x\Delta y\Delta z$ and letting $\Delta x \to 0$, $\Delta y \to 0$, $\Delta z \to 0$ and $\Delta t \to 0$ eqn (5.20) yields the following partial differential equation,

$$\frac{\partial T}{\partial t} = \alpha[\frac{\partial^2 T}{\partial x^2} + \frac{\partial^2 T}{\partial y^2} + \frac{\partial^2 T}{\partial z^2}] + \frac{\lambda(x,y,z,t)}{c\rho} \tag{5.21}$$

where $\alpha = \frac{k}{c\rho}$ is called the coefficient of thermal diffusion or simply the thermal diffusivity. Equation (5.21) is the equation commonly known as the equation of conduction of heat in three-dimensions. To find the dimension of diffusivity α we note that, writing $[Q]$ and $[\theta]$ for those of quantity of heat and temperature, respectively,

$$\begin{aligned}
[k] &= [Q][L^{-1}][T^{-1}][\theta^{-1}]\\
[c] &= [Q][M^{-1}][\theta^{-1}]\\
[\rho] &= [M][L^{-3}]
\end{aligned}$$

so that

$$[\alpha] = [L^2][T^{-1}].$$

In many important applications, heat is neither generated nor lost in the body by any external agent. In that situation

$$\lambda(x, y, z, t) = 0.$$

Hence, eqn (5.21) reduces to

$$\frac{\partial T}{\partial t} = \alpha[\frac{\partial^2 T}{\partial x^2} + \frac{\partial^2 T}{\partial y^2} + \frac{\partial^2 T}{\partial z^2}]. \quad (5.22)$$

In case of steady temperature in which T does not vary with time, it becomes Laplace's equation,

$$\nabla^2 T = \frac{\partial^2 T}{\partial x^2} + \frac{\partial^2 T}{\partial y^2} + \frac{\partial^2 T}{\partial z^2} = 0. \quad (5.23)$$

In the cases of one and two dimensional heat flow problems, eqn (5.22) reduces to the following partial differential equations, respectively:

One dimensional:

$$\frac{\partial T}{\partial t} = \alpha \frac{\partial^2 T}{\partial x^2}. \quad (5.24)$$

Two dimensional:

$$\frac{\partial T}{\partial t} = \alpha[\frac{\partial^2 T}{\partial x^2} + \frac{\partial^2 T}{\partial y^2}]. \quad (5.25)$$

The transfer of heat by conduction is due to random molecular motions; the greater the speed the higher the temperature will be. Also the flow of heat by convection from places of higher to lower temperature is due to the spreading or diffusion of such particles from places where their concentration or density is high to places where it is low. Thus diffusion is the process by which matter is transported from one part of the system to another as a result of random molecular motions. This was first recognized by Fick, who first put diffusion on a quantitative basis by adopting the mathematical equation of heat conduction developed by Fourier a few years earlier. Thus there is an obvious analogy between these two processes and the same equations derived above for heat conduction can be used for diffusion. The only essential difference is that T should be replaced by the symbol C to denote the concentration of the diffusing substance, and the symbol for thermal diffusivity α should be interpreted in this case as the molecular diffusivity or eddy diffusivity.

III. Problems involving Laplace's equation

We have already seen occurrence of Laplace's equation in connection with the heat conduction equation derived in the preceding section. The Laplace equation plays a very important role as a mathematical model in the problems involving electrical or gravitational potential. To demonstrate the usefulness of this equation, consider a three-dimensional region D as in Fig. 5.5, which may represent a continuous distribution of electric charges or a continuous distribution of mass. Let us consider that ρ denotes the charge per unit volume or mass per unit volume. ρ may be a function of position and time. Hence, the quantity ρ is the charge density or the mass density. The electrical potential at P due to charge q at Q or the gravitational potential at P due to mass m at Q is defined to be,

$$\frac{q}{r} \quad \text{or} \quad \frac{m}{r}$$

where r is the distance between P and Q.

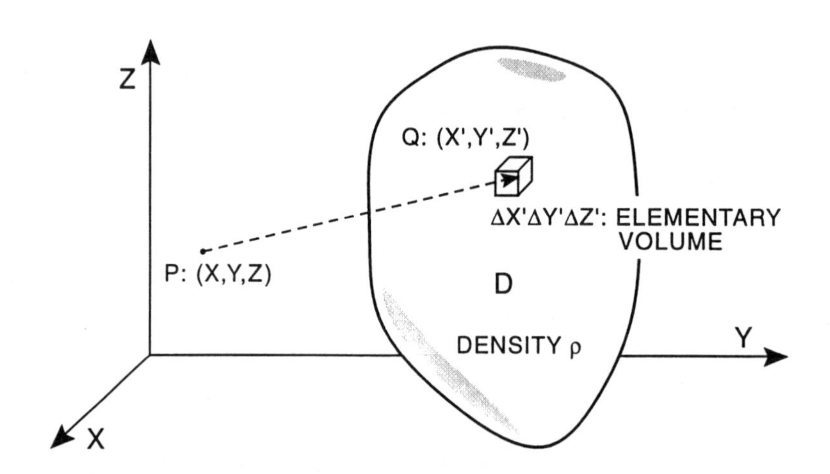

Figure 5.5. Potential at P due to the charge or mass density ρ at Q

Now r is given by

$$r = \sqrt{(x-x')^2 + (y-y')^2 + (z-z')^2}.$$

Let dV be the potential, electrical or gravitational, then due to the charge or mass given by $\rho\, dx'dy'dz'$ we then have,

$$\begin{aligned} dV &= \frac{\rho\, dx'dy'dz'}{r} \\ &= \frac{\rho\, dx'dy'dz'}{\sqrt{(x-x')^2 + (y-y')^2 + (z-z')^2}}. \end{aligned} \tag{5.26}$$

Thus the total potential V due to the entire charge or mass distribution in the region D is found by integrating (5.26) over the region D to obtain,

$$V = \int\int\int_D \frac{\rho\, dx'dy'dz'}{\sqrt{(x-x')^2 + (y-y')^2 + (z-z')^2}}. \tag{5.27}$$

It can be easily verified that V satisfies Laplace's equation

$$\nabla^2 V = \frac{\partial^2 V}{\partial x^2} + \frac{\partial^2 V}{\partial y^2} + \frac{\partial^2 V}{\partial z^2} = 0. \tag{5.28}$$

This result can be accomplished by taking the ∇^2 of both sides of eqn (5.27). By interchanging the order of differentiation and integration which can be justified, the results amount to showing that the Laplacian of $\frac{1}{r}$ is zero.

Thus, by partial differentiation, $\frac{\partial^2 u}{\partial x^2} = -\frac{1}{r^3} + \frac{3(x-x')^2}{r^5}$. Similar expressions for $\frac{\partial^2 u}{\partial y^2}$ and $\frac{\partial^2 u}{\partial z^2}$ can be obtained replacing x for y and z respectively.

Therefore,

$$\frac{\partial^2 u}{\partial x^2} + \frac{\partial^2 u}{\partial y^2} + \frac{\partial^2 u}{\partial z^2} = -\frac{3}{r^3} + \frac{3}{r^3} = 0.$$

Hence, the required result follows.

In deriving Laplace's eqn (5.28) it was assumed that the potential to be found at points is not occupied by matter or electric charge. To find the potential at points occupied by matter or charge it turns out the equation is given by

$$\nabla^2 V = -4\pi\rho$$

which is called Poisson's equation.

IV. Problems involving flow of electricity in a transmission line

Partial differential equations play a very important role in the problems involving flow of electricity in a long cable or transmission line. It is of paramount importance to be able to predict instantaneous current and instantaneous voltage drop at any point in a transmission line or in a long cable. In the following we shall derive the partial differential equations governing this physical system.

Consider the flow of electricity in a long cable or transmission line as shown in Fig. 5.6.

We assume the cable is imperfectly insulated so that there is ground leakage to both capacitance and current. We know from the physics of the problem that at certain instant t,

$$\begin{aligned} \text{the potential at } Q \;&=\; \text{the potential at } P \\ &-\; \text{potential drop due to resistance along the element } PQ \\ &-\; \text{potential drop due to inductance along the element } PQ. \end{aligned} \tag{5.29}$$

Also, we know that,

$$\begin{aligned}\text{The Current at } Q &= \text{the current at } P \\ &- \text{the current loss on the element through leakage to ground} \\ &- \text{the apparent current loss due to varying charges stored in the capacitor on the element.}\end{aligned} \tag{5.30}$$

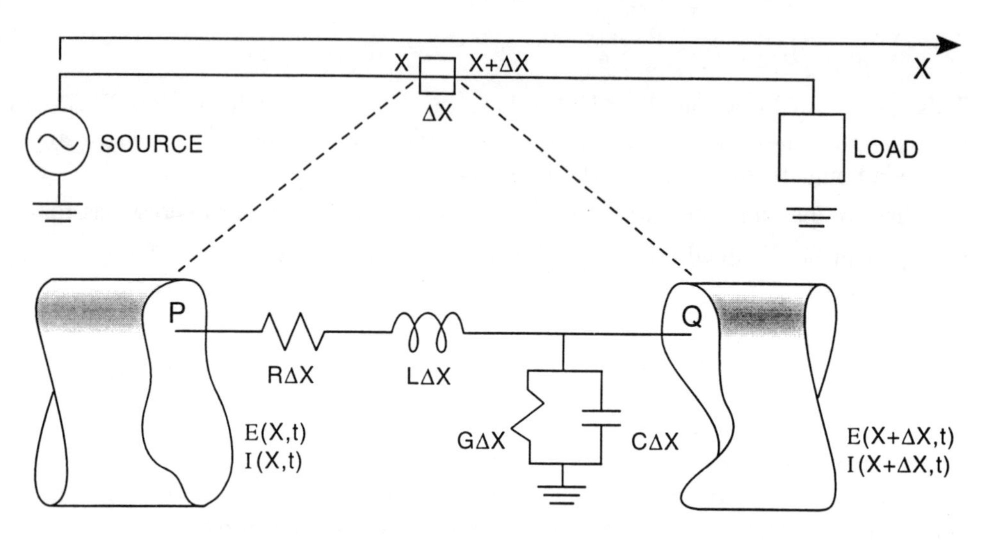

Figure 5.6. A typical element of a transmission line

The eqns (5.29) and (5.30) can be expressed as,

$$E(x+\Delta x,t) = E(x,t) - (R\,\Delta x)I(x,t) - (L\,\Delta x)\frac{\partial I(x,t)}{\partial t} \tag{5.31}$$

$$I(x+\Delta x,t) = I(x,t) - (G\,\Delta x)E(x,t) - (C\,\Delta x)\frac{\partial E(x,t)}{\partial t} \tag{5.32}$$

where

x	=	the distance from the sending end of the cable
$E(x,t)$	=	the potential at any point on the cable at any time
$I(x,t)$	=	the current at any point on the cable at any time
R	=	the resistance of the cable per unit length
L	=	the inductance of the cable per unit length
G	=	the conductance to ground per unit length
C	=	the capacitance to ground per unit length.

Rewriting eqns (5.31) and (5.32) and dividing by Δx throughout,

$$\frac{E(x+\Delta x,t)-E(x,t)}{\Delta x} = -RI(x,t) - L\frac{\partial I(x,t)}{\partial t} \tag{5.33}$$

$$\frac{I(x+\Delta x,t)-I(x,t)}{\Delta x} = -GE(x,t) - C\frac{\partial E(x,t)}{\partial t}. \tag{5.34}$$

Let $\Delta x \to 0$, then (5.33) and (5.34) can be written as

$$\frac{\partial E}{\partial x} = -RI - L\frac{\partial I}{\partial t} \tag{5.35}$$

$$\frac{\partial I}{\partial x} = -GE - C\frac{\partial E}{\partial t}. \tag{5.36}$$

These two simultaneous partial differential equations are called the transmission line equations.

Differentiating (5.35) with respect to x and (5.36) with respect to t, we obtain

$$\begin{aligned}\frac{\partial^2 E}{\partial x^2} &= -R\frac{\partial I}{\partial x} - L\frac{\partial^2 I}{\partial x \partial t} \\ \frac{\partial^2 I}{\partial x \partial t} &= -G\frac{\partial E}{\partial t} - C\frac{\partial^2 E}{\partial t^2}.\end{aligned}$$

Elimination of $\frac{\partial^2 I}{\partial x \partial t}$ between these two equations yields

$$\frac{\partial^2 E}{\partial x^2} = -R\frac{\partial I}{\partial x} + LG\frac{\partial E}{\partial t} + LC\frac{\partial^2 E}{\partial t^2}.$$

Then substitute $\frac{\partial I}{\partial x}$ from eqn (5.36), the above equation becomes

$$\frac{\partial^2 E}{\partial x^2} = LC\frac{\partial^2 E}{\partial t^2} + (RC + LG)\frac{\partial E}{\partial t} + RGE. \tag{5.37}$$

Similarly by differentiating eqn (5.35) with respect to t and (5.36) with respect to x and then eliminating the variable E, we obtain,

$$\frac{\partial^2 I}{\partial x^2} = LC\frac{\partial^2 I}{\partial t^2} + (RC + LG)\frac{\partial I}{\partial t} + RGI. \tag{5.38}$$

Equations (5.37) and (5.38) are called the Telephone Equations. Note that both E and I satisfy the same partial differential equation.

Special cases

1. If ground leakage and inductance are negligible, that is, if $G = 0$ and $L = 0$, as they are for co-axial cables, eqns (5.37) and (5.38) reduce respectively to

$$\frac{\partial^2 E}{\partial x^2} = RC\frac{\partial E}{\partial t} \tag{5.39}$$

$$\frac{\partial^2 I}{\partial x^2} = RC\frac{\partial I}{\partial t}. \tag{5.40}$$

   These equations are called Telegraph equations, which correspond to the one-dimensional heat conduction equation.

2. With reference to the frequency σ of some sine or cosine wave, we have the relationship between time period T and frequency as $T = \frac{2\pi}{\sigma}$.

At high frequencies (see eqn (5.37)),

$$\begin{aligned}
\frac{\partial^2 E}{\partial t^2} &\sim 0(\frac{E}{T^2}) = 0(\frac{E\sigma^2}{4\pi^2}) \\
\frac{\partial E}{\partial t} &\sim 0(\frac{E}{T}) = 0(\frac{E\sigma}{2\pi}) \\
E &= E
\end{aligned}$$

where 0 represents the measure of order of some quantity.

Hence, when $\sigma \to \infty$, terms $\frac{\partial E}{\partial t}, E$ are insignificant in comparison with term $\frac{\partial^2 E}{\partial t^2}$. Similar analysis follows in the case of term $\frac{\partial^2 I}{\partial t^2}$ in Equation (5.38). In this situation eqns (5.37) and (5.38) can be approximated to yield,

$$\frac{\partial^2 E}{\partial x^2} = LC\frac{\partial^2 E}{\partial t^2} \tag{5.41}$$

$$\frac{\partial^2 I}{\partial x^2} = LC\frac{\partial^2 I}{\partial t^2}. \tag{5.42}$$

These two equations are examples of one-dimensional wave equations. Here $1\sqrt{LC}$ has the dimension of velocity.

Equations (5.41) and (5.42) can also be obtained at any frequency provided the leakage to the ground and the resistance of the cable are both zero, that is $G = 0$ and $R = 0$.

Note

In this section we have derived the mathematical equations corresponding to the physical situation of the problems. Note that the partial differential equations will remain the same from problem to problem. As for example, the partial differential equations derived for the heat conduction problem can be applied to the diffusion problem. In actual practice, one physical problem will be different from the other provided the boundary conditions are clearly defined. Thus in the succeeding section we shall deal with some boundary value problems.

5.3 Classification of PDE: Method of characteristics

In the preceding section we formulated partial differential equations governing the physical situation of problems. Those equations are the particular cases of a general class of partial differential equations. In general, the second order partial differential equations may be written as

$$A(x,y)u_{xx} + 2B(x,y)u_{xy} + C(x,y)u_{yy} = f(x,y,u,u_x,u_y) \tag{5.43}$$

where $u(x,y)$ is a dependent variable and x, y are independent variables. The coefficients A, B, and C are in general, functions of x and y. The function f on the right hand side of eqn (5.43) includes all the terms like, u, u_x, u_y, etc. Here the subscripts are defined as the partial derivatives as for example,

$$\begin{aligned} u_x &= \frac{\partial u}{\partial x} \\ u_{xy} &= \frac{\partial^2 u}{\partial x \partial y}, \text{ etc.} \end{aligned}$$

For simplicity, let us consider first the possibility of obtaining solutions of the form

$$u = g(x + \lambda y) \tag{5.44}$$

where $(x + \lambda y)$ is the argument of g, for the equation

$$A u_{xx} + 2B u_{xy} + C u_{yy} = 0 \tag{5.45}$$

where A, B and C are assumed constants. If (5.44) is a solution of (5.45), then substituting, we obtain,

$$A g''(x + \lambda y) + 2B\lambda g''(x + \lambda y) + C\lambda^2 g''(x + \lambda y) = 0$$

which will be an identity if and only if

$$C\lambda^2 + 2B\lambda + A = 0. \tag{5.46}$$

This is a quadratic equation in λ, and solving for λ, we have

$$\lambda = \frac{-B \pm \sqrt{B^2 - CA}}{C}. \tag{5.47}$$

Thus,

(i) λ has two real distinct roots if $B^2 - CA > 0$.

(ii) λ has two equal roots if $B^2 - CA = 0$.

(iii) λ has two complex roots if $B^2 - CA < 0$.

Hence, there are two, one or no real values of λ for which solutions of the form $g(x+\lambda y)$ exist, according to the discriminant $B^2 - AC >=< 0$.

By analogy with the criterion for a conic to be hyperbolic, parabolic or elliptic, eqn (5.45) is said to be of

$$\begin{aligned} \text{Hyperbolic type if } B^2 - CA &> 0 \\ \text{Parabolic type if } B^2 - CA &= 0 \\ \text{Elliptic type if } B^2 - CA &< 0. \end{aligned}$$

Now in each case, the loci

$$\begin{aligned} x + \lambda_1 y &= C_1 \\ x + \lambda_2 y &= C_2 \end{aligned} \tag{5.48}$$

where λ_1 and λ_2 are the two roots of (5.47), are called the **characteristic curves** or simply the **characteristics** of eqn (5.45). These two characteristics define two families of parallel lines in the xy plane.

Consider the characteristic in its simplest form, $x + \lambda y = c$ such that $\frac{dy}{dx} = -\frac{1}{\lambda}$. Now eliminating λ in eqn (5.46) by the slope $(\frac{dy}{dx})$, we obtain the following differential equation

$$A(\frac{dy}{dx})^2 - 2B(\frac{dy}{dx}) + C = 0. \tag{5.49}$$

This differential equation must be satisfied by any characteristics of the form $x+\lambda y = c$ which are found by the solution of (5.49). This equation is usually known as the characteristic differential equation.

Following the treatment presented in a more advanced book, eqn (5.43) may be defined to be

$$\begin{aligned} \text{Hyperbolic if } B^2(x,y) - A(x,y)C(x,y) &> 0 \\ \text{Parabolic if } B^2(x,y) - A(x,y)C(x,y) &= 0 \\ \text{Elliptic if } B^2(x,y) - A(x,y)C(x,y) &< 0 \end{aligned}$$

at all points in a region D. Note that A, B and C are in general functions of x and y. Furthermore, generalizing the property expressed in eqn (5.49), we obtain

$$A(x,y)(\frac{dy}{dx})^2 - 2B(x,y)(\frac{dy}{dx}) + C(x,y) = 0 \tag{5.50}$$

which is known as the characteristic equation. Solving in terms of $\frac{dy}{dx}$,

$$\frac{dy}{dx} = \frac{B \pm \sqrt{B^2 - AC}}{A}. \tag{5.51}$$

Equation (5.51) represents a pair of ordinary differential equations. The solution of which exists in the following forms:

$$\begin{aligned} \phi(x,y) &= C_1 = \xi \\ \psi(x,y) &= C_2 = \eta. \end{aligned} \tag{5.52}$$

These two solutions are usually said to be characteristics of the partial differential equations (5.42) (see Fig. 5.7).

The simplest examples of hyperbolic, parabolic and elliptic partial differential equations are, respectively

$$\begin{aligned} \text{the wave equation, } u_{tt} &= c^2 u_{xx} \\ \text{the heat conduction equation, } u_t &= \alpha u_{xx} \\ \text{Laplace's equation, } u_{xx} + u_{yy} &= 0. \end{aligned}$$

Using the concept of characteristics, a general second order partial differential equation can be reduced to one of the above simple equations. Thus in the following we summarize the techniques to reduce this type of equation into simple form.

Case (I): Hyperbolic Class

In this case, eqn (5.43) can be classified as a hyperbolic type if $B^2 - CA > 0$. The characteristic will be given by the solution of the following pair of ordinary differential equations,

$$\frac{dy}{dx} = \frac{B \pm \sqrt{B^2 - CA}}{A}.$$

Integrating the above pair of differential equations, we get $\phi(x, y) = c_1$ and $\psi(x, y) = c_2$.

Thus, the characteristics are defined as

$$\begin{aligned} \xi &= \phi(x, y) \\ \eta &= \psi(x, y). \end{aligned} \tag{5.53}$$

Under this transformation, the original eqn (5.43) can be reduced to the standard form

$$u_{\xi\eta} = h(\xi, \eta, u, u_\xi, u_\eta). \tag{5.54}$$

Case (II): Parabolic Class

In this case, eqn (5.43) can be classified as of parabolic type if $B^2 = CA$. The characteristics will then be given by the solution of the following ordinary differential equation $\frac{dy}{dx} = \frac{B}{A}$ and upon integrating, we get $\phi(x, y) = c_1$.

Note that we will get only one characteristic from the solution of the above equation, say, $\eta = \phi(x, y)$. The other characteristics may be defined as $\xi = x$. Therefore, under the transformation

$$\begin{aligned} \xi &= x \\ \eta &= \phi(x, y). \end{aligned} \tag{5.55}$$

Equation (5.43) may be reduced to the standard form

$$u_{\xi\xi} = h(\xi, \eta, u, u_\xi, u_\eta). \tag{5.56}$$

Case (III): Elliptic Class

In this case, eqn (5.43) can be classified as an Elliptic type provided $B^2 - AC < 0$.

The characteristics will be obtained by solving the following pair of ordinary differential equations $\frac{dy}{dx} = \frac{B \pm i\sqrt{CA - B^2}}{A}$. Taking the positive sign and integrating we get $\phi(x, y) = C$ where $\phi(x, y)$ is a complex function.

The characteristics in this case are given by

$$\begin{aligned} \xi &= Re\{\phi(x, y)\} \\ \eta &= Im\{\phi(x, y)\}. \end{aligned} \tag{5.57}$$

Under this transformation, eqn (5.43) reduces to the standard form

$$u_{\xi\xi} + u_{\eta\eta} = h(\xi, \eta, u, u_\xi, u_\eta). \tag{5.58}$$

Alternatively, considering both positive and negative signs and then solving the differential equations, we obtain $\phi(x, y) = C_1$ and $\psi(x, y) = C_2$.

These two functions are complex conjugates. The characteristics are then given by

$$\begin{aligned} \xi &= \phi(x, y) \\ \eta &= \psi(x, y) = \bar{\phi}(x, y). \end{aligned} \tag{5.59}$$

Both ξ and η are complex conjugates. Under this transformation, eqn (5.43) reduces to the standard form

$$u_{\xi\eta} = h(\xi, \eta, u, u_\xi, u_\eta). \tag{5.60}$$

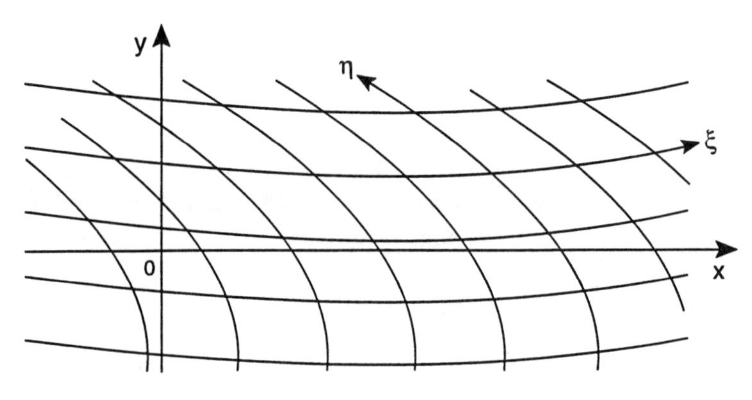

Figure 5.7. Characteristic plane (ξ, η)

We have already seen how to classify a given second order partial differential equation. General solutions of a wide range of partial differential equations can be obtained by using the characteristic variables which usually transform the equation into one of standard form. The method of finding these characteristics and subsequently the solution of the equation is known as the method of characteristics. In the following we will demonstrate this powerful method with some examples.

Example 1

Find the characteristics of the following equation and reduce it to the appropriate standard form and then obtain the general solution.

$$3u_{xx} + 10u_{xy} + 3u_{yy} = 0.$$

Solution

Comparing this equation with eqn (5.43), we have $A = 3$, $B = 5$ and $C = 3$. Then, $B^2 - CA = 25 - 9 = 16 > 0$. Therefore the equation is of hyperbolic type.

The characteristics are given by the solution of the ordinary differential equation

$$A(\frac{dy}{dx})^2 - 2B(\frac{dy}{dx}) + C = 0$$

i.e., $\frac{dy}{dx} = \frac{5\pm4}{3} = 3, \frac{1}{3}$. Integrating these, we get $3x - y = C_1$ and $x - 3y = C_2$ and hence the characteristics are defined by $\xi = 3x - y$ and $\eta = x - 3y$.

Thus under this transformation,

$$\begin{aligned} u_x &= u_\xi \xi_x + u_\eta \eta_x = 3u_\xi + u_\eta \\ u_y &= u_\xi \xi_y + u_\eta \eta_y = -u_\xi - 3u_\eta \end{aligned}$$

since

$$\begin{aligned} \xi_x &= 3 \qquad & \eta_x &= 1 \\ \xi_y &= -1 \qquad & \eta_y &= -3, \end{aligned}$$

$$\begin{aligned} u_{xx} &= 3\{u_{\xi\xi}\xi_x + u_{\xi\eta}\eta_x\} + \{u_{\eta\xi}\xi_x + u_{\eta\eta}\eta_x\} \\ &= 9u_{\xi\xi} + 6u_{\xi\eta} + u_{\eta\eta} \end{aligned}$$

$$\begin{aligned} u_{xy} &= 3\{u_{\xi\xi}\xi_y + u_{\xi\eta}\eta_y\} + \{u_{\eta\xi}\xi_y + u_{\eta\eta}\eta_y\} \\ &= -3u_{\xi\xi} - 10u_{\xi\eta} - 3u_{\eta\eta} \end{aligned}$$

$$\begin{aligned} u_{yy} &= -\{u_{\xi\xi}\xi_x + u_{\xi\eta}\eta_y\} - 3\{u_{\eta\xi}\xi_y + u_{\eta\eta}\eta_y\} \\ &= u_{\xi\xi} + 6u_{\xi\eta} + 9u_{\eta\eta}. \end{aligned}$$

Substituting these values into the given equation, we have

$$\begin{aligned} &3\{9u_{\xi\xi} + 6u_{\xi\eta} + u_{\eta\eta}\} + 10\{-3u_{\xi\xi} - 10u_{\xi\eta} - 3u_{\eta\eta}\} \\ &\quad +3\{u_{\xi\xi} + 6u_{\xi\eta} + 9u_{\eta\eta}\} = 0. \end{aligned}$$

Simplifying, we obtain $u_{\xi\eta} = 0$. Now integrating with respect to ξ partially $u_\eta = f'(\eta)$ where f is an arbitrary function and a prime denotes differentiation. Again integrating with respect to η partially, we obtain

$$u = f(\eta) + g(\xi).$$

Then the solution of the original equation is

$$u = f(x - 3y) + g(3x - y)$$

where f and g are arbitrary functions.

5.4 The D'Alembert solution of the wave equation

In the last section we saw that a general solution of the second order partial differential equation can easily be found by using the method of characteristics. However, in applied sciences and engineering problems, the "method of characteristics" is of little importance in parabolic and elliptic equations. On the other hand, the two families of characteristics of hyperbolic equations, being real and distinct, are of considerable practical value. Especially, in one-dimensional progressive wave propagation, consideration of the characteristics can give us a good deal of information about the propagation of wave fronts. Thus in the following we are going to discuss the progressive wave solution of one-dimensional wave equations. This solution was first discovered by a French mathematician, Jean Le Rond D'Alembert (1717–1783). The D'Alembert solution of the wave equation is actually not a special method, but a special application of the method of characteristics.

Now consider the one-dimensional wave equation

$$u_{tt} = c^2 u_{xx} \tag{5.61}$$

where u is the dependent variable, x and t are the independent variables and c is a parameter and its dimension is the speed.

Comparing this equation with (5.43), we have $A = 1$, $B = 0$ and $C = -c^2$. Then $B^2 - CA = 0 - (-c^2)(1) = c^2 > 0$. Hence eqn (5.61) is of hyperbolic type. The characteristics are given by the solution of the ordinary differential equation $A(\frac{dx}{dt})^2 - 2B(\frac{dx}{dt}) + C = 0$ such that $(\frac{dx}{dt}) = \pm c$. Integrating, we get

$$\begin{aligned} x - ct &= c_1 \\ x + ct &= c_2 \end{aligned} \tag{5.62}$$

and hence the characteristics are defined by

$$\begin{aligned} \xi &= x - ct \\ \eta &= x + ct. \end{aligned} \tag{5.63}$$

Under this transformation

$$u_x = u_\xi \xi_x + u_\eta \eta_x = u_\xi + u_\eta$$

$$u_t = u_\xi \xi_t + u_\eta \eta_t = -cu_\xi + cu_\eta$$

$$\begin{aligned} u_{xx} &= (u_{\xi\xi}\xi_x + u_{\xi\eta}\eta_x) + (u_{\eta\xi}\xi_x + u_{\eta\eta}\eta_x) \\ &= u_{\xi\xi} + 2u_{\xi\eta} + u_{\eta\eta} \end{aligned}$$

$$\begin{aligned} u_{tt} &= -c(u_{\xi\xi}\xi_t + u_{\xi\eta}\eta_t) + c(u_{\eta\xi}\xi_t + u_{\eta\eta}\eta_t) \\ &= c^2(u_{\xi\xi} - 2u_{\xi\eta} + u_{\eta\eta}). \end{aligned}$$

Substituting these values into the given equation, we obtain

$$u_{\xi\eta} = 0. \tag{5.64}$$

Integrating with respect to η, partially

$$u_\xi = f'(\xi).$$

Again integrating with respect to ξ, partially, we obtain

$$u = f(\xi) + g(\eta).$$

Therefore our solution in physical variables is

$$u(x,t) = f(x-ct) + g(x+ct) \tag{5.65}$$

where f and g are arbitrary functions.

By partial differentiations of f, we obtain

$$\begin{aligned}
\frac{\partial f(x-ct)}{\partial x} &= f'(x-ct) \\
\frac{\partial^2 f(x-ct)}{\partial x^2} &= f''(x-ct) \\
\frac{\partial f(x-ct)}{\partial t} &= -cf'(x-ct) \\
\frac{\partial^2 f(x-ct)}{\partial t^2} &= c^2 f''(x-ct).
\end{aligned}$$

Using these results, it is evident that $u = f(x-ct)$ satisfies the equation

$$u_{tt} = c^2 u_{xx}.$$

Similarly, if g is a double differentiable function, then $g(x+ct)$ is likewise a solution of (5.61). Because (5.61) is a linear equation, it follows that the sum

$$u = f(x-ct) + g(x+ct)$$

is also a solution.

The physical interpretation of these functions is quite interesting. The functions f and g represent two progressive waves travelling in opposite directions with the speed c. To see this let us first consider the solution $u = f(x-ct)$. At $t = 0$, it defines the curve $u = f(x)$, and after time $t = t_1$, it defines the curve $u = f(x - ct_1)$. But these curves are identical (see Fig. 5.8) except that the latter is translated to the right a distance equal to ct_1.

Thus the entire configuration moves along the positive direction of x-axis a distance of ct_1 in time t_1. The velocity with which the wave is propagated is, therefore,

$$v = \frac{ct_1}{t_1} = c.$$

Similarly, the function $g(x+ct)$ defines a wave progressing in the negative direction of x-axis with constant velocity c. The total solution is, therefore, the algebraic sum of these two travelling waves.

Solution (5.65) is a very convenient representation for progressive waves which travel large distances through a uniform medium.

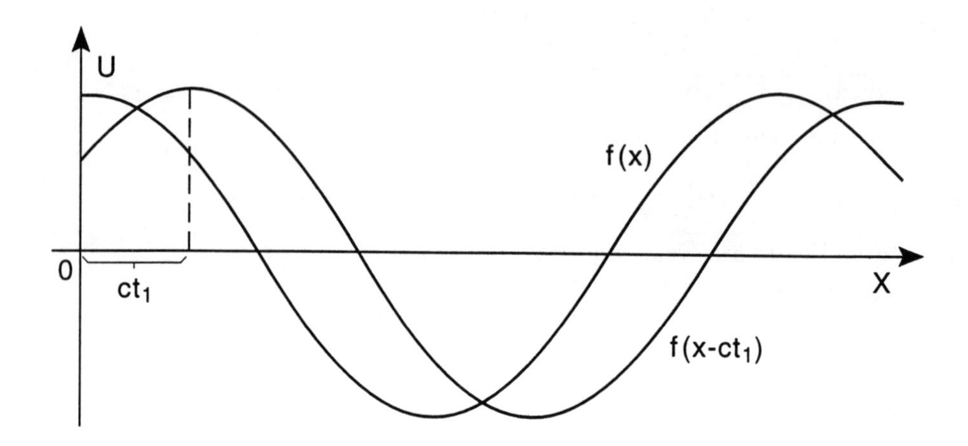

Figure 5.8. A progressive wave

Let us consider the following two initial conditions for a uniform medium extending from $-\infty < x < \infty$.

$$\begin{aligned} \text{Displacement} : u(x,0) &= \phi(x) \\ \text{Velocity} : u_t(x,0) &= \psi(x). \end{aligned} \tag{5.66}$$

Then in solution (5.64) we find that

$$\begin{aligned} f(x) + g(x) &= \phi(x) \\ -cf'(x) + cg'(x) &= \psi(x) \end{aligned} \tag{5.67}$$

for all values of x.

Integrating the second equation with respect to x, we have

$$-f(x) + g(x) = \frac{1}{c}\int_{x_0}^{x} \psi(s)ds + A.$$

Combining this and the first of (5.67), we obtain

$$f(x) = \frac{1}{2}[\phi(x) - \frac{1}{c}\int_{x_0}^{x} \psi(s)ds - A] \tag{5.68}$$

$$g(x) = \frac{1}{2}[\phi(x) + \frac{1}{c}\int_{x_0}^{x} \psi(s)ds + A] \tag{5.69}$$

where A is an integration constant and s is a dummy variable.

Substituting these expressions into (5.65), we obtain

$$u = \frac{1}{2}[\phi(x-ct) + \phi(x+ct)] + \frac{1}{2c}\{\int_{x-ct}^{x+ct} \psi(s)ds\}. \tag{5.70}$$

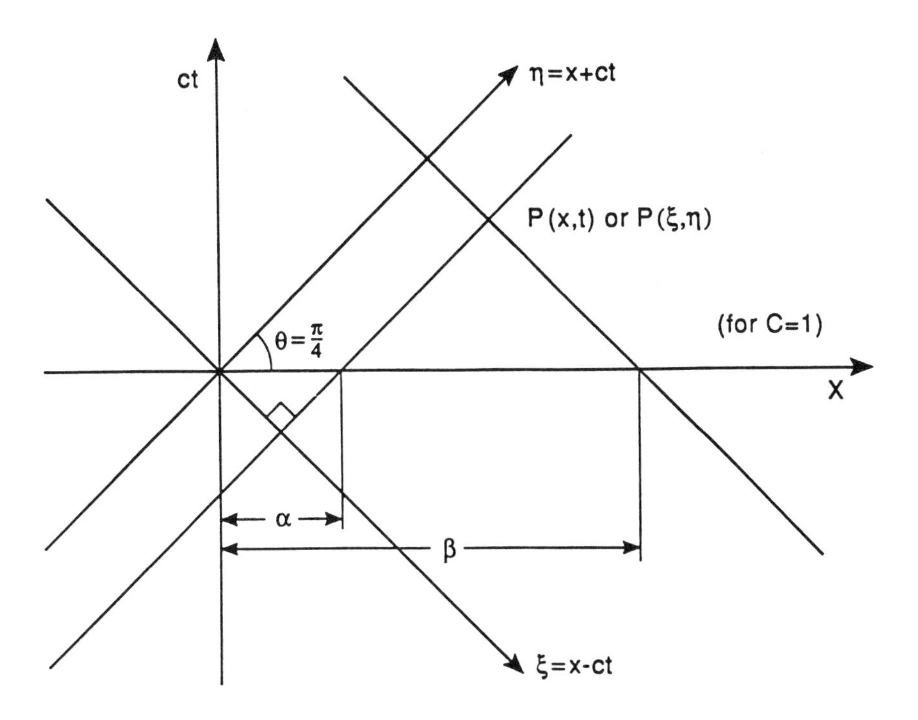

Figure 5.9. Relationship between a characteristic plane and a physical plane

This is D'Alembert's solution of the wave equation. Thus for a given initial displacement and velocity in the vertical direction, the wave equation is completely solved. This solution is usually called the progressive wave solution. Note that the use of the string problem to demonstrate the solution of the wave problem is a matter of convenience. However, any variable satisfying the wave equation possess the mathematical properties developed for the string. We shall see the standing wave solution by the method of separation of variables when we consider this problem in the next section.

It is clear that the wave equation can be handled very easily by introducing the characteristic variables (ξ, η). The relationship between the physical plane and the characteristic plane for this particular example can be visualized in Fig. 5.9.

Equation (5.65) represents the solution as the sum of two progressive waves; one going to the right, and the other to the left. The wave velocity is c. For each of the two progressive waves, we can also follow the wave motion by observing that in the xt plane $\frac{1}{2}f(x-ct)$ is constant along each line $x-ct=$ constant and similarly, $\frac{1}{2}g(x+ct)$ is constant along each line $(x+ct)=$ constant. Thus there are two families of parallel lines called the characteristics, along which waves are propagated.

Furthermore, along the x-axis the values of $u(x,0)$ and $u_t(x,0)$ are given as initial conditions of displacement and velocity and they determine the constant values of f and g along the individual characteristic. The characteristics therefore represent the paths in xt plane along which disturbances in the medium are propagated.

Finally, since the solution of a wave equation is $u=f(x-ct)+g(x+ct)$ the value of u at any point in the xt plane is the sum of the values of f and g on the respective characteristics which pass through that point.

Suppose we need to find the wave function at $P(x,t)$, then draw two characteristics passing through that point. Measure the x intercepts subtended by these two characteristics. Then using the formula (5.70), we obtain

$$u(x,t)=\frac{1}{2}[\phi(\alpha)+\phi(\beta)]+\frac{1}{2c}\int_{\alpha}^{\beta}\psi(s)ds$$

where $\alpha=x-ct$ and $\beta=x+ct$, which is as required.

5.5 The method of separation of variables

In this section, we shall discuss the solution technique of partial differential equations by the method of separation of variables. This appears to appears to be the most powerful method for a wide class of boundary value problems in engineering and physics. The main idea of this method is to convert the given partial differential equations into several ordinary differential equations and then obtain the solutions by familiar solution techniques. We shall illustrate this method by considering examples, namely, the wave equation, the heat equation and Laplace's equation.

Example 2: Displacement of a vibrating string

Consider the wave equation which governs the displacement of a vibrating string. The knowledge gained in this classical problem can be applied to other problems of physical interest such as oceanic waves, electromagnetic waves, sound waves, etc. We saw the progressive wave solution and its physical implication by the method of characteristics. The separation of variables method will lead to the standing wave solution and we shall find later how a standing wave solution can be expressed as the progressive wave solution.

Solution

The partial differential equation, together with the boundary and initial conditions, are respectively given by

$$u_{tt}=c^2u_{xx}. \tag{5.71}$$

Boundary Conditions:

$$\begin{aligned} x = 0 &: \ u(0,t) = 0 \\ x = \ell &: \ u(\ell,t) = 0. \end{aligned} \tag{5.72}$$

Initial Conditions:

$$\begin{aligned} t = 0 \ : \quad u(x,0) &= \phi(x) \\ u_t(x,0) &= \psi(x). \end{aligned} \tag{5.73}$$

Here, u represents the displacement of the string at any position at any time, u_t represents the velocity of the wave, produced by the string, t represents the time, x is the distance from the fixed position 0 and ℓ is the length of the string.

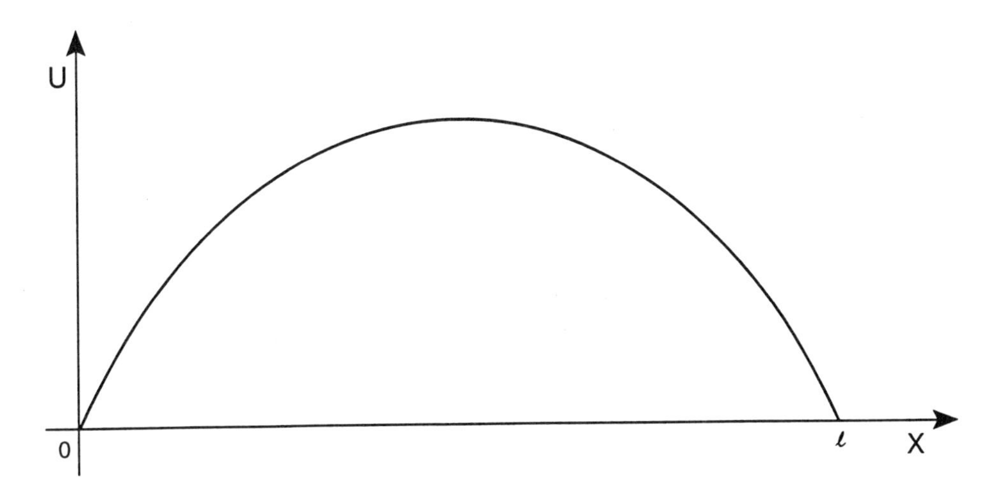

Figure 5.10. Displacement of a string

By the method of separation of variables, we mean that the dependent variable $u(x,t)$ may be expressed as a product solution $u(x,t) = X(x)T(t)$ where $X(x)$ and $T(t)$ are functions of x and t respectively. Substituting these into the wave equation and after a little reduction, we obtain

$$\frac{T''}{T} = c^2 \frac{X''}{X}. \tag{5.74}$$

The left hand member is a function of t and the right hand member is a function of x alone. Therefore,

$$\frac{T''}{T} = c^2 \frac{X''}{X} = k \tag{5.75}$$

where k is known as a separation constant. Therefore, we obtain

$$T'' - kT = 0 \tag{5.76}$$

$$X'' - \frac{k}{c^2} X = 0. \tag{5.77}$$

Thus the solution of the original partial differential equation will be determined by the solutions of the ordinary differential equations stated above.

Assuming that we need to consider only physical solutions, there are three cases to investigate:

$$\begin{array}{lcl} \text{Case (I)} & : & k = 0 \\ \text{Case (II)} & : & k > 0 \\ \text{Case (III)} & : & k < 0. \end{array}$$

Case (I)

If $k = 0$, the equations and their solutions are

$$\begin{array}{ll} T'' = 0 & X'' = 0 \\ T = At + B & X = Cx + D \end{array}$$

and hence,

$$\begin{aligned} u(x,t) &= X(x)T(t) \\ &= (Cx + D)(At + B). \end{aligned}$$

This solution cannot describe the undamped vibration of a system because it is not periodic. Hence, although the product solution exists for $k = 0$, the solution does not have any physical significance. Therefore, we reject this solution. On the other hand, it can be easily verified that the boundary conditions (5.72) produce a trivial solution, i.e., $u(x,t) = 0$ always.

Case (II)

Let us write

$$k = \lambda^2 > 0.$$

Then the two differential equations and their solutions are

$$\begin{array}{ll} T'' - \lambda^2 T = 0 & X'' - (\frac{\lambda}{c})^2 X = 0 \\ T = Ae^{\lambda t} + Be^{-\lambda t} & X = Ce^{\frac{\lambda}{c}x} + De^{-\frac{\lambda}{c}x} \end{array}$$

and hence

$$\begin{aligned} u(x,t) &= X(x)T(t) \\ &= (Ce^{\frac{\lambda}{c}x} + De^{-\frac{\lambda}{c}x})(Ae^{\lambda t} + Be^{-\lambda t}). \end{aligned}$$

This solution also cannot describe the undamped oscillatory motion of a system because the solution is not periodic. Therefore from the physical ground, the case $k = \lambda^2 > 0$ must be rejected. On the other hand, it can be easily verified that the boundary conditions (5.72) produce a trivial solution.

Case (III)

Let us write

$$k = -\lambda^2 < 0.$$

Then the two differential equations and their solutions are

$$T'' + \lambda^2 T = 0 \qquad X'' + (\tfrac{\lambda}{c})^2 X = 0$$
$$T = A\cos\lambda t + B\sin\lambda t \qquad X = C\cos(\tfrac{\lambda}{c}x) + D\sin(\tfrac{\lambda}{c}x)$$

and the solution becomes

$$\begin{aligned} u(x,t) &= X(x)T(t) \\ &= (C\cos\frac{\lambda}{c}x + D\sin\frac{\lambda}{c}x)(A\cos\lambda t + B\sin\lambda t). \end{aligned} \tag{5.78}$$

This solution is clearly periodic of period $\frac{2\pi}{\lambda}$. That means this solution represents a vibratory motion with period $\frac{2\pi}{\lambda}$ or frequency $\frac{\lambda}{2\pi}$. In the following we shall determine the value of λ and the constants A, B, C, and D.

By using the boundary condition at $x = 0$, we have

$$u(0,t) = 0 = C(A\cos\lambda t + B\sin\lambda t).$$

This condition can obviously be satisfied if both A and B are zero, which leads to the trivial solution only, i.e., $u(x,t) = 0$ for all the time and at all positions of the string, in which we are not interested. Hence, we are led to the alternative that $C = 0$, which reduces eqn (5.79) to the form

$$u(x,t) = (D\sin\frac{\lambda}{c}x)(A\cos\lambda t + B\sin\lambda t).$$

The second boundary condition at $x = \ell$ requires that

$$u(\ell,t) = 0 = (D\sin\frac{\lambda\ell}{c})(A\cos\lambda t + B\sin\lambda t).$$

As before, we reject the possibility that $A = 0 = B$. At the same time, we cannot allow D to vanish, because in that case we will get a trivial solution. The only possibility is that

$$\sin\frac{\lambda\ell}{c} = 0$$

or,

$$\lambda_n = \frac{n\pi c}{\ell}, \qquad n = 1, 2, 3, \cdots$$

Here λ_n is called the eigenvalue and corresponding to any λ_n, the solution is called the eigensolution. Then there are an infinite number of values of λ_n for each of which the product solution of the wave equation exists. Thus,

$$\begin{aligned} u_n(x,t) &= (\sin\frac{\lambda_n x}{c})(A_n\cos\lambda_n t + B_n\sin\lambda_n t) \\ &= \sin(\frac{n\pi x}{\ell})(A_n\cos\frac{n\pi ct}{\ell} + B_n\sin\frac{n\pi ct}{\ell}) \end{aligned} \tag{5.79}$$

for $n = 1, 2, 3, \cdots$ is a possible mode of vibrations of the string. Since the wave equation is linear, the solution of the equations is the superposition of all the linear solutions. Hence our solution exists in the Fourier series expansion in the following way

$$\begin{aligned} u(x,t) &= \sum_{n=1}^{\infty} u_n(x,t) \\ &= \sum_{n=1}^{\infty} \sin\frac{n\pi x}{\ell}(A_n \cos\frac{n\pi ct}{\ell} + B_n \sin\frac{n\pi ct}{\ell}) \end{aligned} \tag{5.80}$$

where A_n and B_n are defined as the Fourier coefficients and they are given in terms of the initial conditions (5.73).

To Obtain A_n

The initial displacement condition (5.73) at $t = 0$ yields

$$u(x,0) = \phi(x) = \sum_{n=1}^{\infty} A_n \sin\frac{n\pi x}{\ell}. \tag{5.81}$$

This series can be recognized as the half range sine expansion of a periodic function $\phi(x)$ defined in the range $(0, \ell)$. Now A_n can be obtained, multiplying the series (5.81) by $\sin\frac{n\pi x}{\ell}$ and integrating with respect to x from 0 to ℓ, which yields

$$A_n = \frac{2}{\ell}\int_0^{\ell} \phi(x) \sin\frac{n\pi x}{\ell} dx \quad n = 1, 2, 3, \cdots \tag{5.82}$$

To Determine B_n

The initial velocity condition (5.73) at $t = 0$ must be applied. To do this, we note that

$$u_t(x,t) = \sum_{n=1}^{\infty} \sin\frac{n\pi x}{\ell}\{-A_n \sin\frac{n\pi ct}{\ell} + B_n \cos\frac{n\pi ct}{\ell}\}(\frac{n\pi c}{\ell}).$$

Hence, at $t = 0$

$$u_t(x,0) = \psi(x) = \sum_{n=1}^{\infty} B_n \frac{n\pi c}{\ell} \sin\frac{n\pi x}{\ell}. \tag{5.83}$$

Now multiplying both sides of (5.84) by $\sin\frac{n\pi x}{\ell}$ and integrating with respect to x from 0 to ℓ, we obtain

$$B_n = \frac{2}{n\pi c}\int_0^{\ell} \psi(x) \sin\frac{n\pi x}{\ell} dx \quad n = 1, 2, 3, \cdots \tag{5.84}$$

Therefore, eqn (5.80) is the required solution where the Fourier coefficients A_n and B_n are given by (5.83) and (5.85).

The displacement (5.80) is referred to as the nth eigenfunction or nth normal mode of the vibrating string. Note that the normal mode corresponding to $n = 0$ is $u_0 = 0$

and the normal modes corresponding to the negative integers can be omitted as they are identical to those which correspond to the positive integers.

The nth normal mode vibrates with a period of $\frac{2\ell}{nc}$ seconds which corresponds to a frequency of $\frac{nc}{2\ell}$ cycles per second. Since $c^2 = (\frac{gT}{\omega})$, where g is the acceleration due to gravity, T is the tension and ω is the weight of the string per unit length, the frequency is

$$\frac{n}{2\ell}(\frac{gT}{\omega})^{1/2}.$$

Hence if a string on a musical instrument is vibrating in a normal mode, its pitch may be sharpened (frequency increased) by either decreasing the length ℓ of the string or increasing the tension T in the string.

The first normal mode $n = 1$ vibrates with the lowest frequency

$$\frac{1}{2\ell}(\frac{gT}{\omega})^{1/2}$$

and this is called the *fundamental frequency* of the string. If the string can be made to vibrate in a higher normal mode the frequency is increased by an integer multiple; this corresponds to the production of a musical harmonic or overtone.

For a given normal mode at the input $t = t_0$, the shape of the string is the sine curve presented by (5.80) with t replaced by t_0. The shape functions of the first four normal modes, together with their frequencies, are shown below in Fig. 5.11. Also, it can be observed from eqn (5.80) that the point $x = x_0$ of the string undergoes a periodic transverse oscillation with amplitude

$$\sqrt{A_n^2 + B_n^2}\sin(\frac{n\pi x_0}{\ell}).$$

The solution presented in eqn (5.81) may be regarded as the *standing* wave solution, which also represents eigenfunctions corresponding to the respective eigenvalues. For an arbitrary initial displacement $\phi(x)$ of the string, the motion is more complicated since it represents a combination in general of all the modes of vibrations and thus all frequencies. However, all the harmonic frequencies or *overtones* are integer multiples of the *fundamental frequency.* Whenever such a situation arises in a vibrating system, as in a violin or piano string, we say that we have *music*, provided, of course, that sounds are produced, i.e., the frequencies are audible ranges. If a vibrating system produces sounds where frequencies are not integer multiples of some *fundamental frequency*, we say that we have *noise.*

The *progressive wave* solution has been presented in the last section. It is interesting to observe that the standing wave solution (5.81) can be expressed in the form of a *progressive wave* solution (5.65).

Since we have

$$\begin{aligned} u(x,t) &= \sum_{n=1}^{\infty}\{A_n \sin\frac{n\pi x}{\ell}\cos\frac{n\pi ct}{\ell} + B_n \sin\frac{n\pi x}{\ell}\sin\frac{n\pi ct}{\ell}\} \\ &= \sum_{n=1}^{\infty}\{\frac{A_n}{2}\sin\frac{n\pi(x+ct)}{\ell} - \frac{B_n}{2}\cos\frac{n\pi(x+ct)}{\ell}\} \end{aligned}$$

$$+\sum_{n=1}^{\infty}\{\frac{A_n}{2}\sin\frac{n\pi(x-ct)}{\ell}+\frac{B_n}{2}\cos\frac{n\pi(x-ct)}{\ell}\}$$

which can be expressed as $u(x,t) = f(x+ct) + g(x-ct)$. This is known as the progressive wave form of a solution.

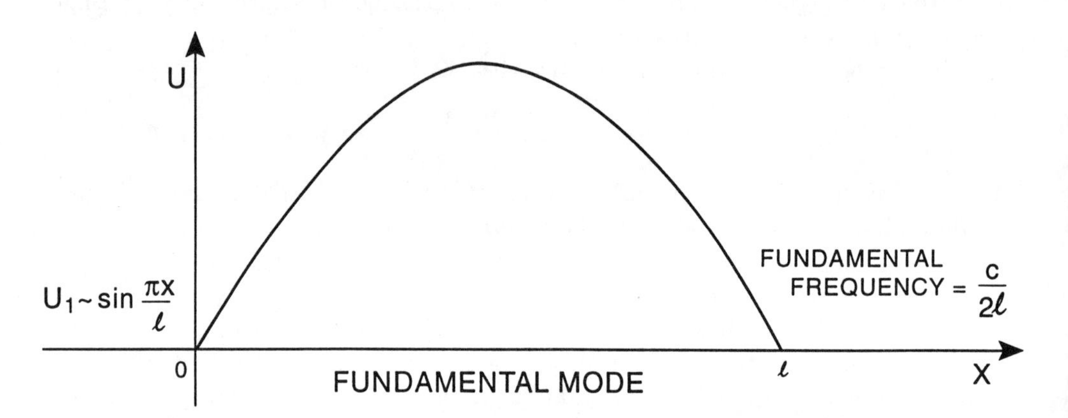

Figure 5.11a. Fundamental mode of a vibrating string

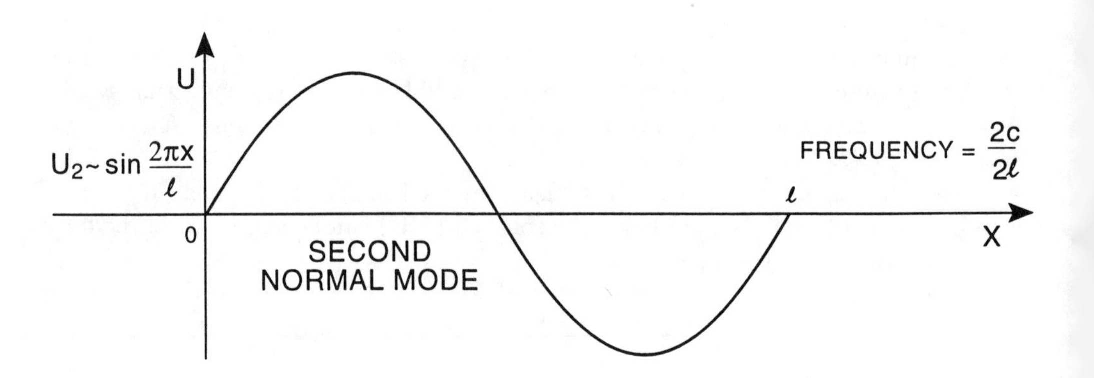

Figure 5.11b. Second normal mode of a vibrating string

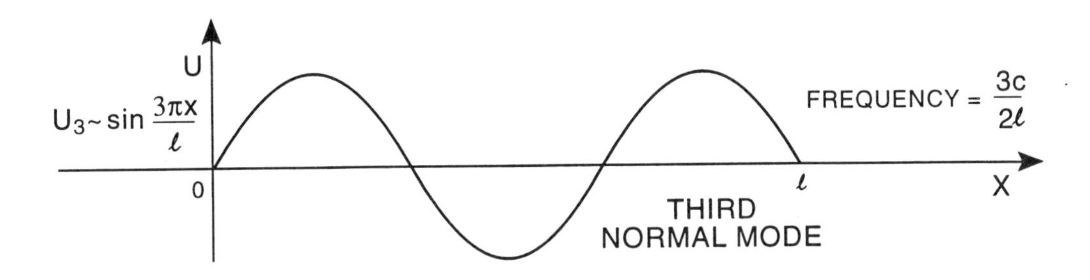

Figure 5.11c. Third normal mode of a vibrating string

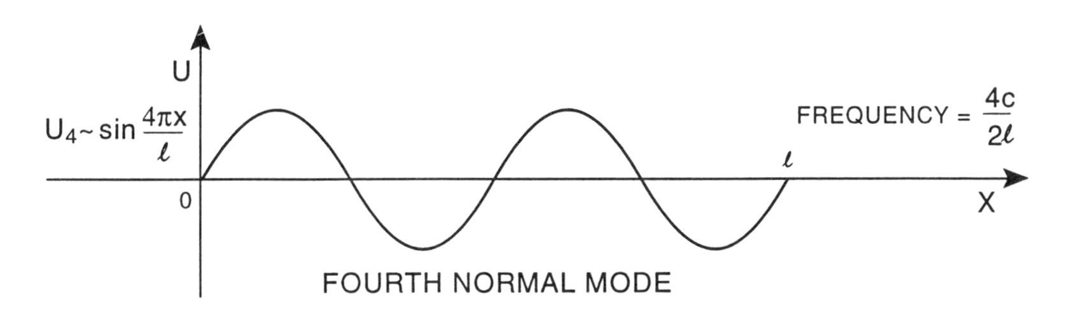

Figure 5.11d. Fourth normal mode of a vibrating string

Note

Consider now the three-dimensional wave equation

$$u_{tt} = c^2(u_{xx} + u_{yy} + u_{zz}).$$

Solutions of this equation in Cartesian coordinates x, y, z and time t behave similiarly to a one-dimensional wave equation. This equation governs a wide variety of wave and vibration phenomena. Therefore, it is natural to enquire whether the three-dimensional wave equation possesses solutions similar to D'Alembert's solution of the one-dimensional wave equation. It can be easily verified that $u = \lambda(\ell x + my + nz - ct)$, where λ is an arbitrary differentiable function, satisfies the above wave equation whenever

$$\ell^2 + m^2 + n^2 = 1$$

so that ℓ, m, n may be interpreted as the direction cosines, and c is the wave speed.

Similarly,

$$u = \mu(\ell x + my + nz + ct)$$

also satisfies the wave equation when ℓ, m, n are the direction cosines. The physical interpretation of these two solutions may be defined to be the plane waves travelling in opposite senses along the direction of the unit vector $\vec{n} = \vec{i}\ell + \vec{j}m + \vec{k}n$. We know the equation of a plane with the unit normal $\vec{n} = \vec{i}\ell + \vec{j}m + \vec{k}n$ is given by $\ell x + my + nz - p = 0$. The line through the origin along the unit normal direction $\vec{n}$ intersects the plane at a distance p from the origin. Suppose the point of intersection moves along the normal line with a speed c and passes through the origin at $t = 0$, so that $p = ct$, then a plane moving along its fixed normal direction $\vec{n}$ with the speed c is given by the equation $\ell x + my + nz - ct = 0$. The surface $\ell x + my + nz - ct = \xi$, where ξ is a constant, represents a parallel plane, moving with the speed c along the normal direction $\vec{n}$ which passes through the origin at time $t = -\xi/c$.

Now considering the solution

$$u = \lambda(\ell x + my + nz - ct) = \lambda(\xi),$$

we shall see that this corresponds to the 'signal' $u = \lambda(\xi_0)$, for $\xi = \xi_0$, which is carried by the surfaces $\ell x + my + nz - ct = \xi_0$. Consequently, all points on the moving plane which pass through the origin at the time $t = -\xi_0/c$ and which move along its normal direction $\vec{n}$ with speed c, carries the same signal $u = \lambda(\xi_0)$. Because of this the solutions of the wave equations in three-dimensions are called the plane waves.

At a fixed position (x_0, y_0, z_0) in space, the signals $u = \lambda(\ell x_0 + my_0 + nz_0 - ct)$ are received when t varies. This corresponds to a wave train of signals $u = \lambda(\xi)$ passing through this point as $\xi = \ell x_0 + my_0 + nz_0 - ct$ varies with the time t.

Similarly, the wave solution $u = \mu(\ell x + my + nz + ct)$ corresponds to a plane wave of arbitrary profile moving with speed c along the direction $-\vec{n}$. Thus any superposition of these waves which satisfies the wave equation is the general solution.

Example 3: Vibrations of a rectangular membrane

We shall consider in this example the transverse vibrations of a rectangular membrane stretched over a plane frame under an all-round tension, which satisfies the two-dimensional wave equation

$$u_{tt} = c^2(u_{xx} + u_{yy}) \tag{5.85}$$

where the plane of the frame is $z = 0$, and the displacement of the membrane normal to the plane $z = 0$ is $u(x, y, t)$. The wave speed c is given by

$$c^2 = (\frac{gT}{\omega})$$

where T is the all-round tension per unit length, g is the acceleration due to gravity and ω is the weight per unit area.

Solution

Consider the rectangular membrane $0 \leq x \leq a$, $0 \leq y \leq b$ which is stretched over a rectangular frame under an all-round tension, which is clamped around its edges so that the boundary conditions are

$$\begin{array}{ll} u(0, y, t) = 0 & u(a, y, t) = 0 \\ u(x, 0, t) = 0 & u(x, b, t) = 0. \end{array} \tag{5.86}$$

Consider the initial conditions as

$$t = 0: \quad \begin{array}{r} u(x, y, 0) = \phi(x, y) \\ u_t(x, y, 0) = \psi(x, y). \end{array} \tag{5.87}$$

Assume that the solution can be expressed as

$$u(x, y, t) = X(x)Y(y)T(t). \tag{5.88}$$

On substituting this solution into wave eqn (5.86), we find that

$$\frac{X''}{X} + \frac{Y''}{Y} = \frac{1}{c^2}\frac{T''}{T}.$$

Since x, y and t are independent variables, each of these terms is a constant. Thus,

$$\frac{X''}{X} = p, \frac{Y''}{Y} = q, \frac{T''}{T} = c^2(p + q) \tag{5.89}$$

where p and q are the separation constants. We have seen in the transverse vibration of a string that an undamped vibratory motion of the membrane can be generated provided

$$\begin{aligned} p &= -\lambda^2 < 0 \\ q &= -\mu^2 < 0 \end{aligned}$$

where λ and μ are two real parameters. Then the three ordinary differential equations are

$$X'' + \lambda^2 X = 0 \tag{5.90}$$

$$Y'' + \mu^2 Y = 0 \tag{5.91}$$

$$T'' + c^2(\lambda^2 + \mu^2)T = 0. \tag{5.92}$$

Solving these equations, we obtain

$$\begin{aligned} X &= A\cos\lambda x + B\sin\lambda x \\ Y &= C\cos\mu y + D\sin\mu y \\ T &= E\cos(c\sqrt{\lambda^2+\mu^2})t + F\sin(c\sqrt{\lambda^2+\mu^2})t \end{aligned}$$

and hence,

$$\begin{aligned} u(x,y,t) &= (A\cos\lambda x + B\sin\lambda x)(C\cos\mu y + D\sin\mu y) \\ &\quad [E\cos(c\sqrt{\lambda^2+\mu^2})t + F\sin(c\sqrt{\lambda^2+\mu^2})t]. \end{aligned} \tag{5.93}$$

To satisfy the boundary conditions (5.87), we observe that these conditions may be satisfied provided

$$\begin{aligned} \lambda_m &= \tfrac{m\pi}{a} \quad , \quad m = 1,2,3,\cdots \\ \mu_n &= \tfrac{n\pi}{b} \quad , \quad n = 1,2,3,\cdots \end{aligned} \tag{5.94}$$

These are called the eigenvalues. The eigenfunctions or the normal modes corresponding to these eigenvalues are

$$\begin{aligned} u_{mn}(x,y,t) &= \sin\frac{m\pi x}{a}\sin\frac{n\pi y}{b}\{A_{mn}\cos c(\frac{m^2}{a^2}+\frac{n^2}{b^2})^{1/2}\pi t \\ &\quad + B_{mn}\sin c(\frac{m^2}{a^2}+\frac{n^2}{b^2})^{1/2}\pi t\}. \end{aligned} \tag{5.95}$$

The natural frequency for the mth and nth normal mode is

$$\frac{c}{2}(\frac{m^2}{a^2}+\frac{n^2}{b^2})^{1/2} \quad \text{cycles per second.}$$

The fundamental frequency is $\frac{c}{2}(\frac{1}{a^2}+\frac{1}{b^2})^{1/2}$ and corresponds to $m=1, n=1$ normal mode.

Note that these frequencies are all higher than those for an elastic string of length ℓ with the same wave speed c. The string frequencies $\frac{mc}{2\ell}$ are obtained in the plate as the dimension $b \to \infty$, $a \to \ell$, that is when the plate resembles a long strip.

Since the wave equation is a linear equation, the linear combination of all the normal modes is also a solution which can be expressed as

$$\begin{aligned} u &= \sum_{m=1}^{\infty}\sum_{n=1}^{\infty} u_{mn}(x,y,t) \\ &= \sum_{m=1}^{\infty}\sum_{n=1}^{\infty}\sin\frac{m\pi x}{a}\sin\frac{n\pi y}{b}[A_{mn}\cos c(\frac{m^2}{a^2}+\frac{n^2}{b^2})^{1/2}\pi t \\ &\quad + B_{mn}\sin c(\frac{m^2}{a^2}+\frac{n^2}{b^2})^{1/2}\pi t)] \end{aligned} \tag{5.96}$$

where A_{mn}, B_{mn} are arbitrary constants for $m = 1,2,3,\cdots$, $n = 1,2,3,\cdots$. The initial displacement condition (5.88) yields

$$\sum_{m=1}^{\infty}\sum_{n=1}^{\infty} A_{mn}\sin\frac{m\pi x}{a}\sin\frac{n\pi y}{b} = \phi(x,y) \quad 0 \le x \le a, 0 \le y \le b. \tag{5.97}$$

In this case (5.98) is simply a double Fourier series representation.

Now rearranging the terms

$$\phi(x, y) = \sum_{m=1}^{\infty} H_m(y) \sin \frac{m\pi x}{a} \tag{5.98}$$

where

$$H_m(y) = \sum_{n=1}^{\infty} A_{mn} \sin \frac{n\pi y}{b}. \tag{5.99}$$

Using Euler's formulas,

$$H_m(y) = \frac{2}{a} \int_0^a \phi(x, y) \sin \frac{m\pi x}{a} dx \tag{5.100}$$

and subsequently

$$\begin{aligned} A_{mn} &= \frac{2}{b} \int_0^b H_m(y) \sin \frac{n\pi y}{b} dy \\ &= \frac{4}{ab} \int_0^a \int_0^b \phi(x, y) \sin \frac{n\pi y}{b} \sin \frac{m\pi x}{a} dy dx. \end{aligned} \tag{5.101}$$

Also the initial velocity condition yields

$$u_t(x, y, 0) = \psi(x, y) = \sum_{m=1}^{\infty} \sum_{n=1}^{\infty} B_{mn} (c\sqrt{\frac{m^2}{a^2} + \frac{n^2}{b^2}}\pi) \sin \frac{m\pi x}{a} \sin \frac{n\pi y}{b}.$$

which is a double Fourier series expansion. Using the same technique as above,

$$B_{mn} = \frac{4}{ab\pi c (\frac{m^2}{a^2} + \frac{n^2}{b^2})^{1/2}} \int_0^a \int_0^b \psi(x, y) \sin \frac{n\pi y}{b} \sin \frac{m\pi x}{a} dy dx. \tag{5.102}$$

Thus the displacement of the membrane is known provided A_{mn} and B_{mn} are known from eqns (5.102) and (5.103).

Example 4

Consider the heat conduction in a thin metal bar of length ℓ with insulated sides. Let us suppose that the ends $x = 0$ and $x = \ell$ are held at temperature $u = 0^oC$ for all time $t > 0$. In addition, let us suppose that the temperature distribution at $t = 0$ is $u(x, 0) = f(x)$, $0 \leq x \leq \ell$. Determine the temperature distribution in the bar at some subsequent time $t > 0$.

Solution

The boundary value problem is to solve the heat conduction equation

$$u_t = \alpha u_{xx} \tag{5.103}$$

subject to the boundary conditions

$$\left.\begin{array}{ll} x = 0: & u(0,t) = 0 \\ x = \ell: & u(\ell,t) = 0 \end{array}\right\} \quad \text{for all } t > 0. \tag{5.104}$$

and the initial condition

$$u(x,0) = f(x) \quad , \quad 0 \le x \le \ell. \tag{5.105}$$

By the method of separation of variables,

$$u(x,t) = X(x)T(t) \tag{5.106}$$

and substituting this solution into (5.104) and after separating variables, we obtain

$$T' - k\alpha T = 0 \tag{5.107}$$

$$X'' - kX = 0. \tag{5.108}$$

where k is a separation constant.

Assuming that we need to consider only physical solutions, there are three cases to investigate:

$$\begin{array}{ll} \text{Case (I)}: & k = 0 \\ \text{Case (II)}: & k > 0 \\ \text{Case (III)}: & k < 0. \end{array}$$

Case (I)

If $k = 0$, the equations and their solutions are

$$\begin{array}{ll} T' = 0 & X'' = 0 \\ T = A & X = Cx + D \end{array}$$

and hence

$$u(x,t) = A(Cx + D).$$

This solution cannot describe a damped oscillatory solution. Also, applying the boundary conditions it can be seen that we get a trivial solution $u(x,t) = 0$. Hence this solution should be rejected.

Case (II)

If $k = \lambda^2 > 0$, the two diferential equations and their solutions are

$$\begin{array}{ll} T' - \lambda^2 \alpha T = 0 & X'' - \lambda^2 X = 0 \\ T = Ae^{\lambda^2 \alpha t} & X = Ce^{\lambda x} + De^{-\lambda x} \end{array}$$

and, hence

$$u(x,t) = (Ce^{\lambda x} + De^{-\lambda x})(Ae^{\lambda^2 \alpha t}).$$

This solution also cannot describe damped oscillatory motion of the system because it is not periodic with respect to x and not damped with respect to the time t. Thus on the physical grounds the case $k = \lambda^2$ must be rejected. Also, applying the boundary conditions, it can be easily seen that we get a trivial solution as before.

Case (III)

If $k = -\lambda^2 < 0$, for real λ, the two differential equations and their solutions are

$$\begin{array}{ll} T' + \lambda^2 \alpha T = 0 & X'' + \lambda^2 X = 0 \\ T = Ae^{-\lambda^2 \alpha t} & X = C\cos\lambda x + D\sin\lambda x \end{array}$$

and hence,

$$\begin{aligned} u(x,t) &= X(x)T(t) \\ &= (C\cos\lambda x + D\sin\lambda x)Ae^{-\lambda^2 \alpha t}. \end{aligned} \tag{5.109}$$

This solution is clearly a damped oscillatory solution. Because it is damped with respect to the time and oscillatory with respect to the x coordinates. The solution is of periodic of period $\frac{2\pi}{\lambda}$ or frequency $\frac{\lambda}{2\pi}$ cycles per unit length. This solution corresponds to the physics of the problem. In the following we shall determine the value of λ and the constants A, B and C.

Now the boundary condition at $x = 0$ yields

$$u(0,t) = 0 = C(Ae^{-\lambda^2 \alpha t}).$$

This condition can obviously be satisfied if either $A = 0$ or $C = 0$. $A = 0$ leads to the trivial solution $u(x,t) = 0$ in which we are not interested. Hence, we are led to the alternative that $C = 0$ which reduces eqn (5.110) to the form

$$u(x,t) = (D\sin\lambda x)(Ae^{-\lambda^2 \alpha t}). \tag{5.110}$$

The second boundary condition at $x = \ell$ requires that

$$u(\ell,t) = 0 = (D\sin\lambda\ell)(Ae^{-\lambda^2 \alpha t}).$$

As before, we reject the possibility that $A = 0$. Likewise D cannot equal zero because it will result in a trivial solution. The only alternative is that $\sin\lambda\ell = 0$ or, $\lambda\ell = n\pi$. Therefore, $\lambda_n = \frac{n\pi}{\ell}$, $n = 1, 2, 3, \cdots$.

As before, λ_n is defined to be the eigenvalues and the function $u_n(x,t)$ corresponding to any λ_n is called the eigenfunction. Thus, eqn (5.111) yields

$$\begin{aligned} u_n(x,t) &= B_n \sin(\lambda_n x) e^{-\lambda_n^2 \alpha t} \\ &= B_n \sin(\frac{n\pi x}{\ell}) e^{-\frac{n^2\pi^2\alpha}{\ell^2}t} \\ &\quad \text{for } n = 1, 2, 3, \cdots \end{aligned}$$

Now through superposition of all the linear solutions, we get the general solution,

$$\begin{aligned} u(x,t) &= \sum_{n=1}^{\infty} u_n(x,t) \\ &= \sum_{n=1}^{\infty} B_n \sin(\frac{n\pi x}{\ell}) e^{-\frac{n^2\pi^2\alpha}{\ell^2}t}. \end{aligned} \tag{5.111}$$

This is the Fourier half-range sine series and B_n is the Fourier coefficient and can be determined from the given initial condition.

Thus, using the initial condition (5.107), we obtain

$$u(x,0) = f(x) = \sum_{n=1}^{\infty} B_n \sin \frac{n\pi x}{\ell}. \tag{5.112}$$

Now, multiplying both sides by $\sin \frac{n\pi x}{\ell}$, and integrating with respect to x from $x = 0$ to $x = \ell$, we obtain

$$B_n = \frac{2}{\ell} \int_0^{\ell} f(x) \sin \frac{n\pi x}{\ell} dx \quad \text{for } n = 1, 2, 3, \cdots \tag{5.113}$$

Thus (5.112) is our required solution provided B_n is known from (5.114).

Example 5

Consider the heat conduction in a thin metal bar of length ℓ with insulated sides. Let us suppose that the end $x = 0$ is held at u_0 degrees celsius and the end $x = \ell$ is held at u_ℓ degrees celsius for all time $t > 0$. Let us suppose that the temperature distribution at $t = 0$ is $u(x,0) = f(x)$, $0 \leq x \leq \ell$. Determine the temperature distribution in the bar at any position at any time $t > 0$.

Solution

The mathematical problem is the following:

The heat conduction equation

$$u_t = \alpha u_{xx} \tag{5.114}$$

is to be solved subject to the boundary conditions

$$\left.\begin{aligned} x = 0: \quad & u(0,t) = u_0 \\ x = \ell: \quad & u(\ell,t) = u_\ell \end{aligned}\right\} \quad \text{for all } t > 0 \tag{5.115}$$

and the intitial condition

$$u(x,0) = f(x). \tag{5.116}$$

By the method of separation of variables, the solution can be written as

$$u(x,t) = (C\cos\lambda x + D\sin\lambda x)(Ae^{-\lambda^2\alpha t})$$

as given in eqn (5.110).

Using the nonhomogeneous boundary conditions (5.116), it can be seen that

$$\begin{aligned} C(Ae^{-\lambda^2\alpha t}) &= u_0 \\ (C\cos\lambda\ell + D\sin\lambda\ell)(Ae^{-\lambda^2\alpha t}) &= u_\ell. \end{aligned}$$

It is obvious that it will be difficult to satisfy the boundary conditions because no conclusion can be drawn unless u_0 and u_ℓ are both zero. After some thought as to how this problem might be resolved, we may be led to the idea of looking for a solution of the form

$$u(x,t) = \omega(x,t) + \phi(x). \tag{5.117}$$

Then

$$\begin{aligned} u_t &= \omega_t \\ u_{xx} &= \omega_{xx} + \phi''(x). \end{aligned}$$

Substituting these into (5.116), we obtain

$$\omega_t = \alpha[\omega_{xx} + \phi''(x)]. \tag{5.118}$$

The boundary conditions are:

$$\begin{aligned} \omega(0,t) &= u_0 - \phi(0) \\ \omega(\ell,t) &= u_\ell - \phi(\ell) \end{aligned} \tag{5.119}$$

and the initial condition is

$$\omega(x,0) = f(x) - \phi(x). \tag{5.120}$$

Now we would like to use the method of separation of variables on (5.119). To achieve this, we set

$$\phi'' = 0 \tag{5.121}$$

or $\phi = Ax + B$ so that (5.119) yields,

$$\omega_t = \alpha\omega_{xx}. \tag{5.122}$$

Now, if we choose the right hand sides of boundary condition (5.120) to be zero, then

$$\begin{aligned} \omega(0,t) &= 0 \\ \omega(\ell,t) &= 0 \end{aligned} \tag{5.123}$$

which leads to the following two boundary conditions which must satisfy $\phi(x)$:

$$\begin{aligned}\phi(0) &= u_0 \\ \phi(\ell) &= u_\ell.\end{aligned} \tag{5.124}$$

The initial condition for ω is given by (5.121). Now consider (5.122) and (5.125):

$$\begin{aligned}&x = 0: \qquad u_0 = B \\ &x = \ell: \qquad u_\ell = A\ell + B = A\ell + u_0.\end{aligned}$$

Hence, $A = (\frac{u_\ell - u_0}{\ell})$.

Therefore, the solution is

$$\phi(x) = (\frac{u_\ell - u_0}{\ell})x + u_0. \tag{5.125}$$

Now to solve the boundary value problem given by eqns (5.123), (5.124) and (5.121), we use the method of separation of variables illustrated in the previous examples.

The solution is thus

$$\omega(x,t) = (C\cos\lambda x + D\sin\lambda x)(Ae^{-\lambda^2\alpha t}). \tag{5.126}$$

The boundary conditions (5.124) lead to

$$C = 0 \quad \text{and} \quad \lambda_n = \frac{n\pi}{\ell} \quad n = 1, 2, 3, \cdots$$

so that

$$\omega_n(x,t) = B_n \sin\frac{n\pi x}{\ell} e^{-\frac{n^2\pi^2\alpha}{\ell^2}t} \quad n = 1, 2, 3, \cdots \tag{5.127}$$

To satisfy the initial condition (5.121), we first superimpose all the linear solutions to obtain

$$\begin{aligned}\omega(x,t) &= \sum_{n=1}^{\infty} \omega_n(x,t) \\ &= \sum_{n=1}^{\infty} B_n \sin\frac{n\pi x}{\ell} e^{-\frac{n^2\pi^2\alpha}{\ell^2}t}.\end{aligned} \tag{5.128}$$

Applying the initial condition (5.121), we get

$$\omega(x,0) = f(x) - \phi(x) = \sum_{n=1}^{\infty} B_n \sin\frac{n\pi x}{\ell}$$

where the Fourier coefficients B_n can be obtained by

$$B_n = \frac{2}{\ell}\int_0^\ell \{f(x) - \phi(x)\} \sin\frac{n\pi x}{\ell} dx \quad n = 1, 2, 3, \cdots \tag{5.129}$$

Hence the required solution is

$$u(x,t) = u_0 + (\frac{u_\ell - u_0}{\ell})x + \sum_{n=1}^{\infty} B_n \sin\frac{n\pi x}{\ell} e^{-\frac{n^2\pi^2\alpha}{\ell^2}t} \tag{5.130}$$

where B_n is given by (5.130).

In the particular case of a bar at zero initial temperature $f(x) = 0$, which is heated to a temperature $u = 100^o C$ applied at the end $x = 0$ for all times $t > 0$, the other end being held at $u = 0^o C$, the solution is given by

$$u(x,t) = 100(\frac{\ell - x}{\ell}) + \sum_{n=1}^{\infty} B_n \sin\frac{n\pi x}{\ell} e^{-\frac{n^2\pi^2\alpha}{\ell^2}t}$$

where

$$\begin{aligned} B_n &= \frac{2}{\ell}\int_0^\ell [-\phi(x)] \sin\frac{n\pi x}{\ell} dx \\ &= \frac{-200}{\ell^2}\int_0^\ell (\ell - x) \sin\frac{n\pi x}{\ell} dx \\ &= \frac{-200}{\ell^2}(\frac{\ell}{n\pi})\ell - 0 \\ &= -\frac{200}{n\pi} \qquad n = 1, 2, 3, \cdots \end{aligned}$$

Therefore, the solution is

$$u(x,t) = 100(\frac{\ell - x}{\ell}) - \frac{200}{\pi}\sum_{n=1}^{\infty}\frac{1}{n}\sin(\frac{n\pi x}{\ell})e^{-\frac{n^2\pi^2\alpha}{\ell^2}t}. \tag{5.131}$$

It is interesting to note that the above solution can be defined as the sum of the steady-state (the first part) plus the transient solution (second part) when $t \to \infty$, the transient part dies out leaving only the steady-state solution, i.e.,

$$\lim_{t\to\infty} u(x,t) = 100(\frac{\ell - x}{\ell}). \tag{5.132}$$

The method of separation of variables can be used to solve any intitial value problems for the heat conduction equation on $0 \le x \le \ell$ provided that the boundary conditions at $x = 0$ and $x = \ell$ are constants. If the boundary conditions are time dependent, then this method is not suitable. For such a problem the solution may be found by using the Laplace Transform and Fourier Transforms and this is discussed in the next section.

Example 6

Consider Laplace's equation,

$$u_{xx} + u_{yy} = 0 \tag{5.133}$$

where $u(x, y)$ represents the velocity potential of a fluid particle in a certain domain. For example, we need to determine $u(x, y)$ inside a unit circle, $x^2 + y^2 < 1$, when its values on the circumference $x^2 + y^2 = 1$, are prescribed.

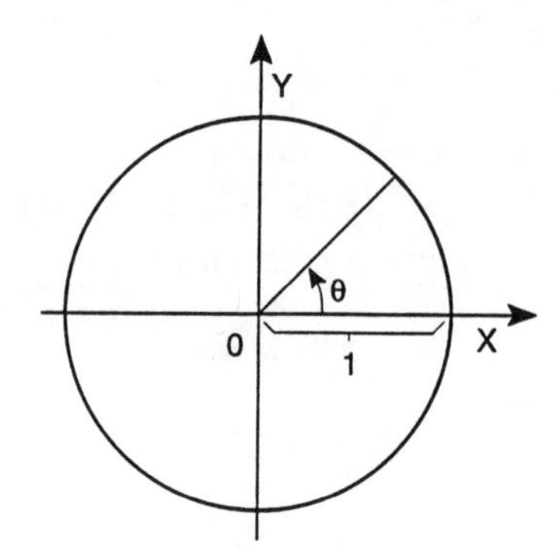

Figure 5.12. A unit circle as a domain

Solution

This boundary value problem for Laplace's equation is defined as the DIRICHLET Problem. This problem usually requires a different geometry and lesser number of arbitrary functions in the boundary conditions which distinguishes it from the initial-boundary value problems discussed above.

In obtaining the solution of (5.133), we make the following transformations, $x = r\cos\theta$ and $y = r\sin\theta$. Then Laplace's eqn (5.133) can be reduced to,

$$u_{rr} + \frac{1}{r}u_r + \frac{1}{r^2}u_{\theta\theta} = 0 \tag{5.134}$$

where $u = u(r, \theta)$.

The boundary condition for (5.134) is given by

$$u(1, \theta) = f(\theta). \tag{5.135}$$

By using the method of separation of variables, $u(r, \theta) = R(r)\Theta(\theta)$, eqn (5.134) can be written in the following two ordinary differential equations:

$$r^2R'' + rR' - n^2R = 0 \tag{5.136}$$

$$\theta'' + n^2\theta = 0 \quad n = 0, 1, 2, \cdots \tag{5.137}$$

Equation (5.136) is a Cauchy-Euler type equation and its solution can be obtained by changing the independent variable as $r = e^z$ or $\ell nr = z$, solution of which is

$$R = Ar^n + Br^{-n}. \tag{5.138}$$

Then the solution of (5.137) is

$$\Theta = C\cos n\theta + D\sin n\theta \tag{5.139}$$

and hence,

$$u = (Ar^n + Br^{-n})(C \cos n\theta + D \sin n\theta). \tag{5.140}$$

From a physical point of view, we are looking for a bounded solution and (5.140) is bounded provided $B = 0$.

Therefore,

$$u_n = (A_n \cos n\theta + B_n \sin n\theta)r^n \quad n = 0, 1, 2, 3, \cdots \tag{5.141}$$

Hence the most general bounded solution can be written as

$$\begin{aligned} u &= \sum_{n=0}^{\infty} u_n \\ &= \frac{a_0}{2} + \sum_{n=1}^{\infty} (a_n \cos n\theta + b_n \sin n\theta) r^n. \end{aligned} \tag{5.142}$$

By applying the boundary condition (5.135), the Fourier coefficients a_n and b_n can be obtained as

$$\begin{aligned} a_n &= \frac{1}{\pi} \int_0^{2\pi} f(\theta) \cos n\theta d\theta \quad , \quad n = 0, 1, 2, 3, \cdots \\ b_n &= \frac{1}{\pi} \int_0^{2\pi} f(\theta) \sin n\theta d\theta \quad , \quad n = 1, 2, 3, \cdots \end{aligned}$$

Replacing the Fourier coefficients into eqn (5.142), the solution $u(r, \theta)$ can be obtained as

$$\begin{aligned} u(r, \theta) &= \frac{1}{2\pi} \int_0^{2\pi} \{1 + 2 \sum_1^{\infty} r^n \cos n(\beta - \theta)\} f(\beta) d\beta \\ &= \frac{1}{2\pi} \int_0^{2\pi} \{\frac{1 - r^2}{1 - 2r \cos(\theta - \beta) + r^2}\} f(\beta) d\beta \end{aligned}$$

which is known as Poisson's integral equation. This equation has many practical applications.

Example 7

A transmission line of negligible resistance and conductance has its sending end at $x = 0$ and its receiving end at $x = \ell$. A constant voltage E_0 is applied at the sending end while an open circuit is maintained at the receiving end so that the current there is zero. Assuming the initial voltage and current are zero, find the voltage and current at any position x at any time $t > 0$.

Solution

The transmission line equation is

$$\frac{\partial^2 E}{\partial x^2} = LC\frac{\partial^2 E}{\partial t^2} + (RC + LG)\frac{\partial E}{\partial t} + RGE. \tag{5.143}$$

It is given that $R = 0$ and $G = 0$.

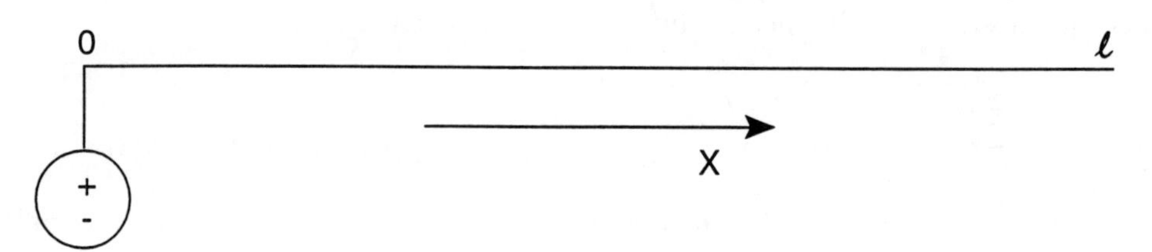

Figure 5.13. A transmission line of length ℓ

Thus the given equation reduces to

$$\frac{\partial^2 E}{\partial x^2} = LC\frac{\partial^2 E}{\partial t^2}. \tag{5.144}$$

Boundary conditions are

$$\begin{array}{ll} x = 0: & E(0,t) = E_0 \\ x = \ell: & I(\ell,t) = 0. \end{array}$$

The second boundary condition can be approximately written as

$$\frac{\partial E}{\partial x} = -L\frac{\partial I}{\partial t} \simeq 0.$$

Therefore, $\frac{\partial E(\ell,t)}{\partial x} = 0$. The initial conditions are

$$\text{at t} = 0: \quad \begin{array}{l} E(x,0) = 0 \\ I(x,0) = 0. \end{array}$$

The second initial condition may be approximately written as

$$\frac{\partial I}{\partial x} = -C\frac{\partial E}{\partial t} \simeq 0.$$

Therefore, $\frac{\partial E(x,0)}{\partial t} = 0$. Thus the mathematical problem (5.144) is to be solved with the following conditions:

Boundary conditions:

$$\begin{aligned} E(0,t) &= E_0 \\ \frac{\partial E}{\partial x}(\ell,t) &= 0. \end{aligned} \tag{5.145}$$

Initial conditions:

$$\begin{aligned} E(x,0) &= 0 \\ \frac{\partial E}{\partial t}(x,0) &= 0. \end{aligned} \tag{5.146}$$

By the method of separation of variables, the solution can be written as

$$E(x,t) = X(x)T(t)$$

such that (5.144) yields

$$\frac{X''}{X} = LC\frac{T''}{T} = -\lambda^2 \tag{5.147}$$

where λ^2 is a separation parameter. The minus sign is needed to obtain oscillatory solutions of physical interest.

Thus the two differential equations and their solutions are

$$\begin{array}{ll} X'' + \lambda^2 X = 0 & T'' + \frac{\lambda^2}{LC}T = 0 \\ X = A\cos\lambda x + B\sin\lambda x & T = C\cos\frac{\lambda t}{\sqrt{LC}} + D\sin\frac{\lambda t}{\sqrt{LC}}. \end{array}$$

and hence

$$\begin{aligned} E(x,t) &= X(x)T(t) \\ &= (A\cos\lambda x + B\sin\lambda x)(C\cos\frac{\lambda t}{\sqrt{LC}} + D\sin\frac{\lambda t}{\sqrt{LC}}). \end{aligned}$$

Using the nonhomogeneous boundary condition it can be seen that

$$A(C\cos\frac{\lambda t}{\sqrt{LC}} + D\sin\frac{\lambda t}{\sqrt{LC}}) = E_0.$$

It is obvious that it will be difficult to satisfy this boundary condition. Although it may be possible to satisfy the other homogeneous boundary condition, no conclusion can be drawn unless E_0 is zero. With the previous experience gained in the heat conduction problem, we may be led to the idea of looking for a solution of the form

$$E(x,t) = u(x,t) + \phi(x). \tag{5.148}$$

Then eqn (5.144) may be written as

$$LCu_{tt} = [u_{xx} + \phi''(x)]. \tag{5.149}$$

The boundary conditions are:

$$\begin{array}{ll} x = 0: & u(0,t) = E_0 - \phi(0) \\ x = \ell: & u_x(\ell,t) = -\phi'(\ell), \end{array} \tag{5.150}$$

and the initial conditions are:

$$\begin{array}{ll} t = 0: & u(x,0) = -\phi(x) \\ & u_t(x,0) = 0. \end{array} \tag{5.151}$$

The boundary value problem (5.149) to (5.151) can be split up into two boundary value problems in the following manner:

$$LCu_{tt} = u_{xx}. \tag{5.152}$$

Boundary conditions:

$$\begin{aligned} u(0,t) &= 0 \\ u_x(\ell,t) &= 0. \end{aligned} \tag{5.153}$$

Initial conditions:

$$\begin{aligned} u(x,0) &= -\phi(x) \\ u_t(x,0) &= 0, \end{aligned} \tag{5.154}$$

and

$$\phi''(x) = 0. \tag{5.155}$$

Boundary conditions:

$$\begin{aligned} \phi(0) &= E_0 \\ \phi'(\ell) &= 0. \end{aligned} \tag{5.156}$$

Now solving the boundary value problem (5.155) to (5.156), we obtain

$$\phi(x) = E_0. \tag{5.157}$$

Now to solve the boundary value problem given by (5.152) to (5.154), we use the method of separation of variables which yields the solution as

$$u(x,t) = (A\cos\lambda x + B\sin\lambda x)(C\cos\frac{\lambda t}{\sqrt{LC}} + D\sin\frac{\lambda t}{\sqrt{LC}}).$$

The boundary conditions (5.153) lead to

$$A = 0$$

and

$$\begin{aligned} \cos\lambda\ell &= 0 \\ \lambda_n &= \frac{(2n-1)\pi}{2\ell} \qquad n = 1, 2, 3, \cdots \end{aligned}$$

so that the solution becomes

$$\begin{aligned} u_n(x,t) &= (\sin\lambda_n x)(A_n\cos\frac{\lambda_n t}{\sqrt{LC}} + B_n\sin\frac{\lambda_n t}{\sqrt{LC}}) \\ &= (\sin\frac{(2n-1)\pi x}{2\ell})(A_n\cos\frac{(2n-1)\pi t}{2\ell\sqrt{LC}} + B_n\sin\frac{(2n-1)\pi t}{2\ell\sqrt{LC}}). \end{aligned} \tag{5.158}$$

To satisfy the initial conditions (5.154), we first superimpose all the linear solutions to obtain

$$u(x,t) = \sum_{n=1}^{\infty} (\sin \frac{(2n-1)\pi x}{2\ell})(A_n \cos \frac{(2n-1)\pi t}{2\ell\sqrt{LC}} + B_n \sin \frac{(2n-1)\pi t}{2\ell\sqrt{LC}}). \quad (5.159)$$

Applying the initial conditions, we obtain

$$u(x,0) = -E_0 = \sum_{n=1}^{\infty} A_n \sin \frac{(2n-1)\pi x}{2\ell}$$

where

$$\begin{aligned} A_n &= \frac{2}{\ell}\int_0^{\ell} (-E_0) \sin \frac{(2n-1)\pi x}{2\ell} dx \\ &= \frac{-4E_0}{(2n-1)\pi} \qquad n = 1, 2, 3, \cdots \end{aligned}$$

Similarly, applying the initial conditions $u_t(x,0) = 0$, we must have

$$0 = \sum_{n=1}^{\infty} B_n \sin \frac{(2n-1)\pi x}{2\ell} . \frac{(2n-1)\pi}{2\ell\sqrt{LC}}$$

which implies that

$$B_n = 0.$$

Hence the required solution is

$$u(x,t) = \frac{-4E_0}{\pi} \sum_{n=1}^{\infty} \frac{1}{(2n-1)} \sin \frac{(2n-1)\pi x}{2\ell} \cos \frac{(2n-1)\pi t}{2\ell\sqrt{LC}}.$$

Therefore, the voltage drop at any position and at any time is given by

$$E(x,t) = E_0[1 - \frac{4}{\pi} \sum_{n=1}^{\infty} \frac{1}{2n-1} \sin \frac{(2n-1)\pi x}{2\ell} \cos \frac{(2n-1)\pi t}{2\ell\sqrt{LC}}]. \quad (5.160)$$

5.6 Laplace and Fourier transform methods

Methods of Laplace and Fourier transforms are powerful whenever a boundary value problem is to be solved over an infinite or semi-infinite domain. The philosophy of these methods lies in the fact that by taking the transform of the given partial differential equations, we can reduce the number of independent variables by one.

As for example, we have seen how a Laplace transformation converted a linear ordinary differential equation with constant coefficient into a linear algebraic equation from which the dependent variable can be obtained easily. In a similar manner, the Laplace and Fourier transforms can be used to solve linear partial differential equations with constant coefficients. In the case of two independent variables, the transform methods convert a partial differential equation to an ordinary differential equation and provide the solution in the transform of the dependent variable.

The formal step is that the given partial differential equation with its boundary and initial conditions is transformed with respect to one of its independent variables. In doing so, we assume that the differential operation can be interchanged with the Laplace or Fourier transform operator. For example, for a function $u(x,t)$ we form the Laplace transform with respect to t,

$$\mathcal{L}\{u(x,t)\} = \int_0^\infty e^{-st}u(x,t)dt.$$

Then to obtain the Laplace transform of u_x and u_{xx}, we simply take the transforms as follows:

$$\begin{aligned}
\mathcal{L}\{u_x\} &= \int_0^\infty \frac{\partial u}{\partial x}e^{-st}dt \\
&= \frac{\partial}{\partial x}\int_0^\infty e^{-st}u\,dt \\
&= \frac{d}{dx}\mathcal{L}\{u(x,t)\}
\end{aligned}$$

$$\begin{aligned}
\mathcal{L}\{u_{xx}\} &= \int_0^\infty \frac{\partial^2}{\partial x^2}u\,e^{-st}dt \\
&= \frac{\partial^2}{\partial x^2}\int_0^\infty u\,e^{-st}dt \\
&= \frac{d}{dx^2}\mathcal{L}\{u(x,t)\}
\end{aligned}$$

and

$$\begin{aligned}
\mathcal{L}\{u_t\} &= \int_0^\infty e^{-st}\frac{\partial u}{\partial t}dt = [u\,e^{-st}]_0^\infty + s\int_0^\infty e^{-st}u\,dt \\
&= s\,\mathcal{L}\{u\} - u(x,0).
\end{aligned}$$

Similarly we can form the Fourier transform of $u(x,t)$, $u_x(x,t)$ and $u_{xx}(x,t)$, with respect to t as follows:

$$\mathcal{F}\{u(x,t)\} = \int_{-\infty}^\infty u(x,t)e^{-i\sigma t}dt$$

$$\begin{aligned}
\mathcal{F}\{u_x\} &= \int_{-\infty}^\infty \frac{\partial}{\partial x}u(x,t)e^{-i\sigma t}dt \\
&= \frac{\partial}{\partial x}\int_{-\infty}^\infty u(x,t)e^{-i\sigma t}dt \\
&= \frac{d}{dx}\mathcal{F}\{u(x,t)\}
\end{aligned}$$

$$\mathcal{F}\{u_{xx}\} = \int_{-\infty}^\infty \frac{\partial^2}{\partial x^2}u(x,t)e^{-i\sigma t}$$

$$= \frac{d^2}{dx^2}\int_{-\infty}^{\infty} u(x,t)e^{-i\sigma t}dt$$
$$= \frac{d^2}{dx^2}\mathcal{F}\{u(x,t)\}$$

and

$$\mathcal{F}\{u_t\} = \int_{-\infty}^{\infty} \frac{\partial u}{\partial t} e^{-i\sigma t}dt = [e^{-i\sigma t}u]_{-\infty}^{\infty} + i\sigma \int_{-\infty}^{\infty} u\, e^{-i\sigma t}dt.$$

If $u(\pm\infty) = 0$, then $\mathcal{F}\{u_t\} = i\sigma\mathcal{F}\{u\}$.

Note that Fourier transform is useful when the domain of definition with respect to x and t are $-\infty < x < \infty$ and $-\infty < t < \infty$. And for semi-infinite range, i.e., $0 < x < \infty$ and $0 < t < \infty$, a Laplace transform is very useful. Also, it requires that $u(x,t) \to 0$ as $x \to \infty$, and we assume that this implies that $\mathcal{L}\{u(x,t)\} \to 0$ as $x \to \infty$. In the following we shall consider some examples which will demonstrate the applications of these transforms.

Example 8

A transmission line of negligible resistance and capacitance has its sending end at $x = 0$ and its receiving end at $x = \ell$. A constant voltage E_0 is applied at the sending end while an open circuit is maintained at the receiving end, so that the current there is zero. Assuming the initial voltage and current are zero, determine the voltage and current at any position x at any time $t > 0$.

Solution

The telephone equation can be written as

$$\frac{\partial^2 E}{\partial x^2} = LC\frac{\partial^2 E}{\partial t^2} + (RC + LG)\frac{\partial E}{\partial t} + RGE.$$

It is given that $R = 0$ and $C = 0$.

Then the telephone equation reduces to

$$\frac{\partial^2 E}{\partial x^2} = LG\frac{\partial E}{\partial t}. \tag{5.161}$$

The boundary conditions are given by

$$\begin{aligned} x = 0: &\quad E(0,t) = E_0 \\ x = \ell: &\quad \tfrac{\partial E}{\partial x}(\ell,t) = 0 \end{aligned} \tag{5.162}$$

and the initial condition is

$$t = 0: \quad E(x,0) = 0. \tag{5.163}$$

We will use the Laplace transform method in this example.

Taking the Laplace transform of eqn (5.161) with respect to time t, we obtain

$$\begin{aligned}\mathcal{L}\{\frac{\partial^2 E}{\partial x^2}\} &= LG\mathcal{L}(\frac{\partial E}{\partial t}) \\ \frac{d^2}{dx^2}\mathcal{L}\{E\} &= LG[s\mathcal{L}\{E\} - E(x,0)].\end{aligned}$$

Now, using the initial condition (5.164),

$$\frac{d^2}{dx^2}\mathcal{L}\{E\} - LGs\mathcal{L}\{E\} = 0. \tag{5.164}$$

Solving this ordinary differential equation for $\mathcal{L}\{E(x,t)\}$, we find that

$$\mathcal{L}\{E(x,t)\} = A(s)e^{\sqrt{LGs}x} + B(s)e^{-\sqrt{LGs}x}. \tag{5.165}$$

To determine the integration constants $A(s)$ and $B(s)$, we first need to transform the boundary conditions (5.162). The transformed boundary conditions are:

$$\begin{aligned} x = 0: &\quad \mathcal{L}\{E(0,t)\} = \tfrac{E_0}{s} \\ x = \ell: &\quad \tfrac{d}{dx}\mathcal{L}\{E(\ell,t)\} = 0. \end{aligned} \tag{5.166}$$

The application of boundary conditions (5.166) yields

$$A(s) + B(s) = \frac{E_0}{s} \tag{5.167}$$

$$A(s)e^{\sqrt{LGs}\ell} - B(s)e^{-\sqrt{LGs}\ell} = 0. \tag{5.168}$$

Solving these two equations for the unknowns $A(s)$ and $B(s)$, we obtain

$$A(s) = \frac{E_0}{s(1 + e^{2\ell\sqrt{LGS}})}$$

and

$$B(s) = \frac{E_0 e^{2\ell\sqrt{LGs}}}{s(1 + e^{2\ell\sqrt{LGS}})}.$$

Then substituting values of $A(s)$ and $B(s)$ into (5.165) and after simplification, we obtain

$$\mathcal{L}\{E(x,t)\} = \frac{E_0 \cosh\sqrt{LGs}(\ell - x)}{s\cosh\sqrt{LGs}\ell}. \tag{5.169}$$

Using the inverse Laplace transform table, the inversion can be obtained as

$$E(x,t) = E_0[1 + \frac{4}{\pi}\sum_{n=1}^{\infty}\frac{(-1)^n}{2n-1}e^{-\frac{(2n-1)^2\pi^2}{4LG\ell^2}t}\cos\frac{(2n-1)(\ell - x)\pi}{2\ell}]. \tag{5.170}$$

To find the current, we use the following transmission line equation,

$$\begin{aligned}\frac{\partial}{\partial x}I(x,t) &= -GE(x,t)\\ &= -GE_0[1+\frac{4}{\pi}\sum_{n=1}^{\infty}\frac{(-1)^n}{2n-1}e^{-\frac{(2n-1)^2\pi^2}{4LG\ell^2}t}\cos\frac{(2n-1)\pi(\ell-x)}{2\ell}].\end{aligned}$$

Integrating partially with respect to x

$$\begin{aligned}I(x,t) &= -GE_0[x+\frac{4}{\pi}\sum_{n=1}^{\infty}\frac{(-1)^n}{2n-1}e^{-\frac{(2n-1)^2\pi^2}{4LG\ell^2}t}\sin\frac{(2n-1)\pi(\ell-x)}{2\ell}\\ &\quad\times\frac{-2\ell}{2(n-1)\pi}]+K(t).\end{aligned}$$

where $K(t)$ is an arbitrary function of t. From the boundary conditions at $x=\ell$, i.e. $I(\ell,t)=0$, we obtain $K=GE_0\ell$.

Therefore, the current is obtained as

$$\begin{aligned}I(x,t) &= GE_0(\ell-x)+\frac{8\ell GE_0}{\pi^2}\sum_{n=1}^{\infty}\frac{(-1)^n}{(2n-1)^2}e^{-\frac{(2n-1)^2\pi^2}{4LG\ell^2}t}\\ &\quad\times\sin\frac{(2n-1)\pi(\ell-x)}{2\ell}.\end{aligned} \tag{5.171}$$

To see the initial condition, $I(x,0)=0$ is satisfied, put $t=0$ in the above expression which yields

$$\begin{aligned}\ell-x &= -\frac{8\ell}{\pi^2}\sum_{n=1}^{\infty}\frac{(-1)^n}{(2n-1)^2}\sin\frac{(2n-1)\pi(\ell-x)}{2\ell}\\ &= \sum_{n=1}^{\infty}B_n\sin\frac{(2n-1)\pi(\ell-x)}{2\ell}\end{aligned}$$

where

$$\begin{aligned}B_n &= \frac{2}{\ell}\int_0^{\ell}(\ell-x)\sin\frac{(2n-1)\pi(\ell-x)}{2\ell}dx\\ &= -\frac{8\ell}{\pi^2}.\frac{(-1)^n}{(2n-1)^2}.\end{aligned}$$

It is obvious that initial condition $I(x,0)$ is satisfied.

Example 9

An infinitely long string having one end at $x=0$ is initially at rest on the x-axis. At $t=0$, the end $x=0$ begins to move along the y-axis in a manner described by $y(0,t)=a\cos\sigma t$. Find the displacement $y(x,t)$ of the string at any point at any time.

Solution

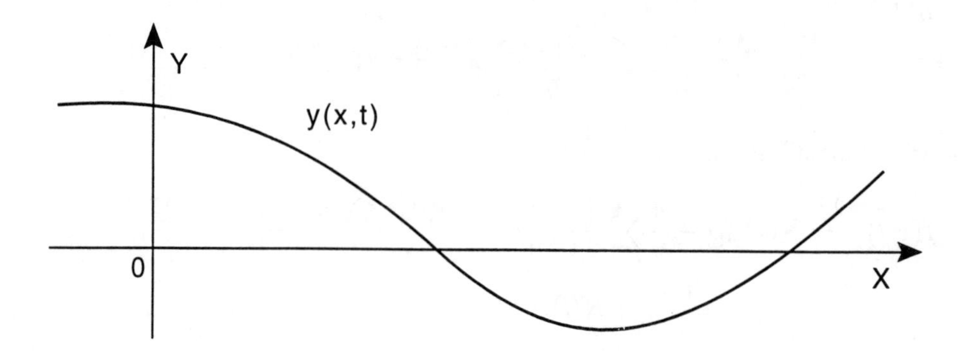

Figure 5.14. A vibrating string

The partial differential equation to be solved is the one-dimensional wave equation

$$y_{tt} = c^2 y_{xx} \tag{5.172}$$

subject to the boundary conditions:

$$x = 0: \qquad y(0,t) = a\cos\sigma t \tag{5.173}$$

$$y(x,t) \text{ bounded as } x \to \infty \tag{5.174}$$

and the initial conditions are

$$y(x,0) = 0 \tag{5.175}$$

$$y_t(x,0) = 0. \tag{5.176}$$

Taking the Laplace transform of (5.172) with respect to t, we obtain

$$\begin{aligned} s^2\mathcal{L}\{y(x,t)\} &- sy(x,0) - y_t(x,0) = c^2\mathcal{L}\{y_{xx}\} \\ &= c^2\frac{d^2}{dx^2}\mathcal{L}\{y(x,t)\}. \end{aligned}$$

Using the initial conditions (5.175) and (5.176)

$$\frac{d^2}{dx^2}\mathcal{L}\{y(x,t)\} - (\frac{s}{c})^2\mathcal{L}\{y(x,t)\} = 0. \tag{5.177}$$

Solving this ordinary differential equation for $\mathcal{L}\{y(x,t)\}$, we find that

$$\mathcal{L}\{y(x,t)\} = A(s)e^{(\frac{s}{c})x} + B(s)e^{-(\frac{s}{c})x}. \tag{5.178}$$

To determine $A(s)$ and $B(s)$, we observe first that if $y(x,t)$ remains finite as $x \to \infty$, so must $\mathcal{L}\{y(x,t)\}$. Hence, $A(s)$ must be zero. Therefore,

$$\mathcal{L}\{y(x,t)\} = B(s)e^{-(\frac{s}{c})x}. \tag{5.179}$$

Now Laplace transform of (5.173) yields

$$\mathcal{L}\{y(0,t)\} = \frac{a\,s}{s^2+\sigma^2}. \tag{5.180}$$

Therefore, applying this condition to (5.179), we obtain

$$\mathcal{L}\{y(x,t)\} = \frac{a\,s}{s^2+\sigma^2}e^{-(\frac{s}{c})x}.$$

The inverse of this can be found immediately by using the standard Laplace transform table which yields

$$y(x,t) = a\cos\sigma(t-\frac{x}{c})u(t-\frac{x}{c}) \tag{5.181}$$

where $u(t)$ represents the unit step function.

This result represents a wave travelling to the right along the string with velocity c. The physical meaning of this result is that a point x of the string stays at rest until the time $t = x/c$. Thereafter, it undergoes motion identical with that of the end $x = 0$ but lags behind it in time by the amount x/c.

Example 10

A semi-infinite transmission line of negligible inductance and conductance per unit length has its voltage and current equal to zero. At $t = 0$, a constant voltage E_0 is applied at the sending end, $x = 0$. Find the voltage and current at any point at any subsequent time.

Solution

The telephone equation is

$$\frac{\partial^2 E}{\partial x^2} = LC\frac{\partial^2 E}{\partial t^2} + (RC+LG)\frac{\partial E}{\partial t} + RGE.$$

It is given that $L = 0$ and $G = 0$.

Thus the above equation reduces to

$$\frac{\partial^2 E}{\partial x^2} = RC\frac{\partial E}{\partial t}. \tag{5.182}$$

The boundary conditions are:

$$x = 0: \qquad E(0,t) = E_0 \tag{5.183}$$

$$x \to \infty: \qquad |E(x,t)| < M \tag{5.184}$$

and the initial condition is:

$$E(x,0) = 0. \tag{5.185}$$

Taking the Laplace transform of (5.182) with respect to time t, we obtain

$$\begin{aligned} \mathcal{L}\{E_{xx}\} &= RC\mathcal{L}\{E_t\} \\ \frac{d^2}{dx^2}\mathcal{L}\{E(x,t)\} &= RC[s\mathcal{L}\{E(x,t)\} - E(x,0)]. \end{aligned}$$

Now using the initial condition (5.185)

$$\frac{d^2}{dx^2}\mathcal{L}\{E(x,t)\} - RCs\mathcal{L}\{E(x,t)\} = 0. \tag{5.186}$$

Solving this ordinary differential equation for $\mathcal{L}\{E(x,t)\}$, we obtain

$$\mathcal{L}\{E(x,t)\} = A(s)e^{\sqrt{RCs}x} + B(s)e^{-\sqrt{RCs}x}. \tag{5.187}$$

From the boundedness condition on $E(x,t)$ as $x \to \infty$, we have that $A(s) = 0$. Then

$$\mathcal{L}\{E(x,t)\} = B(s)e^{-\sqrt{RCs}x}. \tag{5.188}$$

Now the Laplace transform of (5.183) yields

$$\mathcal{L}\{E(0,t)\} = \frac{E_0}{s}. \tag{5.189}$$

Therefore, applying this condition to (5.188), we obtain

$$\mathcal{L}\{E(x,t)\} = \frac{E_0}{s}e^{-\sqrt{RCs}x}. \tag{5.190}$$

The inverse of this can be found at once from the standard table, and is given by

$$E(x,t) = E_0 erfc(x\sqrt{RC}/2\sqrt{t}) \tag{5.191}$$

where $erfc(\theta)$ is defined as

$$\begin{aligned} erfc(\theta) &= 1 - erf(\theta) \\ &= 1 - \frac{2}{\sqrt{\pi}}\int_0^\theta e^{-\eta^2}d\eta \end{aligned}$$

in which $erf(\theta) = \frac{2}{\sqrt{\pi}}\int_0^\theta e^{-\eta^2}d\eta$ is called the error-function, a tabulated function which can be found in most handbooks of mathematical tables.

To find the current, we use the transmission line equation,

$$\frac{\partial E}{\partial x} = -RI(x,t).$$

Therefore,

$$\begin{aligned} I(x,t) &= -\frac{1}{R}\frac{\partial E}{\partial x} \\ &= -\frac{E_0}{R}[\frac{\partial}{\partial x}\{1 - \frac{2}{\sqrt{\pi}}\int_0^{\frac{x\sqrt{RC}}{2\sqrt{t}}} e^{-\eta^2}d\eta\}] \\ &= E_0\sqrt{\frac{C}{Rt\pi}}e^{-\frac{RCx^2}{4t}}. \end{aligned} \tag{5.192}$$

Note

In evaluating the result (5.192), we have used the Leibnitz rule of differentiating a function under the integral sign which is

$$\begin{aligned}
&\frac{d}{dt}\int_{a(t)}^{b(t)} f(x,t)dx \\
&= \int_{a(t)}^{b(t)} \frac{\partial f}{\partial t}(x,t)dx + f[b(t),t]\frac{db(t)}{dt} - f[a(t),t]\frac{da(t)}{dt}.
\end{aligned} \tag{5.193}$$

Example 11

An infinitely long string extending from $-\infty < x < \infty$ under tension is displaced into the curve $y = f(x)$ and lets go from rest. Find the displacement $y(x,t)$ of the string at any point at any subsequent time.

Solution

The partial differential equation to be solved is the one-dimensional wave equation

$$y_{tt} = c^2 y_{xx} \qquad -\infty < x < \infty \quad , \quad t > 0. \tag{5.194}$$

subject to the initial conditions

$$\left.\begin{aligned} y(x,0) &= f(x) \\ y_t(x,0) &= 0 \end{aligned}\right\} \quad \text{at } t = 0 \quad -\infty < x < \infty. \tag{5.195}$$

We write the complex Fourier transform with respect to x, as

$$\mathcal{F}\{y(x,t)\} = \int_{-\infty}^{\infty} e^{-i\sigma x} y(x,t)dx. \tag{5.196}$$

The Fourier transform of y_{xx} is

$$\begin{aligned}
\mathcal{F}\{y_{xx}\} &= \int_{-\infty}^{\infty} e^{-i\sigma x} y_{xx} dx \\
&= [y_x e^{-i\sigma x}]_{-\infty}^{\infty} + i\sigma \int_{-\infty}^{\infty} e^{-i\sigma x} y_x dx \\
&= (i\sigma)[y\, e^{-i\sigma x}]_{-\infty}^{\infty} + (i\sigma)^2 \int_{-\infty}^{\infty} e^{-i\sigma x} y\, dx \\
&= -\sigma^2 \int_{-\infty}^{\infty} e^{-i\sigma x} y(x,t)dx \\
&= -\sigma^2 \mathcal{F}\{y(x,t)\}.
\end{aligned} \tag{5.197}$$

In evaluating this we have made use of the fact that both y and y_x are zero at both ends of the string.

Then taking the Fourier transform of (5.194) with respect to x, we obtain

$$\mathcal{F}\{y_{tt}\} = c^2 \mathcal{F}\{y_{xx}\}$$

or

$$\frac{d^2}{dt^2}\mathcal{F}\{y(x,t)\} = -(c\sigma)^2\mathcal{F}\{y(x,t)\}. \tag{5.198}$$

The transforms of initial conditions (5.195) are then

$$\mathcal{F}\{y(x',0)\} = \mathcal{F}\{f(x')\} = \int_{-\infty}^{\infty} e^{-i\sigma x'} f(x')dx' \tag{5.199}$$

and

$$\mathcal{F}\{y_t(x',0)\} = \frac{d}{dt}\mathcal{F}\{y(x',0)\} = 0. \tag{5.200}$$

The dashes have been introduced in the integral to avoid later confusion with x. The solution of (5.198) is given by

$$\mathcal{F}\{y(x,t)\} = A(\sigma)\cos(c\sigma t) + B(\sigma)\sin(c\sigma t). \tag{5.201}$$

Applying the conditions (5.199) and (5.200), we obtain

$$B(\sigma) = 0 \quad , \quad A(\sigma) = \mathcal{F}\{f(x)\}$$

and hence

$$\mathcal{F}\{y(x,t)\} = \mathcal{F}\{f(x')\}\cos(c\sigma t). \tag{5.202}$$

Taking the inversion of the Fourier transform (5.202), yields

$$y(x,t) = \frac{1}{2\pi}\int_{-\infty}^{\infty}[\mathcal{F}\{f(x')\}\cos(c\sigma t)]e^{i\sigma x}d\sigma. \tag{5.203}$$

Using the expression for $\mathcal{F}\{f(x')\}$ given in (5.199), this can be written as

$$\begin{aligned}
y(x,t) &= \frac{1}{2\pi}\int_{-\infty}^{\infty}\{\int_{-\infty}^{\infty} e^{-i\sigma x'} f(x')dx' \cos(c\sigma t)\}e^{i\sigma x}d\sigma \\
&= \frac{1}{2\pi}\int_{-\infty}^{\infty}(\frac{e^{ic\sigma t}+e^{-ic\sigma t}}{2})e^{i\sigma x}d\sigma \int_{-\infty}^{\infty} f(x')e^{-i\sigma x'}dx' \\
&= \frac{1}{4\pi}\int_{-\infty}^{\infty}[e^{(x-ct)i\sigma}+e^{(x+ct)i\sigma}]d\sigma \int_{-\infty}^{\infty} f(x')e^{-i\sigma x'}dx' \\
&= \frac{1}{2}[\{\frac{1}{2\pi}\int_{-\infty}^{\infty} e^{(x-ct)i\sigma}d\sigma\}\{\int_{-\infty}^{\infty} f(x')e^{-i\sigma x'}dx'\} \\
&\quad +\{\frac{1}{2\pi}\int_{-\infty}^{\infty} e^{(x+ct)i\sigma}d\sigma\}\{\int_{-\infty}^{\infty} f(x')e^{-i\sigma x'}dx'\}] \\
&= \frac{1}{2}[f(x-ct)+f(x+ct)].
\end{aligned} \tag{5.204}$$

In obtaining this result we have used the fact that the Fourier integral representation of the function $f(x)$ can be written as

$$f(x) = \frac{1}{2\pi}\int_{-\infty}^{\infty} e^{i\sigma x}d\sigma \int_{-\infty}^{\infty} f(x')e^{-i\sigma x'}dx'. \tag{5.205}$$

Result (5.204) can be recognized to be D'Alembert's solution to the wave equation.

Example 12

Use the complex form of the Fourier transform to show that

$$u(x,t) = \frac{1}{2\sqrt{\pi t \alpha}} \int_{-\infty}^{\infty} f(x') e^{-\frac{(x-x')^2}{4t\alpha}} dx' \tag{5.206}$$

is the solution of the boundary value problem governing the heat conduction in a very long metal bar which extends from $-\infty < x < \infty$.

$$\begin{aligned} u_t &= \alpha\, u_{xx}, \qquad -\infty < x < \infty & (5.207) \\ u(x,0) &= f(x), \qquad \text{when } t = 0. & (5.208) \end{aligned}$$

Solution

We write the complex Fourier transform with respect to x as

$$\mathcal{F}\{u(x,t)\} = \int_{-\infty}^{\infty} e^{-i\sigma x} u(x,t) dx.$$

Then

$$\begin{aligned} \mathcal{F}\{u_{xx}\} &= \int_{-\infty}^{\infty} e^{-i\sigma x} u_{xx} dx \\ &= -\sigma^2 \mathcal{F}\{u(x,t)\}. \end{aligned}$$

Hence, taking the complex Fourier transform of (5.207) with respect to x, we obtain

$$\mathcal{F}\{u_t\} = \alpha \mathcal{F}\{u_{xx}\}$$

or

$$\frac{d}{dt}\mathcal{F}\{u(x,t)\} = -\sigma^2 \alpha \mathcal{F}\{u(x,t)\}. \tag{5.209}$$

The transform of the initial condition (5.208) is then

$$\mathcal{F}\{u(x',0)\} = \mathcal{F}\{f(x')\} = \int_{-\infty}^{\infty} e^{-i\sigma x'} f(x') dx'. \tag{5.210}$$

The dashes have been introduced in the integral to avoid later confusion with x.

Now, solving eqn (5.209), we have

$$\mathcal{F}\{u(x,t)\} = A(\sigma) e^{-\alpha\sigma^2 t}. \tag{5.211}$$

Using the initial condition (5.210) in (5.211),

$$A(\sigma) = \mathcal{F}\{f(x')\}$$

and hence

$$\mathcal{F}\{u(x,t)\} = \mathcal{F}\{f(x')\} e^{-\alpha\sigma^2 t}. \tag{5.212}$$

Fourier inverse of (5.212) yields

$$u(x,t) = \frac{1}{2\pi}\int_{-\infty}^{\infty}[\mathcal{F}\{f(x')\}e^{-\alpha\sigma^2 t}]e^{i\sigma x}d\sigma$$

which can be written as

$$\begin{aligned} u(x,t) &= \frac{1}{2\pi}\int_{-\infty}^{\infty}[\int_{-\infty}^{\infty} e^{-i\sigma x'}f(x')dx'e^{-\alpha\sigma^2 t}]e^{i\sigma x}d\sigma \\ &= \frac{1}{2\pi}\int_{-\infty}^{\infty} f(x')[\int_{-\infty}^{\infty} e^{i\sigma x - i\sigma x' - \alpha\sigma^2 t}d\sigma]dx' \\ &= \frac{1}{2\pi}\int_{-\infty}^{\infty} f(x')[\int_{-\infty}^{\infty} e^{-\alpha t(\sigma - \frac{i(x-x')}{2\alpha t})^2} e^{-\frac{(x-x')^2}{4\alpha t}}d\sigma]dx' \\ &= \frac{1}{2\pi}\int_{-\infty}^{\infty} f(x')e^{-\frac{(x-x')^2}{4\alpha t}}[\int_{-\infty}^{\infty} e^{-\alpha t(\sigma - \frac{i(x-x')}{2\alpha t})^2}d\sigma]dx'. \end{aligned}$$

Now consider

$$I = \int_{-\infty}^{\infty} e^{-\alpha t(\sigma - \frac{i(x-x')}{2\alpha t})^2}d\sigma.$$

In this integral, put $\sigma - \frac{i(x-x')}{2\alpha t} = z$, such that $d\sigma = dz$. Substituting $\alpha t z^2 = \theta$ we obtain,

$$\begin{aligned} I &= 2\int_0^{\infty}\frac{1}{2\sqrt{\alpha t}}e^{-\theta}\theta^{-\frac{1}{2}}d\theta = \frac{1}{\sqrt{\alpha t}}\Gamma(\frac{1}{2}) \\ &= \frac{\sqrt{\pi}}{\sqrt{\alpha t}}. \end{aligned}$$

Hence, the solution is

$$u(x,t) = \frac{1}{2\sqrt{\pi\alpha t}}\int_{-\infty}^{\infty} f(x')e^{-\frac{(x-x')^2}{4\alpha t}}dx' \qquad (5.213)$$

which is therefore the required result.

Sine and cosine Fourier transforms

The sine and cosine Fourier transforms can be employed when the range of the variable selected extends from 0 to ∞. The choice of sine or cosine transform is decided by the form of the boundary conditions at the lower limit of the variable selected. As for example, consider the one-dimensional heat conduction in a semi-infinite metal bar $0 \leq x \leq \infty$. The partial differential equation which governs this process can be written as

$$u_t = \alpha u_{xx}, \qquad \begin{cases} 0 \leq x < \infty \\ t > 0. \end{cases}$$

Now, if a sine transform is being used, we multiply the differential equation by $\sin\sigma x$ and integrate with respect to x from 0 to ∞. Thus the sine transform of the term u_{xx} can be written as follows

$$\int_0^{\infty}\sin\sigma x(u_{xx})dx = [(u_x)\sin\sigma x]_{x=0}^{\infty} - \sigma\int_0^{\infty}\cos\sigma x(u_x)dx.$$

The first term of the right hand side vanishes at the lower limit through the sine term. It vanishes also at the upper limit if $u(x,t)$ is such that $u_x \to 0$ as $x \to \infty$. This is usually the case in physical problems.

Then a second integration gives

$$\int_0^\infty \sin\sigma x(u_{xx})dx = -\sigma[\cos\sigma x u(x,t)]_{x=0}^\infty - \sigma^2 \int_0^\infty u \sin\sigma x dx.$$

Now, if we assume that $u(x,t) \to 0$ as $x \to \infty$, then

$$\int_0^\infty \sin\sigma x(u_{xx})dx = \sigma[u(0,t)] - \sigma^2 \mathcal{F}_s\{u\}$$

where

$$\mathcal{F}_s\{u\} = \int_0^\infty u(x,t)\sin\sigma x dx$$

and $u(0,t)$ is the value of u at $x = 0$.

In a similar manner, the cosine transform of the term u_{xx} may be written as

$$\begin{aligned}\int_0^\infty \cos\sigma x(u_{xx})dx &= [\cos\sigma x u_x]_0^\infty + \sigma\int_0^\infty \sin\sigma x u_x dx\\ &= -u_x(0,t) + \sigma[\sin\sigma x u]_0^\infty - \sigma^2\int_0^\infty u\cos\sigma x dx\\ &= -u_x(0,t) - \sigma^2\mathcal{F}_c\{u\}\end{aligned}$$

where

$$\mathcal{F}_c\{u\} = \int_0^\infty u(x,t)\cos\sigma x dx.$$

In evaluating these integrals we have assumed that $u(x,t)$ and $u_x(x,t)$ both approach zero as $x \to \infty$.

Thus the successful use of the sine and cosine Fourier transform lies in the fact that in removing the term u_{xx} from the differential equation requires a knowledge of $u(0,t)$ and $u_x(0,t)$ at $x = 0$ for sine and cosine transforms, respectively. Thus with this knowledge in mind we solve the following heat conduction problem.

Example 13

If $u(x,t)$ is the temperature at time t and α the thermal diffusivity of the metal bar, find $u(x,t)$ from the partial differential equation

$$u_t = \alpha\, u_{xx}, \qquad x > 0 \quad \text{and} \quad t > 0 \tag{5.214}$$

with the boundary condition

$$u(0,t) = u_0, \qquad t > 0 \tag{5.215}$$

and the initial condition

$$u(x,0) = 0, \qquad x > 0. \tag{5.216}$$

Solution

Since the temperature $u(0,t) = u_0$ is given when $x = 0$, the sine transform

$$\bar{U} = \int_0^\infty u(x,t) \sin \sigma x dx \tag{5.217}$$

is appropriate.

Then, taking the sine transform of (5.214), we obtain

$$\int_0^\infty u_t \sin \sigma x dx = \alpha \int_0^\infty u_{xx} \sin \sigma x dx$$

or

$$\frac{d\bar{U}}{dt} = \alpha[\sigma u(0,t) - \sigma^2 \bar{U}].$$

Therefore,

$$\frac{d\bar{U}}{dt} + \alpha\sigma^2 \bar{U} = \alpha\sigma u_0 \tag{5.218}$$

which is an ordinary differential equation, the solution of which exists in the following form

$$\bar{U} = A\, e^{-\alpha\sigma^2 t} + \frac{1}{\sigma} u_0, \qquad \text{at } t = 0 \text{ and } \bar{U} = 0. \tag{5.219}$$

Therefore, we have

$$\bar{U} = \frac{u_0}{\sigma}[1 - e^{-\alpha\sigma^2 t}]. \tag{5.220}$$

The inversion formula yields

$$u(x,t) = \frac{2u_0}{\pi} \int_0^\infty (1 - e^{-\alpha\sigma^2 t}) \frac{\sin \sigma x}{\sigma} d\sigma.$$

Since

$$\int_0^\infty \frac{\sin \sigma x}{\sigma} d\sigma = \frac{\pi}{2},$$

this can be written as

$$u(x,t) = u_0[1 - \frac{2}{\pi} \int_0^\infty e^{-\sigma^2 \alpha t} \frac{\sin \sigma x}{\sigma} d\sigma].$$

The infinite integral can be expressed in terms of the error function by using the well-known result

$$\int_0^\infty e^{-\lambda^2 x^2} \cos 2\mu x dx = (\frac{\sqrt{\pi}}{2\lambda}) e^{-\frac{\mu^2}{\lambda^2}}$$

and integrating with respect to μ from 0 to μ. Therefore,

$$u(x,t) = u_0 erfc(\frac{x}{2\sqrt{\alpha t}}). \tag{5.221}$$

5.7 Similarity technique

We saw in the previous sections the solution techniques for linear partial differential equations arising in engineering fields. The classical nature of solutions such as the method of characteristics, separation of variables, Laplace transforms, Fourier transforms, etc. can be very successfully applied to a wide class of different problems when the partial differential equations are linear. However, there are a number of problems in which solutions cannot be found by these methods. This is particularly true when the partial differential equations are nonlinear. As for example, the system of equations of motion for a viscous fluid flow presents formidable obstacles to classical analyses because the equations are nonlinear. In that situation, solutions are obtained by employing transformations that reduce a system of partial differential equations in two independent variables to a system of ordinary differential equations in one independent variable. These solutions are usually called "similarity solutions." For detailed information about this technique readers should turn their attention to standard textbooks. As for example, A.G. Hansen gives a lucid description about this technique in *Similarity Analysis of Boundary Value Problems in Engineering*, Prentice Hall, 1964, Englewood Cliffs. In the following, we shall demonstrate this powerful technique to obtain solutions for two classical problems connected with heat flow and viscous flow in which the first one is linear and the second one is nonlinear.

Example 14: Similarity solution of the diffusion equation

Consider the heat conduction in a semi-infinite medium. The partial differential equation to be resolved is

$$u_t = \alpha\, u_{xx}, \qquad 0 < x < \infty \quad \text{and} \quad t > 0 \tag{5.222}$$

subject to the boundary conditions

$$x = 0: \qquad u(0,t) = u_0 \tag{5.223}$$

$$x \to \infty: \qquad u(x,t) \to 0 \tag{5.224}$$

and the initial condition

$$t = 0: \quad u(x,t) = 0, \qquad x > 0. \tag{5.225}$$

Solution

Depending on the physical significance attached to the variables, this mathematical model described in (5.222) to (5.225) can be applied to a variety of fluid mechanics problems such as fluid flow parallel to a plate suddenly set into motion.

We saw the solution to this problem by using classical methods in the previous sections. Here we shall demonstrate the similarity technique using this diffusion problem.

Let us define a new independent variable called the similarity variable, η, in the following manner

$$\eta = x\, b(t) \tag{5.226}$$

where $b(t)$ is unknown.

Then we can directly see that when $x = 0$, $\eta = 0$; and when $x \to \infty$ then $\eta \to \infty$ for fixed t.

It will be seen in our subsequent calculations that the form of $b(t)$ will be such that when $t = 0, \eta \to \infty$ for fixed x. Thus both the space and time boundaries can be combined to yield two boundaries, one at $\eta = 0$ and the other at $\eta \to \infty$.

Consider the dependent variable in the following manner

$$u = u_0 f(\eta).$$

Now we can obtain the values of partial derivatives of u as follows:

$$\begin{aligned} u_x &= u_0 f_\eta \eta_x = u_0 b(t) f_\eta \\ u_{xx} &= u_0 b^2(t) f_{\eta\eta} \\ u_t &= u_0 f_\eta \eta_t = u_0 x\, b'(t) f_\eta \\ &= u_0 \{\frac{b'(t)}{b(t)}\} \eta\, f_\eta. \end{aligned}$$

Then substituting these values into eqn (5.223), we obtain

$$u_0 \{\frac{b'(t)}{b(t)}\} \eta\, f_\eta = u_0\, \alpha b^2(t) f_{\eta\eta}$$

or

$$\{\frac{b'}{\alpha b^3}\} \eta\, f_\eta = f_{\eta\eta}. \tag{5.227}$$

This equation will be completely in terms of independent variable η provided we assume that

$$\frac{b'}{\alpha b^3} = c \tag{5.228}$$

where c is a constant. Now, integrating with respect to t, $-\frac{1}{2\alpha b^2} = ct +$constant. Since we will be interested in the form of $b(t)$, set the constant to be zero. Then we get, $b^2(t) = \frac{1}{-2c\alpha t}$. Now, choose $c = -2$, such that

$$b(t) = \frac{1}{2\sqrt{\alpha t}}. \tag{5.229}$$

Therefore, the similarity variable $\eta(x, t)$ can be written as

$$\eta(x, t) = \frac{x}{2\sqrt{\alpha t}}. \tag{5.230}$$

Equation (5.227) takes the form

$$f_{\eta\eta} + 2\eta\, f_\eta = 0$$

or

$$\frac{d^2 f}{d\eta^2} + 2\eta \frac{df}{d\eta} = 0 \tag{5.231}$$

which is an ordinary differential equation. The boundary conditions are:

$$\eta = 0: \qquad f = 1 \tag{5.232}$$

$$\eta \to \infty: \qquad f = 0. \tag{5.233}$$

Integrating eqn (5.231) with respect to η once, we get $f' = c_1 e^{-\eta^2}$. Again integrating,

$$f = c_1 \int_0^\eta e^{-z^2} dz + c_2 \tag{5.234}$$

where c_1 and c_2 are two constants. Using the boundary condition at $\eta = 0$ $c_2 = 1$ and using the condition when $\eta \to \infty$, we have $c_1 = -\frac{2}{\sqrt{\pi}}$.

The solution is then

$$\begin{aligned} f &= \frac{u}{u_0} = 1 - \frac{2}{\sqrt{\pi}} \int_0^\eta e^{-z^2} dz = 1 - erf(\eta) \\ &= erfc(\eta). \end{aligned}$$

Therefore,

$$u = u_0 erfc(\frac{x}{2\sqrt{\alpha t}}). \tag{5.235}$$

Note that the error function is defined as

$$erf(\eta) = \frac{2}{\sqrt{\pi}} \int_0^\eta e^{-z^2} dz$$

and the error complementary function is defined as

$$erfc(\eta) = 1 - erf(\eta).$$

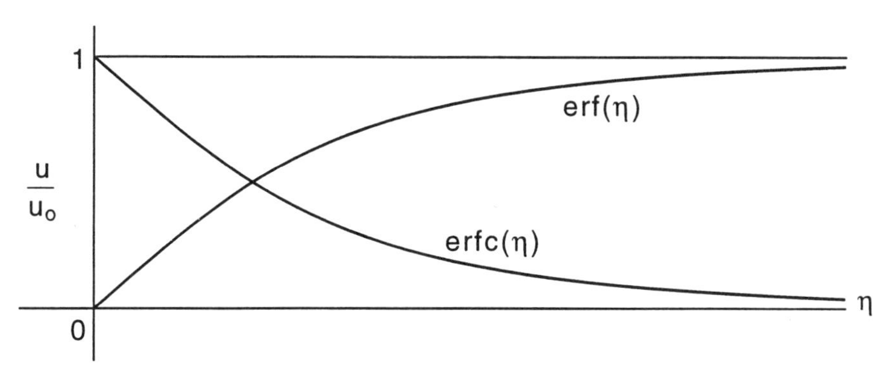

Figure 5.15. A similarity solution

Example 15: Boundary-layer equations over a flat plate

A semi-infinite flat plate is immersed in the steady uniform stream of an incompressible fluid with a viscosity ν. This flow behaviour about the flat plate can be predicted by using the boundary-layer flow equations in two dimensions.

The mathematical theory of boundary layer flow was first discussed by Ludwig Prandtl in 1904. Using the flow configuration illustrated by Fig. 5.16, the boundary layer equations in two dimensions can be written as
Momentum:

$$u\, u_x + v\, u_y = \nu\, u_{yy} \tag{5.236}$$

Continuity:

$$u_x + v_y = 0. \tag{5.237}$$

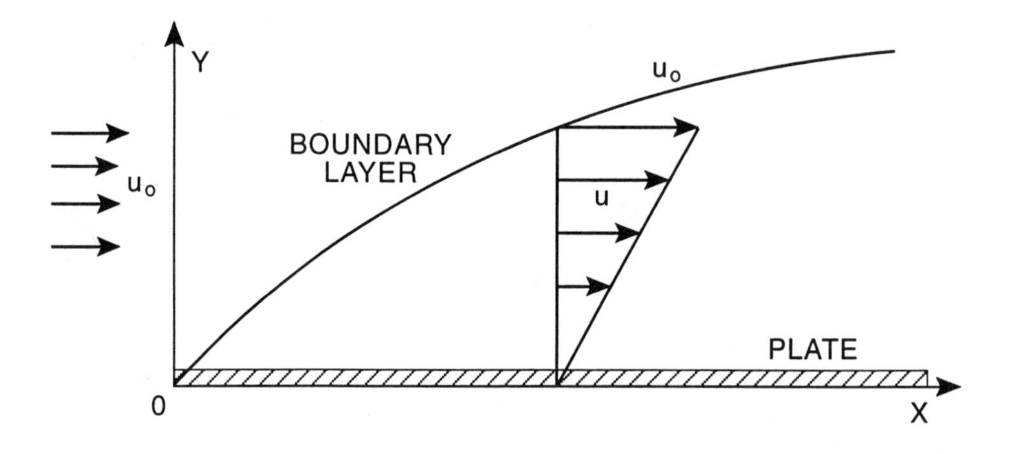

Figure 5.16. Flat plate immersed in a uniform stream

The boundary conditions are

$$\begin{array}{l} u = 0 \\ v = 0 \end{array} \text{ at } y = 0 \tag{5.238}$$

$$\lim_{y \to \infty} u = u_0. \tag{5.239}$$

The solution of the boundary-layer equation for uniform flow over a flat plate was obtained by H. Blassius (1908). Here we shall show the technique briefly by using the concept of stream function.

Solution

From (5.237) we see that there exists a function $\psi(x, y)$ called the stream function such that

$$u = \frac{\partial \psi}{\partial y}$$
$$v = -\frac{\partial \psi}{\partial x}. \tag{5.240}$$

Then the momentum equation (5.236) reduces to

$$\psi_y \psi_{xy} - \psi_x \psi_{yy} = \nu \psi_{yyy}. \tag{5.241}$$

Now, consider a new independent variable $\eta(x, y)$ defined by

$$\eta = y\, b(x) \tag{5.242}$$

and a dependent variable defined by

$$\psi(x, y) = u_0 g(x) f(\eta). \tag{5.243}$$

Then,

$$\begin{aligned} -v = \psi_x &= u_0[g'(x)f(\eta) + g(x)f'(\eta)\eta_x] \\ &= u_0[g'(x)f(\eta) + \frac{b'(x)}{b} g(x)\eta f'(\eta)] \end{aligned}$$

$$\begin{aligned} \psi_{xy} &= u_0[g'(x)f'(\eta)\eta_y + \frac{b'}{b} g(x)f'(\eta) + \frac{b'}{b} g(x)\eta f''(\eta) b] \\ &= u_0[b\, g' f'(\eta) + b'\, g\, f'(\eta) + b' g\eta f''(\eta)] \end{aligned}$$

$$u = \psi_y = u_0 g\, f'(\eta)\eta_y = u_0 g\, b\, f'(\eta)$$

$$\psi_{yy} = u_0 b^2 g f''(\eta)$$

$$\psi_{yyy} = u_0 b^3 g f'''(\eta).$$

Substituting these values into (5.241) and simplifying, we obtain

$$\frac{u_0}{\nu\, b^2}[bg' + b'g] f'^2 - \frac{u_0 g'}{\nu\, b} f f'' = f'''.$$

This equation will be defined in terms of the independent variable η, provided $\frac{u_0 g'}{\nu\, b} = C_1 = \text{constant}$ and $\frac{u_0}{\nu}[\frac{g'}{b} + \frac{gb'}{b^2}] = C_2 = \text{constant}$. Let us choose $C_2 = 0$, then $d(gb) = 0$ and upon integration $gb = C_3$. By choosing $C_3 = 1$ and $b = \frac{1}{g}$, we have $\frac{u_0}{\nu} gg' = C_1$, or, $gg' = \frac{\nu}{u_0} C_1$ which upon integration yields $g^2 = \frac{2\nu x}{u_0} C_1$. Without loss of generality, choose $C_1 = 1$. Thus, $g = \sqrt{\frac{2\nu x}{u_0}}$.

Hence the similarity variable η is given by

$$\eta = y\sqrt{\frac{u_0}{2\nu x}} \tag{5.244}$$

and

$$\psi(x,y) = u_0\sqrt{\frac{2\nu x}{u_0}}f(\eta) \tag{5.245}$$

and the differential eqn (5.241) takes the following simple form

$$f''' + ff'' = 0 \tag{5.246}$$

which is the ordinary differential equation to be solved subject to the following boundary conditions:

As $y = 0$ implies $\eta = 0$ and as $\lim_{y\to\infty} \eta = \infty$, we can transform the boundary conditions as follows:

$$\begin{aligned} u(x,0) &= 0 \to f'(0) = 0 && (5.247)\\ v(x,0) &= 0 \to f(0) = 0 && (5.248)\\ \lim_{y\to\infty} u(x,y) &= u_0 \to \lim_{\eta\to\infty} f'(0) = 1. && (5.249) \end{aligned}$$

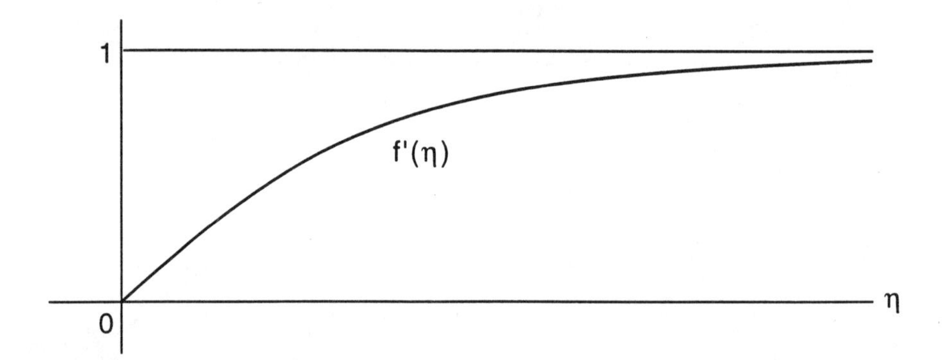

Figure 5.17. $\frac{u}{u_0}$ versus η for flow over a flat plate

The graph shows the numerical solution obtained by using Runge-Kutta method to the boundary value problem

$$\begin{aligned} f''' + ff'' &= 0\\ f(0) &= 0\\ f'(0) &= 0\\ \lim_{\eta\to\infty} f'(\eta) &= 1. \end{aligned}$$

5.8 Applications to miscellaneous problems

We saw in the previous sections the solution techniques for linear and nonlinear partial differential equations governing a physical system. In this section we shall consider some problems in partial differential equations which need special types of solution techniques. However, we shall not investigate in detail the theoretical developments of such problems. Rather, we shall demonstrate the extension of the existing methods to some practical problems.

Example 16

Confirm that the telegraph equation has a solution of the form:

$$e^{-kx}\cos\{\lambda(x-ct)\}.$$

Show that the wave speed c and the damping length k^{-1} depend on the wave length λ through the relations

$$\begin{aligned} c^2 &= \frac{4(RC+\lambda^2)}{(RC+LG)^2+4LC\lambda^2} \\ k &= \frac{1}{2}(RC+LG)c. \end{aligned}$$

Solution

The telegraph equation can be written as

$$\frac{\partial^2 E}{\partial x^2} = LC\frac{\partial^2 E}{\partial t^2} + (RC+LG)\frac{\partial E}{\partial t} + RGE. \tag{5.250}$$

The trial solution is

$$E = e^{-kx}\cos\{\lambda(x-ct)\}. \tag{5.251}$$

Then,

$$\begin{aligned} \frac{\partial E}{\partial x} &= -ke^{-kx}\cos\{\lambda(x-ct)\} - \lambda\, e^{-kx}\sin\{\lambda(x-ct)\} \\ \frac{\partial^2 E}{\partial x^2} &= k^2e^{-kx}\cos\{\lambda(x-ct)\} + 2\lambda\, k\, e^{-kx}\sin\{\lambda(x-ct)\} \\ &\quad -\lambda^2 e^{-kx}\cos\{\lambda(x-ct)\} \\ \frac{\partial E}{\partial t} &= \lambda\, c\, e^{-kx}\sin\{\lambda(x-ct)\} \\ \frac{\partial^2 E}{\partial t^2} &= -\lambda^2 c^2 e^{-kx}\cos\{\lambda(x-ct)\}. \end{aligned}$$

Substituting these values into (5.250) yields

$$\begin{aligned} &(k^2-\lambda^2)e^{-kx}\cos\{\lambda(x-ct)\} + 2\lambda\, k\, e^{-kx}\sin\{\lambda(x-ct)\} \\ &= -(LC)(\lambda^2c^2)e^{-kx}\cos\{\lambda(x-ct)\} + (RC+LG)(\lambda c)e^{-kx}\sin\{\lambda(x-ct)\} \\ &\quad +RC\, e^{-kx}\cos\{\lambda(x-ct)\}. \end{aligned} \tag{5.252}$$

Equating the coefficients of cosine and sine, we obtain

$$(k^2 - \lambda^2) = -(LC)(\lambda^2 c^2) + RG \tag{5.253}$$
$$2\lambda k = (RC + LG)(\lambda c). \tag{5.254}$$

Thus, $k = \frac{1}{2}(RC + LG)c$ and $\frac{1}{4}(RC + LG)^2 c^2 - \lambda^2 = RC - \lambda^2 c^2(LC)$.
Therefore,

$$c^2 = \frac{4(RG + \lambda^2)}{(RC + LG)^2 + 4LC\lambda^2}$$

as asserted.

Example 17

Find the solution of Laplace's equation in the rectangular region $0 \le x \le a$, $0 \le y \le b$, subject to the boundary conditions

$$\left.\begin{array}{l} u(0,y) = 0 \\ u(a,y) = 0 \end{array}\right\} \quad 0 \le y \le b \tag{5.255}$$

$$\left.\begin{array}{l} u(x,0) = f(x) \\ u(x,b) = 0 \end{array}\right\} \quad 0 \le x \le a. \tag{5.256}$$

Solution

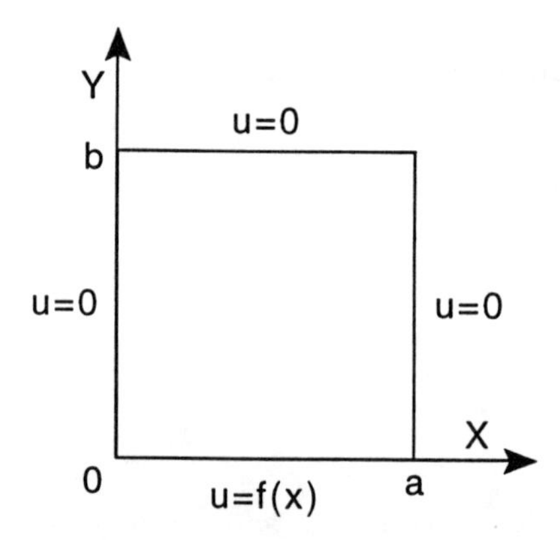

Figure 5.18. Boundary conditions on a rectangular region

Laplace's equation in $2D$ is

$$u_{xx} + u_{yy} = 0. \tag{5.257}$$

We assume the product solution $u(x,y) = X(x)Y(y)$ and thus eqn (5.257) becomes $\frac{X''}{X} = -\frac{Y''}{Y} = -\lambda^2$ where λ^2 is a separation constant. Then the two differential equations with their solutions are

$$X'' + \lambda^2 X = 0 \qquad Y'' - \lambda^2 Y = 0$$
$$X = A\cos\lambda x + B\sin\lambda x \qquad Y = C\cosh\lambda y + D\sinh\lambda y$$

and hence

$$u(x,y) = (A\cos\lambda x + B\sin\lambda x)(C\cosh\lambda y + D\sinh\lambda y) \tag{5.258}$$

where A, B, C, D and λ are all constants.

Using the boundary conditions (5.255) at $x = 0, a$, we find that

$$A = 0$$

and

$$\begin{aligned} \sin\lambda a &= 0 \\ \lambda_n &= \frac{n\pi}{a}, \quad n = 1, 2, 3, \cdots \end{aligned}$$

Therefore,

$$u_n(x,y) = \sin\frac{n\pi x}{a}\left(A_n\cosh\frac{n\pi y}{a} + B_n\sinh\frac{n\pi y}{a}\right). \tag{5.259}$$

Superimposing all the linear solutions yields

$$\begin{aligned} u(x,y) &= \sum_{n=1}^{\infty} u_n(x,y) \\ &= \sum_{n=1}^{\infty} \sin\frac{n\pi x}{a}\left\{A_n\cosh\frac{n\pi y}{a} + B_n\sinh\frac{n\pi y}{a}\right\}. \end{aligned} \tag{5.260}$$

The conditions at $y = 0, b$ confirm that

$$u(x,0) = f(x) = \sum_{n=1}^{\infty} A_n \sin\frac{n\pi x}{a}$$

from which

$$A_n = \frac{2}{a}\int_0^a f(x)\sin\frac{n\pi x}{a}\,dx \tag{5.261}$$

and

$$u(x,b) = 0 = \sum_{n=1}^{\infty} \sin\frac{n\pi x}{a}\left[A_n\cosh\frac{n\pi b}{a} + B_n\sinh\frac{n\pi b}{a}\right]$$

from which we obtain $B_n = -A_n\frac{\cosh\frac{n\pi b}{a}}{\sinh\frac{n\pi b}{a}}$. Thus solution (5.260) becomes

$$u(x,y) = \sum_{n=1}^{\infty} \frac{A_n\sin\frac{n\pi x}{a}}{\sinh\frac{n\pi b}{a}}\left\{\sinh\frac{n\pi}{a}(b-y)\right\} \tag{5.262}$$

which is the required solution.

Example 18

Determine the solution of Laplace's equation in the rectangular region $0 \leq x \leq a$, $0 \leq y \leq b$, subject to the following boundary conditions:

$$u(x,0) = f(x), u(x,b) = g(x) \quad \text{for } 0 \leq x \leq a \tag{5.263}$$

$$\left.\begin{array}{l} \frac{\partial u}{\partial x}(0,y) = 0 \\ \frac{\partial u}{\partial x}(a,y) = 0 \end{array}\right. \quad \text{for } 0 \leq y \leq b. \tag{5.264}$$

Solution

Laplace's equation is

$$u_{xx} + u_{yy} = 0. \tag{5.265}$$

Using the separation of variables method, $u(x,y) = X(x)Y(y)$ we obtain as before $\frac{X''}{X} = -\frac{Y''}{Y} = -\lambda^2$ where λ^2 is a separation constant. Then the two ordinary differential equations and their solutions are

$$\begin{array}{ll} X'' + \lambda^2 X = 0 & Y'' - \lambda^2 Y = 0 \\ X = A\cos\lambda x + B\sin\lambda x & Y = C\cosh\lambda y + D\sinh\lambda y \end{array}$$

and hence

$$u(x,y) = (A\cos\lambda x + B\sin\lambda x)(C\cosh\lambda y + D\sinh\lambda y). \tag{5.266}$$

Note that because of the presence of two nonhomogeneous boundary conditions at $y = 0, b$, the separation of variables method should be dealt with care. This is evident in the following analysis.

Satisfying the two homogeneous conditions at $x = 0, a$, we obtain $B = 0$ and $\lambda_n = \frac{n\pi}{a} \quad n = 0, 1, 2, \cdots$ Then (5.266) reduces to

$$u_n(x,y) = \cos\frac{n\pi x}{a}[A_n\cosh\frac{n\pi y}{a} + B_n\sinh\frac{n\pi y}{a}].$$

Corresponding to $n = 0$, we have $\lambda = 0$, and therefore $X'' = 0$ and $Y'' = 0$. The solutions can be written as

$$\begin{aligned} u_0(x,y) &= (A_0x + B_0)(C_0y + D_0) \\ \frac{\partial u_0}{\partial x} &= A_0(C_0y + D_0). \end{aligned}$$

Satisfying the condition (5.264), we find that $A_0 = 0$. Thus,

$$u_0(x,y) = \alpha y + \beta.$$

Superimposing all the linear solutions, we have,

$$\begin{aligned} u(x,y) &= u_0(x,y) + \sum_{n=1}^{\infty} u_n(x,y) \\ &= \alpha y + \beta + \sum_{n=1}^{\infty}\cos\frac{n\pi x}{a}\{A_n\cosh\frac{n\pi y}{a} + B_n\sin\frac{n\pi y}{a}\}. \end{aligned} \tag{5.267}$$

Now, at $y = 0$:

$$u(x,0) = f(x) = \beta + \sum_{n=1}^{\infty} A_n \cos\frac{n\pi x}{a}$$

and from which we obtain

$$\begin{aligned} \beta &= \frac{1}{a}\int_0^a f(x)dx \\ A_n &= \frac{2}{a}\int_0^a f(x)\cos\frac{n\pi x}{a}dx. \end{aligned} \tag{5.268}$$

Then, at $y = b$:

$$\begin{aligned} u(x,b) = g(x) &= \alpha b + \beta + \sum_{n=1}^{\infty} \cos\frac{n\pi x}{a}\{A_n \cosh\frac{n\pi b}{a} + B_n \sinh\frac{n\pi b}{a}\} \\ &= \alpha b + \beta + \sum_{n=1}^{\infty} C_n \cos\frac{n\pi x}{a} \end{aligned}$$

where

$$C_n = A_n \cosh\frac{n\pi b}{a} + B_n \sinh\frac{n\pi b}{a}.$$

Then

$$(\alpha b + \beta) = \frac{1}{a}\int_0^a g(x)dx$$

and

$$C_n = \frac{2}{a}\int_0^a g(x)\cos\frac{n\pi x}{a}dx. \tag{5.269}$$

Hence

$$\begin{aligned} \alpha &= \frac{1}{ab}\int_0^a [g(x) - f(x)]dx \\ B_n &= \frac{C_n}{\sinh\frac{n\pi b}{a}} - A_n\frac{\cosh\frac{n\pi b}{a}}{\sinh\frac{n\pi b}{a}}. \end{aligned} \tag{5.270}$$

Hence the problem is completely solved.

In practice, the boundary conditions are much more complicated than those considered in the last two problems. If, for example, u is prescribed in all 4 edges of a rectangular plate such as for $u(x,0) = f(x)$, $u(x,b) = g(x)$, $u(0,y) = h(y)$ and $u(a,y) = k(y)$, then we have to split the problem into two subproblems in each of which the function $u(x,y)$ is zero on two edges and takes its specific values on the other two edges. The problem can then be solved by the method illustrated in Example 18.

Example 19

By using the Laplace transform method, solve the following boundary value problem

$$u_{tt} = c^2 u_{xx} + \sin\frac{\pi x}{\ell}\sin\sigma t, \qquad 0 < x < \ell, \qquad t > 0. \tag{5.271}$$

Boundary conditions:

$$u(0,t) = 0, \quad u(\ell,t) = 0, \qquad t > 0$$

Initial conditions:

$$u(x,0) = 0, \quad u_t(x,0) = 0, \qquad 0 < x < \ell. \tag{5.272}$$

Solution

Define the Laplace transform of $u(x,t)$ with respect to t

$$\mathcal{L}\{u(x,t)\} = \int_0^\infty e^{-st}u(x,t)dt.$$

Taking the Laplace transform of the given partial differential equation, we have

$$\begin{aligned} & s^2\mathcal{L}\{u(x,t)\} - s\,u(x,0) - u_t(x,0) \\ = \; & c^2\frac{d^2}{dx^2}\mathcal{L}\{u(x,t)\} + \sin(\frac{\pi x}{\ell})\frac{\sigma}{s^2+\sigma^2}. \end{aligned}$$

Using the given initial conditions

$$\frac{d^2}{dx^2}\mathcal{L}\{u(x,t)\} - \frac{s^2}{c^2}\mathcal{L}\{u(x,t)\} = -\frac{\sigma}{c^2(s^2+\sigma^2)}\sin(\frac{\pi x}{\ell}).$$

The solution of which may be written as

$$\begin{aligned} \mathcal{L}\{u(x,t)\} &= A(s)\cosh\frac{sx}{c} + B(s)\sinh(\frac{sx}{c}) \\ & \quad -\frac{\sigma}{c^2(s^2+\sigma^2)}.\frac{1}{D^2-\frac{s^2}{c^2}}\sin(\frac{\pi x}{\ell}) \end{aligned}$$

where $D = \frac{d}{dx}$ is a differential operator.

Thus we have

$$\mathcal{L}\{u(x,t)\} = A(s)\cosh(\frac{sx}{c}) + B(s)\sinh(\frac{sx}{c}) + \frac{(\sigma)\sin(\frac{\pi x}{\ell})}{(s^2+\sigma^2)(s^2+\frac{\pi^2}{\ell^2}c^2)}.$$

The transformed boundary conditions are

$$\begin{aligned} \mathcal{L}\{u(0,t)\} &= 0 \\ \mathcal{L}\{u(\ell,t)\} &= 0. \end{aligned}$$

Therefore, using these boundary conditions, we obtain $A(s) = 0$ and $B(s) = 0$.

Hence

$$\mathcal{L}\{u(x,t)\} = \frac{(\sigma)\sin(\frac{\pi x}{\ell})}{(s^2+\sigma^2)(s^2+\frac{\pi^2c^2}{\ell^2})}.$$

Taking the inverse of Laplace transform

$$\begin{aligned}
u(x,t) &= \mathcal{L}^{-1}\{\frac{1}{(s^2+\sigma^2)(s^2+(\frac{\pi c}{\ell})^2)}\}(\sigma)\sin\frac{\pi x}{\ell} \\
&= \frac{1}{\sigma^2-\frac{\pi^2c^2}{\ell^2}}\sigma\sin\frac{\pi x}{\ell}\mathcal{L}^{-1}[\frac{1}{s^2+\frac{\pi^2c^2}{\ell^2}} - \frac{1}{s^2+\sigma^2}] \\
&= \frac{(\sigma)\sin\frac{\pi x}{\ell}}{\sigma^2-\frac{\pi^2c^2}{\ell^2}}[\frac{\ell}{\pi c}\sin(\frac{\pi ct}{\ell}) - \frac{1}{\sigma}\sin\sigma t] \\
&= \frac{\sin\frac{\pi x}{\ell}}{\sigma^2-\frac{\pi^2c^2}{\ell^2}}[\frac{\ell\sigma}{c\pi}\sin(\frac{\pi ct}{\ell}) - \sin\sigma t]
\end{aligned} \tag{5.273}$$

which is the required solution that satisfies all boundary and initial conditions. Thus, when the initial conditions are given to be homogeneous, the Laplace transform method is very powerful.

Example 20

A string vibrates in a vertical plane. The differential equation describing the small vibration of the string if gravitational force is considered is given by

$$y_{tt} = c^2 y_{xx} - g \tag{5.274}$$

where g is the acceleration due to gravity.

Find the displacement of the string at any position and at any time subject to the following boundary and initial conditions:

$$\left.\begin{aligned} y(0,t) &= 0 \\ y(\ell,t) &= 0 \end{aligned}\right. \quad t>0 \tag{5.275}$$

$$\left.\begin{aligned} y(x,0) &= f(x) \\ y_t(x,0) &= h(x) \end{aligned}\right. \quad 0 \le x \le \ell. \tag{5.276}$$

Solution

We use the separation of variables method here rather than the Laplace transform method because the initial conditions are nonhomogeneous. Due to the presence of g in (5.274), it is obvious that direct application of the method of separation of variables will fail.

So, we assume our solution in the following form:

$$y(x,t) = u(x,t) + \phi(x) \tag{5.277}$$

where $u(x,t)$ is the dependent variable and $\phi(x)$ is an unknown function of x alone. Now, substituting (5.277) into (5.274), we obtain

$$u_{tt} = c^2[u_{xx} + \phi''(x)] - g \tag{5.278}$$

subject to the conditions:

$$\begin{aligned} u(0,t) + \phi(0) = 0 \\ u(\ell,0) + \phi(\ell) = 0 \end{aligned} \tag{5.279}$$

$$\begin{aligned} u(x,0) + \phi(x) = f(x) \\ u_t(x,0) = h(x). \end{aligned} \tag{5.280}$$

Consider the two boundary value problems in the following manner:

$$\begin{aligned} u_{tt} &= c^2 u_{xx} \\ u(0,t) &= 0 \\ u(\ell,t) &= 0 \\ u(x,0) &= f(x) - \phi(x) \\ u_t(x,0) &= h(x) \end{aligned} \tag{5.281}$$

and

$$\begin{aligned} c^2\phi''(x) &= g \\ \phi(0) &= 0 \\ \phi(\ell) &= 0. \end{aligned} \tag{5.282}$$

Solving (5.282), we obtain

$$\phi(x) = \frac{gx^2}{2c^2} + Ax + B \tag{5.283}$$

and using the boundary conditions, yields

$$\phi(x) = -\frac{g(\ell - x)x}{2c^2}. \tag{5.284}$$

Thus the boundary value problem (5.281) can now be solved by the method of separation of variables.

We assume

$$u(x,t) = X(x)T(t) \tag{5.285}$$

and then the physical solution can be written as

$$u(x,t) = (A\cos\frac{\lambda x}{c} + B\sin\frac{\lambda x}{c})(C\cos\lambda t + D\sin\lambda t).$$

Boundary conditions at $x = 0, \ell$ yields $A = 0$ and $\sin\frac{\lambda\ell}{c} = 0$ such that the eigen values are given by $\lambda_n = \frac{n\pi c}{\ell} \quad n = 1, 2, 3, \cdots$

Thus,

$$u_n(x,t) = \sin\frac{n\pi x}{\ell}[A_n \cos\frac{n\pi ct}{\ell} + B_n \sin\frac{n\pi ct}{\ell}].$$

Superimposing all the linear solutions, we have

$$u(x,t) = \sum_{n=1}^{\infty} \sin\frac{n\pi x}{\ell}[A_n \cos\frac{n\pi ct}{\ell} + B_n \sin\frac{n\pi ct}{\ell}]. \tag{5.286}$$

Using the displacement condition at $t = 0$,

$$u(x,0) = f(x) - \phi(x) = \sum_{n=1}^{\infty} A_n \sin\frac{n\pi x}{\ell}$$

from which

$$A_n = \frac{2}{\ell}\int_0^{\ell}\{f(x) - \phi(x)\}\sin\frac{n\pi x}{\ell}dx. \tag{5.287}$$

Using the initial velocity condition at $t = 0$,

$$u_t(x,0) = h(x) = \sum_{n=1}^{\infty} \frac{n\pi c}{\ell} B_n \sin\frac{n\pi x}{\ell}$$

from which

$$B_n = \frac{2}{n\pi c}\int_0^{\ell} h(x)\sin\frac{n\pi x}{\ell}dx. \tag{5.288}$$

Thus the problem of the vibrating string under gravity is completely solved.

Example 21

Find the solution of the boundary value problem describing the heat conduction in a semi-infinite thin rod:

$$u_t = \alpha\, u_{xx}, \qquad 0 \le x < \infty,\ t > 0 \tag{5.289}$$
$$u(0,t) = 0, \qquad t > 0 \tag{5.290}$$
$$u(x,0) = f(x), \qquad x > 0. \tag{5.291}$$

Solution

We cannot use the Laplace transform method here because the initial condition is prescribed in a functional form. Thus in this situation we will use the method of separation of variables.

Assume a product solution $u(x,t) = X(x)T(t)$.

Then the given differential equation can be written as

$$\frac{T'}{T} = \alpha\frac{X'}{X} = -\lambda^2 \tag{5.292}$$

where λ^2 is a separation constant. Hence, the two ordinary differential equations with their solutions are:

$$\begin{array}{ll} T' + \lambda^2 \alpha T = 0 & X'' + \lambda^2 X = 0 \\ T = A\, e^{-\lambda^2 \alpha t} & X = B \cos \lambda x + C \sin \lambda x \end{array}$$

and thus

$$u(x,t) = (B \cos \lambda x + C \sin \lambda x) A\, e^{-\lambda^2 \alpha t}.$$

Now, satisfying the boundary condition at $x = 0$, we have that $B = 0$ and therefore,

$$u(x,t) = D \sin \lambda x\, e^{-\lambda^2 \alpha t}. \tag{5.293}$$

Due to the lack of a second boundary condition at a finite x, we have no further restriction on λ. Hence, instead of having a discrete eigenvalue λ_n, with corresponding eigensolution

$$u_n(x,t) = D_n \sin \lambda_n x\, e^{-\lambda_n^2 \alpha t}$$

we must have a continuous family of solutions given by

$$u_\lambda(x,t) = D(\lambda) \sin \lambda x\, e^{-\lambda^2 \alpha t}$$

where the arbitrary constant D is now associated, not with n, but with λ, which is a continuous parameter. Thus, instead of adding all linear solutions, we shall integrate with respect to λ to get the general solution,

$$u(x,t) = \int_0^\infty D(\lambda) e^{-\lambda^2 \alpha t} \sin \lambda x d\lambda. \tag{5.294}$$

By direct substitution into the given equation, it can be verified that this is a solution.

Using the initial condition at $t = 0$,

$$u(x,0) = f(x) = \int_0^\infty D(\lambda) \sin \lambda x d\lambda \tag{5.295}$$

where $D(\lambda)$ can be obtained from

$$D(\lambda) = \frac{2}{\pi} \int_0^\infty f(x) \sin \lambda x dx \tag{5.296}$$

and hence the solution is

$$\begin{aligned} u(x,t) &= \int_0^\infty D(\lambda) e^{-\lambda^2 \alpha t} \sin \lambda x d\lambda \\ &= \frac{2}{\pi} \int_0^\infty [\int_0^\infty f(s) \sin \lambda s ds] e^{-\lambda^2 \alpha t} \sin \lambda x d\lambda \\ &= \frac{2}{\pi} \int_0^\infty \int_0^\infty e^{-\lambda^2 \alpha t} f(s) \sin \lambda s \sin \lambda x ds d\lambda. \end{aligned} \tag{5.297}$$

Note that (5.296) and (5.295) turn out to be Fourier sine transform pairs. Now,

$$\begin{aligned} u(x,t) &= \frac{2}{\pi}\int_0^\infty f(s)[\int_{\lambda=0}^\infty e^{-\lambda^2\alpha t}\sin\lambda s\sin\lambda x d\lambda]ds \\ &= \frac{1}{\pi}\int_{s=0}^\infty f(s)[\int_{\lambda=0}^\infty e^{-\lambda^2\alpha t}\{\cos\lambda(s-x)-\cos\lambda(s+x)\}d\lambda]ds. \end{aligned}$$

We know that

$$\int_0^\infty e^{-\lambda^2x^2}\cos 2\nu x dx = \frac{\sqrt{\pi}}{2\lambda}e^{-\frac{\nu^2}{\lambda^2}}.$$

Hence, using this result in the above equation, we obtain

$$\begin{aligned} u(x,t) &= \frac{1}{\pi}\int_0^\infty f(s)[\frac{\sqrt{\pi}}{2\sqrt{\alpha t}}\{e^{-\frac{(s-x)^2}{4\alpha t}} - e^{-\frac{(s+x)^2}{4\alpha t}}\}]ds \\ &= \frac{1}{\sqrt{4\pi\alpha t}}\int_0^\infty f(s)\{e^{-\frac{(s-x)^2}{4\alpha t}} - e^{-\frac{(s+x)^2}{4\alpha t}}\}ds. \end{aligned} \tag{5.298}$$

Substituting $\eta = \frac{s-x}{2\sqrt{\alpha t}}$ in the first integral and $\eta = \frac{s+x}{2\sqrt{\alpha t}}$ in the second integral, we obtain,

$$\begin{aligned} u(x,t) &= \frac{1}{\sqrt{\pi}}\int_{-\frac{x}{2\sqrt{\alpha t}}}^\infty f(x+2\eta\sqrt{\alpha t})e^{-\eta^2}d\eta \\ &\quad -\frac{1}{\sqrt{\pi}}\int_{\frac{x}{2\sqrt{\alpha t}}}^\infty f(-x+2\eta\sqrt{\alpha t})e^{-\eta^2}d\eta. \end{aligned} \tag{5.299}$$

Thus, if the initial temperature is a constant, u_0 say, then

$$\begin{aligned} u(x,t) &= \frac{1}{\sqrt{\pi}}\int_{-\frac{x}{2\sqrt{\alpha t}}}^{\frac{x}{2\sqrt{\alpha t}}} u_0 e^{-\eta^2}d\eta \\ &= u_0\frac{2}{\sqrt{\pi}}\int_0^{\frac{x}{2\sqrt{\alpha t}}} e^{-\eta^2}d\eta \\ &= u_0 erf(\frac{x}{2\sqrt{\alpha t}}) \end{aligned} \tag{5.300}$$

where

$$erf\, z = \frac{2}{\sqrt{\pi}}\int_0^z e^{-\eta^2}d\eta.$$

Note

The function

$$u(x,t) = u_0[1 - erf(\frac{x}{2\sqrt{\alpha t}})]$$

will therefore have the property that $u(x,0) = 0$. Furthermore, $u(0,t) = u_0$.

Thus the function

$$u(x,t,t') = g(t')[1 - erf(\frac{x}{2\sqrt{\alpha t}})] \tag{5.301}$$

satisfies the one-dimensional diffusion equation and the conditions $u(x,0,t')=0$ and $u(0,t,t')=g(t')$.

By applying Duhamel's theorem[1] it follows that the solution of the boundary value problem

$$u(x,0)=0 \quad \text{and} \quad u(0,t)=g(t) \tag{5.302}$$

is

$$\begin{aligned} u(x,t) &= \frac{\partial}{\partial t}\int_0^t g(t')dt' \int_{\frac{x}{2\sqrt{\alpha t-\alpha t'}}}^{\infty} e^{-\eta^2} d\eta \\ &= \frac{x}{2\sqrt{\pi\alpha}}\int_0^t g(t')\frac{e^{-\frac{x^2}{4\alpha(t-t')}}}{(t-t')^{3/2}}dt'. \end{aligned} \tag{5.303}$$

Changing the variable of integration for t' to ζ, where

$$t' = t - \frac{x^2}{4\alpha\zeta^2}$$

we see that the solution may be written as

$$u(x,t) = \frac{2}{\sqrt{\pi}}\int_{\frac{x}{2\sqrt{\alpha t}}}^{\infty} g(t-\frac{x^2}{4\alpha\zeta^2})e^{-\zeta^2}d\zeta. \tag{5.304}$$

Another short cut approach

$$\begin{array}{ll} u_t = \alpha\, u_{xx} & 0 \le x < \infty \\ & t > 0 \\ u(0,t) = g(t) & t > 0 \\ u(x,0) = 0 & x > 0 \\ x \to \infty & |u| < M. \end{array} \tag{5.305}$$

In this case, take the Laplace transform with respect to time, then the given equation transforms to

$$s\mathcal{L}\{u\} - u(x,0) = \alpha\frac{d^2}{dx^2}\mathcal{L}\{u\}$$

or,

$$\frac{d^2}{dx^2}\mathcal{L}\{u\} - \frac{s}{\alpha}\mathcal{L}\{u\} = 0.$$

The solution is

$$\mathcal{L}\{u\} = A\, e^{\sqrt{\frac{s}{\alpha}}x} + B\, e^{-\sqrt{\frac{s}{\alpha}}x}. \tag{5.306}$$

The solution must be bounded when $x \to \infty$, therefore, we set $A=0$, and hence

$$\mathcal{L}\{u\} = B(s)e^{-\sqrt{\frac{s}{\alpha}}x}. \tag{5.307}$$

[1] R.C.F. Bartels and R.V. Churchill, *Bull. Am. Math. Soc.*, **48**, 176, (1942).

However,

$$\mathcal{L}\{u(0,t)\} = \mathcal{L}\{g(t)\}.$$

But, from the Laplace transform table[2] we know that

$$\mathcal{L}\{\frac{x}{2\sqrt{\pi\alpha t^3}}e^{-\frac{x^2}{4\alpha t}}\} = e^{-\sqrt{s/\alpha}x}.$$

Hence, using the convolution theorem of Laplace transform, we obtain

$$u(x,t) = \int_0^t g(\tau)\frac{x\, e^{-\frac{x^2}{4\alpha(t-\tau)}}}{2\sqrt{\pi\alpha(t-\tau)^3}}d\tau.$$

Changing the variable of integration from τ to ζ, where $\tau = t - \frac{x^2}{4\alpha\zeta^2}$ we see that

$$u(x,t) = \frac{2}{\sqrt{\pi}}\int_{\frac{x}{2\sqrt{\alpha t}}}^{\infty} g(t - \frac{x^2}{4\alpha\zeta^2})e^{-\zeta^2}d\zeta$$

which is the same as in (5.304).

Thus, as a rule, if the problem of heat conduction is given by, suppose

$$\begin{array}{ll} u_t = \alpha\, u_{xx} & 0 < x < \infty \\ & t > 0 \\ u(x,0) = f(x) & x > 0 \\ u(0,t) = g(t) & t > 0 \\ x \to \infty & |u| < M. \end{array} \tag{5.308}$$

then the problem can be split into two linear problems as follows:

$$u = u_1 + u_2 \tag{5.309}$$

$$\begin{aligned} \frac{\partial u_1}{\partial t} &= \alpha\frac{\partial^2 u_1}{\partial x^2} \\ u_1(x,0) &= f(x) \\ u_1(0,t) &= 0 \end{aligned} \tag{5.310}$$

and

$$|u_1| < M.$$

Also,

$$\begin{aligned} \frac{\partial u_2}{\partial t} &= \alpha\frac{\partial^2 u_2}{\partial x^2} \\ u_2(x,0) &= 0 \\ u_2(0,t) &= g(t) \end{aligned} \tag{5.311}$$

[2]G. Doetsch *Guide to the Application of Laplace Transform*, D. Van Nostrand (1961).

and

$$|u_2| < M.$$

As we know, the solutions of (5.310) is given by (5.299) and that of (5.311) by (5.304), the solution of (5.308) may be immediately written as

$$\begin{aligned} u &= u_1 + u_2 \\ &= \frac{1}{\sqrt{\pi}} \int_{-\frac{x}{2\sqrt{\alpha t}}}^{\infty} f(x + 2\eta\sqrt{\alpha t}) e^{-\eta^2} d\eta - \frac{1}{\sqrt{\pi}} \int_{\frac{x}{2\sqrt{\alpha t}}}^{\infty} f(-x + 2\eta\sqrt{\alpha t}) e^{-\eta^2} d\eta \\ &\quad + \frac{2}{\sqrt{\pi}} \int_{\frac{x}{2\sqrt{\alpha t}}}^{\infty} g(t - \frac{x^2}{4\alpha\zeta^2}) e^{-\zeta^2} d\zeta. \end{aligned} \tag{5.312}$$

We shall use this solution technique in the next example of a transmission line problem.

Example 22

A semi-infinite transmission line of negligible inductance and negligible ground conductance per unit length has a voltage applied to its sending end $x = 0$ given by $E(0, t) = E_0 \cos \omega t$ for $t > 0$. Assuming the initial voltage and current are zero, find the voltage and current at any point $x > 0$ at any time $t > 0$.

Solution

We know that the telephone equation is given by

$$E_{xx} = LC\, E_{tt} + (RC + LG)E_t + RGE. \tag{5.313}$$

It is given that inductance and conductance are negligible. Hence $L = 0$ and $G = 0$. Thus, (5.313) reduces to

$$E_{xx} = RC\, E_t. \tag{5.314}$$

The boundary and initial conditions are, respectively

$$x = 0: \quad E(0, t) = E_0 \cos \omega t \tag{5.315}$$

$$x \to \infty: \quad |E(x, t)| < M \quad \text{bounded} \tag{5.316}$$

and at

$$t = 0: \quad E(x, 0) = 0. \tag{5.317}$$

We shall use here the method of Laplace transform rather than that of separation of variables. Taking the Laplace transform of (5.314) with respect to t, we obtain

$$\mathcal{L}\{E_{xx}\} = RC\mathcal{L}\{E_t\}$$

or,

$$\frac{d^2}{dx^2}\mathcal{L}\{E(x, t)\} = RC[s\mathcal{L}\{E\} - E(x, 0)].$$

Using the initial condition, $E(x,0)=0$, we have

$$\frac{d^2}{dx^2}\mathcal{L}\{E\} - RC\, s\mathcal{L}\{E\} = 0 \tag{5.318}$$

the solution of which can be written as

$$\mathcal{L}\{E\} = A(s)e^{\sqrt{RCs}x} + B(s)e^{-\sqrt{RCs}x}. \tag{5.319}$$

From condition (5.316), we have that $|\mathcal{L}\{E\}| < M$ must be bounded as x goes to infinity, and hence $A(s)=0$. Thus

$$\mathcal{L}\{E\} = B(s)e^{-\sqrt{RCs}x}. \tag{5.320}$$

The Laplace transform of (5.315) is

$$\mathcal{L}\{E(0,t)\} = E_0\frac{s}{s^2+\omega^2} \tag{5.321}$$

and so

$$B(s) = E_0\frac{s}{s^2+\omega^2}. \tag{5.322}$$

Hence (5.320) yields

$$\mathcal{L}\{E(x,t)\} = E_0\left(\frac{s}{s^2+\omega^2}\right)\left(e^{-\sqrt{RCs}x}\right). \tag{5.323}$$

We know from the Laplace transform table

$$\mathcal{L}\left\{\frac{x}{2\sqrt{\pi\alpha t^3}}e^{-\frac{x^2}{4\alpha t}}\right\} = e^{-\sqrt{s/\alpha}x} \quad \text{where } \alpha = 1/RC.$$

Therefore,

$$\mathcal{L}\left\{\frac{\sqrt{RC}x}{2\sqrt{\pi t^3}}e^{-\frac{RC\,x^2}{4t}}\right\} = e^{-\sqrt{RCs}x}.$$

Using the convolution theorem of Laplace transform, we obtain

$$E(x,t) = E_0\int_0^t\left\{\frac{\sqrt{RC}x}{2\sqrt{\pi\tau^3}}e^{-\frac{RC\,x^2}{4\tau}}\right\}\cos(\omega t-\omega\tau)d\tau. \tag{5.324}$$

By substituting, $v^2 = \frac{RC\,x^2}{4\tau}$, the above expression can be written as

$$E(x,t) = \frac{2E_0}{\sqrt{\pi}}\int_{\frac{x\sqrt{RC}}{2\sqrt{t}}}^{\infty} e^{-v^2}\cos\left\{\omega t - \frac{\omega RC\,x^2}{4v^2}\right\}dv \tag{5.325}$$

which is the required voltage. The current is given by

$$\begin{aligned}
I(x,t) &= -\frac{1}{R}\frac{\partial E}{\partial x} \\
&= -\frac{2E_0}{R\sqrt{\pi}}\left[\int_{\frac{x}{2}\sqrt{\frac{RC}{t}}}^{\infty}\frac{\partial}{\partial x}\{e^{-v^2}\cos(\omega t-\frac{\omega RCx^2}{4v^2})\}dv-\frac{1}{2}\sqrt{\frac{RC}{t}}e^{-\frac{RCx^2}{4t}}\right] \\
&= E_0\sqrt{\frac{C}{R\pi t}}e^{-\frac{RCx^2}{4t}}-\frac{E_0\omega Cx}{\sqrt{\pi}}\int_{\frac{x}{2}\sqrt{\frac{RC}{t}}}^{\infty}v^{-2}e^{-v^2} \\
&\quad \sin(\omega t-\frac{\omega RCx^2}{4v^2})dv. \qquad (5.326)
\end{aligned}$$

In this differentiation, we have used the Leibnitz rule of differentiation under the integral sign.

Example 23

Find the solution of the following boundary value problem:

$$u_{tt}=c^2u_{xx}+F(x,t) \quad 0<x<\ell,\ t>0.$$

Boundary conditions:

$$\left\{\begin{array}{l} u(0,t)=0 \\ u(\ell,t)=0 \end{array}\right\} \quad t>0.$$

Initial conditions:

$$\left\{\begin{array}{l} u(x,0)=f(x) \\ u_t(x,0)=g(x) \end{array}\right\} \quad 0<x<\ell.$$

Solution

Assume that

$$u(x,t)=v(x,t)+\omega(x,t)$$

is a solution of the given partial differential equation. Then the above boundary value problem can be stated as follows:

$$(v_{tt}+\omega_{tt})=c^2(v_{xx}+\omega_{xx})+F(x,t).$$

Boundary conditions:

$$\begin{aligned}
v(0,t)+\omega(0,t) &= 0 \\
v(\ell,t)+\omega(\ell,t) &= 0.
\end{aligned}$$

Initial conditions:

$$\begin{aligned}
v(x,0)+\omega(x,0) &= f(x) \\
v_t(x,0)+\omega_t(x,0) &= g(x).
\end{aligned}$$

Because the problem is linear, we can split this problem into two boundary value problems:

$$\begin{aligned} v_{tt} &= c^2 v_{xx} + F(x,t) \\ v(0,t) &= 0 \\ v(\ell,t) &= 0 \end{aligned}$$

and

$$\begin{aligned} \omega_{tt} &= c^2 \omega_{xx} \\ \omega(0,t) &= 0 \\ \omega(\ell,t) &= 0 \\ \omega(x,0) &= f(x) - v(x,0) \\ \omega_t(x,0) &= g(x) - v_t(x,0). \end{aligned}$$

Consider that $F(x,t) = c^2 \sin x \cos \omega t$. Then we have

$$\begin{aligned} v_{tt} &= c^2 v_{xx} + c^2 \sin x \cos \omega t \\ v(0,t) &= 0 = v(\ell,t). \end{aligned}$$

Assume

$$v = \phi(x) \cos \omega t$$

then

$$\begin{aligned} -\omega^2 \phi \cos(\omega t) &= c^2(\phi'' \cos \omega t + \sin x \cos \omega t) \\ -\omega^2 \phi &= c^2(\phi'' + \sin x) \\ \phi'' + \frac{\omega^2}{c^2}\phi &= -\sin x. \end{aligned}$$

The boundary conditions are

$$\begin{aligned} \phi(0) &= 0 \\ \phi(\ell) &= 0. \end{aligned}$$

Solving

$$\phi = A \cos \frac{\omega x}{c} + B \sin \frac{\omega x}{c} - \frac{\sin x}{-1 + \omega^2/c^2}.$$

using the boundary conditions, at

$$\begin{aligned} x = 0: &\qquad A = 0 \\ x = \ell: &\qquad B = \tfrac{c^2 \sin \ell}{(\sin \frac{\omega \ell}{c})(\omega^2 - c^2)}. \end{aligned}$$

Therefore,

$$\begin{aligned}\phi &= \frac{c^2 \sin\ell \sin\frac{\omega x}{c}}{(\omega^2 - c^2)\sin\frac{\omega\ell}{c}} - \frac{c^2 \sin x}{(\omega^2 - c^2)} \\ &= \frac{c^2}{\omega^2 - c^2} \times \frac{1}{\sin\frac{\omega\ell}{c}}[\sin\ell \sin\frac{\omega x}{c} - \sin x \sin\frac{\omega\ell}{c}]\end{aligned}$$

and hence

$$\begin{aligned}v(x,t) &= \phi(x)\cos\omega t \\ &= \frac{c^2}{(\omega^2 - c^2)\sin\frac{\omega\ell}{c}}[\sin\ell\sin\frac{\omega x}{c} - \sin x \sin\frac{\omega\ell}{c}]\cos\omega t.\end{aligned}$$

The boundary value problem in $\omega(x,t)$ can now be solved by the usual method of separation of variables.

The solution can be written directly as

$$\omega(x,t) = \sum_{n=1}^{\infty} \sin\frac{n\pi x}{\ell}\{A_n \cos\frac{n\pi ct}{\ell} + B_n \sin\frac{n\pi ct}{\ell}\}$$

where

$$\begin{aligned}A_n &= \frac{2}{\ell}\int_0^{\ell}[f(x) - v(x,0)]\sin\frac{n\pi x}{\ell}dx \\ B_n &= \frac{2}{n\pi c}\int_0^{\ell}[g(x) - v_t(x,0)]\sin\frac{n\pi x}{\ell}dx \quad \text{where } n = 1,2,3,\cdots\end{aligned}$$

and hence the solution to the original problem can be explicitly written as

$$\begin{aligned}u(x,t) &= \sum_{n=1}^{\infty}\sin\frac{n\pi x}{\ell}\{A_n\cos\frac{n\pi ct}{\ell} + B_n \sin\frac{n\pi ct}{\ell}\} \\ &\quad + \frac{c^2}{\sin\frac{\omega\ell}{c}(\omega^2 - c^2)}\{\sin\ell\sin\frac{\omega x}{c} - \sin x\sin\frac{\omega\ell}{c}\}\cos\omega t\end{aligned}$$

where A_n and B_n are given by above.

Example 24

A thin metal bar of length ℓ has its curved surface perfectly insulated against the flow of heat. Its left end is maintained at a constant temperature $u = 0^oC$, and its right end radiates freely into the air at a constant temperature of $u = 0^oC$. If the initial temperature of the bar is given by

$$u(x,0) = f(x),$$

find the temperature at any point of the bar at any subsequent time.

Solution

This problem is similar to the heat conduction problem with the radiation boundary condition.

The problem can be mathematically described as

$$u_t = \alpha\, u_{xx}. \tag{5.327}$$

Boundary conditions:

$$x = 0: \quad u(0,t) = 0 \tag{5.328}$$

$$x = \ell: \quad u_x + h\,u = 0. \tag{5.329}$$

Initial condition:

$$t = 0: \quad u(x,0) = f(x). \tag{5.330}$$

Here h is defined to be a radiation parameter.

We use here the method of separation of variables,

$$u(x,t) = X\,T$$

and then substituting back into (5.327), we obtain

$$\frac{1}{\alpha}\frac{T'}{T} = \frac{X''}{X} = -\lambda^2$$

where λ^2 is a separation constant in which a minus sign is chosen to obtain the physical solution of the problem.

Then we get two ordinary differential equations and their solution as follows:

$$\begin{array}{ll} T' + \lambda^2 \alpha T = 0 & X'' + \lambda^2 X = 0 \\ T = A\,e^{-\lambda^2 \alpha t} & X = B\cos\lambda x + C\sin\lambda x. \end{array}$$

Thus,

$$u(x,t) = (B\cos\lambda x + C\sin\lambda x)A\,e^{-\lambda^2\alpha t}. \tag{5.331}$$

Satisfying the left hand condition at $x = 0$, we must have

$$u(0,t) = 0 = B(A\,e^{-\lambda^2\alpha t})$$

and hence $B = 0$ and the equation reduces to

$$u(x,t) = (C\sin\lambda x)(A\,e^{-\lambda^2\alpha t}). \tag{5.332}$$

To satisfy the radiation condition at $x = \ell$, we must have

$$[(C\lambda\cos\lambda x)(A\,e^{-\lambda^2\alpha t}) + h(C\sin\lambda x)(A\,e^{-\lambda^2\alpha t})]_{x=\ell} = 0$$

or

$$CA(\lambda\cos\lambda\ell + h\,\sin\lambda\ell) = 0.$$

If $C = 0$ or $A = 0$, then the solution is trivial. Hence, another alternative is

$$\lambda \cos \lambda\ell + h \sin \lambda\ell = 0$$

so that

$$\tan \lambda\ell = -\frac{\lambda}{h}.$$

Substituting $\lambda\ell = z$, yields

$$\tan z = -\frac{z}{\ell h}.$$

The roots of this equation can be obtained by the graphical method which is shown below.

The roots are given by the abscissas of the points of intersection of the curves

$$y_1 = \tan z \quad \text{and} \quad y_2 = -\frac{z}{\ell h}.$$

Obviously, there are an infinite number of roots z_n. However, unlike the roots of $\sin \lambda\ell = 0$ and $\cos \lambda\ell = 0$, these roots are not evenly spaced, although as Fig. 5.20 indicates, the interval between successive values of z_n approaches π as $n \to \infty$.

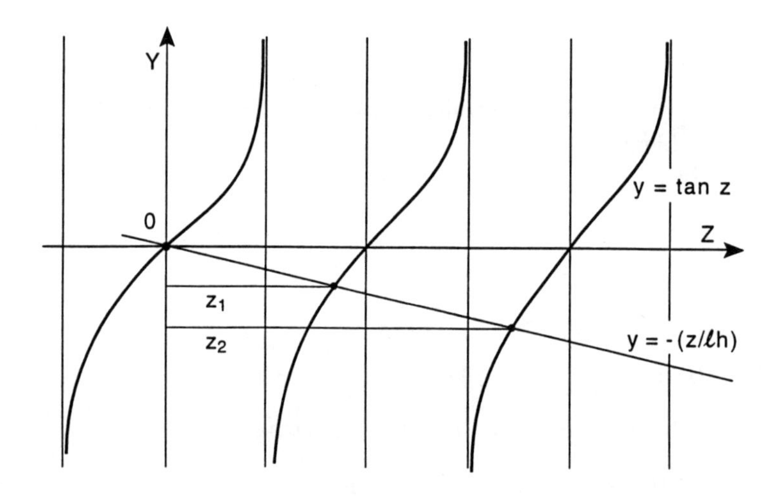

Figure 5.19: Graphical solution of $\tan z = -\frac{z}{\ell h}$

Thus, we have

$$\begin{aligned} z_n &= \lambda_n \ell \\ \lambda_n &= \frac{z_n}{\ell} \end{aligned}$$

and the associated product solution can be written as

$$u_n(x,t) = A_n \sin \lambda_n x e^{-\lambda_n^2 \alpha t}. \tag{5.333}$$

Hence the general solution is

$$\begin{aligned} u(x,t) &= \sum_{n=1}^{\infty} u_n(x,t) \\ &= \sum_{n=1}^{\infty} A_n \sin \lambda_n x\, e^{-\lambda_n^2 \alpha t}. \end{aligned} \tag{5.334}$$

Finally, using the initial condition, we have

$$u(x,0) = f(x) = \sum_{n=1}^{\infty} A_n \sin \lambda_n x \tag{5.335}$$

which requires the expansion of $f(x)$ into a series of trigonometric functions. The series (5.335) bears the resemblance to the Fourier sine series except for the fact that the values of λ_n are not equally spaced whereas the corresponding values $\lambda_n = n\pi/\ell$ in the Fourier series are equally spaced. Evidently, Fourier series can be classified as the special case of a more general and more fundamental system of functions which will be defined as Orthogonal set of functions over some interval. In the next section we shall try to explain this notion elaborately by citing some important theorems. Until then we shall accept the orthogonal property as defined below

$$\int_0^{\ell} \sin \lambda_m x \sin \lambda_n x dx = 0 \qquad \lambda_n \neq \lambda_n. \tag{5.336}$$

This can be easily verified, of course, by direct integration which is shown below:

$$\begin{aligned} \frac{1}{2}\{\int_0^{\ell} 2 \sin \lambda_m x \sin \lambda_n x dx\} &= \frac{1}{2}\int_0^{\ell} \{\cos(\lambda_m - \lambda_n)x - \cos(\lambda_m + \lambda_n)x\} dx \\ &= \frac{1}{2}[\frac{\sin(\lambda_m - \lambda_n)x}{\lambda_m - \lambda_n} - \frac{\sin(\lambda_m + \lambda_n)x}{\lambda_m + \lambda_n}]_0^{\ell} \\ &= \frac{\cos \lambda_m \ell \cos \lambda_n \ell}{\lambda_m^2 - \lambda_n^2}[-\lambda_m \tan \lambda_n \ell + \lambda_n \tan \lambda_m \ell] \\ &= \frac{\cos \lambda_m \ell \cos \lambda_n \ell}{h(\lambda_m^2 - \lambda_n^2)}[\lambda_m \lambda_n - \lambda_m \lambda_n] = 0. \end{aligned}$$

This result has been obtained using the relations $\tan \lambda_m \ell = -\frac{\lambda_m}{h}$ and $\tan \lambda_n \ell = -\frac{\lambda_n}{h}$.
Thus

$$\int_0^{\ell} \sin \lambda_m x \sin \lambda_n x dx = 0 \qquad \lambda_m \neq \lambda_n.$$

Thus, multiplying (5.335) by $\sin \lambda_m x$ and then integrating with respect to x from $x = 0$ to $x = \ell$, we find

$$\begin{aligned} \int_0^{\ell} f(x) \sin \lambda_m x dx &= \sum_{n=1}^{\infty} A_n \int_0^{\ell} \sin \lambda_m x \sin \lambda_n x dx \\ &= A_m \int_0^{\ell} \sin^2 \lambda_m x dx. \end{aligned}$$

Therefore, because of (5.336)

$$A_n = \frac{\int_0^\ell f(x)\sin\lambda_n x dx}{\int_0^\ell \sin^2\lambda_n x dx} \qquad n = 1, 2, 3, \cdots \tag{5.337}$$

Now

$$\begin{aligned}\int_0^\ell \sin^2\lambda_n x dx &= \frac{1}{2}\int_0^\ell [1-\cos 2\lambda_n x]dx \\ &= \frac{1}{2}[\ell - \frac{\sin 2\lambda_n\ell}{2\lambda_n}] \\ &= \frac{1}{2h}(\ell h + \cos^2\lambda_n\ell).\end{aligned}$$

This result is obtained using the relation $\tan\lambda_n\ell = -\frac{\lambda_n}{h}$

$$A_n = \frac{2h}{\ell h + \cos^2\lambda_n\ell}\int_0^\ell f(x)\sin\lambda_n x dx.$$

Thus, with the determination of A_n, the whole problem is solved.

5.9 Sturm-Liouville problems

5.9.1 Orthogonal functions

In this section we shall discuss the notion of orthogonality of functions and Sturm-Liouville-type boundary value problems. We saw the orthogonal property of sine and cosine functions arising in connection with the Fourier series expansion. These trigonometric functions are by no means the only functions from which sets can be constructed which having the property that the integral between suitable limits of the product of two distinct members of the set is zero. In fact, the trigonometric functions which appear in the Fourier series expansions are merely one of the simplest examples of infinitely many such systems of functions. In the following, we shall give some definitions and theorems connected with this topic.

Definition 1

If a sequence of real functions $\{\phi_n(x)\}$, $n = 1, 2, 3, \cdots$, which are defined over some interval $a \le x \le b$, has the property that

$$\begin{aligned}\textstyle\int_a^b \phi_m(x)\phi_n(x)dx &= 0 \qquad m \ne n \\ &\ne 0 \qquad m = n\end{aligned}$$

then the set of functions is said to form an orthogonal set over that interval.

Definition 2

If the functions of an orthogonal set $\{\phi_n(x)\}$ have the property that

$$\int_a^b \phi_n^2(x)dx = 1 \quad \text{for all } x$$

then the functions are said to be orthonormal on the interval $a \leq x \leq b$. As, for example, suppose

$$\int_a^b \phi_n^2(x)dx = k_n > 0$$

then

$$\frac{\phi_1}{\sqrt{k_1}}, \frac{\phi_2}{\sqrt{k_2}}, \frac{\phi_3}{\sqrt{k_3}}, \cdots \text{ is called a set of orthonormal functions.}$$

Thus, for the set $\{\sin\frac{n\pi x}{\ell}\}$, $n = 1, 2, 3, \cdots$, defined over the interval $0 \leq x \leq \ell$, we obtain

$$\begin{aligned}\int_0^\ell \sin^2 \frac{n\pi x}{\ell} dx &= \frac{1}{2}(\ell) \\ &= \frac{\ell}{2}.\end{aligned}$$

Therefore, the orthonormal set can be written as

$$\frac{\sqrt{2}\sin\frac{\pi x}{\ell}}{\sqrt{\ell}}, \frac{\sqrt{2}\sin\frac{2\pi x}{\ell}}{\sqrt{\ell}}, \frac{\sin\frac{3\pi x}{\ell}}{\sqrt{\ell}}, \cdots$$

Definition 3

If a set of real functions $\{\phi_n(x)\}$ has the property that over some interval $a \leq x \leq b$,

$$\begin{aligned}\textstyle\int_a^b \omega(x)\phi_m(x)\phi_n(x)dx &= 0 \qquad m \neq n \\ &\neq 0 \qquad m = n\end{aligned}$$

then the functions are called orthogonal with respect to the weight function $\omega(x)$ on that interval.

Suppose $f(x)$ has a formal expansion analogous to a Fourier series expansion and can be expressed in terms of any set of functions $\{\phi_n(x)\}$ orthogonal over an interval (a, b), then we can write

$$f(x) = \sum_{n=1}^{\infty} A_n \phi_n(x). \tag{5.338}$$

Then multiplying both sides by $\phi_n(x)$ and integrating with respect to x between a and b, we obtain

$$\int_a^b f(x)\phi_n(x)dx = A_n \int_a^b \phi_n^2(x)dx$$

and hence

$$A_n = \frac{\int_0^b f(x)\phi_n(x)dx}{\int_a^b \phi_n^2(x)dx} \qquad n = 1, 2, 3, \cdots$$

Definition 4

A real function $f(x)$ is said to be a null function on the interval (a, b) provided

$$\int_a^b f^2(x)dx = 0.$$

Hence if $f(x)$ is identically zero, it is obviously a null function.

Definition 5

A set of orthogonal functions $\{\phi_n(x)\}$ is said to be complete if the relation

$$\int_a^b f(x)\phi_n(x)dx = 0$$

is true for all values of n only if $f(x)$ is a null function.

5.9.2 Sturm-Liouville problems

Orthogonal functions arise in many types of pure and applied mathematics problems. In practice, solutions obtained from the second-order linear ordinary differential equation, subject to two point boundary conditions, can be shown to form orthogonal sets of functions. As for example, a boundary value problem that has the form

$$\frac{d}{dx}[P(x)\frac{dy}{dx}] + [Q(x) + \lambda R(x)]y = 0 \tag{5.339}$$

$$\begin{aligned} a_1 y(a) + a_2 y'(a) = 0 \\ b_1 y(b) + b_2 y'(b) = 0 \end{aligned} \tag{5.340}$$

is called the Sturm-Liouville system where $P(x)$ and $Q(x)$ are real and continuous on the closed interval $a \leq x \leq b$ and $R(x)$ be continuous and positive. The system is named after a Swiss mathematician, Jacques C.F. Sturm (1803–1855) and a French mathematician, Joseph Liouville (1809–1882) who investigated its properties. A Sturm-Liouville problem consists of finding the values of λ and the corresponding values of y that satisfy the system. The Sturm-Liouville problem is called regular provided λ is a constant, a_1 and a_2 are not both zero and that b_1 and b_2 are not both zero and that $P(x) > 0$ for $a \leq x \leq b$.

To solve a regular Sturm-Liouville problem constitutes finding values of λ which are called eigenvalues or characteristic values and the corresponding values of y called eigenfunctions or characteristic functions. The following properties can be cited in connection with the solution of a Sturm-Liouville problem:

(a) There are an infinite number of eigenvalues that are real and distinct and can be ordered $\lambda_1 < \lambda_2 < \lambda_3 < \cdots$

(b) For each eigenvalue there corresponds an eigenfunction and all eigenfunctions belonging to different eigenvalues are linearly independent.

(c) The eigenfunctions contribute an orthogonal set on the interval $a < x < b$ with respect to the weight function $R(x)$.

To see the orthogonal property of the set of eigenfunctions arising in the Sturm-Liouville problem described mathematically by (5.343) and (5.344), we follow the following steps.

Let y_m and y_n be the two solutions associated with two distinct values of λ, say, λ_m and λ_n. This means that

$$\begin{aligned} \frac{d}{dx}(Py_m') + (Q + \lambda_m R)y_m &= 0 \\ \frac{d}{dx}(Py_n') + (Q + \lambda_n R)y_n &= 0. \end{aligned}$$

If y_n times the first of these equations is subtracted from y_m times the second, we obtain

$$y_m(Py_n')' - y_n(Py_m')' = (\lambda_m - \lambda_n)R\, y_m y_n.$$

Now it is easy to verify that

$$\frac{d}{dx}[P(y_m y_n' - y_m' y_n)] = y_m(Py_n')' - y_n(Py_m')'$$

and subsequently, we have

$$\frac{d}{dx}[P(y_m y_n' - y_m' y_n)] = (\lambda_m - \lambda_n)R\, y_m y_n. \tag{5.341}$$

Integrating from a to b

$$[P(y_m y_n' - y_m' y_n)]_a^b = (\lambda_m - \lambda_n)\int_a^b R\, y_m y_n dx$$

or,

$$\begin{aligned} P(b)[W(y_m, y_n)]_{x=b} &- P(a)[W(y_m, y_n]_{x=a} \\ = (\lambda_m &- \lambda_n)\int_a^b R\, y_m y_n dx \end{aligned} \tag{5.342}$$

where $W(y_m, y_n)$ denotes the Wronskian of y_m and y_n. Now, from the given boundary conditions

$$\begin{aligned} b_1 y_m(b) + b_2 y_m'(b) &= 0 \\ b_1 y_n(b) + b_2 y_n'(b) &= 0 \end{aligned}$$

if b_1 and b_2 are both not zero, then a nontrivial solution exists if

$$W(y_m, y_n) = \begin{vmatrix} y_m(b) & y_m'(b) \\ y_n(b) & y_n'(b) \end{vmatrix} = 0.$$

Similarly, from the given boundary conditions

$$\begin{aligned} a_1 y_m(a) + a_2 y'_m(a) &= 0 \\ a_1 y_n(a) + a_2 y'_n(a) &= 0 \end{aligned}$$

if a_1 and a_2 are both not zero, then a nontrivial solution exists if

$$[W(y_m, y_n)]_{x=a} = 0$$

and consequently

$$(\lambda_m - \lambda_n) \int_a^b R\, y_m y_n dx = 0.$$

Since λ_m and λ_n are two distinct values of λ, and hence

$$\lambda_m - \lambda_n \neq 0 \quad m \neq n$$

thus

$$\int_a^b R\, y_m y_n dx = 0 \quad m \neq n.$$

Therefore, the sets $\{y_m\}$ and $\{y_n\}$ are mutually orthogonal over the interval $a \leq x \leq b$ with respect to the weight function $R(x)$. We shall see some applications of orthogonal functions when we discuss Bessel functions and Legendre polynomials in Chapter 6. In the following we demonstrate some simple problems which can give rise to an orthogonal set of functions in the solutions of boundary value problems.

Example 25

Show that the general linear second-order differential equation

$$a(x)y'' + b(x)y' + c(x)y = -\lambda y$$

can be reduced to an equation of the Sturm-Liouville form by multiplying it by the factor

$$\frac{1}{a(x)} e^{\{\int_{x_0}^{x} \frac{b(x)}{a(x)} dx\}}.$$

Solution

Multiplying the given equation by the given factor we obtain

$$\begin{aligned} y'' e^{\int_{x_0}^{x} \frac{b(s)}{a(s)} ds} &+ \frac{b(x)}{a(x)} y' e^{\int_{x_0}^{x} \frac{b(s)}{a(s)} ds} + \frac{c(x)}{a(x)} y\, e^{\int_{x_0}^{x} \frac{b(s)}{a(s)} ds} \\ &= -\frac{\lambda y}{a(x)} e^{\int_{x_0}^{x} \frac{b(s)}{a(s)} ds} \end{aligned}$$

which can be written as

$$\frac{d}{dx}[y' e^{\int_{x_0}^{x} \frac{b(s)}{a(s)} ds}] + [\frac{c(x)}{a(x)} e^{\int_{0}^{x} \frac{b(s)}{a(s)} ds} + \frac{\lambda}{a(s)} e^{\int_{x_0}^{x} \frac{b(s)}{a(s)} ds}]y = 0$$

and subsequently can be written as

$$\frac{d}{dx}[P(x)y'] + [Q(x) + \lambda R(x)]y = 0.$$

This is called the Sturm-Liouville equation. Here,

$$\begin{aligned} P(x) &= e^{\{\int_{x_0}^{x} \frac{b(s)}{a(s)} ds\}} \\ Q(x) &= \frac{c(x)}{a(x)} e^{\{\int_{x_0}^{x} \frac{b(s)}{a(s)} ds\}} \\ R(x) &= \frac{1}{a(x)} e^{\{\int_{x_0}^{x} \frac{b(s)}{a(s)} ds\}}. \end{aligned}$$

Example 26

Given the boundary-value problem

$$\begin{aligned} x^2 y'' + 2xy' + \mu y &= 0 \\ y(1) = 0, y(e) &= 0. \end{aligned}$$

(a) Show that it is a Sturm-Liouville problem.

(b) Find the eigenvalues and eigenfunctions.

(c) Obtain a corresponding set of functions which are mutually orthogonal in the interval $1 \leq x \leq e$.

(d) Obtain a corresponding set of functions orthonormal in the interval $1 \leq x \leq e$.

Solution

(a) The given equation can be written as

$$y'' + \frac{2}{x} y' + \frac{\mu}{x^2} y = 0.$$

This equation can be put into Sturm-Liouville form by multiplying throughout by the integrating factor x^2 such that $\frac{d}{dx}(x^2 y') + \mu y = 0$. Comparing with

$$\frac{d}{dx}[P(x)y'] + [Q(x) + \mu R(x)]y' = 0$$

we have $P(x) = x^2$, $Q(x) = 0$ and $R(x) = 1$. The weight function here is $R(x) = 1$, so that the given equation is a Sturm-Liouville problem.

(b) $$x^2y'' + 2xy' + \mu y = 0$$

is of Euler-Cauchy type equation. Putting $x = e^z$ or $z = \ell n x$ and with the following notations

$$\begin{aligned} x\frac{d}{dx} &= \frac{d}{dz} = \theta \\ x^2\frac{d^2}{dx^2} &= \frac{d^2}{dz^2} - \frac{d}{dz} = \theta(\theta - 1). \end{aligned}$$

the given equation becomes $(\theta^2 + \theta + \mu)y = 0$. The auxiliary equation is $m^2 + m + \mu = 0$ and solving, we have

$$m = \frac{-1 \pm \sqrt{1 - 4\mu}}{2}.$$

Therefore, the solution is

$$y = A\,e^{\frac{-1+\sqrt{1-4\mu}}{2}}z + B\,e^{-\frac{1+\sqrt{1-4\mu}}{2}}z \quad \text{for } \mu < \frac{1}{4}$$

and

$$y = e^{-\frac{z}{2}}\{A\cos\frac{\sqrt{4\mu - 1}}{2}z + B\sin\frac{\sqrt{4\mu - 1}}{2}z\} \quad \text{for } \mu > \frac{1}{4}$$

where the nontrivial solution can be obtained for $\mu > \frac{1}{4}$ only.

The boundary conditions are: when $x = 1$, $z = 0$; and when $x = e$, $z = 1$.

Hence

$$\begin{aligned} y(0) &= 0 = A \\ y(1) &= 0 = e^{-1/2}B\sin\frac{\sqrt{4\mu - 1}}{2} \end{aligned}$$

which implies that

$$\begin{aligned} \sin\frac{\sqrt{4\mu - 1}}{2} &= 0 \\ \lambda_n = \frac{\sqrt{4\mu - 1}}{2} &= n\pi \\ 4\mu - 1 &= 4n^2\pi^2 \quad n = 1, 2, \cdots \end{aligned}$$

Therefore, corresponding to the eigenvalue λ_n, the eigenfunction is

$$y_n = B_n e^{-z/2}\sin n\pi z = B_n\sin(n\pi\ell n x)/\sqrt{x}.$$

(c) To show the orthogonal property,

$$\begin{aligned}\int_1^e R(x)y_m y_n dx &= \int_1^e (1)\{B_m \frac{\sin(m\pi \ell n x)}{\sqrt{x}}\}\{B_n \frac{\sin(n\pi \ell n x)}{\sqrt{x}}\}dx \\ &= \int_1^e B_m B_n \frac{1}{x}\sin(m\pi \ell n x)\sin(n\pi \ell n x)dx.\end{aligned}$$

Put

$$\begin{aligned}\ell n x &= z \\ \frac{1}{x}dx &= dz.\end{aligned}$$

Therefore, the above integral becomes

$$\begin{aligned}\int_1^e R(x)y_m y_n &= B_m B_n \int_0^1 \sin(m\pi z)\sin(n\pi z)dz \\ &= \frac{B_m B_n}{2}\int_0^1 [\cos(m-n)\pi z - \cos(m+n)\pi z]dz \\ &= \frac{B_m B_n}{2}[\frac{\sin(m-n)\pi z}{(m-n)\pi} - \frac{\sin(m+n)\pi z}{(m+n)\pi}]_0^1 = 0 \quad m \neq n.\end{aligned}$$

Hence the set $\{y_n\}$ is orthogonal.

(d) To find the orthonormal set, we must determine the constant B_n so that the scalar product of y_n with itself is 1. This leads to

$$\begin{aligned}\int_0^1 B_n^2 \sin^2 m\pi z dz &= 1 \\ \frac{B_n^2}{2} &= 1.\end{aligned}$$

Therefore, $B_n = \sqrt{2}$. Hence, the orthonormal set is

$$y_n = \sqrt{2}\sin(n\pi \ell n x)/\sqrt{x} \quad n = 1, 2, 3, \cdots$$

Chapter 6

Bessel functions and Legendre polynomials

6.1 Introduction

Bessel functions and Legendre polynomials, like many other branches of mathematics, had their origin in the solution of physical problems. We saw the extensive use of Bessel functions in applied problems, including almost all applications involving partial differential equations, such as the wave equation or the heat equation, in regions possessing circular symmetry. Bessel functions were first used by the German mathematician and astronomer, B.W. Bessel (1784–1846), and so are named after him; Legendre's polynomials were first used by, and are named after the French mathematician, A.M. Legendre (1752–1833).

We shall first give brief descriptions of the origins of both Bessel functions and Legendre's polynomials. We will see that the solution of Laplace's equation in cylindrical coordinates by the method of separation of variables leads to Bessel's equation; in much the same way, when we apply the method of separation of variables to Laplace's equation in spherical coordinates, one of the differential equations results in the well known Legendre's equation.

Consider Laplace's equation in cylindrical polar coordinates (r, θ, z)

$$\nabla^2 u = \frac{1}{r}\frac{\partial}{\partial r}(r\frac{\partial u}{\partial r}) + \frac{1}{r^2}\frac{\partial^2 u}{\partial \theta^2} + \frac{\partial^2 u}{\partial z^2} = 0. \tag{6.1}$$

Any solution $u(r, \theta, z)$ of this equation is known as a cylindrical harmonic. We apply the separation of variables method by assuming that u is a product of functions of r, θ and z, i.e.,

$$u = R(r)\Theta(\theta)Z(z)$$

and substitute the appropriate derivatives into the partial differential equation (6.1). We obtain

$$\frac{\Theta Z}{r}\frac{d}{dr}(r\frac{dR}{dr}) + \frac{RZ}{r^2}\frac{d^2\Theta}{d\Theta^2} + R\Theta\frac{d^2 Z}{dz^2} = 0$$

and dividing throughout by $\frac{R\Theta Z}{r^2}$ yields,

$$\frac{r}{R}\frac{d}{dr}(r\frac{dR}{dr}) + \frac{r^2}{Z}\frac{d^2Z}{dz^2} = -\frac{1}{\Theta}\frac{d^2\Theta}{d\theta^2}.$$

Since the left member of this equation is independent of θ, the equation can be satisfied if both members are equal to a constant.

Hence we have

$$-\frac{1}{\Theta}\frac{d^2\Theta}{d\theta^2} = \nu^2.$$

A second separation yields

$$\frac{1}{rR}\frac{d}{dr}(r\frac{dR}{dr}) - \frac{\nu^2}{r^2} = -\frac{1}{Z}\frac{d^2Z}{dz^2} = -\lambda^2.$$

We call the first separation constant ν^2 because this will force Θ (and u) to be periodic in θ. This is due to the physical situation of some applied problems. We call the second separation constant λ^2 because we do not want Z (and u) to be periodic in z.

Therefore, by separation of variables, we have reduced Laplace's equation to the following ordinary differential equation using prime notations:

$$Z'' - \lambda^2 Z = 0 \quad (6.2)$$

$$\Theta'' + \nu^2\Theta = 0 \quad (6.3)$$

$$R'' + \frac{1}{r}R' + (\lambda^2 - \frac{\nu^2}{r^2})R = 0. \quad (6.4)$$

The solutions to the first two equations are straight-forward and are given by

$$Z = A\,e^{\lambda z} + B\,e^{-\lambda z} \quad \text{and} \quad \Theta = C\cos\nu\theta + D\sin\nu\theta \quad (6.5)$$

respectively.

Equation (6.4) is simply known as the Bessel's equation of order ν with a parameter λ, and solutions of this equation are called Bessel functions.

Consider Laplace's equation in spherical coordinates (r, θ, ϕ)

$$\nabla^2 u = \frac{\partial^2 u}{\partial r^2} + \frac{2}{r}\frac{\partial u}{\partial r} + \frac{1}{r^2\sin^2\theta}\frac{\partial^2 u}{\partial\phi^2} + \frac{1}{r^2}\frac{\partial^2 u}{\partial\theta^2} + \frac{\cot\theta}{r^2}\frac{\partial u}{\partial\theta} = 0. \quad (6.6)$$

An equivalent form of this equation is

$$\frac{1}{r^2}\frac{\partial}{\partial r}(r^2\frac{\partial u}{\partial r}) + \frac{1}{r^2\sin\theta}\frac{\partial}{\partial\theta}(\sin\theta\frac{\partial u}{\partial\theta}) + \frac{1}{r^2\sin^2\theta}\frac{\partial^2 u}{\partial\phi^2} = 0. \quad (6.7)$$

Any solution $u(r, \theta, \phi)$ of this equation is known as a spherical harmonic.

We seek a solution by the method of separation of variables

$$u(r, \theta, \phi) = R(r)\Theta(\theta)\Phi(\phi).$$

Substituting this into (3.7), yields

$$\frac{\Phi\Theta}{r^2}\frac{d}{dr}(r^2\frac{dR}{dr}) + \frac{R\Phi}{r^2\sin\theta}\frac{d}{d\theta}(\sin\theta\frac{d\Theta}{d\theta}) + \frac{R\Theta}{r^2\sin^2\theta}\frac{d^2\Phi}{d\phi^2} = 0. \tag{6.8}$$

Dividing throughout by $\frac{R\Theta\Phi}{r^2\sin^2\theta}$, we obtain

$$\frac{\sin^2\Theta}{R}\frac{d}{dr}(r^2\frac{dR}{dr}) + \frac{\sin\theta}{\Theta}\frac{d}{d\theta}(\sin\theta\frac{d\Theta}{d\theta}) = -\frac{1}{\Phi}\frac{d^2\Phi}{d\phi^2}.$$

Since the left hand member of this equation is independent of ϕ, the equation can be satisfied if both members are equal to a constant.

Hence we have

$$-\frac{1}{\Phi}\frac{d^2\Phi}{d\phi^2} = m^2 \quad , \quad m = 0, 1, 2, \cdots \tag{6.9}$$

where the first separation constant is chosen to be a non-negative integer in order that the function Φ, and also u, are periodic in ϕ.

Separating variables again, we obtain

$$\frac{1}{R}\frac{d}{dr}(r^2\frac{dR}{dr}) = -\{\frac{1}{\Theta\sin\theta}\frac{d}{d\theta}(\sin\theta\frac{d\Theta}{d\theta}) - \frac{m^2}{\sin^2\theta}\} = \mu$$

where we call the second separation constant μ. At this point, nothing further is known about this quantity.

Thus, we have reduced Laplace's equation to the following three second-order ordinary differential equations:

$$\frac{d^2\Phi}{d\phi^2} + m^2\Phi = 0 \tag{6.10}$$

$$\frac{d}{dr}(r^2\frac{dR}{dr}) - \mu R = 0 \tag{6.11}$$

$$\frac{1}{\sin\theta}\frac{d}{d\theta}(\sin\theta\frac{d\Theta}{d\theta}) + (\mu - \frac{m^2}{\sin^2\theta})\Theta = 0. \tag{6.12}$$

The solution of (6.10) is

$$\Phi_m = A_m\cos m\phi + B_m\sin m\phi \quad m = 0, 1, 2, \cdots \tag{6.13}$$

Equation (6.11) is an Euler-Cauchy type equation

$$r^2\frac{d^2R}{dr^2} + 2r\frac{dR}{dr} - \mu R = 0. \tag{6.14}$$

It can be solved by making the following substitution

$$R = r^k$$

in which case the differential equation (6.14) becomes

$$r^2k(k-1)r^{k-2} + 2r\,kr^{k-1} - \mu r^k = 0.$$

After simplification this equation yields

$$(k^2 + k - \mu)r^k = 0.$$

Hence, r^k is a solution, provided

$$k^2 + k - \mu = 0.$$

To find a second linearly independent solution may not be so simple. If we choose $k = n$, a non-negative integer, then $\mu = n(n+1)$. Or, if we choose $k = -(n+1)$, then $\mu = n(n+1)$ also. Thus, with $\mu = n(n+1)$, the equation has two independent solutions r^n and $r^{-(n+1)}$ so that the general solution can be written as

$$R_n(r) = C_n r^n + D_n r^{-(n+1)}. \tag{6.15}$$

In order to solve equation (6.12), we make the following substitutions:

$$\begin{aligned} x &= \cos\theta \\ y(x) &= \Theta(\theta) \\ \frac{d}{d\theta} &= \frac{dx}{d\theta}\frac{d}{dx} = -\sin\theta\frac{d}{dx} \end{aligned}$$

Then

$$\begin{aligned} \frac{d}{d\theta}(\sin\theta\frac{d\Theta}{d\theta}) &= -\sin\theta\frac{d}{dx}(\sin\theta\frac{d\Theta}{dx}\frac{dx}{d\theta}) \\ &= \sin\theta\frac{d}{dx}(-\sin^2\theta\frac{d\Theta}{dx}) \\ &= \sin\theta\frac{d}{dx}((1-x^2)\frac{dy}{dx}). \end{aligned}$$

Therefore the equation (6.12) can be written as

$$\frac{d}{dx}\{(1-x^2)\frac{dy}{dx}\} + \{n(n+1) - \frac{m^2}{1-x^2}\}y = 0 \tag{6.16}$$

or, equivalently, we can write

$$(1-x^2)\frac{d^2y}{dx^2} - 2x\frac{dy}{dx} + \{n(n+1) - \frac{m^2}{1-x^2}\}y = 0. \tag{6.17}$$

This is the algebraic form of the associated Legendre equation. If $m = 0$, that is, if the solution of the original equation is independent of the longitudinal angle ϕ, then (6.17) reduces to

$$(1-x^2)\frac{d^2y}{dx^2} - 2x\frac{dy}{dx} + n(n+1)y = 0 \tag{6.18}$$

which is known as Legendre's equation. Its solutions are called Legendre Polynomials.

6.2 Series solution of Bessel's equation

The Bessel equation (6.4) can be written with familiar notation

$$x^2\frac{d^2y}{dx^2} + x\frac{dy}{dx} + (\lambda^2x^2 - \nu^2)y = 0. \tag{6.19}$$

By the substitution $z = \lambda x$, equation (6.19) is reduced to a simpler form,

$$z^2\frac{d^2y}{dz^2} + z\frac{dy}{dz} + (z^2 - \nu^2)y = 0, \tag{6.20}$$

which is known as Bessel's equation of order ν.

Comparing this equation with the normal standard equation,

$$\frac{d^2y}{dz^2} + P(z)\frac{dy}{dz} + Q(z)y = 0, \tag{6.21}$$

we obtain

$$P(z) = \frac{1}{z} \quad \text{and} \quad Q(z) = \frac{z^2 - \nu^2}{z^2}.$$

Hence, $z = 0$ is a singular point of the differential equation, and $z = \infty$ is also a singular point of the differential equation; all other values of z are ordinary points. $z = 0$ is a regular singular point because $\lim_{z\to 0} z\,P(z) = 1$ and $\lim_{z\to 0} z^2Q(z) = -\nu^2$.

Because these two limits exist, $z = 0$ is a regular singular point. A series solution exists around $z = 0$ and the infinite series will be convergent for $|z| < \infty$, i.e., for all real values of z.

By the method of Frobenius, we are led to try a series solution of the form

$$y = \sum_{k=0}^{\infty} a_k z^{\gamma+k}, \qquad a_0 \neq 0. \tag{6.22}$$

Hence,

$$\frac{dy}{dz} = \sum_{k=0}^{\infty} a_k(\gamma + k)z^{\gamma+k-1} \tag{6.23}$$

$$\frac{d^2y}{dz^2} = \sum_{k=0}^{\infty} a_k(\gamma + k)(\gamma + k - 1)z^{\gamma+k-2}. \tag{6.24}$$

By substitution of these expressions into (6.20), we obtain

$$z^2\sum_{k=0}^{\infty} a_k(\gamma + k)(\gamma + k - 1)z^{\gamma+k-2} + z\sum_{k=0}^{\infty} a_k(\gamma + k)z^{\gamma+k-1} + (z^2 - \nu^2)\sum_{k=0}^{\infty} a_k z^{\gamma+k} = 0.$$

Bringing the coefficients of like powers under the same brackets,

$$\sum_{k=0}^{\infty} a_k[(\gamma+k)(\gamma+k-1)+(\gamma+k)-\nu^2]z^{\gamma+k}+\sum_{k=0}^{\infty} a_k z^{\gamma+k+2}=0$$

or

$$\sum_{k=0}^{\infty} a_k((\gamma+k)^2-\nu^2)z^{\gamma+k}+\sum_{k=0}^{\infty} a_k z^{\gamma+k+2}=0.$$

The index of summation in the second sum is changed from k to $k-2$, and hence

$$\sum_{k=0}^{\infty} a_k((\gamma+k)^2-\nu^2)z^{\gamma+k}+\sum_{k=2}^{\infty} a_{k-2} z^{\gamma+k}=0.$$

Equating the coefficients of z^γ, $z^{\gamma+1}$, $z^{\gamma+2}$, etc., we obtain

$$k = 0: \quad a_0(\gamma^2-\nu^2)=0 \tag{6.25}$$

$$k = 1: \quad a_1((\gamma+1)^2-\nu^2)=0 \tag{6.26}$$

$$k \geq 2: \quad a_k((\gamma+k)^2-\nu^2)+a_{k-2}=0. \tag{6.27}$$

Equation (6.25), which corresponds to $k=0$, is called the indicial equation (that is, $\gamma^2-\nu^2=0$, provided $a_0 \neq 0$). Therefore $\gamma=\pm\nu$ are the two indicial roots. First consider the root $\gamma=\nu$ (we shall consider the root $\gamma=-\nu$ later). From (6.26), we see that

$$a_1(2\nu+1)=0. \tag{6.28}$$

Because we impose the restriction $\nu \geq 0$, it is clear that $a_1=0$. Then, from (6.27), it follows that $a_3=a_5=\cdots=a_{2m+1}=0$.

Equation (6.27) can be written in a simplified form when $\gamma=\nu$, giving

$$k(2\nu+k)a_k+a_{k-2}=0, \qquad k\geq 2.$$

Therefore,

$$a_2 = -\frac{a_0}{2(2\nu+2)} = -\frac{a_0}{2^2.(\nu+1).1!}$$

$$a_4 = \frac{a_2}{4(2\nu+4)} = -\frac{a_2}{2^2.2(\nu+2)} = \frac{a_0}{2^4.2!(\nu+1)(\nu+2)}$$

$$a_6 = -\frac{a_4}{6(2\nu+6)} = -\frac{a_4}{2^2.3(\nu+3)} = \frac{-a_0}{2^6.3!(\nu+1)(\nu+2)(\nu+3)}$$

and, in general,

$$a_{2m} = \frac{(-1)^m a_0}{2^{2m}.m!(\nu+m)(\nu+m-1)\cdots(\nu+2)(\nu+1)}, \qquad m=1,2,3,\cdots$$

Now a_{2m} is the coefficient of $z^{\nu+2m}$ in the series (3.22) for y. We can simplify this coefficient considerably if we multiply both the numerator and the denominator of the right hand side by the factor $2^{\nu}\Gamma(\nu+1)$, giving

$$a_{2m} = \frac{(-1)^m 2^{\nu}\Gamma(\nu+1)a_0}{2^{\nu+2m}m!(\nu+m)(\nu+m-1)\cdots(\nu+1)\Gamma(\nu+1)}.$$

Using the recurrence relation of the gamma function,

$$\begin{aligned}(\nu+\jmath)\Gamma(\nu+\jmath) &= \Gamma(\nu+\jmath+1), \\ a_{2m} &= \frac{(-1)^m 2^{\nu}\Gamma(\nu+1)a_0}{2^{\nu+2m}m!\Gamma(\nu+m+1)}.\end{aligned}$$

Note that here we used the gamma function instead of the factorial because ν is not necessarily an integer.

As we are looking for a particular solution, and a_0 is an arbitrary constant, we can choose

$$a_0 = \frac{1}{2^{\nu}\Gamma(v+1)}$$

such that

$$a_{2m} = \frac{(-1)^m}{2^{\nu+2m}m!\Gamma(\nu+m+1)}, \qquad m = 0, 1, 2, \cdots$$

We are now in a position to define a Bessel function of the first kind of order ν, which is denoted by the symbol $J_{\nu}(z)$. Thus,

$$\begin{aligned}J_{\nu}(z) &= z^{\nu}[\frac{1}{2^{\nu}\Gamma(\nu+1)} - \frac{z^2}{2^{\nu+2}\Gamma(\nu+2)} + \frac{z^4}{2^{\nu+4}2!\Gamma(\nu+3)} - \cdots] \\ &= \sum_{m=0}^{\infty} \frac{(-1)^m z^{\nu+2m}}{2^{\nu+2m}m!\Gamma(\nu+m+1)}. \qquad (6.29)\end{aligned}$$

As we mentioned earlier, this series converges for all values of $z \geq 0$. In Fig. 6.1, we have graphed $J_0(z)$ and $J_1(z)$ to get an idea of the shape of Bessel functions. These graphs resemble the graphs of $e^{-z}\sin z$ and $e^{-z}\cos z$.

Let us consider the series corresponding to the other root ($\gamma = -\nu$) of the indicial equation. Without going into the formal calculations, as we did in the case of $J_{\nu}(z)$, we can form a second particular solution of Bessel's equation (6.20) by replacing ν with $-\nu$, such that the gamma functions appearing in the denominators of the various forms are all defined. When ν is not an integer, the gamma functions are all well-defined, and so the function

$$J_{-\nu}(z) = \sum_{m=0}^{\infty} \frac{(-1)^m z^{-\nu+2m}}{2^{-\nu+2m}m!\Gamma(-\nu+m+1)} \qquad (6.30)$$

is a second particular solution of Bessel's equation of order ν. This function is unbounded at $z = 0$ because of the negative power of z, while $J_{\nu}(z)$ remains finite.

Hence, for non-integral values of ν, $J_\nu(z)$ and $J_{-\nu}(z)$ are two linearly independent solutions; a complete solution of Bessel's equation, when ν is not an integer, is then

$$y(z) = c_1 J_\nu(z) + c_2 J_{-\nu}(z) \tag{6.31}$$

where c_1 and c_2 are arbitrary constants.

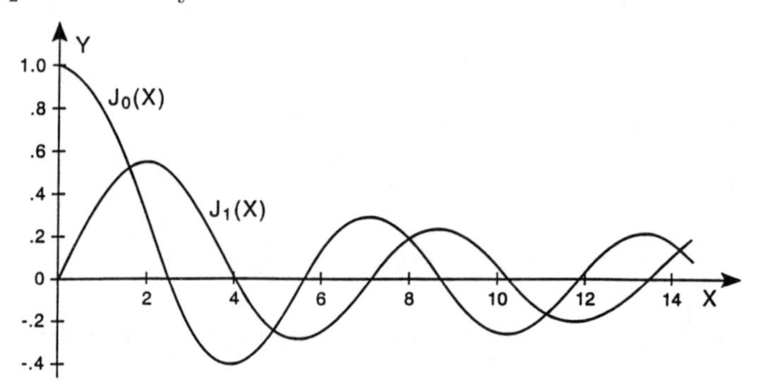

Figure 6.1. Bessel functions of the first kind $J_0(z)$ and $J_1(z)$

To construct a second solution, it is convenient to take a linear combination of $J_\nu(z)$ and $J_{-\nu}(z)$ such that

$$Y_\nu(z) = \alpha J_\nu(z) + \beta J_{-\nu}(z) \tag{6.32}$$

where α and β are non-zero arbitrary contants. But, for standardization, it is expedient that α, β have particular values. The most suitable ones are found to be $\alpha = \cot \pi\nu$ and $\beta = -cosec\pi\nu$, and, when ν is non-integral, the second solution, as defined by Weber, is

$$Y_\nu(z) = \frac{\cos \pi\nu J_\nu(z) - J_{-\nu}(z)}{\sin \pi\nu}. \tag{6.33}$$

This $Y_\nu(z)$ is known as the Bessel function of the second kind of order ν. Accordingly, for ν non-integral, an alternative complete solution is

$$y(z) = C_1 J_\nu(z) + C_2 Y_\nu(z). \tag{6.34}$$

If $C_1 = c_1 + c_2 \cos \nu\pi$, $C_2 = -c_2 \sin \nu\pi$, then (6.34) and (6.31) are identical. Again, in some applied problems, it is convenient to use another form of the general solution of Bessel's equation based upon the following particular solutions

$$\begin{aligned} H_\nu^{(1)}(z) &= J_\nu(z) + \imath Y_\nu(z) \\ H_\nu^{(2)}(z) &= J_\nu(z) - \imath Y_\nu(z) \end{aligned} \tag{6.35}$$

where ν is not an integer.

These are known as the Hankel functions (named after the German mathematician Hermann Hankel (1839–1873)) of the first and second kinds of order ν respectively, or, simply, as Bessel functions of the third kind of order ν. Note that they are conjugate complex functions.

A complete solution of Bessel's equation in terms of these two functions, is

$$y(z) = A\, H_\nu^{(1)}(z) + B\, H_\nu^{(2)}(z). \tag{6.36}$$

It is interesting that the expressions (6.31), (6.34) and (6.36) are the true solutions of Bessel's equation (6.20), even if ν is an odd multiple of $\frac{1}{2}$ and the roots of the indicial equation $\gamma^2 - \nu^2 = 0$ differ by an integer.

The situation is somewhat different when ν is an integer, say, $\nu = n$. Again, the roots of the indicial equation differ by an integer, namely $2n$. When $\nu = n$ (an integer), solution (6.29) of Bessel's equation is valid. It is

$$J_n(z) = \sum_{m=0}^{\infty} \frac{(-1)^m z^{n+2m}}{2^{n+2m} m!(n+m)!}. \tag{6.37}$$

But, when $\nu = -n$ (an integer), it can be easily verified that $J_{-n}(z)$ is not an independent solution and that, in fact, $J_{-n}(z) = (-1)^n J_n(z)$. To see this point, let us write the series expression for $J_{-n}(z)$:

$$\begin{aligned} J_{-n}(z) &= \sum_{m=0}^{\infty} \frac{(-1)^m z^{-n+2m}}{2^{-n+2m} m!(-n+m)!} \\ &= \sum_{m=n}^{\infty} \frac{(-1)^m z^{-n+2m}}{2^{-n+2m} m!(-n+m)!} \end{aligned}$$

because for $m < n$, $(-n+m)! = \infty$.

Now, replacing m by $n + \jmath$, we obtain

$$\begin{aligned} J_{-n}(z) &= \sum_{\jmath=0}^{\infty} \frac{(-1)^{n+\jmath} z^{-n+2n+2\jmath}}{2^{-n+2n+2\jmath}(n+\jmath)!\jmath!} \\ &= (-1)^n \sum_{\jmath=0}^{\infty} \frac{(-1)^\jmath z^{n+2\jmath}}{2^{n+2\jmath}\jmath!(n+\jmath)!} \\ &= (-1)^n J_n(z). \end{aligned}$$

Thus, when ν is an integer, the function $J_{-n}(z)$ is proportional to $J_n(z)$; therefore, $y(z) = c_1 J_n(z) + c_2 J_{-n}(z)$ is no longer a complete solution of Bessel's equation.

To find a second independent solution when ν is an integer, we can use Wronskian's property. The result is

$$y(z) = c_1 J_n(z) + c_2 J_n(z) \int \frac{dz}{z J_n^2(z)}.$$

Alternately, we can follow Weber's definition to find a second solution

$$\begin{aligned} Y_n(z) &= \lim_{\nu \to n} Y_\nu(z) \\ &= \lim_{\nu \to n} \left(\frac{\cos \nu\pi J_\nu(z) - J_{-\nu}(z)}{\sin \pi\nu} \right), \end{aligned} \tag{6.38}$$

which takes the indeterminate form 0/0. We use L'Hospital's rule to find the limit

$$
\begin{aligned}
Y_n(z) &= \lim_{\nu\to n} \frac{\frac{\partial}{\partial\nu}(\cos\pi\nu J_\nu(z)) - \frac{\partial}{\partial\nu}J_{-\nu}(z)}{\frac{\partial}{\partial\nu}(\sin\pi\nu)} \\
&= \frac{[\cos n\pi \frac{\partial}{\partial\nu}J_\nu(z) - \frac{\partial}{\partial\nu}J_{-\nu}]_{\nu=n}}{\pi\cos n\pi} \\
&= \frac{(-1)^n}{\pi}[(-1)^n\frac{\partial}{\partial\nu}J_\nu(z) - \frac{\partial}{\partial\nu}J_\nu]_{\nu=n} \\
&= \frac{1}{\pi}[\frac{\partial}{\partial\nu}J_\nu(z) + (-1)^{n+1}\frac{\partial}{\partial\nu}J_{-\nu}(z)]_{\nu=n}.
\end{aligned} \tag{6.39}
$$

Detailed differentiations are given below:

$$
\begin{aligned}
J_\nu(z) &= \sum_{m=0}^{\infty} \frac{(-1)^m z^{\nu+2m}}{2^{\nu+2m}m!\Gamma(\nu+m+1)} \\
&= \sum_{m=0}^{\infty} \frac{(-1)^m(\frac{z}{2})^{\nu+2m}}{m!\Gamma(\nu+m+1)} \\
&= (\frac{z}{2})^\nu \sum_{m=0}^{\infty} \frac{(-1)^m(\frac{z}{2})^{2m}}{m!\Gamma(\nu+m+1)} \\
[\frac{\partial}{\partial\nu}J_\nu(z)]_{\nu=n} &= [(\frac{z}{2})^\nu \ell n(\frac{z}{2}) \sum_{m=1}^{\infty} \frac{(-1)^m(\frac{z}{2})^{2m}}{m!\Gamma(\nu+m+1)} \\
&\quad + \sum_{m=0}^{\infty} \frac{(-1)^m(\frac{z}{2})^{\nu+2m}}{m!} \frac{\partial}{\partial\nu}(\frac{1}{\Gamma(\nu+m+1)})]_{\nu=n} \\
[\frac{\partial}{\partial\nu}J_\nu(z)]_{\nu=n} &= \ell n(\frac{z}{2})J_n(z) \\
&\quad - \sum_{m=0}^{\infty} \frac{(-1)^m(\frac{z}{2})^{n+2m}\psi^{(n+m+1)}}{m!\Gamma(n+m+1)}
\end{aligned}
$$

where

$$\frac{d}{d\nu}(\frac{1}{\Gamma(\nu+m+1)}) = -(\Gamma'/\Gamma)/\Gamma = -\psi/\Gamma,$$

and

$$\psi(n+m+1) = \{1 + \frac{1}{2} + \frac{1}{3} + \cdots + \frac{1}{n+m}\} - \gamma.$$

Here γ is Euler's constant and is defined by

$$\gamma = \lim_{r\to\infty}(1 + \frac{1}{2} + \frac{1}{3} + \cdots \frac{1}{r} - \ell nr) = 0.5772\cdots$$

Note

$\Gamma(t)$ can be defined as

$$\begin{aligned}
\Gamma(t) &= \lim_{r\to\infty} \frac{r!r^t}{t(t+1)(t+2)\cdots(t+r)} \\
\ell n\Gamma(t) &= \lim_{r\to\infty}[\ell n(r!) + t\ell nr - \ell nt - \ell n(t+1) - \cdots - \ell n(t+r)] \\
\frac{d}{dt}(\ell n\Gamma(t)) &= \lim_{r\to\infty}[0 + \ell nr - \frac{1}{t} - \frac{1}{t+1} - \cdots - \frac{1}{t+r}] \\
(\frac{\Gamma'}{\Gamma}) &= (1 + \frac{1}{2} + \frac{1}{3} + \cdots + \frac{1}{t-1}) - \lim_{r\to\infty}[1 + \frac{1}{2} \\
&\quad + \frac{1}{3} + \cdots + \frac{1}{t+r} - \ell nr]
\end{aligned}$$

Substitute $t = n + m + 1$. Therefore,

$$\begin{aligned}
\psi(n+m+1) &= (1 + \frac{1}{2} + \cdots + \frac{1}{n+m}) - \lim_{r\to\infty}(1 + \frac{1}{2} \\
&\quad + \cdots + \frac{1}{n+m+r+1} - \ell nr).
\end{aligned}$$

Now, coming back to our problem of differentiation, $\frac{\partial}{\partial\nu}J_{-\nu}(z)$, we see that it is advantageous to use another definition of the gamma function, which is

$$\Gamma(t)\Gamma(1-t) = \frac{\pi}{\sin\pi t},$$

such that

$$\frac{1}{\Gamma(1-t)} = \frac{\Gamma(t)\sin\pi t}{\pi},$$

or, replacing t by $\nu - m$ in the above,

$$\frac{1}{\Gamma(-\nu+m+1)} = \frac{\Gamma(\nu-m)\sin(\pi(\nu-m)}{\pi}.$$

Thus, we have

$$\begin{aligned}
J_{-\nu}(z) &= \frac{1}{\pi}\sum_{m=0}^{n-1}\frac{(-1)^m(\frac{z}{2})^{-\nu+2m}}{m!}\Gamma(\nu-m)\sin\pi(\nu-m) \\
&\quad + \sum_{m=n}^{\infty}\frac{(-1)^m(\frac{z}{2})^{-\nu+2m}}{m!\Gamma(-\nu+m+1)}.
\end{aligned} \tag{6.40}$$

Differentiating the first series with respect to ν yields

$$\begin{aligned}
\frac{1}{\pi}\sum_{m=0}^{n-1}\frac{(-1)^m}{m!}[(\frac{z}{2})^{-\nu+2m}\Gamma(\nu &- m)\pi\cos\pi(\nu-m) \\
&+ \sin\pi(\nu-m)\frac{d}{d\nu}((\frac{z}{2})^{-\nu+2m}\Gamma(\nu-m))]
\end{aligned}$$

and, when $\nu = n$, we get

$$\sum_{m=0}^{n-1} \frac{(-1)^m}{m!} (\frac{z}{2})^{-n+2m} \Gamma(n-m)(-1)^{n-m} = (-1)^n \sum_{m=0}^{n-1} \frac{(n-m-1)!}{m!} (\frac{2}{z})^{2m-n}. \quad (6.41)$$

Differentiating the second series in (6.40), we obtain

$$-\ell n(\frac{z}{2}) \sum_{m=n}^{\infty} \frac{(-1)^m (\frac{z}{2})^{-\nu+2m}}{m!\Gamma(-\nu+m+1)} + \sum_{m=n}^{\infty} \frac{(-1)^m (\frac{z}{2})^{-\nu+2m} \psi(-\nu+m+1)}{m!\Gamma(-\nu+m+1)}.$$

When $\nu = n$ and changing m to $k+n$, we obtain

$$(-1)^{n+1} \ell n(\frac{z}{2}) \sum_{k=0}^{\infty} \frac{(-1)^k (\frac{z}{2})^{n+2k}}{k!(k+n)!} + (-1)^n \sum_{k=0}^{\infty} \frac{(-1)^k (\frac{z}{2})^{n+2k} \psi(k+1)}{k!(k+n)!}$$

which is equivalent to

$$(-1)^{n+1} \ell n(\frac{z}{2}) J_n(z) + (-1)^n \sum_{m=0}^{\infty} \frac{(-1)^m (\frac{z}{2})^{n+2m} \psi(m+1)}{m!(m+n)!}. \quad (6.42)$$

Therefore, combining (6.39), (6.40), and (6.41), the second solution, $Y_n(z)$, is

$$\begin{aligned} Y_n &= \frac{2}{\pi} \ell n(\frac{z}{2}) J_n(z) - \frac{1}{\pi} \sum_{m=0}^{n-1} \frac{(n-m-1)!(\frac{2}{z})^{2m-2n}}{m!} \\ &\quad - \frac{1}{\pi} \sum_{m=0}^{\infty} \frac{(-1)^m (\frac{z}{2})^{n+2m} [\psi(n+m+1) + \psi(m+1)]}{m!(m+n)!} \\ &= \frac{2}{\pi}(\gamma + \ell n \frac{z}{2}) J_n(z) - \frac{1}{\pi} \sum_{m=0}^{n-1} \frac{(n-m-1)!}{m!} (\frac{2}{z})^{n-2m} \\ &\quad - \frac{1}{\pi} \sum_{m=0}^{\infty} \frac{(-1)^m (\frac{z}{2})^{n+2m}}{m!(n+m)!} [(1 + \frac{1}{2} + \cdots + \frac{1}{m}) \\ &\quad + (1 + \frac{1}{2} + \cdots + \frac{1}{n+m})]. \end{aligned} \quad (6.43)$$

Note when $n = 0$, this sum is equal to zero. Also, when $m = 0$, the sum is equal to zero.

We are now in a position to define the Hankel functions which correspond to the Bessel functions $J_n(z)$ and $Y_n(z)$:

$$\begin{aligned} H_n^{(1)}(z) &= J_n(z) + \imath Y_n(z) \\ H_n^{(2)}(z) &= J_n(z) - \imath Y_n(z). \end{aligned} \quad (6.44)$$

Note that these two functions are complex conjugates. Thus, with the formulae (6.38) and (6.44), we can now eliminate the restriction that ν is non-integral from (6.34) and (6.35). In fact, we can use the results (6.34) and (6.35), together with the formulae (6.38) and (6.44), for all values of ν, integral as well as non-integral.

In summary, the various forms of the complete solution of (6.20) are

$$y = A\,J_\nu(z) + B\,J_{-\nu}(z) \quad , \quad \text{non-integral} \tag{6.44a}$$
$$y = A\,J_\nu(z) + B\,Y_\nu(z) \quad , \quad \text{always} \tag{6.44b}$$
$$y = A\,J_n(z) + B\,Y_n(z) \quad , \quad n \text{ integral} \tag{6.44c}$$
$$y = A\,H_\nu^{(1)}(z) + B\,H_\nu^{(2)}(z) \quad , \quad \text{always} \tag{6.44d}$$
$$y = A\,H_n^{(1)}(z) + B\,H_n^{(2)}(z) \quad , \quad n \text{ integral.} \tag{6.44e}$$

Depending upon the nature of the problem, we choose the appropriate solutions from the above list. For instance, if $\nu = \frac{1}{3}$, either (6.44a) or (6.44b) may be used. The arbitrary constants are determined from the boundary conditions by choosing the most convenient form of the solution.

In Fig. 6.2, plots of $Y_0(z)$ and $Y_1(z)$ are shown. It is obvious from this figure that $Y_\nu(z)$ is unbounded in the neighbourhood of the origin for all values of ν.

Note

Going back to the original Bessel's equation, (6.19), of order ν, with a parameter λ

$$x^2\frac{d^2y}{dx^2} + x\frac{dy}{dx}(\lambda^2x^2 - \nu^2)y = 0,$$

a complete solution can be written as

$$y(x) = c_1 J_\nu(\lambda x) + c_2 Y_\nu(\lambda x)$$

or as

$$y(x) = c_1 H_\nu^{(1)}(\lambda x) + c_2 H_\nu^{(2)}(\lambda x).$$

If ν is not an integer, a complete solution can also be written as

$$y(x) = c_1 J_\nu(\lambda x) + c_2 J_{-\nu}(\lambda x).$$

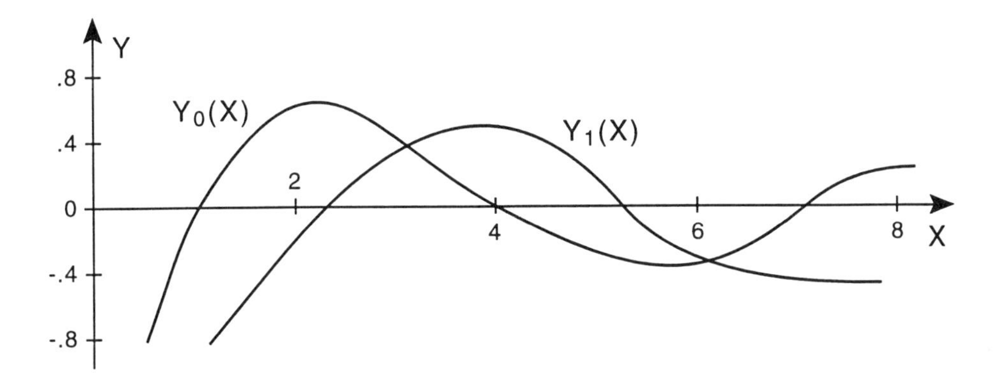

Figure 6.2. Plots of $Y_0(z)$ and $Y_1(z)$

6.3 Modified Bessel functions

In the previous section, it was shown that the Bessel functions $J_\nu(z)$, $Y_\nu(z)$, and $J_{-\nu}(z)$ are solutions of the differential equation

$$\frac{d^2y}{dz^2} + \frac{1}{z}\frac{dy}{dz} + (1 - \frac{\nu^2}{z^2})y = 0.$$

If z is replaced by $\imath z$, then we must write

$$\imath^{-1}\frac{dy}{dz} \quad \text{for} \quad \frac{dy}{dz}$$

and

$$-\frac{d^2y}{dz^2} \quad \text{for} \quad \frac{d^2y}{dz^2}.$$

With these substitutions, the above equation becomes

$$\frac{d^2y}{dz^2} + \frac{1}{z}\frac{dy}{dz} - (1 + \frac{\nu^2}{z^2})y = 0 \tag{6.45}$$

which is known as the Modified Bessel equation of order ν.

The complete solution of (6.45) is

$$y = A\,J_\nu(\imath z) + B\,J_{-\nu}(\imath z) \tag{6.46}$$

when ν is non-integral, and

$$y = A\,J_n(\imath z) + B\,Y_n(\imath z) \tag{6.47}$$

when $\nu = n$, an integer.

It is convenient in applications to give solutions in real variables instead of in complex variable form. Accordingly, we seek those modifications of Bessel functions which will be real functions of real variables. Writing

$$z\,e^{\pm\frac{\pi\imath}{2}} = \pm\imath z \quad \text{for} \quad z \text{ in (6.29)}$$

we obtain

$$\begin{aligned} J_\nu(z\,e^{\pm\frac{\imath\pi}{2}}) &= (e^{\pm\frac{\imath\pi\nu}{2}})\sum_{m=0}^{\infty}\frac{(\frac{z}{2})^{2m+\nu}}{m!\Gamma(\nu+m+1)} \\ &= e^{\pm\frac{\imath\pi\nu}{2}}\sum_{m=0}^{\infty}\frac{(\frac{z}{2})^{2m+\nu}}{m!\Gamma(\nu+m+1)}. \end{aligned}$$

Multiplying throughout by $e^{\pm\frac{\imath\pi\nu}{2}}$ yields

$$e^{\pm\frac{\imath\pi\nu}{2}}J_\nu(z\,e^{\pm\frac{\pi\imath}{2}}) = \sum_{m=0}^{\infty}\frac{(\frac{z}{2})^{\nu+2m}}{m!\Gamma(\nu+m+1)}.$$

Now, defining the function

$$I_\nu(z) = e^{\pm\frac{\imath\pi\nu}{2}} J_\nu(z\, e^{\pm\frac{\pi\imath}{2}}) \tag{6.48}$$

we have

$$I_\nu(z) = \sum_{m=0}^{\infty} \frac{\left(\frac{z}{2}\right)^{\nu+2m}}{m!\Gamma(\nu+m+1)} \tag{6.49}$$

which is known as the modified Bessel function of order ν.

Note this is a completely real function, identical to $J_\nu(z)$ except that its terms are all positive. If ν is not an integer, the function $I_{-\nu}(z)$ obtained from $I_\nu(z)$ by replacing ν by $-\nu$ throughout is a second, independent solution of (6.45). A complete solution can then be written as

$$y = A\, I_\nu(z) + B\, I_{-\nu}(z). \tag{6.50}$$

In applications, it often happens that a bounded solution is needed as $z \to +\infty$. Thus, many writers define the second solution of the modified Bessel equation to be the linear combination of $I_\nu(z)$ and $I_{-\nu}(z)$ as follows:

$$K_\nu(z) = \frac{\pi}{2}\frac{I_{-\nu}(z) - I_\nu(z)}{\sin \nu\pi}, \qquad \nu \text{ not an integer.} \tag{6.51}$$

This is known as the Modified Bessel function of the second kind of order ν.

Corresponding to $\nu = n$, an integer, we can see from (6.48) that

$$I_n(z) = \imath^{-n} J_n(\imath z).$$

Therefore, writing $-n$ for n,

$$I_{-n}(z) = \imath^n J_{-n}(\imath z) = (-1)^n \imath^n J_n(\imath z) = \imath^{-n} J_n(\imath z). \tag{6.52}$$

Therefore

$$I_n(z) = I_{-n}(z). \tag{6.53}$$

According to (6.49),

$$I_n(-z) = (-1)^n I_n(z). \tag{6.54}$$

For $\nu = n$, an integer, (6.51) takes the indeterminate form $0/0$, and so we use L'Hospital's rule

$$\begin{aligned} K_n(z) &= \lim_{\nu\to n} K(z) \\ &= \frac{\pi}{2}\left[\frac{\frac{\partial}{\partial\nu}(I_{-\nu}(z) - I_\nu(z))}{\frac{\partial}{\partial\nu}(\sin \pi\nu)}\right]_{\nu=n}. \end{aligned} \tag{6.55}$$

Proceeding as in the previous section, we find that

$$\begin{aligned} K_n(z) &= (-1)^{n+1}\ell n(\frac{z}{2})I_n(z) + \frac{1}{2}\sum_{m=0}^{n-1}\frac{(-1)^m(n-m-1)!}{m!}(\frac{2}{z})^{n-2m} \\ &+\frac{(-1)^n}{2}\sum_{m=0}^{\infty}\frac{(\frac{z}{2})^{n+2m}}{m!(n+m)!}[\psi(n+m+1)+\psi(m+1)] \qquad (6.56) \\ &= (-1)^{n+1}\{\gamma+\ell n(\frac{z}{2})\}I_n(z) + \frac{1}{2}\sum_{m=0}^{n-1}\frac{(-1)^m(n-m-1)!}{m!}(\frac{2}{z})^{n-2m} \\ &+\frac{(-1)^n}{2}\sum_{m=0}^{\infty}\frac{(\frac{z}{2})^{n+2m}}{m!(n+m)!}[(1+\frac{1}{2}+\frac{1}{3}+\cdots+\frac{1}{m}) \\ &+(1+\frac{1}{2}+\cdots+\frac{1}{n+m})] \qquad (6.57) \end{aligned}$$

which is a solution independent of $I_n(z)$

Plots of $I_0(z)$, $I_1(z)$ and $K_0(z)$, $K_1(z)$ are shown in Fig. 6.3.

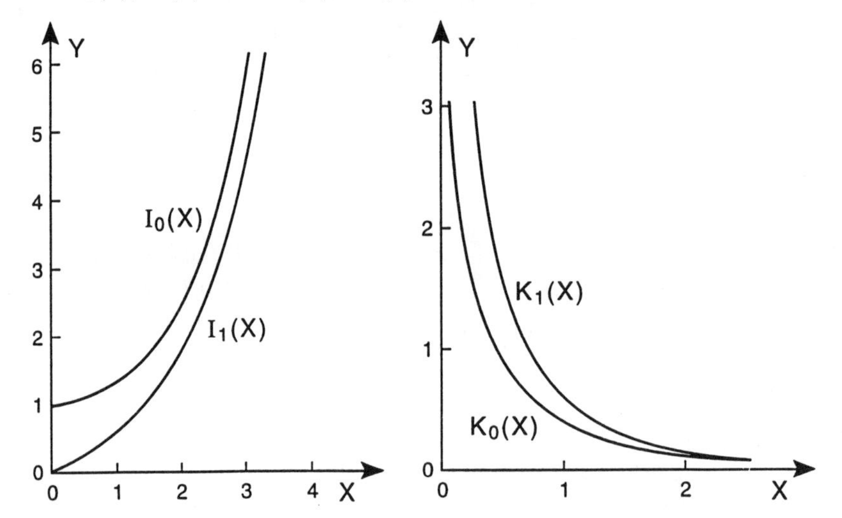

Figure 6.3. Modified Bessel functions: $I_0(z)$, $I_1(z)$, $K_0(z)$ and $K_1(z)$

When z is replaced by λx, we see that (6.45) can be transformed to

$$\frac{d^2y}{dx^2}+\frac{1}{x}\frac{dy}{dx}-(\lambda^2+\frac{\nu^2}{x^2})y=0 \qquad (6.58)$$

where λ is a parameter.

A complete solution of this is, of course,

$$y = c_1 I_\nu(\lambda x) + c_2 K_\nu(\lambda x) \quad , \quad \text{always} \qquad (6.58a)$$
$$y = c_1 I_\nu(\lambda x) + c_2 I_{-\nu}(\lambda x) \quad , \quad \nu \text{ non-integral} \qquad (6.58b)$$
$$y = c_1 I_n(\lambda x) + c_2 K_n(\lambda x) \quad , \quad \nu = n, \text{ an integral} \qquad (6.58c)$$

From Fig. 6.3, it can be observed that if $\nu \geq 0$, $z > 0$, all I-functions are monotonic increasing (without oscillation) to $+\infty$, where they are mutually asymptotic. All K-functions are monotonic decreasing to zero, where they are mutually asymptotic.

6.4 Ber, Bei, Ker, and Kei functions

In certain engineering problems, it is necessary to solve the differential equation

$$x^2\frac{d^2y}{dx^2}+x\frac{dy}{dx}+(-\imath x^2-\nu^2)y=0. \tag{6.59}$$

This equation is similar to Bessel's equation and so can be regarded as Bessel's equation of order ν with parameter $\lambda=\sqrt{-\imath}$ or as the modified Bessel equation of order ν with parameter $\lambda=\sqrt{\imath}$. In the first instance, a complete solution is

$$y=c_1J_\nu(\sqrt{-\imath}x)+c_2Y_\nu(\sqrt{-\imath}x), \tag{6.60}$$

and, in the second instance, the solution is

$$y=A\,I_\nu(\sqrt{\imath}x)+B\,K_\nu(\sqrt{\imath}x). \tag{6.61}$$

As we know, a complete solution can be constructed from any pair of independent particular solutions. Here we choose $J_\nu(\sqrt{-\imath}x)$ and $K_\nu(\sqrt{\imath}x)$ to form the complete solution

$$\begin{aligned} y &= A\,J_\nu(\sqrt{-\imath}x)+B\,K_\nu(\sqrt{\imath}x)\\ &= A\,J_\nu(\imath^{3/2}x)+B\,K_\nu(\imath^{1/2}x) \end{aligned} \tag{6.62}$$

where $-\imath=\imath^3$ and $(-\imath)^{1/2}=\imath^{3/2}$.
Now consider

$$J_\nu(\imath^{3/2}x)=\sum_{m=0}^{\infty}\frac{(-1)^m(\imath^{3/2}x)^{\nu+2m}}{2^{\nu+2m}m!\Gamma(\nu+m+1)}=\imath^{\frac{3\nu}{2}}\sum_{m=0}^{\infty}\frac{(-1)^m\imath^{3m}x^{\nu+2m}}{2^{\nu+2m}m!\Gamma(\nu+m+1)}.$$

It can be easily observed that $\imath^{3m}$ can take on only one of the following values

$$\begin{array}{rl} 1 & \text{when}\quad m=0,4,8,\cdots\\ -\imath & \text{when}\quad m=1,5,9,\cdots\\ -1 & \text{when}\quad m=2,6,10,\cdots\\ \imath & \text{when}\quad m=3,7,11,\cdots \end{array}$$

Hence, for the values $m=0,2,4,6,\cdots$, the terms in the series for $J_\nu(\imath^{3/2}x)$ are real and when $m=1,3,5,\cdots$, these terms are imaginary.

Thus, when $m=2\jmath$, $\jmath=0,1,2,3,\cdots$, then

$$\begin{aligned} (-1)^m\imath^{3m}=(-1)^{2\jmath}\imath^{6\jmath} &= \imath^{4\jmath}\imath^{6\jmath}\\ &= \imath^{10\jmath}\\ &= (-1)^{5\jmath}=(-1)^{\jmath} \end{aligned}$$

and, when $m=2\jmath+1$, $\jmath=0,1,2,3,\cdots$, then

$$\begin{aligned} (-1)^m\imath^{3m} &= (-1)^{2\jmath+1}\imath^{6\jmath+3}\\ &= \imath^{4\jmath+2}\imath^{6\jmath+3}\\ &= \imath^5\imath^{10\jmath}\\ &= \imath(-1)^{\jmath}. \end{aligned}$$

Using these observations to separate the series into its real and imaginary parts, we obtain, after changing $\jmath$ to m again,

$$\begin{aligned} J_\nu(\imath^{3/2}x) &= \imath^{\frac{3\nu}{2}}[\sum_{m=0}^{\infty} \frac{(-1)^m x^{\nu+4m}}{2^{\nu+4m}(2m)!\Gamma(\nu+2m+1)} \\ &+\imath \sum_{m=0}^{\infty} \frac{(-1)^m x^{\nu+2+4m}}{2^{\nu+2+4m}(2m+1)!\Gamma(\nu+2m+2)}]. \end{aligned}$$

Therefore

$$\begin{aligned} J_\nu(\imath^{3/2}x) &= \imath^{3\nu/2}(\Sigma Re + \imath\,\Sigma Im) \\ &= (\cos\frac{3\pi\nu}{4} + \imath\sin\frac{3\pi\nu}{4})(\Sigma Re + \imath\,\Sigma Im) \\ &= (\cos\frac{3\pi\nu}{4}\Sigma Re - \sin\frac{3\pi\nu}{4}\Sigma Im) \\ &+\imath(\cos\frac{3\pi\nu}{4}\Sigma Im + \sin\frac{3\pi\nu}{4}\Sigma Re). \end{aligned} \tag{6.63}$$

Thus, we are in a position to define the following quantities.

The real part of $J_\nu(\imath^{3/2}x)$ is defined as the function $ber\,x$ and the imaginary part of $J_\nu(\imath^{3/2}x)$ is defined as the function $bei\,x$. The letters '*be*' suggest the relation between these new functions and the Bessel functions themselves. The terminal letters 'r' and '$\imath$' stand for the adjectives real and imaginary.

In engineering applications, an important case arises when $\nu = 0$. In this situation, we have

$$ber_0 x \equiv ber\,x = \sum_{m=0}^{\infty} \frac{(-1)^m x^{4m}}{2^{4m}[(2m)!]^2} \tag{6.64}$$

$$bei_0 x \equiv bei\,x = \sum_{m=0}^{\infty} \frac{(-1)^m x^{4m+2}}{2^{4m+2}[(2m+1)!]^2}. \tag{6.65}$$

Similarly, we can define $ker_\nu x$ and $kei_\nu x$ as

$$Re[K_\nu(\imath^{1/2}x)] = ker_\nu x \tag{6.66}$$

$$Im[K_\nu(\imath^{1/2}x)] = kei_\nu x. \tag{6.67}$$

Here, the letters '*ke*' are derived from the name of the British mathematical physicist Lord Kelvin (1824–1907).

For the case $\nu = 0$, we get

$$\begin{aligned} K_0(\imath^{1/2}x) &= -\{\gamma + \ell n(\frac{1}{2}x\,\imath^{1/2})\}I_0(\imath^{1/2}x) \\ &+ \sum_{m=1}^{\infty} \frac{(\frac{1}{2}x\,\imath^{1/2})^{2m}}{(m!)^2}\{1+\frac{1}{2}+\cdots+\frac{1}{m}\} \end{aligned} \tag{6.68}$$

$$\begin{aligned} &= -\{\gamma + \ell n(\frac{x}{2}) + \frac{1}{4}\pi\imath\}(ber\,x + \imath\,bei\,x) \\ &+\imath[(\frac{x}{2})^2 - \frac{(\frac{x}{2})^6}{(3!)^2}(1+\frac{1}{2}+\frac{1}{3}) + \cdots] \\ &-[\frac{(\frac{x}{2})^4}{(2!)^2}(1+\frac{1}{2}) - \frac{(\frac{x}{2})^8}{(4!)^2}(1+\frac{1}{2}+\frac{1}{3}+\frac{1}{4}) + \cdots]. \end{aligned} \tag{6.69}$$

Since

$$J_0(\imath^{3/2}x) = I_0(\imath^{1/2}x) = ber\, x + \imath\, bei\, x,$$

equating real and imaginary parts leads to

$$\begin{aligned} ker\, x &= (\ell n2 - \gamma - \ell n x) ber\, x + \frac{\pi}{4} bei\, x \\ &\quad - \frac{(\frac{x}{2})^4}{(2!)^2}(1 + \frac{1}{2}) + \frac{(\frac{x}{2})^8}{(4!)^2}(1 + \frac{1}{2} + \frac{1}{3} + \frac{1}{4}) \cdots \end{aligned} \tag{6.70}$$

and

$$\begin{aligned} kei\, x &= (\ell n2 - \gamma - \ell n x) bei\, x - \frac{\pi}{4} ber\, x \\ &\quad + (\frac{x}{2})^2 - \frac{(\frac{x}{2})^6}{(3!)^2}(1 + \frac{1}{2} + \frac{1}{3}) \cdots \end{aligned} \tag{6.71}$$

where $\ell n2 - \gamma = 0.1159\cdots$ and $\gamma = 0.5772\cdots$.

In Fig. 6.4, $ber\, x$ and $bei\, x$ are plotted. The graphs show the oscillations with increasing amplitudes.

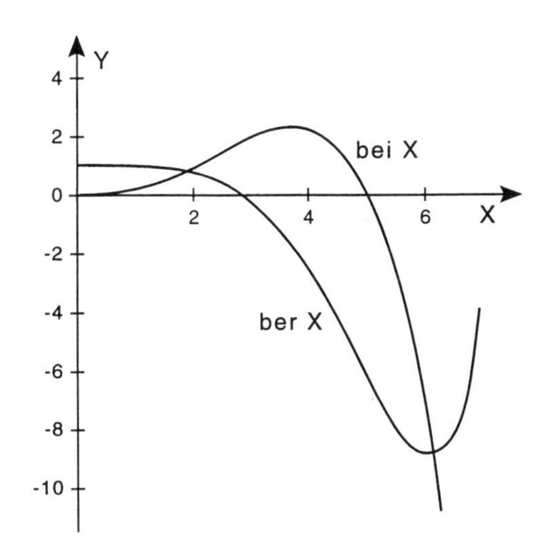

Figure 6.4. The functions $ber\, x$ and $bei\, x$

A complete solution of (6.59) is then written as

$$y = c_1(ber_\nu x + \imath\, bei_\nu x) + c_2(ker_\nu x + \imath\, kei_\nu x). \tag{6.72}$$

The functions $ber_\nu x$ and $bei_\nu x$ are finite at the origin but become infinite when x approaches infinity; however, the functions $ker_\nu x$ and $kei_\nu x$ are infinite at the origin but approach zero when x approaches infinity.

As we know, for real x, the quantity $ber_\nu x + \imath\, bei_\nu x$ is a complex number and hence

$$ber_\nu x + \imath\, bei_\nu x = M(x)e^{\imath\Theta}\nu^{(x)}$$

where

$$M_\nu(x) = \sqrt{ber_\nu^2 x + bei_\nu^2 x}$$

and

$$\theta_\nu(x) = \tan^{-1}\{\frac{bei_\nu x}{ber_\nu x}\}.$$

Similarly,

$$ker_\nu x + \imath\, kei_\nu x = N_\nu(x)e^{\imath\phi_\nu(x)}$$

where

$$\begin{aligned} N_\nu(x) &= \sqrt{ker_\nu^2 x + kei_\nu^2 x} \\ \phi_\nu(x) &= \tan^{-1}\{\frac{kei_\nu x}{ker_\nu x}\}. \end{aligned}$$

Thus, we have

$$\begin{aligned} ber_\nu x &= M_\nu(x)\cos\theta_\nu(x) \\ bei_\nu x &= M_\nu(x)\sin\theta_\nu(x) \\ ker_\nu x &= N_\nu(x)\cos\phi_\nu(x) \\ kei_\nu x &= N_\nu(x)\sin\phi_\nu(x). \end{aligned}$$

In particular, when $\nu = 0$,

$$\begin{aligned} ber\, x &= M_0(x)\cos\theta_0(x) \\ bei\, x &= M_0(x)\sin\theta_0(x) \\ ker\, x &= N_0(x)\cos\theta_0(x) \\ kei\, x &= N_0(x)\sin\theta_0(x). \end{aligned}$$

$M_0(x)$, $\theta_0(x)$; $N_0(x)$, $\phi_0(x)$ are shown in Figs. 6.5 and 6.6.

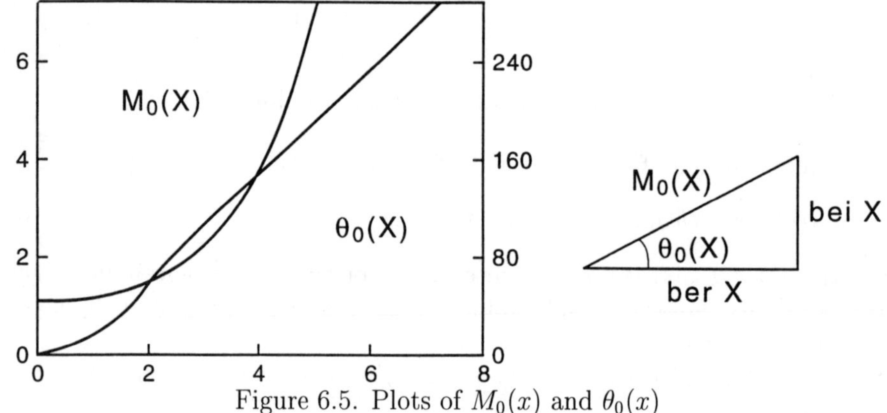

Figure 6.5. Plots of $M_0(x)$ and $\theta_0(x)$

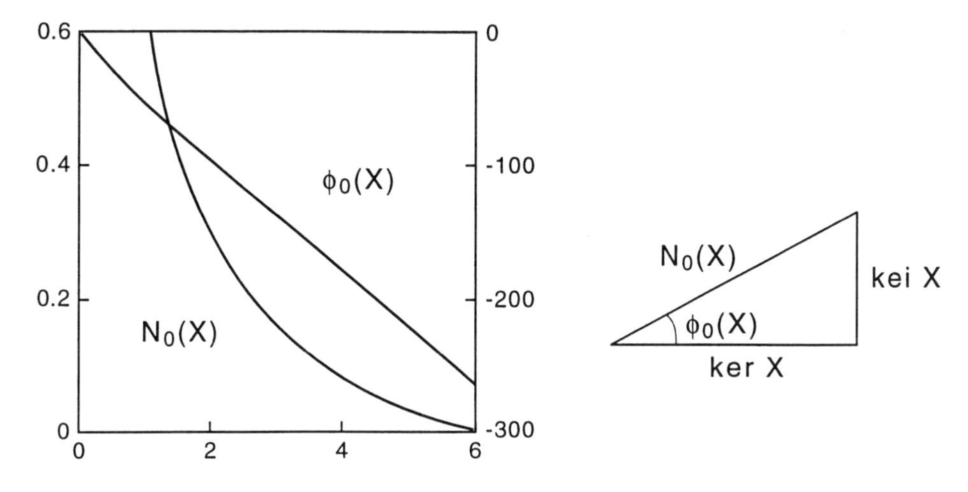

Figure 6.6. Plots of $N_0(x)$ and $\phi_0(x)$

Example 1

Show that $\int_0^x x\, ber\, x\, dx = x\, bei' x$, where the prime denotes differentiation.

Solution

Using the expression for $ber\, x$ from (6.64), we have

$$\begin{aligned}
&\int_0^x x\, ber\, x\, dx \\
&= \int_0^x [x - \frac{x^5}{2^2.4^2} + \frac{x^9}{2^2.4^2.6^2.8^2} - \cdots]dx \\
&= \frac{x^2}{2} - \frac{x^6}{2^2.4^2.6} + \frac{x^{10}}{2^2.4^2.6^2.8^2.10} - \cdots \\
&= x\, bei' x
\end{aligned}$$

or $x\, ber\, x = \frac{d}{dx}(x\, bei' x)$. Similarly, we can show that $\int_0^x x\, bei\, x\, dx = -x\, ber' x$ or $x\, bei\, x = -\frac{d}{dx}(x\, ber' x)$.

6.5 Equations solvable in terms of Bessel functions

There are some second-order ordinary differential equations which can be transformed into the Bessel equation through suitable transformations. The solutions can then be expressed as Bessel functions. Here we shall discuss only four important types:

Type I

Consider the differential equation

$$\frac{d^2y}{dz^2} + \frac{1-2\nu}{z}\frac{dy}{dz} + k^2 y = 0. \tag{6.73}$$

Substituting $y = vz^\nu$, we obtain

$$\frac{d^2v}{dz^2} + \frac{1}{z}\frac{dv}{dz} + (k^2 - \frac{\nu^2}{z^2})v = 0. \tag{6.74}$$

We know a complete solution of (6.74) is

$$v = A\,J_\nu(kz) + B\,Y_\nu(kz). \tag{6.75}$$

Therefore, a complete solution of (6.73) is

$$y = z^\nu(A\,J_\nu(kz) + B\,Y_\nu(kz)). \tag{6.76}$$

Type II

If in the differential equation (6.74) we substitute $z = x^\beta$, we obtain

$$\frac{d^2v}{dx^2} + \frac{1}{x}\frac{dv}{dx} + (\beta^2 k^2 x^{2\beta-2} - \frac{\beta^2\nu^2}{x^2})v = 0. \tag{6.77}$$

Hence a solution of (6.77) can be written as

$$v = A\,J_\nu(kx^\beta) + B\,Y_\nu(kx^\beta). \tag{6.78}$$

Type III

If in the differential equation (6.77) a transformation is effected by putting $v = u\,x^{\alpha-\beta\nu}$, we get

$$\frac{d^2u}{dx^2} + (\frac{2\alpha - 2\beta\nu + 1}{x})\frac{du}{dx} + \{\beta^2 k^2 x^{2\beta-2} + \frac{\alpha(\alpha-2\beta\nu)}{x^2}\}u = 0 \tag{6.79}$$

and hence a complete solution of (6.79) is given by

$$u = x^{\beta\nu-\alpha}\{A\,J_\nu(k\,x^\beta) + B\,Y_\nu(k\,x^\beta)\}. \tag{6.80}$$

Type IV

If $(1 - a^2) \geq 4c$ and if neither d, p, nor q is zero, then, except when it reduces to Euler-Cauchy type equation, the differential equation

$$\frac{d^2y}{dx^2} + \frac{a + 2b\,x^p}{x}\frac{dy}{dx} + [\frac{c + d\,x^{2q} + b(a+p-1)x^p + b^2\,x^{2p}}{x^2}]y = 0 \tag{6.81}$$

has a complete solution

$$y = x^{\alpha}e^{-\beta x^p}[c_1 J_\nu(\lambda x^q) + c_2 Y_\nu(\lambda x^q)] \tag{6.82}$$

where the parameters α, β, λ and ν are related to

$$\alpha = \frac{1-a}{2} \quad , \quad \beta = \frac{b}{p} \quad , \quad \lambda = \frac{\sqrt{|d|}}{q} \quad \text{and} \quad \nu = \frac{\sqrt{(1-a^2) - 4c}}{2q}.$$

When $d < 0$, J_ν and Y_ν are replaced by I_ν and K_ν respectively; and if ν is not an integer, Y_ν and K_ν are replaced by $J_{-\nu}$ and $I_{-\nu}$.

It can be seen by the transformations

$$y = x^{\alpha}\,e^{-\beta x^p}\,v \quad \text{and} \quad x^q = \frac{qz}{\sqrt{|d|}}$$

that equation (6.81) can be reduced to Bessel's equation. The proof is left to the reader.

Example 2

Find a complete solution of the equation

$$x\,y'' - y' + 4x^5 y = 0.$$

Solution

The given equation can be written as

$$y'' - \frac{1}{x}y' + 4x^4 y = 0. \tag{1}$$

Comparing (1) with

$$y'' + \frac{2\alpha - 2\beta\nu + 1}{x}y' + [\beta^2 k^2 x^{2\beta-2} + \frac{\alpha(\alpha - 2\beta\nu)}{x^2}]y = 0, \tag{2}$$

we have $\alpha(\alpha - 2\beta\nu) = 0$, $2\beta - 2 = 4$, $\beta^2 k^2 = 4$ and $2\alpha - 2\beta\nu + 1 = -1$.

From the first equation, $\alpha = 0$ or $\alpha = \beta\nu$. Corresponding to $\alpha = 0$, we have $\beta = 3$, $k = \pm 2/3$ and $\nu = 1/3$. Corresponding to $\alpha = -2$, we have $\beta = 3$, $k = \pm 2/3$ and $\nu = -1/3$.

Therefore, a complete solution of (1) is

$$\begin{aligned} y &= x^{\beta\nu - \alpha}[A\,J_\nu(k\,x^\beta) + B\,J_{-\nu}(k\,x^\beta)] \\ &= x[A\,J_{1/3}(\frac{2}{3}x^3) + B\,J_{-1/3}(\frac{2}{3}x^3)]. \end{aligned}$$

Example 3

Find a complete solution of the equation

$$y'' + x^m y = 0.$$

Solution

Comparing the above equation with (6.79), we have $\alpha(\alpha - 2\beta\nu) = 0, 2\alpha - 2\beta\nu + 1 = 0$, $2\beta - 2 = m$ and $\beta^2 k^2 = 1$. Solving the third and fourth equations, we have $\beta = \frac{m+2}{2}$ and $k = (\frac{2}{2+m})$. From the second, we have $\beta\nu - \alpha = \frac{1}{2}$ and, from the first, $\alpha = 0$ or $\alpha = 2\beta\nu$. Now, if $\alpha = 0$, then $\nu = \frac{1}{m+2}$. Also, if $\alpha = 2\beta\nu$, then $\nu = \frac{-1}{m+2}$. Hence a complete solution is

$$\begin{aligned} y &= x^{\beta\nu - \alpha}[A\,J_\nu(k\,x^\beta) + B\,Y_\nu(k\,x^\beta)] \\ &= x^{1/2}[A\,J_{(\frac{1}{2+m})}(\frac{2}{2+m}x^{\frac{2+m}{2}}) + B\,Y_{(\frac{1}{2+m})}(\frac{2}{2+m}x^{\frac{2+m}{2}})]. \end{aligned}$$

Example 4

Find a complete solution of the equation

$$x^2 y'' + (2x^2 + x)y' + (x^2 + 3x - 1)y = 0. \quad (1)$$

Solution

Compare (1) with

$$x^2 y'' + x[a + 2bx^p]y' + [c + d\,x^{2q} + b(a + p - 1)x^p + b^2 x^{2p}]y = 0. \quad (2)$$

Then, we have $a = 1, b = 1, p = 1, c = -1, q = \frac{1}{2}, d = 2, \alpha = 0, \beta = 1, \lambda = \frac{\sqrt{|d|}}{q} = 2\sqrt{2}$, and $\nu = \frac{\sqrt{(1-a)^2 - 4c}}{2q} = 2$

Hence, a complete solution is

$$\begin{aligned} y &= x^\alpha e^{-\beta x^p}[(A\,J_\nu(\lambda x^q) + B\,Y_\nu(\lambda x^q)] \\ &= e^{-x}[A\,J_2(2\sqrt{2x}) + B\,Y_2(2\sqrt{2x})]. \end{aligned}$$

6.6 Recurrence relations of Bessel functions

There are many Bessel function identities; among these, the following eight are the most important:

Formula 1 : $\frac{d}{dx}(x^\nu J_\nu(x)) = x^\nu J_{\nu-1}(x).$

Formula 2 : $\frac{d}{dx}(x^{-\nu} J_\nu(x)) = -x^{-\nu} J_{\nu+1}(x).$

Formula 3 : $\frac{d}{dx}(x^{\nu} I_{\nu}(x)) = x^{\nu} I_{\nu-1}(x).$

Formula 4 : $\frac{d}{dx}(x^{-\nu} I_{\nu}(x)) = -x^{-\nu} I_{\nu+1}(x).$

Formula 5 : $\frac{d}{dx}(x^{\nu} Y_{\nu}(x)) = x^{\nu} Y_{\nu-1}(x).$

Formula 6 : $\frac{d}{dx}(x^{-\nu} Y_{\nu}(x)) = -x^{-\nu} Y_{\nu+1}(x).$

Formula 7 : $\frac{d}{dx}(x^{\nu} K_{\nu}(x)) = x^{\nu} K_{\nu-1}(x).$

Formula 8 : $\frac{d}{dx}(x^{-\nu} K_{\nu}(x)) = -x^{-\nu} K_{\nu+1}(x).$

In a similar way, we can extend the Hankel functions $H_{\nu}^{(1)}(x)$ and $H_{\nu}^{(2)}(x)$ such that

Formula 9 : $\frac{d}{dx}(x^{\nu} H_{\nu}^{(1)}(x)) = x^{\nu} H_{\nu-1}^{(1)}(x)$

Formula 10 : $\frac{d}{dx}(x^{\nu} H_{\nu}^{(2)}(x)) = x^{\nu} H_{\nu-1}^{(2)}(x)$

We demonstrate the proof of Formula 1 and Formula 2 below:

Proof of Formula 1

We know that

$$
\begin{aligned}
J_{\nu}(x) &= \sum_{m=0}^{\infty} \frac{(-1)^m x^{\nu+2m}}{2^{\nu+2m} m! \Gamma(\nu+m+1)} \\
x^{\nu} J_{\nu}(x) &= \sum_{m=0}^{\infty} \frac{(-1)^m x^{2(\nu+m)}}{2^{\nu+2m} m! \Gamma(\nu+m+1)} \\
\frac{d}{dx}(x^{\nu} J_{\nu}(x)) &= \sum_{m=0}^{\infty} \frac{(-1)^m 2(\nu+m) x^{2(\nu+m)-1}}{2^{\nu+2m} m! \Gamma(\nu+m+1)} \\
&= x^{\nu} \sum_{m=0}^{\infty} \frac{(-1)^m x^{\nu+2m-1}}{2^{\nu+2m-1} (m!) \Gamma(\nu+m)} \\
&= x^{\nu} J_{\nu-1}(x).
\end{aligned}
$$

Proof of Formula 2

$$
\begin{aligned}
x^{-\nu} J_{\nu}(x) &= \sum_{m=0}^{\infty} \frac{(-1)^m x^{2m}}{2^{\nu+2m} m! \Gamma(\nu+m+1)} \\
\frac{d}{dx} x^{-\nu} J_{\nu}(x) &= \sum_{m=0}^{\infty} \frac{(-1)^m (2m) x^{2m-1}}{2^{\nu+2m} m! \Gamma(\nu+m+1)}
\end{aligned}
$$

$$\begin{aligned}
&= -x^{-\nu}\sum_{m=0}^{\infty}\frac{(-1)^{m+1}(2m)x^{\nu+2m-1}}{2^{\nu+2m}m!\Gamma(\nu+m+1)}\\
&= -x^{-\nu}\sum_{m=0}^{\infty}\frac{(-1)^{m+1}x^{\nu+2m-1}}{2^{\nu+2m-1}(m-1)!\Gamma(\nu+m+1)}\\
\frac{d}{dx}(x^{-\nu}J_\nu(x)) &= -x^{-\nu}\sum_{m=0}^{\infty}\frac{(-1)^{m+1}x^{\nu+2m-1}}{2^{\nu+2m-m}(m-1)!\Gamma(\nu+m+1)}\\
&= -x^{-\nu}\sum_{m=0}^{\infty}\frac{(-1)^{m}x^{\nu+2m+1}}{2^{\nu+2m+1}(m!)\Gamma(\nu+m+2)}\\
&= -x^{-\nu}J_{\nu+1}(x).
\end{aligned}$$

Now, from Formula 1 and Formula 2, after differentiating, we obtain

$$\begin{aligned}
x^\nu J_\nu'(x) + \nu x^{\nu-1}J_\nu(x) &= x^\nu J_{\nu-1}(x)\\
x^{-\nu}J_\nu'(x) - \nu x^{-\nu-1}J_\nu(x) &= -x^\nu J_{\nu+1}(x).
\end{aligned}$$

Then, dividing the first equation by x^ν and the second by $x^{-\nu}$, we have

$$J_\nu'(x) = J_{\nu-1}(x) - \frac{\nu}{x}J_\nu(x) \tag{6.83}$$

$$J_\nu'(x) = \frac{\nu}{x}J_\nu(x) - J_{\nu+1}(x). \tag{6.84}$$

Adding

$$2J_\nu'(x) = J_{\nu-1}(x) - J_{\nu+1}(x). \tag{6.85}$$

Subtracting

$$J_{\nu+1}(x) = \frac{2\nu}{x}J_\nu(x) - J_{\nu-1}(x). \tag{6.86}$$

Similarly we can obtain many more identities.

Use of a generating function to obtain Bessel function identities

Consider the expansion of the function

$$\begin{aligned}
e^{\{\frac{x}{2}(t-\frac{1}{t})\}} &= e^{\frac{xt}{2}}.e^{-\frac{x}{2t}}\\
&= [1+\frac{xt}{2}+\frac{1}{2!}(\frac{xt}{2})^2+\frac{1}{3!}(\frac{xt}{2})^3+\cdots]\\
&\quad\times[1-\frac{x}{2t}+\frac{1}{2!}(\frac{x}{2t})^2-\frac{1}{3!}(\frac{x}{2t})^3+\cdots]\\
&= (\sum_{m=0}^{\infty}\frac{(\frac{xt}{2})^m}{m!})(\sum_{n=0}^{\infty}\frac{(\frac{x}{2t})^n(-1)^n}{n!})\\
&= \sum_{m=0}^{\infty}\sum_{n=0}^{\infty}\frac{(\frac{x}{2})^{m+n}t^{m-n}(-1)^n}{m!n!}.
\end{aligned} \tag{6.87}$$

Now, to collect the coefficient of t^ℓ, we substitute $m - n = \ell$ in the form $m = n + \ell$ which yields

$$\sum_{n=0}^{\infty} \frac{(\frac{x}{2})^{2n+\ell}(-1)^n}{n!(n+\ell)!} = J_\ell(x). \tag{6.88}$$

Similarly, to collect the coefficient of $t^{-\ell}$, we have to substitute $m - n = -\ell$ in the form $n = m + \ell$, which then yields

$$\begin{aligned} &\sum_{m=0}^{\infty} \frac{(\frac{x}{2})^{2m+\ell}(-1)^{m+\ell}}{m!(m+\ell)!} \\ &= (-1)^\ell \sum_{m=0}^{\infty} \frac{(\frac{x}{2})^{2m+\ell}(-1)^m}{m!(m+\ell)!} \\ &= (-1)^\ell J_\ell(x). \end{aligned} \tag{6.89}$$

Hence

$$\begin{aligned} &e^{\{\frac{x}{2}(t-\frac{1}{t})\}} \\ &= J_0(x) + \sum_{\ell=1}^{\infty} J_\ell(x)\{t^\ell + (-1)^\ell t^{-\ell}\}. \end{aligned}$$

Put

$$t = e^{i\theta}, \qquad \frac{1}{2}(t - \frac{1}{t}) = \frac{e^{i\theta} - e^{-i\theta}}{2} = i\,\sin\theta$$

and

$$\begin{aligned} t^\ell + (-1)^\ell \frac{1}{t^\ell} &= e^{\ell i\theta} + (-1)^\ell e^{-i\ell\theta} \\ &= e^{2ik\theta} + e^{-2ik\theta}, \qquad \text{when } \ell = 2k, \text{ even} \\ &= 2\cos(2k\theta). \end{aligned}$$

Also

$$\begin{aligned} t^\ell + (-1)^\ell t^{-\ell} &= e^{\ell i\theta} + (-1)^\ell e^{-i\ell\theta} \\ &= e^{(2k-1)i\theta} + (-1)^{(2k-1)} e^{-(2k-1)i\theta}, \qquad \text{when } \ell = 2k-1, \text{ odd} \\ &= e^{(2k-1)i\theta} - e^{-(2k-1)i\theta} \\ &= 2i\sin(2k-1)\theta \end{aligned}$$

Therefore, from (6.87), we have

$$e^{(ix\sin\theta)} = J_0(x) + 2\sum_{k=1}^{\infty} J_{2k}(x)\cos(2k\theta) + 2i\sum_{k=1}^{\infty} J_{(2k-1)}\sin(2k-1)\theta. \tag{6.90}$$

Equating the real and imaginary parts, we obtain

$$\cos(x\sin\theta) = J_0(x) + 2\sum_{k=1}^{\infty} J_{2k}(x)\cos(2k\theta) \tag{6.91}$$

$$\sin(x\sin\theta) = 2\sum_{k=1}^{\infty} J_{2k-1}\sin(2k-1)\theta. \tag{6.92}$$

The series on the right in (6.91) and (6.92) appear to be the Fourier cosine and sine expansion respectively.

Multiply both sides of (6.91) by $\cos n\theta$ and both sides of (6.92) by $\sin n\theta$ and integrate each identity with respect to θ from 0 to π. Since

$$\int_0^\pi \cos m\theta \cos n\theta d\theta = 0 \quad m \neq n$$

and

$$\begin{aligned} \int_0^\pi \sin m\theta \sin n\theta d\theta &= 0 \\ \int_0^\pi \cos^2 n\theta d\theta = \int_0^\pi \sin^2 n\theta d\theta &= \frac{\pi}{2}, \end{aligned}$$

it follows that

$$\int_0^\pi \cos(x \sin\theta) \cos n\theta d\theta = \begin{cases} \pi J_n(x) & n \text{ even} \\ 0 & n \text{ odd} \end{cases} \tag{6.93}$$

$$\int_0^\pi \sin(x \sin\theta) \sin n\theta d\theta = \begin{cases} 0 & n \text{ even} \\ \pi J_n(x) & n \text{ odd.} \end{cases} \tag{6.94}$$

Hence, by adding (6.93) and (6.94), we have, for all positive integral values of n,

$$\int_0^\pi (\cos(x \sin\theta) \cos n\theta + \sin(x \sin\theta) \sin n\theta) d\theta = \pi\, J_n(x). \tag{6.95}$$

The integrand in (6.95) may be simplified, and we obtain

$$\int_0^\pi \cos(n\theta - x \sin\theta) d\theta = \pi\, J_n(x). \tag{6.96}$$

In general we obtain $\frac{1}{2\pi}\int_{-\pi}^{\pi} e^{i(n\theta - \lambda r \sin(\theta+\phi))} d\theta = e^{-in\phi} J_n(\lambda r)$.

Writing $(\frac{\pi}{2} - \theta)$ for θ in (6.90), yields

$$\begin{aligned} e^{ix\cos\theta} &= J_0(x) + 2\sum_{k=1}^{\infty} J_{2k}(x) \cos 2k\theta \\ &\quad + 2i \sum_{k=1}^{\infty} (-1)^{k-1} J_{2k-1} \cos(2k-1)\theta \\ &= \sum_{k=0}^{\infty} \epsilon_k i^k J_k \cos k\theta \end{aligned} \tag{6.97}$$

where $\epsilon = 1$, $\epsilon_n = 2$, $n \geq 1$.

Now, equating real and imaginary parts, gives

$$\cos(x\cos\theta) = J_0(x) + 2\sum_{k=1}^{\infty} (-1)^k J_2 k(x) \cos 2k\theta \tag{6.98}$$

$$\sin(x\cos\theta) = \sum_{k=1}^{\infty} (-1)^{k-1} J_{2k-1}(x) \cos(2k-1)\theta. \tag{6.99}$$

When $\theta = 0$, we obtain

$$\cos x = J_0(x) + 2\sum_{k=1}^{\infty}(-1)^k J_{2k}(x) \tag{6.100}$$

$$\sin x = 2\sum_{k=1}^{\infty}(-1)^{k-1} J_{2k-1}(x). \tag{6.101}$$

These formulas show a relationship between circular functions and Bessel functions of the first kind.

Example 5

If ν is not an integer, show that

$$J_\nu(x)J'_{-\nu}(x) - J'_\nu(x)J_{-\nu}(x) = -\frac{2}{\pi x}\sin \pi\nu.$$

Proof

Since J_ν and $J_{-\nu}$ are two independent solutions of Bessel's equation,

$$J''_\nu + \frac{1}{x}J'_\nu + (1 - \frac{\nu^2}{x^2})J_\nu = 0 \tag{1}$$

$$J''_{-\nu} + \frac{1}{x}J'_{-\nu} + (1 - \frac{\nu^2}{x^2})J_{-\nu} = 0. \tag{2}$$

Multiplying (1) by $J_{-\nu}$ and (2) by J_ν and subtracting one from the other,

$$(J_\nu J''_{-\nu} - J''_\nu) + \frac{1}{x}(J_\nu J'_{-\nu} - J'_\nu J_{-\nu}) = 0.$$

Substituting $W = J_\nu J'_{-\nu} - J'_\nu J_{-\nu}$ in the above equation, we have $\frac{dW}{dx} + \frac{W}{x} = 0$, and then integrating, we obtain $W = \frac{A}{x}$. Therefore,

$$J_\nu J'_{-\nu} - J'_\nu J_{-\nu} = \frac{A}{x}.$$

Now we need to find A. By series expansion, we know

$$J_\nu = \sum_{m=0}^{\infty} \frac{(-1)^m x^{\nu+2m}}{2^{\nu+2m}(m!)\Gamma(\nu+m+1)}$$

and

$$J_{-\nu} = \sum_{m=0}^{\infty} \frac{(-1)^m x^{-\nu+2m}}{2^{-\nu+2m}m!\Gamma(-\nu+m+1)}$$

$$J'_\nu = \sum_{m=0}^{\infty} \frac{(-1)^m(\nu+2m)x^{\nu+2m-1}}{2^{\nu+2m}m!\Gamma(\nu+m+1)}$$

$$J'_{-\nu} = \sum_{m=0}^{\infty} \frac{(-1)^m(-\nu+2m)x^{-\nu+2m-1}}{2^{-\nu+2m}m!\Gamma(-\nu+m+1)}.$$

Now,

$$\begin{aligned} J_\nu J'_{-\nu} &= \sum_{m=0}^{\infty}\sum_{n=0}^{\infty}\left(\frac{(-1)^m x^{\nu+2m}}{2^{\nu+2m}m!\Gamma(\nu+m+1)}\right)\left(\frac{(-1)^n(-\nu+2n)x^{-\nu+2n-1}}{2^{-\nu+2n}n!\Gamma(-\nu+n+1)}\right) \\ &= \sum_{m=0}^{\infty}\sum_{n=0}^{\infty}\frac{(-1)^{m+n}(-\nu+2n)x^{2(m+n)-1}}{2^{2(m+n)}m!n!\Gamma(\nu+m+1)\Gamma(-\nu+n+1)}. \end{aligned}$$

The coefficient of $\frac{1}{x}$ is

$$\frac{-\nu}{\Gamma(1+\nu)\Gamma(1-\nu)}.$$

Also,

$$J_{-\nu}J'_\nu = \sum_{m=0}^{\infty}\sum_{n=0}^{\infty}\frac{(-1)^{m+n}(\nu+2m)x^{2(m+n)-1}}{2^{2(m+n)}m!n!\Gamma(\nu+m+1)\Gamma(-\nu+n+1)}.$$

The coefficient of $\frac{1}{x}$ is

$$\frac{\nu}{\Gamma(1+\nu)\Gamma(1-\nu)}.$$

Thus,

$$\begin{aligned} J_\nu J'_{-\nu} - J'_\nu J_{-\nu} &= \frac{A}{x} \\ &= \frac{-2\nu}{x\Gamma(1+\nu)\Gamma(1-\nu)} \\ &= \frac{-2}{x\Gamma(\nu)\Gamma(1-\nu)}. \end{aligned}$$

But, we know that

$$\Gamma(\nu)\Gamma(1-\nu) = \frac{\pi}{\sin\nu\pi}.$$

Hence

$$J_\nu J'_{-\nu} - J'_\nu J_{-\nu} = -\frac{2}{\pi x}\sin\nu\pi$$

as required.

Example 6

Show that

$$\begin{aligned} J_0(x) &= \frac{1}{\pi}\int_0^\pi \cos(x\cos\theta)d\theta \\ &= \frac{1}{2\pi}\int_0^{2\pi} e^{\pm ix\cos\theta}d\theta. \end{aligned}$$

Proof

We know from (6.97) that

$$\begin{aligned} e^{\pm ix\cos\theta} &= J_0(x) + 2\sum_{k=1}^{\infty}(-1)^k J_{2k}(x)\cos 2k\theta \\ &\quad \pm 2i\sum_{k=1}^{\infty}(-1)^{k-1}J_{2k-1}(x)\cos(2k-1)\theta. \end{aligned}$$

Then, integrating with respect to θ from 0 to 2π, we see

$$\begin{aligned} \int_0^{2\pi} e^{\pm ix\cos\theta}d\theta &= J_0(x)\int_0^{2\pi} d\theta = 2\pi J_0(x) \\ J_0(x) &= \frac{1}{2\pi}\int_0^{2\pi} e^{\pm ix\cos\theta}d\theta \\ &= \frac{1}{2}[\int_0^{\pi} e^{\pm ix\cos\theta}d\theta + \int_{\pi}^{2\pi} e^{\pm ix\cos\theta}d\theta]. \end{aligned}$$

Replacing θ by $2\pi - \theta$ in the second integral, we have

$$\frac{1}{2\pi}\int_{\pi}^{0} e^{\pm ix\cos\theta}(-d\theta) = \frac{1}{2\pi}\int_0^{\pi} e^{\pm ix\cos\theta}d\theta.$$

Therefore,

$$\begin{aligned} J_0(x) &= \frac{1}{2\pi}[\int_0^{2\pi} e^{\pm ix\cos\theta}d\theta] \\ &= \frac{1}{\pi}\int_0^{\pi} e^{\pm ix\cos\theta}d\theta \\ &= \frac{1}{\pi}[\int_0^{\pi}(\cos(x\cos\theta) \pm i\sin(x\cos\theta))d\theta]. \end{aligned}$$

Equating real and imaginary parts

$$J_0(x) = \frac{1}{\pi}\int_0^{\pi}\cos(x\cos\theta)d\theta.$$

This is the required proof.

Example 7

Show that

$$\frac{d}{dx}[x^2 J_{\nu-1}(x)J_{\nu+1}(x)] = 2x^2 J_\nu(x)\frac{d}{dx}J_\nu(x).$$

Then, using the result, show that

$$\int x^2 J_n^2(x)dx = \frac{x^2}{2}[J_n^2(x) - J_{n-1}(x)J_{n+1}(x)] + C.$$

Proof

$$\begin{aligned}
\text{Left hand side} &= x^2 J'_{\nu-1}(x)J_{\nu+1}(x) + x^2 J_{\nu-1}(x)J'_{\nu+1}(x) \\
&\quad +2xJ_{\nu-1}(x)J_{\nu+1}(x) \\
&= x^2[\frac{\nu-1}{x}J_{\nu-1}(x) - J_\nu(x)]J_{\nu+1}(x) \\
&\quad +x^2[J_\nu(x) - \frac{\nu+1}{x}J_{\nu+1}(x)]J_{\nu-1}(x) \\
&\quad +2xJ_{\nu-1}(x)J_{\nu+1}(x) \\
&= x(\nu-1)J_{\nu-1}(x)J_{\nu+1}(x) - x^2 J_\nu(x)J_{\nu+1}(x) \\
&\quad +x^2 J_\nu(x)J_{\nu-1}(x) - x(\nu+1)J_{\nu+1}(x)J_{\nu-1}(x) \\
&\quad +2xJ_{\nu-1}(x)J_{\nu+1}(x) \\
&= x^2 J_\nu(x)(J_{\nu-1}(x) - J_{\nu+1}(x)) \\
&= 2x^2 J_\nu(x)J'_\nu.
\end{aligned}$$

Therefore,

$$\frac{d}{dx}[x^2 J_{\nu-1}(x)J_{\nu+1}(x)] = 2x^2 J_\nu(x)J'_\nu(x).$$

Integrating with respect to x, we obtain

$$\begin{aligned}
x^2 J_{\nu-1}(x)J_{\nu+1}(x) &= \int x^2 \frac{d}{dx}J_\nu^2(x)dx + C \\
&= x^2 J_\nu^2(x) - 2\int xJ_\nu^2(x)dx + C.
\end{aligned}$$

Therefore, we have

$$\int xJ_\nu^2(x)dx = \frac{x^2}{2}[J_\nu^2(x) - J_{\nu-1}(x)J_{\nu+1}(x)] + C.$$

Hence the proof.

6.7 Orthogonality of Bessel functions

Bessel's equation of order ν with parameter λ is given by

$$x^2\frac{d^2y}{dx^2} + x\frac{dy}{dx} + (\lambda^2x^2 - \nu^2)y = 0$$

which can subsequently be written as

$$x\frac{d^2y}{dx^2} + \frac{dy}{dx} + (\lambda^2 x - \frac{\nu^2}{x})y = 0. \tag{6.102}$$

Equation (6.102) can easily be cast into Sturm-Liouville form as follows:

$$\frac{d}{dx}(x\frac{dy}{dx}) + (\lambda^2 x - \frac{\nu^2}{x})y = 0. \tag{6.103}$$

Now, comparing (6.103) with the Sturm-Liouville equation

$$\frac{d}{dx}(P(x)\frac{dy}{dx}) + (R(x)\lambda^2 + Q(x))y = 0, \tag{6.104}$$

we see that

$$\begin{aligned} P(x) &= x \\ Q(x) &= -\tfrac{\nu^2}{x} \\ R(x) &= x. \end{aligned} \tag{6.105}$$

Therefore, if solutions of Bessel functions satisfy the conditions of the form

$$\begin{aligned} a_1 y(a) + a_2 y'(a) &= 0 \\ b_1 y(b) + b_2 y'(b) &= 0, \end{aligned} \tag{6.106}$$

they must form an orthogonal set with respect to the weight function $R(x) = X$ over the interval (a, b) such that

$$\int_a^b R(x) y_m y_n dx = 0 \quad m \neq n \tag{6.107}$$

and

$$\int_a^b R(x) y_n^2(x) dx \neq 0 \quad m = n. \tag{6.108}$$

Note as one of the solutions of Bessel's equation is unbounded in the neighbourhood of the origin, then the condition $a_1 y(0) + a_2 y'(0) = 0$ at $a = 0$ is replaced by the requirement that y is bounded in the neighbourhood of the origin.

We can establish the relationships (6.107) and (6.108) from the Bessel equation given in (6.103). Assuming $m \neq n$, consider

$$\begin{aligned} & y_m(x) = J_\nu(\lambda_m x) \\ \text{and} \quad & y_n(x) = J_\nu(\lambda_n x) \end{aligned} \tag{6.109}$$

two solutions of (6.103) corresponding to the eigenvalues λ_m and λ_n respectively. The forms that the boundary conditions $y_m(x)$ and $y_n(x)$ must satisfy are

$$\begin{aligned} a_1 y_m(a) + a_2 y_m'(a) &= 0 \\ b_1 y_m(b) + b_2 y_m'(b) &= 0 \end{aligned} \tag{6.110}$$

and

$$\begin{aligned} a_1 y_n(a) + a_2 y_n'(a) &= 0 \\ b_1 y_n(b) + b_2 y_n'(b) &= 0. \end{aligned} \tag{6.111}$$

$y_m(x)$ and $y_n(x)$ satisfy (6.103), therefore

$$(x\, y_m')' + (\lambda_m^2 x - \frac{\nu^2}{x}) y_m = 0 \tag{6.112}$$

$$(x\, y_n')' + (\lambda_n^2 x - \frac{\nu^2}{x}) y_n = 0. \tag{6.113}$$

Multiply (6.112) by y_n and (6.113) by y_m and subtract one from the other,

$$y_m(x\,y_n')' - y_n(x\,y_m')' = (\lambda_m^2 - \lambda_n^2)x\,y_m y_n.$$

That is,

$$\frac{d}{dx}x(y_m y_n' - y_m' y_n) = (\lambda_m^2 - \lambda_n^2)x\,y_m y_n. \tag{6.114}$$

Hence, integrating (6.114) with respect to x from a to b, we get

$$\begin{aligned}(\lambda_m^2 - \lambda_n^2)\int_a^b x\,y_m y_n dx &= [x(y_m y_n' - y_m' y_n)]_a^b \\ &= b[y_m(b)y_n'(b) - y_m'(b)y_n(b)] \\ &\quad -a[y_m(a)y_n'(a) - y_m'(a)y_n(a)] \\ &= b[W(y_m, y_n)]_{x=b} - a[W(y_m, y_n)]_{x=a}\end{aligned}$$

where $W(y_m, y_n)$ denotes the Wronskian of y_m and y_n.

For nontrivial solutions, these Wronskians vanish and hence we have

$$(\lambda_m^2 - \lambda_n^2)\int_a^b x\,y_m y_n dx = 0 \quad m \neq n.$$

Since $\lambda_m^2 \neq \lambda_n^2$ for $m \neq n$, we have

$$\int_a^b x\,y_m y_n dx = 0$$

or

$$\int_a^b x\,J_\nu(\lambda_m x)J_\nu(\lambda_n x)dx = 0.$$

Hence (6.107) is established.

To establish (6.108), we multiply (6.113) by $2y_n'$ and we obtain

$$2y_n'(x\,y_n')' + 2(\lambda_n^2 x - \tfrac{\nu^2}{x})y_n y_n' = 0$$

or

$$2x^2 y_n' y_n'' + 2x\,y_n'^2 + 2(\lambda_n^2 x^2 - \nu^2)y_n y_n' = 0.$$

It is easy to verify that the above equation can be written as

$$\frac{d}{dx}[x^2 y_n'^2 + (\lambda_n^2 x^2 - \nu^2)y_n^2] = 2\lambda_n^2 x\,y_n^2. \tag{6.115}$$

Now, integrating with respect to x from a to b, we obtain

$$2\lambda_n^2\int_a^b x\,y_n^2 dx = [x^2 y_n'^2 + (\lambda_n^2 x^2 - \nu^2)y_n^2]_a^b$$

or

$$\int_a^b x\,y_n^2 dx = \frac{1}{2\lambda_n^2}[x^2 y_n'^2 + (\lambda_n^2 x^2 - \nu^2)y_n^2]_{x=a}^{x=b}.$$

Now, if $y_n(x) = J_\nu(\lambda_n x)$, then $y_n'(x) = \lambda_n J_\nu'(\lambda_n x)$.

Therefore,

$$\begin{aligned}\int_a^b x\, J_\nu^2(\lambda_n x)dx &= [\frac{\lambda_n^2 x^2}{2\lambda_n^2} J_\nu'^2(\lambda_n x) + (\frac{\lambda_n^2 x^2 - \nu^2}{2\lambda_n^2}) J_\nu^2(\lambda_n x)]_{x=a}^{x=b} \\ &= [\frac{x^2}{2}\{J_\nu'^2(\lambda_n x) + (1 - \frac{\nu^2}{\lambda_n^2 x^2}) J_\nu^2(\lambda_n x)\}]_a^b \qquad (6.116)\end{aligned}$$

provided $Re(\nu) > -1$ to ensure convergence at $a = 0$.

Using the following recurrence relations

$$\begin{aligned} J_\nu'(\lambda_n x) &= \frac{\nu}{\lambda_n x} J_\nu(\lambda_n x) - J_{\nu+1}(\lambda_n x) \\ J_\nu'(\lambda_n x) &= -\frac{\nu}{\lambda_n x} J_\nu(\lambda_n x) - J_{\nu-1}(\lambda_n x) \qquad (6.117) \\ \frac{2\nu}{\lambda_n x} J_\nu(\lambda_n x) &= J_\nu + 1(\lambda_n x) + J_{\nu-1}(x), \end{aligned}$$

we have

$$\begin{aligned}\int_a^b x\, J_\nu^2(\lambda_n x)dx &= [\frac{x^2}{2}(J_\nu^2(\lambda_n x) - J_{\nu-1}(\lambda_n x) J_{\nu+1}(\lambda_n x))]_{x=a}^{x=b} \\ &= \frac{b^2}{2}\,(J_\nu^2(\lambda_n b) - J_{\nu-1}(\lambda_n b) J_{\nu+1}(\lambda_n b)) \\ &\quad -\frac{a^2}{2}(J_\nu^2(\lambda_n a) - J_{\nu-1}(\lambda_n a) J_{\nu+1}(\lambda_n a)). \qquad (6.118)\end{aligned}$$

If the lower limit $a \neq 0$, then in formulas (6.117) and (6.118), $J_\nu(\lambda_n x)$ may be replaced with $Y_\nu(\lambda_n x)$.

Example 8

The power transmitted in a circular wave guide is obtained by evaluating an integral of the type

$$I = \int_{v=0}^{v=ka} \{J_n'^2(v) + \frac{n^2 J_n^2(v)}{v^2}\} v dv$$

where $v = kr$, r is the radius of an annular cross section, a the internal radius of the cylinder, k a constant, and $J_n(ka) = 0$. Evaluate this integral.

Solution

We know from formula (6.83) that

$$J_n'(v) + \frac{n}{v} J_n(v) = J_{n-1}(v).$$

Squaring both sides

$$J_n'^2(v) + \frac{n^2}{v^2} J_n^2(v) = J_{n-1}^2(v) - \frac{2n}{v} J_n J_n'.$$

Using this in the given integral yields

$$I = \int_0^{ka} J_{n-1}^2(v)v dv - 2n \int_0^{ka} J_n(v) d\, J_n(v).$$

Now, applying (6.118) here we have

$$\begin{aligned} I &= [\frac{v^2}{2}(J_{n-1}^2(v) - J_{n-2}(v)J_n(v))]_{v=0}^{ka} - n[J_n^2(v)]_0^{ka} \\ &= \frac{(ka)^2}{2} J_{n-1}^2(ka), \end{aligned}$$

since $J_n(ka) = 0$ by hypothesis and $J_n(0) = 0$ always $n > 0$.
This is the required result.

Note

Let $J_n'(ka) = 0$. Then

$$I = \int_0^{ka} v\, J_{n-1}^2(v) dv - n[J_n^2(v)]_0^{ka}.$$

Using formula (6.117),

$$\begin{aligned} I &= [\frac{v^2}{2}(J_{n-1}'^2(v) + (1 - \frac{n^2}{v^2})J_{n-1}^2(v))]_0^{ka} - n\, J_n^2(ka) \\ &= \frac{(ka)^2}{2}\{J_{n-1}'^2(ka) + (1 - (\frac{n}{ka})^2)J_{n-1}^2(ka)\} - n\, J_n^2(ka) \\ I &= \frac{(ka)^2}{2}\{J_{n-1}'^2(ka) - J_n(ka)J_{n-2}(ka)\} - n\, J_n^2(ka). \end{aligned}$$

Now, if $J_n'(ka) = 0$, then the relation

$$J_n'(ka) + \frac{n}{ka} J_n(ka) = J_{n-1}(ka)$$

becomes

$$\frac{n}{ka} J_n(ka) = J_{n-1}(ka).$$

But,

$$\begin{aligned} J_{n-2}(ka) &= \frac{2(n-1)}{ka} J_{n-1}(ka) - J_n(ka) \\ &= \frac{2(n-1)}{ka} . \frac{n}{ka} J_n(ka) - J_n(ka) \\ &= (\frac{2n^2 - 2n}{(ka)^2} - 1)J_n(ka). \end{aligned}$$

Therefore,

$$\begin{aligned} I &= \frac{(ka)^2}{2}[(\frac{n}{ka})^2 - \frac{2n^2 - 2n}{(ka)^2} + 1]J_n^2(ka) - n\, J_n^2(ka) \\ &= \frac{(ka)^2}{2}(1 - (\frac{n}{ka})^2)J_n^2(ka). \end{aligned}$$

This is the required result.

Note

The basic differential identities of Formula 1 and 2 can be written as the following integration formulae:

$$\text{Formula 11} : x^{\nu} J_{\nu}(x) = \int x^{\nu} J_{\nu-1}(x)dx + c.$$

$$\text{Formula 12} : x^{-\nu} J_{\nu}(x) = -\int x^{-\nu} J_{\nu+1}(x)dx + c.$$

Similar integration formulae can be obtained by considering the other differentiation formulae.

When $\nu = 1$, from Formula 11, we have

$$x\, J_1(x) = \int x\, J_0(x)dx + c$$

and, when $\nu = 0$, from Formula 12, we have

$$J_0(x) = -\int J_1(x)dx + c.$$

For other formulae, however, integration by parts must be used in addition to Formulae 11 and 12.

Example 9

Evaluate

$$\int_0^x x^3 J_0(x)dx.$$

Solution

Integrating by parts yields

$$\begin{aligned}
&\int_0^x x^3 J_0(x)dx \\
&= \int_0^x x^2(x\, J_0(x))dx \\
&= \int_0^x x^2 \frac{d}{dx}(x\, J_1(x))dx \quad \text{by Formula 1} \\
&= x^3 J_1(x) - 2\int_0^x x^2 J_1(x)dx \\
&= x^3 J_1(x) - 2x^2 J_2(x).
\end{aligned}$$

Note

The integral $\int_0^x x^{m+n} J_n(x)dx$ can be reduced and evaluated, provided $m+n$ is an odd positive integer. However, if $m+n$ is even, the reduction is terminated by $\int_0^x J_0(x)dx$.

For example:

$$\begin{aligned}
&\int_0^x x^2 J_0(x)dx \\
&= x^2 J_1(x) - \int_0^x x\, J_1(x)dx \\
&= x^2 J_1(x) + \int_0^x x(\frac{d}{dx} J_0(x))dx \\
&= x^2 J_1(x) + x\, J_0(x) - \int_0^x J_0(x)dx.
\end{aligned}$$

Now to evaluate $\int_0^x J_0(x)dx$, we make repeated use of identity (6.85):

$$\begin{aligned}
\int J_0(x)dx - \int J_2(x)dx &= 2\, J_1(x) \\
\int J_2(x)dx - \int J_4(x)dx &= 2\, J_3(x)
\end{aligned}$$

and so on. By adding both sides separately, we obtain

$$\begin{aligned}
\int_0^x J_0(x)dx &= 2\{J_1(x) + J_3(x) + \cdots + \cdots\} \\
&= 2\sum_{n=0}^{\infty} J_{2n+1}(x).
\end{aligned}$$

Example 10

Expand $f(x) = x^2 \quad , \quad 0 \leq x \leq 1$ in a series of Bessel functions $J_0(\lambda_k x)$ where $\lambda_k, k = 1, 2, 3, \cdots$ are the positive roots of $J_0(\lambda) = 0$.

Solution

We expand a given function $f(x)$ into the Bessel Series having the form

$$\begin{aligned}
f(x) &= c_1 J_\nu(\lambda_1 x) + c_2 J_\nu(\lambda_2 x) + \cdots + c_n J_\nu(\lambda_n x) + \cdots \\
&= \sum_{j=1}^{\infty} c_j J_\nu(\lambda_j x). \qquad (1)
\end{aligned}$$

The method of determining the coefficients $c_1, c_2, \cdots$ consists of multiplying (1) by $x\, J_\nu(\lambda_k x)$, and then integrating from $x = 0$ to $x = 1$ to obtain

$$\int_0^1 x\, f(x) J_\nu(\lambda_k x)dx = \sum_{j=1}^{\infty} c_j \int_0^1 x\, J_\nu(\lambda_k x) J_\nu(\lambda_j x)dx.$$

By orthogonality, the integral on the right hand side except for $j = k$ vanishes so that the infinite series on the right hand side reduces to only one term and we obtain

$$\int_0^1 x\, f(x) J_\nu(\lambda_k x) dx = c_k \int_0^1 x\, J_\nu^2(\lambda_k x) dx$$

or

$$c_k = \frac{\int_0^1 x\, f(x) J_\nu(\lambda_k x) dx}{\int_0^1 x\, J_\nu^2(\lambda_k x) dx} \qquad k = 1, 2, 3, \cdots \tag{2}$$

Using $f(x) = x$, and $\nu = 0$ in this example, we have

$$c_k = \frac{\int_0^1 x^2 J_0(\lambda_k x) dx}{\int_0^1 x\, J_0^2(\lambda_k x) dx}.$$

To evaluate these integrals, we use the results in (6.116) such that

$$\int_0^1 x\, J_0^2(\lambda_k x) dx = \frac{1}{2} J_0'^2(\lambda_k)$$

and

$$\begin{aligned} &\int_0^1 x^2\, J_0(\lambda_k x) dx \\ &= \frac{1}{\lambda_k^3} \int_0^{\lambda_n} z^2 J_0(z) dz \qquad \text{put } \lambda_k x = z \\ &= \frac{1}{\lambda_k^3} [\lambda_k^2 J_1(\lambda_k) + 0(\lambda_k) - 2 \sum_{n=0}^{\infty} J_{2n+1}(\lambda_k)] \\ &= \frac{1}{\lambda_k^3} [\lambda_k^2 J_1(\lambda_k) - 2 \sum_{n=0}^{\infty} J_{2n+1}(\lambda_k)]. \end{aligned}$$

Thus

$$c_k = \frac{2[\lambda_k^2 J_1(\lambda_k) - 2 \sum_{n=0}^{\infty} J_{2n+1}(\lambda_k)]}{\lambda_k^3 J_0'^2(\lambda_k)}.$$

Hence the expansion is

$$x = 2 \sum_{k=1}^{\infty} \frac{\{\lambda_k^2 J_1(\lambda_k) - 2 \sum_{n=0}^{\infty} J_{2n+1}(\lambda_k)\}}{\lambda_k^3 J_0'^2(\lambda_k)} J_0(\lambda_k x). \tag{3}$$

Note

Parseval's identity of the Bessel series expansion in general can be obtained from (1) squaring, multiplying by x, and then integrating from $x = 0$ to $x = 1$ which yields

$$\begin{aligned} \int_0^1 x[f(x)]^2 dx &= \sum_{k=1}^{\infty} c_k \int_0^1 x\, f(x) J_\nu(\lambda_k x) dx \\ &= \sum_{k=1}^{\infty} c_k^2 \int_0^1 x\, J_\nu^2(\lambda_k x) dx \\ &= \sum_{k=1}^{\infty} c_k^2 d_k \end{aligned} \tag{4}$$

where

$$d_k = \int_0^1 x\, J_\nu^2(\lambda_k x) dx.$$

Thus, Parseval's identity for the expansion (3) is

$$\begin{aligned} \int_0^1 x^3 dx &= \sum_{k=1}^{\infty} c_k^2 \int_0^1 x\, J_0^2(\lambda_k x) dx \\ \frac{1}{4} &= \sum_{k=1}^{\infty} c_k^2 \frac{J_0'^2}{2}(\lambda_k). \end{aligned}$$

Simplifying we obtain

$$\frac{1}{4} = \sum_{k=1}^{\infty} \frac{\lambda_k^2 J_1(\lambda_k) - 2\sum_{n=0}^{\infty} J_{2n+1}(\lambda_k)}{\lambda_k^3} c_k.$$

Example 11

(a) Show that for $-1 \leq x \leq 1$

$$\frac{x(x^2-1)}{16} = \sum_{k=1}^{\infty} \frac{J_1(\lambda_k x)}{\lambda_k^3 J_0(\lambda_k)},$$

where $\lambda_k, k = 1, 2, \cdots$ are the positive roots of $J_1(\lambda) = 0$

(b) Deduce that

$$\sum_{k=1}^{\infty} \frac{J_1(\lambda k/2)}{\lambda_k^3 J_0(\lambda_k)} = -\frac{3}{128}.$$

Solution

Expand the function $x(x^2 - 1)$ into Bessel series as

$$x(x^2-1) = \sum_{k=1}^{\infty} c_k J_1(\lambda_k x)$$

where

$$c_k = \frac{\int_0^1 x^2(x^2-1) J_1(\lambda_k x) dx}{\int_0^1 x\, J_1^2(\lambda_k x) dx}.$$

We know from (6.117) $\int_0^1 x\, J_1^2(\lambda_k x) dx = \frac{1}{2} J_1'^2(\lambda_k)$. Because $J_1(\lambda) = 0$, integrating $\int_0^1 (x^4 - x^2) J_1(\lambda_k x) dx$ gives: $\int_0^1 x^4\, J_1(\lambda_k x) dx - \int_0^1 x^2\, J_1(\lambda_k x) dx$.

Substituting $\lambda_k x = z$, the above expression becomes

$$\begin{aligned}
&\frac{1}{\lambda_k^5}\int_0^{\lambda_k} z^4 J_1(z)dz - \frac{1}{\lambda_k^3}[z^2 J_2(z)]_0^{\lambda_k} \\
&= \frac{1}{\lambda_k^5}[z^4 J_2(z) - 2\int z^3 J_2(z)dz]_0^{\lambda_k} - \frac{1}{\lambda_k}J_2(\lambda_k) \\
&= \frac{1}{\lambda_k^5}[\lambda_k^4 J_2(\lambda_k) - 2\lambda_k^3 J_3(\lambda_k)] - \frac{1}{\lambda_k}J_2(\lambda_k) \\
&= -\frac{2}{\lambda_k^2}J_3(\lambda_k).
\end{aligned}$$

However, using the Bessel identities (6.117) we have

$$\frac{1}{2}J_1'^2(\lambda_k) = \frac{1}{2}J_0^2(\lambda_k)$$

and

$$-\frac{2}{\lambda_k^2}J_3(\lambda_k) = \frac{8}{\lambda_k^3}J_0(\lambda_k)$$

and so $c_k = \frac{16}{\lambda_k^3 J_0(\lambda_k)}$. Thus we have

$$\frac{x(x^2-1)}{16} = \sum_{k=1}^{\infty}\frac{J_1(\lambda_k x)}{\lambda_k^3 J_0(\lambda_k)}.$$

This series is also valid for the interval $-1 \leq x \leq 1$ and converges to the given odd function $\frac{x(x^2-1)}{16}$ because $J_1(\lambda_k x)$ is an odd function.

Substituting $x = \frac{1}{2}$ in this series, it can easily be seen that

$$\sum_{k=1}^{\infty}\frac{J_1(\frac{\lambda_k}{2})}{\lambda_k^2 J_0(\lambda_k)} = -\frac{3}{128}.$$

6.8 Legendre polynomials

In Section 6.1, we saw the evolution of Legendre's equation which is given by

$$(1-x^2)\frac{d^2y}{dx^2} - 2x\frac{dy}{dx} + n(n+1)y = 0. \tag{6.119}$$

Comparing this equation with the equation

$$\frac{d^2y}{dx^2} + P(x)\frac{dy}{dx} + Q(x)y = 0, \tag{6.120}$$

we have

$$\begin{aligned}
P(x) &= -\frac{2x}{1-x^2} \\
Q(x) &= \frac{n(n+1)}{1-x^2}.
\end{aligned}$$

We see that $x = \pm 1$ are the singular points of Legendre's equation and that all other points are ordinary points. We use the method of Frobenius to find a series solution of the given differential equation about the origin which is an ordinary point. Because the region of convergence of such a series is $|x| < 1$, the infinite series will be convergent for $-1 < x < 1$.

Assume a series solution of the form

$$y(x) = \sum_{k=0}^{\infty} a_k x^{k+r}. \tag{6.121}$$

Then

$$y'(x) = \sum_{k=0}^{\infty} a_k (k+r) x^{k+r-1} \tag{6.122}$$

and

$$y''(x) = \sum_{k=0}^{\infty} a_k (k+r)(k+r-1) x^{k+r-2}. \tag{6.123}$$

Substituting these values into (6.119), we obtain

$$\begin{aligned}(1-x^2) &\sum_{k=0}^{\infty} a_k (k+r)(k+r-1) x^{k+r-2} \\ &-2x \sum_{k-0}^{\infty} a_k (k+r) x^{k+r-1} \\ &+n(n+1) \sum_{k=0}^{\infty} a_k x^{k+r} = 0\end{aligned}$$

or

$$\begin{aligned}&\sum_{k=0}^{\infty} a_k (k+r)(k+r-1) x^{k+r-2} \\ &\quad + \sum_{k=0}^{\infty} a_k \{n(n+1) - (k+r)(k+r-1) \\ &\quad -2(k+r)\} x^{k+r} = 0,\end{aligned}$$

which can subsequently be reduced to

$$\begin{aligned}&\sum_{k=0}^{\infty} a_k (k+r)(k+r-1) x^{k+r-2} \\ &\quad + \sum_{k=2}^{\infty} a_{k-2} \{n(n+1) - (k+r-2)(k+r-3) \\ &\quad -2(k+r-2)\} x^{k+r-2} = 0.\end{aligned} \tag{6.124}$$

Equating the coefficients of like powers of x, we obtain the following equations:

Corresponding to $k = 0$: $\quad a_0(r-1)r = 0 \qquad (6.125)$

Corresponding to $k = 1$: $\quad a_1(r+1)r = 0 \qquad (6.126)$

and
Corresponding to $k \geq 2$:

$$a_k = \frac{-a_{k-2}\{n(n+1) - (k+r-2)(k+r-3) - 2(k+r-2)\}}{(k+r)(k+r-1)}. \tag{6.127}$$

From (6.125) and (6.126), it is obvious that, if $r = 0$, then both a_0 and a_1 are non-zero arbitrary constants which lead to two independent solutions of (6.119). Of course, this is to be expected because $x = 0$ is an ordinary point of Legendre's equation. If $r = 1$, then $a_0 \neq 0$, but $a_1 = 0$. In this case, we get only one particular solution which can easily be shown to be related to the previous two independent solutions.

Thus, when $r = 0$, (6.127) reduces to

$$a_k = -\frac{(n-k+2)(n+k-1)}{k(k-1)} a_{k-2} \quad k \geq 2.$$

Therefore, we have

$$\begin{array}{rclcrcl} a_0 &=& a_0 & & a_1 &=& a_1 \\ a_2 &=& -\frac{n(n+1)}{2!} a_0 & & a_3 &=& -\frac{(n-1)(n+2)}{3!} a_1 \\ a_4 &=& \frac{n(n-2)(n+1)(n+3)}{4!} a_0 & & a_5 &=& \frac{(n-1)(n-3)(n+2)(n+4)}{5!} a_1 \\ & \cdots\cdots & & \text{etc.} & & \cdots\cdots & \end{array}$$

Hence, a complete solution can be written as

$$\begin{aligned} y(x) &= a_0[1 - \frac{n(n+1)}{2!}x^2 + \frac{n(n-2)(n+1)(n+3)}{4!}x^4 - \cdots] \\ &\quad + a_1[x - \frac{(n-1)(n+2)}{3!}x^3 \\ &\quad + \frac{(n-1)(n-3)(n+2)(n+4)}{5!}x^5 - \cdots]. \end{aligned} \tag{6.128}$$

These infinite series are known as Legendre functions of the second kind. The region of convergence of each series is $-1 < x < 1$.

It is to be noted here that in most engineering applications the parameter n is a positive integer. If n is even, the first series terminates after certain terms, and if n is odd, the second series terminates after certain other terms. In either case, the series which reduces to a finite sum is known as a Legendre polynomial or zonal harmonic of order n.

To obtain the standard form of Legendre polynomials, we substitute the values of a_0 and a_1 in such a way that the coefficients of the highest power of x in each series is equal to

$$\frac{(2n)!}{2^n(n!)^2}.$$

These values are

$$\begin{aligned} a_0 &= (-1)^{\frac{n}{2}} \frac{n!}{2^n[(\frac{n}{2})!]^2} \\ a_1 &= (-1)^{\frac{n-1}{2}} \frac{(n+1)!}{2^n[(\frac{n-1}{2})]![(\frac{n+1}{2})]!} \end{aligned}$$

and the resulting general formula is

$$P_n(x) = \sum_{k=0}^{N} \frac{(-1)^k (2n-2k)!}{2^n k!(n-k)!(n-2k)!} x^{n-2k} \tag{6.129}$$

where

$$\begin{aligned} N &= \frac{n}{2} \quad n \text{ even} \\ N &= \frac{n-1}{2} \quad n \text{ odd.} \end{aligned}$$

This formula then gives

$$\begin{aligned} P_0(x) &= 1 \\ P_1(x) &= x \\ P_2(x) &= \frac{1}{2}(3x^2 - 1) \\ P_3(x) &= \frac{1}{2}(5x^3 - 3x) \\ P_4(x) &= \frac{1}{8}(35x^4 - 30x^2 + 3) \\ P_5(x) &= \frac{1}{8}(63x^5 - 70x^3 + 15x). \end{aligned}$$

We see from these results that

$$P_n(1) = 1 \quad \text{and} \quad P_n(-1) = (-1)^n$$

for all values of n. The Legendre polynomial $P_n(x)$ is finite in the closed interval $-1 \le x \le 1$.

Since we know that $P_n(x)$ is a solution of Legendre's equation, a complete solution can be written as

$$y = A\,P_n(x) + B\,Q_n(x),$$

where

$$Q_n = P_n(x) \int \frac{dx}{(x^2-1)P_n^2(x)}.$$

Note this second solution $Q_n(x)$ has been obtained by using Abel's identity.

In the following, we shall discuss three important identities of Legendre polynomials.

Theorem 1

$$P_n(x) = \frac{1}{2^n n!} \frac{d^n (x^2-1)^n}{dx^n} \qquad n = 0, 1, 2, \cdots$$

This formula, known as Rodrigues' formula, is named after the French mathematician, Olinde Rodrigues (1794–1851).

Proof

The binomial expansion of $(x^2-1)^n$ is given by

$$(x^2-1)^n = \sum_{k=0}^{n}(-1)^k \frac{n!}{(n-k)!k!} x^{2n-2k}.$$

Differentiating n times yields

$$\frac{d^n(x^2-1)^n}{dx^n} = \sum_{k-0}^{N}(-1)^k \frac{n!(2n-2k)!}{k!(n-k)!(n-2k)!} x^{n-2k} \tag{6.130}$$

where the last term is a constant; here $N=\frac{n}{2}$ when n is even and $N=\frac{n-1}{2}$ when n is odd.

Comparing (6.129) and (6.130), we have

$$P_n(x) = \frac{1}{2^n n!}\frac{d^n(x^2-1)^n}{dx^n}.$$

Hence this is the required proof.

Like Bessel functions, Legendre polynomials also have a generating function which is given in the following theorem.

Theorem 2

$$\begin{aligned}\frac{1}{\sqrt{1-2xz+z^2}} &= P_0(x) + z\,P_1(x) + z^2 P_2(x) + \cdots + z^n P_n(x) + \cdots \\ &= \sum_{n=0}^{\infty} z^n P_n(x).\end{aligned}$$

Proof

Expanding the left-hand side binomially, we have

$$\begin{aligned}\{1-z(2x-z)\}^{-1/2} &= 1 + \frac{z}{2}(2x-z) + \frac{1.3}{2^2.2!}z^2(2x-z)^2 + \cdots \\ &\quad + \frac{1.3.5.\cdots(2n-1)}{2^n n!}z^n(2x-z)^n + \cdots \\ &= \sum_{n=0}^{\infty}\frac{1.3.5.\cdots(2n-1)}{2^n n!}z^n(2x-z)^n \\ &= \sum_{n=0}^{\infty}\frac{1.3.5.\cdots(2n-1)}{2^n n!}z^n\sum_{k=0}^{n}\frac{(-1)^k n!(2x)^{n-k}}{(n-k)!k!}z^k \\ &= \sum_{n=0}^{\infty}\sum_{k=0}^{n}\frac{(-1)^k(2n)!}{2^{2n}(n!)^2}.\frac{n!2^{n-k}x^{n-k}}{k!(n-k)!}z^{k+n} \\ &= \sum_{n=0}^{\infty}\sum_{k=0}^{n}\frac{(-1)^k(2n)!x^{n-k}}{2^{n+k}(n!)(k!)(n-k)!}z^{n+k}.\end{aligned}$$

Now consider

$$\sum_{k=0}^{n} \frac{(-1)^k (2n)! x^{n-k}}{2^{n+k} n! k! (n-k)!} z^{n+k}.$$

Changing $n+k$ to m, the above expression may be written as

$$\begin{aligned}
&\sum_{k=0}^{M} \frac{(-1)^k (2m-2k)! x^{m-2k}}{2^m k! (m-k)! (m-2k)!} z^m \\
&= z^m \sum_{k=0}^{M} \frac{(-1)^k (2m-2k)! x^{m-2k}}{2^m k! (m-k)! (m-2k)!} \\
&= z^m P_m(x) \quad \text{where} \quad \begin{cases} M = \frac{m}{2} & m \text{ even} \\ \quad = \frac{m-1}{2} & m \text{ odd.} \end{cases}
\end{aligned}$$

Now, changing m to n again,

$$\{1 - z(2x - z)\}^{-1/2} = \sum_{n=0}^{\infty} z^n P_n(x).$$

as asserted.

In many applications, it is essential that Legendre polynomials be expressed in terms of θ instead of in algebraic form. With the help of Theorem 2, we can accomplish this objective.

Substituting

$$x = \cos\theta = \frac{e^{i\theta} + e^{-i\theta}}{2},$$

then

$$\begin{aligned}
&(1 - 2zx + z^2)^{-1/2} \\
&= [1 - z(e^{i\theta} + e^{-i\theta}) + z^2]^{-1/2} \\
&= [(1 - z\,e^{i\theta})(1 - z\,e^{-i\theta})]^{-1/2} \\
&= (1 - z\,e^{i\theta})^{-1/2}(1 - z\,e^{-i\theta})^{-1/2}.
\end{aligned}$$

Expanding binomially,

$$\begin{aligned}
(1 - 2zx - z^2)^{-1/2} &= \sum_{n=0}^{\infty} \{\frac{(2n)!}{2^{2n}(n!)^2} z^n e^{ni\theta}\} \sum_{m=0}^{\infty} \{\frac{(2m)!}{2^{2m}(m!)^2} z^m e^{-mi\theta} \\
&= \sum_{n=0}^{\infty} \sum_{m=0}^{\infty} \frac{(2n)!(2m)!}{2^{2(m+n)}(n!)^2(m!)^2} z^{n+m} e^{(n-m)i\theta}.
\end{aligned}$$

Now, to collect the coefficient of z^k, we let $n+m=k$. Substituting $n = k-m$ and, subsequently, $m = k-n$ yields two infinite series:

$$\sum_{m=0}^{\infty} \frac{(2k-2m)!(2m)!}{2^{2k}(m!)^2[(k-m)!]^2} z^k e^{i(k-2m)\theta} \tag{6.131}$$

and

$$\sum_{n=0}^{\infty} \frac{(2n)!(2k-2n)!}{2^{2k}(n!)^2[(k-n)!]^2} z^k e^{i(2n-k)\theta}. \tag{6.132}$$

Therefore, combining (6.131) and (6.132), we see that the coefficient of z^k is

$$\begin{aligned}
&\sum_{n=0}^{\infty} \frac{(2n)!(2k-2n)!}{2^{2k}(n!)^2[(k-m)!]^2} \{e^{i(k-2n)\theta} + e^{-i(k-2n)\theta}\} \\
&= \sum_{n=0}^{\infty} \frac{(2n)!(2k-2n)!}{2^{2k}(n!)^2[(k-n)!]^2} (2\cos(k-2n)\theta).
\end{aligned} \tag{6.133}$$

We know that the coefficient of z^k is by definition (see Theorem 2) the Legendre polynomial $P_k(\cos\theta)$; hence

$$P_k(\cos\theta) = \sum_{n=0}^{\infty} \frac{(2n)!(2k-2n)!(2\cos(k-2n)\theta)}{2^{2k}(n!)^2[(k-n)]^2}. \tag{6.134}$$

This infinite series terminates for $n > k$ because of the presence of a negative factorial in the denominator. Therefore,

$$P_k(\cos\theta) = \sum_{n=0}^{K} \frac{(2n)!(2k-2n)!(2\cos(k-2n)\theta)}{2^{2k}(n!)^2((k-n)!)^2} \tag{6.135}$$

where

$$\begin{aligned}
K &= \frac{k}{2}, \quad k \text{ even} \\
&= \frac{k-1}{2}, \quad k \text{ odd}.
\end{aligned}$$

Specifically, we write

$$\begin{aligned}
P_0(\cos\theta) &= 1 \\
P_1(\cos\theta) &= \cos\theta \\
P_2(\cos\theta) &= \frac{3\cos 2\theta + 1}{4} \\
P_3(\cos\theta) &= \frac{5\cos 3\theta + 3\cos\theta}{8} \\
P_4(\cos\theta) &= \frac{35\cos 4\theta + 20\cos 2\theta + 9}{64} \\
&\text{etc.}
\end{aligned}$$

The third identity deals with the orthogonal property of Legendre polynomials. We discuss this in the next theorem.

Theorem 3

Legendre polynomials in algebraic form satisfy the orthogonal properties

$$\int_{-1}^{1} P_m(x)P_n(x)dx = \begin{cases} 0 & m \neq n \\ \frac{2}{2n+1} & m = n. \end{cases}$$

In trigonometric form, Legendre polynomials satisfy the orthogonality properties

$$\int_{0}^{\pi} P_m(\cos\theta)P_n(\cos\theta)\sin\theta d\theta = \begin{cases} 0 & m \neq n \\ \frac{2}{2n+1} & m = n. \end{cases}$$

Proof

The Legendre equation can be written in Sturm-Liouville form as

$$\frac{d}{dx}[(1-x^2)y'] + n(n+1)y = 0$$

where

$$\begin{aligned} P(x) &= (1-x^2) \\ Q(x) &= 0 \\ R(x) &= 1 \quad \text{and} \quad \lambda = n(n+1). \end{aligned}$$

Hence, if solutions of the Legendre equation satisfy suitable boundary conditions, they must be orthogonal. In particular, it can easily be observed that, if any two members of the Legendre polynomials are multiplied and integrated with respect to x between the limits -1 and 1, the integral goes to zero automatically. This is due to the fact that $P(\pm 1) = 0$. That is,

$$\int_{-1}^{1} P_m(x)P_n(x)dx = 0 \quad m \neq n.$$

To find the value $\int_{-1}^{1} P_n^2(x)dx$, we use the generating function. We know

$$\frac{1}{\sqrt{1-2zx+z^2}} = \sum_{n=0}^{\infty} z^n P_n(x).$$

Squaring both sides and integrating with respect to x from -1 to 1:

$$\begin{aligned} \int_{-1}^{1} \frac{dx}{1-2zx+z^2} &= \sum_{n=0}^{\infty}\sum_{m=0}^{\infty} z^{n+m} \int_{-1}^{1} P_n(x)P_m(x)dx \\ &= \sum_{n=0}^{\infty} z^{2n} \int_{-1}^{1} P_n^2(x)dx \end{aligned}$$

because

$$\int_{-1}^{1} P_m P_n dx = 0, \quad m \neq n.$$

But,

$$\begin{aligned}
\int_{-1}^{1} \frac{dx}{1 - 2zx + z^2} &= \frac{-1}{2z}[\ell n(1 - 2xz + z^2)]_{-1}^{1} \\
&= \frac{-1}{2z}[\ell n(1 - 2z + z^2) - \ell n(1 + 2z + z^2)] \\
&= \frac{1}{z}[\ell n(1 + z) - \ell n(1 - z)] \\
&= \frac{1}{z}[z - \frac{z^2}{2} + \frac{z^3}{3} - \cdots + \frac{z^{2n+1}}{2n+1} - \cdots] \\
&\quad + \frac{1}{z}[z + \frac{z^2}{2} + \frac{z^3}{3} - \cdots + \frac{z^{2n+1}}{2n+1}] \\
&= 2(1 + \frac{z^2}{3} + \frac{z^4}{5} + \cdots + \frac{z^{2n}}{2n+1} + \cdots)
\end{aligned}$$

by expanding.
Thus, comparing the coefficient of z^{2n}, we have

$$\int_{-1}^{1} P_n^2(x)dx = \frac{2}{2n+1}.$$

By means of the substitution $x = \cos\theta$, we can easily establish the form of Legendre polynomial identities.

Example 12

Using Rodrigues' formula, prove that

$$P'_{n+1}(x) - P'_{n-1}(x) = (2n+1)P_n(x).$$

Hence, show that

$$\int_{x}^{1} P_n(x)dx = \frac{1}{2n+1}[P_{n-1}(x) - P_{n+1}(x)].$$

Proof

We know from Rodrigues' formula

$$P_n(x) = \frac{1}{2^n n!}\frac{d^n}{dx^n}(x^2 - 1)^n.$$

Therefore,

$$\begin{aligned}
P_{n+1}(x) &= \frac{1}{2^{n+1}(n+1)!}\frac{d^{n+1}}{dx^{n+1}}(x^2 - 1)^{n+1} \\
P'_{n+1}(x) &= \frac{1}{2^{n+1}(n+1)!}\frac{d^{n+2}}{dx^{n+2}}(x^2 - 1)^{n+1}
\end{aligned}$$

$$\begin{aligned}
&= \frac{1}{2^n n!}\frac{d^n}{dx^n}[(x^2-1)^n + 2nx^2(x^2-1)^{n-1}] \\
&= \frac{1}{2^n n!}\frac{d^n}{dx^n}[(x^2-1)^n(2n+1) + 2n(x^2-1)^{n-1}] \\
&= (2n+1)P_n(x) + P'_{n-1}(x).
\end{aligned}$$

Hence,

$$P'_{n+1}(x) - P'_{n-1}(x) = (2n+1)P_n(x)$$

as asserted.

We know

$$P_n(x) = \frac{1}{2n+1}[P'_{n+1}(x) - P'_{n-1}(x)].$$

Integrating the above with respect to x from x to 1, we have

$$\begin{aligned}
\int_x^1 P_n(x)dx &= \frac{1}{2n+1}[\int_x^1 P'_{n+1}(x)dx - \int_x^1 P'_{n-1}(x)dx] \\
&= \frac{1}{2n+1}[P_{n+1}(x) - P_{n-1}(x)]_x^1 \\
&= \frac{1}{2n+1}[P_{n-1}(x) - P_{n+1}(x)]
\end{aligned}$$

because

$$P_{n+1}(1) = P_{n-1}(1) = 1.$$

as asserted.

Example 13

Show that

(a) $$\int_{-1}^1 P_n(x)P'_{n+1}(x)dx = 2.$$

(b) $$\int_{-1}^1 x\,P_n(x)P_{n-1}(x)dx = \frac{2n}{4n^2-1}.$$

Proof

(a) We know

$$(n+1)P_{n+1}(x) - (2n-1)x\,P_n(x) + n\,P_{n-1}(x) = 0 \tag{1}$$

and

$$P'_{n+1}(x) - P'_{n-1}(x) = (2n+1)P_n(x). \tag{2}$$

Multiplying (2) by $P_n(x)$ and integrating between -1 and 1, we have

$$\begin{aligned}\int_{-1}^{1} P_n(x)P'_{n+1}(x)dx - \int_{-1}^{1} P_n(x)P'_{n-1}(x)dx &= (2n+1)\int_{-1}^{1} P_n^2(x)dx \\ &= \frac{2n+1}{(2n+1)}.2 \\ &= 2.\end{aligned}$$

Therefore,

$$\int_{-1}^{1} P_n(x)P'_{n+1}(x)dx = 2$$

provided

$$\int_{-1}^{1} P_n(x)P'_{n-1}dx = 0.$$

(b) Also from (1):

$$\begin{aligned}(2n+1)x\,P_n(x) &= (n+1)P_{n+1}(x) + n\,P_{n-1}(x) \\ \int_{-1}^{1}(2n+1)x\,P_n(x)P_{n-1}(x)dx &= (n+1)\int_{-1}^{1} P_{n-1}(x)P_{n+1}(x)dx \\ &\quad +n\int_{-1}^{1} P_{n-1}^2(x)dx \\ &= (n)\frac{2}{2n-1} \\ \int_{-1}^{1} x\,P_n(x)P_{n-1}(x)dx &= \frac{2n}{4n^2-1}.\end{aligned}$$

Therefore,

$$\int_{-1}^{1} P_{n-1}^2(x)dx = \frac{2}{2(n-1)+1} = \frac{2}{2n-1}$$

as asserted.

Example 14

Expand a function $f(x)$ defined over the interval $-1 \leq x \leq 1$ in a series of Legendre polynomials $P_0(x), P_1(x), P_2(x) \cdots$

Solution

We expand $f(x)$ into a series of Legendre polynomials having the form

$$\begin{aligned} f(x) &= c_0P_0(x) + c_1P_1(x) + c_2P_2(x) + \cdots \\ &= \sum_{j=0}^{\infty} c_jP_j(x) \end{aligned} \quad (1)$$

The series (1) is usually called the Legendre series. Because Legendre polynomials are mutually orthogonal in the interval $-1 \leq x \leq 1$, and can be shown to consitute a complete set, the coefficient c_j's can be determined by multiplying the series by $P_k(x)$ and integrating term by term from $x = -1$ to $x = 1$ with respect to x. We then obtain

$$\int_{-1}^{1} f(x)P_k(x)dx = \sum_{j=0} c_j \int_{-1}^{1} P_k(x)P_j(x)dx.$$

We know from the orthogonal property of Legendre polynomials that

$$\int_{-1}^{1} P_k(x)P_j(x)dx = \begin{cases} 0 & k \neq j \\ \frac{2}{2k+1} & k = j. \end{cases}$$

The series reduces to a single term where $j = k$ so that

$$\int_{-1}^{1} f(x)P_k(x)dx = c_k \int_{-1}^{1} [P_k(x)]^2 dx$$

Simplifying

$$c_k = \frac{\int_{-1}^{1} f(x)P_k(x)dx}{\int_{-1}^{1}[P_k(x)]^2dx} = \frac{2k+1}{2} \int_{-1}^{1} f(x)P_k(x)dx. \quad (2)$$

Therefore the series is

$$f(x) = \sum_{k=0}^{\infty}\{(\frac{2k+1}{2}) \int_{-1}^{1} f(x)P_k(x)dx\}P_k(x). \quad (3)$$

We can write the Parseval identity corresponding to the series (1) as follows: Multiply (1) by $f(x)$ and integrate from $x = -1$ to $x = +1$ which yields

$$\begin{aligned} \int_{-1}^{1}[f(x)]^2dx &= \sum_{k=0}^{\infty} c_k \int_{-1}^{1} f(x)P_k(x)dx \\ \int_{-1}^{1}[f(x)]^2dx &= \sum_{k=0}^{\infty} c_k^2 \int_{-1}^{1} [P_k(x)]^2dx \\ &= \sum_{k=0}^{\infty}(\frac{2}{2k+1})c_k^2 \end{aligned} \quad (4)$$

where the coefficients c_k are given by (2).

Example 15

Express $f(x) = x$, $-1 \leq x \leq 1$ in a series of Legendre polynomials.

Solution

The Legendre Series is given by

$$x = \sum_{k=0}^{\infty} c_k P_k(x)$$

where

$$c_k = \frac{2k+1}{2} \int_{-1}^{1} x\, P_k(x) dx.$$

We know $P_k(x)$ is an even function for even k, i.e., $k = 0, 2, 4, 6, \cdots$, and an odd function for odd k, i.e., $k = 1, 3, 5, \cdots$ such that

$$\int_{-1}^{1} x\, P_k(x) dx = 0 \quad \text{for } k \text{ even}$$

and

$$\int_{-1}^{1} x\, P_k(x) dx = 2 \int_{0}^{1} x\, P_k(x) dx \quad \text{for } k \text{ odd}$$

because x is an odd function. Thus we obtain

$$\begin{aligned} c_1 &= 3 \int_0^1 x^2 dx = 1 \\ c_3 &= 7 \int_0^1 x \frac{1}{2}(5x^3 - 3x) dx = 0 \\ c_5 &= 11 \int_0^1 x\, P_5(x) dx = 11 \int_0^1 \frac{x}{8}[63x^5 - 70x^3 + 15x] dx = 0. \end{aligned}$$

Similarly $c_{2k+1} = 0$ for $k = 1, 2, \cdots$, and hence the expansion is $x = P_1(x)$.

6.9 Applications

In this section, we discuss the applications of Bessel functions and Legendre polynomials to practical problems. We use Bessel functions in the study of configurations possessing circular symmetry and Legendre polynomials in the study of configurations possessing spherical symmetry. The following examples are selected to demonstrate the use of these two important functions.

Example 16

The response of a certain network of tuned circuits to a unit impulse applied in the first circuit is described by the set of differential equations:

$$\begin{array}{rcl} LC\,\ddot{v}_1 + RC\,\dot{v}_1 + v_1 & = & \delta(t) \\ LC\,\ddot{v}_2 + RC\,\dot{v}_2 + v_2 & = & v_1 \\ \hline LC\,\ddot{v}_n + RC\,\dot{v}_n + v_n & = & v_{n-1} \end{array}$$

with initial conditions, $v_n(0) = \dot{v}(0) = 0$ at $t = 0$ for all n, where a dot represents the derivative with respect to time.

Show that $v_n(t)$ is given by

$$v_n(t) = \frac{1}{(LC)^n}\frac{\Gamma(\frac{2n+1}{2})}{\Gamma(2n)}\left(\frac{2t}{\beta}\right)^{\frac{2n-1}{2}} e^{-\alpha t} J_{(\frac{2n-1}{2})}(\beta t)$$

where

$$\alpha = \frac{R}{2L} \quad , \quad \beta = \frac{1}{LC} - \frac{R^2}{4L^2}.$$

Proof

Taking the Laplace transform of the first equation and using the initial conditions, we obtain

$$(LC\,s^2 + RC\,s + 1)\mathcal{L}(v_1) = 1$$

or

$$\begin{aligned} \mathcal{L}(v_1) &= \frac{1}{LC\,s^2 + RC\,s + 1} \\ &= \frac{1}{(LC)\{(s+\alpha)^2 + \beta^2\}} \end{aligned}$$

where

$$\alpha = \frac{R}{2L} \quad \text{and} \quad \beta = \frac{1}{LC} - \frac{R^2}{4L^2}.$$

Taking the Laplace transform of the second equation and using the given initial conditions, we obtain

$$\mathcal{L}(v_2) = \frac{\mathcal{L}(v_1)}{(Lc)\{(s+\alpha)^2 + \beta^2\}} = \frac{1}{(Lc)^2\{(s+\alpha)^2 + \beta^2\}^2}.$$

Proceeding in a similar way, we obtain

$$\mathcal{L}(v_n) = \frac{1}{(Lc)^n\{(s+\alpha)^2 + \beta^2\}^n}.$$

But, we know

$$\mathcal{L}^{-1}\{\frac{1}{(s^2+\lambda^2)^{\frac{2\nu+1}{2}}}\} = \frac{2^\nu\Gamma(\nu+1)}{\lambda^\nu\Gamma(2\nu+1)}t^\nu J_\nu(\lambda t).$$

Hence, using this result in our problem, we obtain

$$v_n(t) = \frac{1}{(Lc)^n}(\frac{2t}{\beta})^{\frac{2n-1}{2}}\frac{\Gamma(\frac{2n+1}{2})}{\Gamma(2n)}e^{-\alpha t}J_{(\frac{2n-1}{2})}(\beta t).$$

Hence the proof.

Example 17

The current density σ in a long cylindrical conductor, which increases from the centre toward the surface, is found to satisfy the differential equation

$$\frac{d^2\sigma}{dr^2}+\frac{1}{r}\frac{d\sigma}{dr}-i\,k^2\sigma=0$$

where

$$k^2=\frac{4\pi\omega}{\rho},$$

ω is the frequency, and ρ is the resistivity of the conductor.
If the current density at the surface of the cylindrical conductor of radius a is σ_0, find $\sigma(r)$.

Solution

We know the solution of the above differential equation may be written as

$$\sigma(r) = A\,M_0(kr)e^{i\,\theta_0(kr)}.$$

Using the boundary condition at $r=a$, we have

$$\sigma(r) = A\,M_0(ka)e^{i\,\theta_0(ka)}.$$

Therefore,

$$\sigma(r)=\sigma_0\frac{M_0(kr)}{M_0(ka)}e^{i\{\theta_0(kr)-\theta_0(ka)\}}$$

where

$$\begin{aligned} M_0(kr) &= \sqrt{Ber^2(kr)+Bei^2(kr)} \\ \theta_0(kr) &= \tan^{-1}\frac{Bei(kr)}{Ber(kr)}. \end{aligned}$$

Example 18

In the problem of tidal waves in estuaries, we assume that the wave $\eta(x,t)$ follows a simple harmonic motion, where $\eta(x,t)$ is proportional to $\cos(\sigma t+\epsilon)$ in which σ is the wave frequency and ϵ is the phase angle. The partial differential equation which governs the tidal elevation is given by

$$\frac{\partial^2\eta}{\partial t^2}=\frac{g}{b}\frac{\partial}{\partial x}(h\,b\frac{\partial\eta}{\partial x}) \tag{1}$$

where g is the acceleration due to gravity, $h(x)$ is the depth, and $b(x)$ is the breadth of the estuary.

Let us assume that the breadth varies as the distance from the end $x=0$, the depth being uniform. Let us suppose that at its mouth $(x=\ell)$, the estuary meets an open sea in which the tidal oscillation $\eta=\eta_0\cos(\sigma+\epsilon)$ is maintained. Find the tidal elevation at any point and at any time in the estuary.

Solution

A schematic diagram of an idealized estuary is given below. Let us assume that $b(x)=b_0\frac{x}{\ell}$ and $h(x)=h_0$ (2)

where b_0 and h_0 are constants. Substituting the values of $b(x)$ and $h(x)$ into the tidal equation, we obtain

$$\frac{\partial^2\eta}{\partial t^2}=g\,h_0[\frac{\partial^2\eta}{\partial x^2}+\frac{1}{x}\frac{\partial\eta}{\partial x}]. \tag{3}$$

Let $y(x)$ be the amplitude at any x in $0\le x\le\ell$. Hence, the vertical elevation η at any time is given by

$$\eta(x,t)=y(x)\cos(\sigma t+\epsilon). \tag{4}$$

Substituting (4) into (3) yields the following ordinary differential equation

$$\frac{d^2y}{dx^2}+\frac{1}{x}\frac{dy}{dx}+\lambda^2y=0 \tag{5}$$

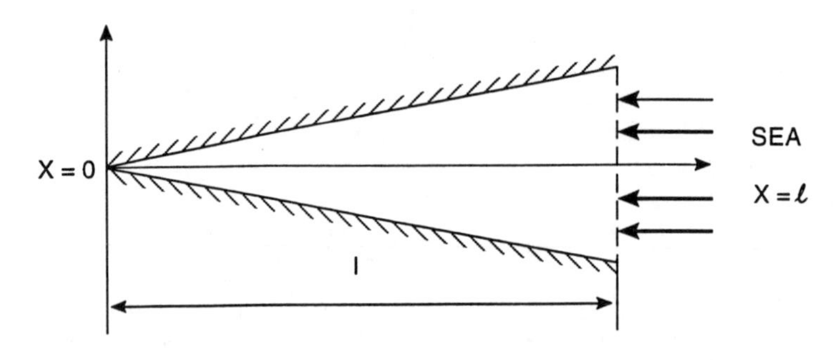

Figure 6.7. An estuary

where $\lambda = \frac{\sigma}{\sqrt{g h_0}} = \frac{\sigma}{c}$, with the dimension ℓ^{-1}. Here c is defined as the speed of the wave front. Equation (5) is a Bessel equation and one of its complete solutions is

$$y = A\,J_0(\lambda x) + B\,Y_0(\lambda x) \tag{6}$$

where A, B are arbitrary constants. The estuary terminates at $x = 0$ where $Y_0(\lambda x)$ ceases to be a solution of (5). Therefore, we are left with

$$y = A\,J_0(\lambda x)$$

and so

$$\eta = A\,J_0(\lambda x)\cos(\sigma t + \epsilon). \tag{7}$$

Now, at $x = \ell$, $\eta = \eta_0 \cos(\sigma t + \epsilon)$. Therefore,

$$A = \frac{\sigma}{J_0(\lambda \ell)} \tag{8}$$

and hence the solution is given by $\eta(x,t) = \eta_0 \frac{J_0(\lambda x)}{J_0(\lambda \ell)} \cos(\sigma t + \epsilon)$. This is the required solution.

Example 19

A metal pipe of infinite length in the form of a circular cylinder of radius a has an initial temperature in the form $f(r)$ at $t = 0$, where r is the distance from the axis of the cylinder. Its surface is kept at $0^o C$ for all time. Find the temperature of the pipe at any point at any time.

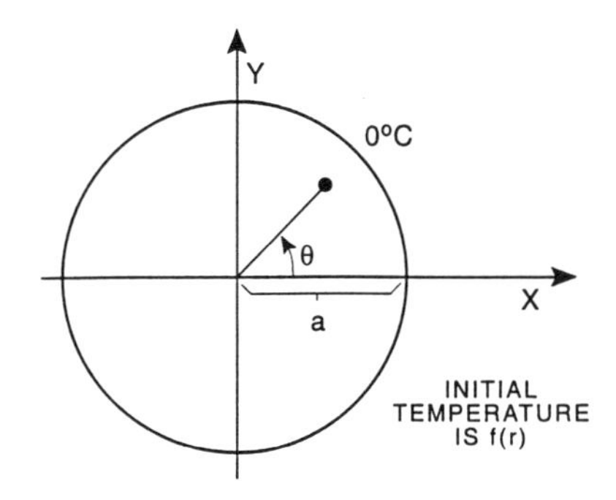

Figure 6.8. Heat conduction in a pipe

Solution

The cross-section of the pipe in the xy plane is depicted in Figure 6.8. We assume that the pipe is symmetrical about its axis in order that the solution will not depend upon θ; at the same time, we have assumed that the pipe is of infinite length and so the solution should be independent of z.

The differential equation which governs this situation is the heat conduction equation, given by

$$\frac{\partial u}{\partial t} = \alpha(\frac{\partial^2 u}{\partial r^2} + \frac{1}{r}\frac{\partial u}{\partial r}) \tag{1}$$

where $u = u(r, t)$ is the temperature, which is dependent on r and t, and α is the thermal diffusivity, which is a constant property of the metal pipe.

We have the boundary condition

$$u(a, t) = 0 \quad \text{at } r = a \tag{2}$$

and the initial condition

$$u(r, 0) = f(r) \quad \text{at } t = 0. \tag{3}$$

Using the method of separation of variables,

$$U(r, t) = R(r)T(t)$$

we see that equation (1) reduces to two ordinary differential equations

$$T' + \alpha\lambda^2 T = 0 \tag{4}$$

$$R'' + \tfrac{1}{r}R' + \lambda^2 R = 0. \tag{5}$$

The solutions of (4) and (5) are respectively $T = A\,e^{-\alpha\lambda^2 t}$ and $R = B\,J_0(\lambda r) + C\,Y_0(\lambda r)$, where J_0 and Y_0 are the Bessel functions of first and second kind with order zero.

Now, a complete solution is given by

$$u(r, t) = A\,e^{-\alpha\lambda^2 t}(B\,J_0(\lambda r) + C\,Y_0(\lambda r)). \tag{6}$$

The solution (6) must be bounded at all points inside the pipe and in particular at $r = 0$. But, $Y_0(\lambda r) \to -\infty$ as $r \to 0$. Therefore, $C = 0$.

Therefore, (6) becomes

$$u(r, t) = A\,e^{-\alpha\lambda^2 t}J_0(\lambda r) \tag{7}$$

where AB is replaced by the new constant A. Using boundary condition (2), we see that

$$J_0(\lambda a) = 0 \tag{8}$$

Equation (8) has an infinite number of positive roots such that

$$\lambda a = \mu_n \quad , \quad n = 1, 2, 3, \cdots \tag{9}$$

Therefore, for each root u_n we obtain the corresponding value of λ which is $\lambda_n = \frac{\mu_n}{a}$ and the product solution (8) becomes

$$u_n(r,t) = A_n e^{-\alpha\lambda_n^2 t} J_0(\lambda_n r).$$

Then, we form a series of these particular solutions

$$u(r,t) = \sum_{n=1}^{\infty} u_n(r,t) = \sum_{n=1}^{\infty} A_n e^{-\alpha\lambda_n^2 t} J_0(\lambda_n r). \tag{10}$$

Using initial condition (3) we have

$$f(r) = \sum_{n=1}^{\infty} A_n J_0(\lambda_n r). \tag{11}$$

The coefficients A_n can be determined by multiplying both sides of (11) by $r\, J_0(\lambda_m r)$ and then integrating with respect to r from 0 to a, obtaining,

$$\int_0^a r\, f(r) J_0(\lambda_m r) dr = \sum_{n=1}^{\infty} A_n \int_0^a r\, J_0(\lambda_m r) J_0(\lambda_n r) dr. \tag{12}$$

We know from the orthogonal property of Bessel functions that

$$\begin{aligned} \int_0^a r\, J_0(\lambda_m r) J_0(\lambda_n r) dr &= 0 \quad m \neq n \\ &\neq 0 \quad m = n. \end{aligned}$$

Thus, we have

$$\int_0^a r\, f(r) J_0(\lambda_n r) dr = A_n \int_0^a r\, J_0^2(\lambda_n r) dr.$$

Hence

$$A_n = \frac{\int_0^a r\, f(r) J_0(\lambda_n r) dr}{\int_0^a r\, J_1^2(\lambda_n r) dr} \qquad n = 1, 2, 3, \cdots$$

or

$$A_n = \frac{2}{a^2 J_0'^2(\lambda_n a)} \int_0^a r\, f(r) J_0(\lambda_n r) dr. \tag{13}$$

We obtain (13) using the formula (6.116). Therefore, (10) is our required solution where A_n is given by (13).

Example 20

A right circular cylinder of radius a and height ℓ has its convex surface and base in the xy plane ($z = 0$) at temperature $0^o C$ while the top end ($z = \ell$) is kept at temperature $f(r)$ degrees celcius. Find the steady-state temperature at any point of the cylinder.

Solution

The schematic diagram of the cylinder is given in Fig. 6.9.

We assume the cylinder is symmetrical about its axis. The differential equation which governs this situation is the steady heat conduction equation, i.e., the Laplace equation in three-dimensional cylindrical coordinates

$$\frac{\partial^2 u}{\partial r^2} + \frac{1}{r}\frac{\partial u}{\partial r} + \frac{\partial^2 u}{\partial z^2} = 0 \tag{1}$$

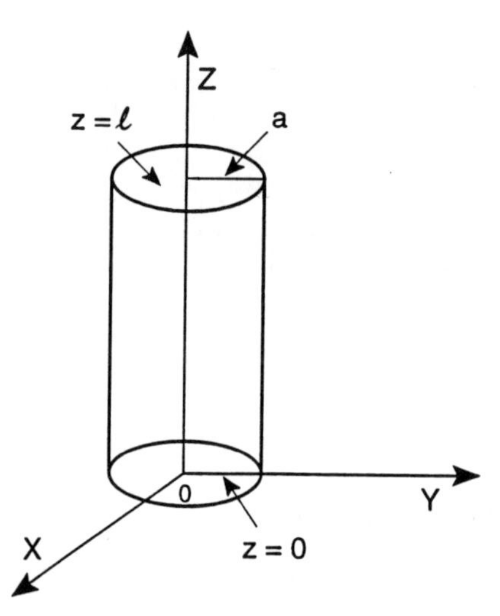

Figure 6.9. Heat conduction in right circular cylinder

where $u = u(r, z)$ is the temperature and depends on r and z. We have the boundary conditions

$$u(a, z) = 0 \tag{2}$$
$$u(r, 0) = 0 \tag{3}$$
$$u(r, \ell) = f(r). \tag{4}$$

By using the separation of variables method, $u(r, z) = R(r)Z(z)$, (1) becomes

$$\begin{aligned} Z(R'' + \frac{1}{r}R') + R\,Z'' &= 0 \\ \frac{R'' + \frac{1}{r}R'}{R} = -\frac{Z''}{Z} &= -\lambda^2. \end{aligned}$$

Therefore, we have two ordinary differential equations:

$$Z'' - \lambda^2 Z = 0 \tag{5}$$
$$R'' + \frac{1}{r}R' + \lambda^2 R = 0 \tag{6}$$

the solutions of which are

$$\begin{aligned} Z &= A\cosh\lambda z + B\sinh\lambda z \\ R &= C\,J_0(\lambda r) + D\,Y_0(\lambda r). \end{aligned}$$

Therefore, a complete solution is

$$u(r,z) = (A\cosh\lambda z + B\sinh\lambda z)(C\,J_0(\lambda r) + D\,Y_0(\lambda r)). \tag{7}$$

But, when $r \to 0$, $Y_0(\lambda r) \to -\infty$. Therefore

$$D = 0.$$

Hence,

$$u(r,z) = J_0(\lambda r)(A\cosh\lambda z + B\sinh\lambda z)$$

where A and B are re-defined constants.

Using boundary condition (2), we have

$$J_0(\lambda a) = 0 \tag{8}$$

which has infinitely many positive real roots such that

$$\lambda a = \mu_n \quad , \quad n = 1, 2, \cdots \tag{9}$$

Therefore, for each root μ_n we obtain the corresponding value of λ

$$\lambda_n = \frac{\mu_n}{a}$$

and the corresponding product solution becomes

$$u_n(r,z) = A_n J_0(\lambda_n r)\cosh\lambda_n z + B_n J_0(\lambda_n r)\sinh\lambda_n z. \tag{10}$$

By superimposing all these solutions, we have

$$u(r,z) = \sum_{n=1}^{\infty} u_n(r,z) = \sum_{n=1}^{\infty}(A_n J_0(\lambda_n r)\cosh\lambda_n z + B_n J_0(\lambda_n r)\sinh\lambda_n z). \tag{11}$$

Using boundary condition (3), it can easily be seen that all of the A's are zero. So, we are left with

$$u(r,z) = \sum_{n=1}^{\infty} B_n J_0(\lambda_n r)\sinh\lambda_n z. \tag{12}$$

Now, using boundary condition (4), we have

$$f(r) = \sum_{n=1}^{\infty}(B_n \sinh\lambda_n \ell) J_0(\lambda_n r).$$

By using the orthogonal property of Bessel functions, we can obtain the coefficients B_n as follows:

$$(B_n \sinh \lambda_n \ell) \int_0^a r\, J_0^2(\lambda_n r) dr = \int_0^a r\, f(r) J_0(\lambda_n r) dr.$$

Therefore,

$$\begin{aligned} B_n &= \frac{\int_0^a r\, f(r) J_0(\lambda_n r) dr}{(\sinh \lambda_n \ell) \int_0^a r\, J_0^2(\lambda_n r) dr} \\ &= \frac{2 \int_0^a r\, f(r) J_0(\lambda_n r) dr}{(\sinh \lambda_n \ell) \int_0^a r\, J_0'^2(\lambda_n r) dr} \\ &= \frac{2 \int_0^a r\, f(r) J_0(\lambda_n r) dr}{a^2 \sinh \lambda_n \ell J_1^2(\lambda_n a)}. \end{aligned} \tag{13}$$

Therefore, (12) is our required solution where B_n is given by (13).

Example 21

A rigid sphere of radius a is moving in a fluid at rest which extends to infinity. The flow is considered axially symmetric. Find the velocity potential of the fluid induced by this moving sphere where the velocity potential satisfies Laplace's equation.

Solution

The schematic diagram of a moving sphere is given in Fig. 6.10.

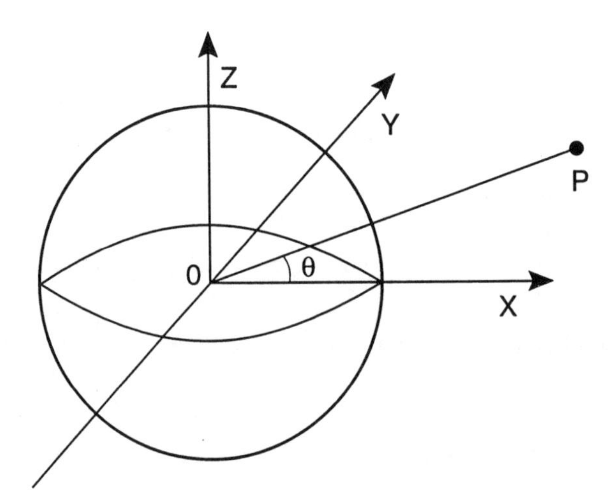

Figure 6.10. Moving sphere

We know that the velocity potential $\phi(r, \theta, \omega)$, which satisfies the Laplace equation in the system of spherical polar coordinates (r, θ, ω), is given by

$$\frac{\partial}{\partial r}(r^2\frac{\partial \phi}{\partial r}) + \frac{1}{\sin\theta}\frac{\partial}{\partial \theta}(\sin\theta\frac{\partial \phi}{\partial \theta}) + \frac{1}{\sin^2\theta}\frac{\partial^2 \phi}{\partial \omega^2} = 0. \tag{1}$$

In an axisymmetric case, the velocity potential ϕ will be independent of ω; in that situation (1) becomes

$$\frac{\partial}{\partial r}(r^2\frac{\partial \phi}{\partial r}) + \frac{1}{\sin\theta}\frac{\partial}{\partial \theta}(\sin\theta\frac{\partial \phi}{\partial \theta}) = 0. \tag{2}$$

By the method of separation of variables, the solution of ϕ exists in the following form

$$\phi = \sum_{n=0}^{\infty}(A_n r^n + B_n r^{-n-1})(C_n P_n(x) + D_n Q_n(x)) \tag{3}$$

where $x = \cos\theta$ and $P_n(x)$ and $Q_n(x)$ are Legendre polynomials. But, for $x = \pm 1$, or $\theta = 0, \pi$, $Q_n(x)$ is unbounded so we choose $D_n = 0$. Also, we know from the boundary condition at $r = \infty$ that $(\phi)_{r\to\infty} = 0$. Therefore, we choose $A_n = 0$. Hence, (3) can be written as

$$\phi = \sum_{n=0}^{\infty} B_n r^{-n-1} P_n(x) \tag{4}$$

where B_n is a re-defined constant.

The boundary condition at the surface of the sphere is

$$(\frac{\partial \phi}{\partial r})_{r=a} = V\cos\theta \tag{5}$$

where V is the velocity of the sphere along the line of reference.

To satisfy condition (5), solution (4) must be in the form

$$\phi = B_1 r^{-2} P_1(x) = B_1 r^{-2}\cos\theta \tag{6}$$

where $P_1(x) = x = \cos\theta$.
Thus,

$$(\frac{\partial \phi}{\partial r})_{r=a} = [-2B_1 r^{-3}\cos\theta]_{r=a} = -2B_1 a^{-3}\cos\theta. \tag{7}$$

Comparing (5) and (7), we have

$$B_1 = -\frac{1}{2}V a^3. \tag{8}$$

Therefore, the solution for ϕ is

$$\phi = -\frac{1}{2}V\frac{a^3}{r^2}\cos\theta.$$

Example 22

A spherical shell has an inner radius r_1 and an outer radius r_2. The temperatures of the inner and outer surfaces are given respectively by $f_1(x)$ and $f_2(x)$, when $x = \cos\theta$. Find the temperature at any point of the shell.

Solution

The partial differential equation governing this physical problem is Laplace's equation in spherical coordinates. For an axisymmetric case, the temperature u must satisfy the following equation

$$\frac{\partial}{\partial r}(r^2\frac{\partial u}{\partial r}) + \frac{1}{\sin\theta}\frac{\partial}{\partial\theta}(\sin\theta\frac{\partial u}{\partial\theta}) = 0. \tag{1}$$

By using the method of separation of variables, we have a complete solution of (1) in the following form

$$u = \sum_{n=0}^{\infty}(A_n r^n + B_n r^{-n-1})P_n(x) \tag{2}$$

where $x = \cos\theta$ and $P_n(x)$ is the Legendre polynomial.

There are two constants, A_n and B_n, which must be evaluated using the boundary conditions

$$\text{at}\, r = r_1 \quad : \quad u(r_1, \theta) = f_1(x) \tag{3}$$
$$\text{at}\, r = r_2 \quad : \quad u(r_2, \theta) = f_2(x). \tag{4}$$

Thus, from (2) we have

$$f_1(x) = \sum_{n=0}^{\infty}(A_n r_1^n + B_n r_1^{-n-1})P_n(x) \tag{5}$$

and

$$f_2(x) = \sum_{n=0}^{\infty}(A_n r_2^n + B_n r_2^{-n-1})P_n(x). \tag{6}$$

Multiplying (5) and (6) by $P_n(x)$, integrating with respect to x from -1 to 1, and using the orthogonal property of Legendre polynomials, we have the following equations

$$(A_n r_1^n + B_n r_1^{-n-1}) = \frac{2n+1}{2}\int_{-1}^{1} f_1(x)P_n(x)dx \tag{7}$$

and

$$(A_n r_2^n + B_n r_2^{-n-1}) = \frac{2n+1}{2}\int_{-1}^{1} f_2(x)P_n(x)dx. \tag{8}$$

The constants A_n and B_n can be evaluated from (7) and (8). Therefore, the temperature distribution is known from (2).

Chapter 7

Applications

We already saw the development of the theories concerning the Fourier Series and integrals and Laplace transforms. In many problems in mechanical, electrical, and industrial engineering, a mathematical model often results in a linear differential e-quation with constant coefficients. In this chapter, we will show how Fourier series and integrals and Laplace transforms can be used to solve these problems.

7.1 Applications of Fourier series

Fourier series play an important role in solving many practical problems. In this section, we will look at some examples which involve the analysis of physical systems subjected to general periodic disturbances.

Example 1

A sinusoidal voltage $E_0 \sin \omega t$ is passed through a half-wave rectifier which clips the negative portion of the wave. Find the Fourier series of the resulting periodic function whose definition in one period is given by

$$f(t) = \begin{cases} 0 & -\pi/\omega < t < 0 \\ E_0 \sin \omega t & 0 < t < \pi/\omega. \end{cases}$$

Solution

Here the period is $\frac{2\pi}{\omega}$. Therefore, the Fourier series can be expressed as

$$f(t) = \frac{a_0}{2} + \sum_{n=1}^{\infty} \{a_n \cos n\omega t + b_n \sin n\omega t\}$$

where $a_n = \frac{\omega}{\pi} \int_{-\pi/\omega}^{\pi/\omega} f(t) \cos n\omega t dt, \quad n = 0, 1, 2, \cdots$ and $b_n = \frac{\omega}{\pi} \int_{-\pi/\omega}^{\pi/\omega} f(t) \sin n\omega t dt, \quad n = 1, 2, \cdots$

Thus the Fourier coefficients are obtained as

$$\begin{aligned} a_n &= \frac{\omega}{\pi}\int_0^{\pi/\omega} E_0 \sin\omega t \cos n\omega t dt \\ &= \frac{2E_0}{\pi(1-n)^2} \quad \text{for n even} \quad \text{and n } \neq 1 \\ a_1 &= \frac{\omega E_0}{2\pi}\int_0^{\pi/\omega} \sin 2\omega t dt = 0. \end{aligned}$$

Also,

$$\begin{aligned} b_n &= \frac{\omega E_0}{\pi}\int_0^{\pi/\omega} \sin\omega t \sin n\omega t = 0 \qquad n \neq 1 \\ b_1 &= \frac{\omega E_0}{2\pi}\int_0^{\pi/\omega} (1-\cos 2\omega t)dt = \frac{E_0}{2}. \end{aligned}$$

Hence, the Fourier series is given by

$$\begin{aligned} f(t) &= \frac{E_0}{\pi} + \frac{E_0}{2}\sin\omega t - \frac{2E_0}{\pi}\sum_{n=2,4,\cdots}^{\infty} \frac{\cos n\omega t}{n^2-1} \\ &= \frac{E_0}{\pi} + \frac{E_0}{2}\sin\omega t - \frac{2E_0}{\pi}\sum_{n=1}^{\infty} \frac{\cos 2n\omega t}{(2n+1)(2n-1)}. \end{aligned}$$

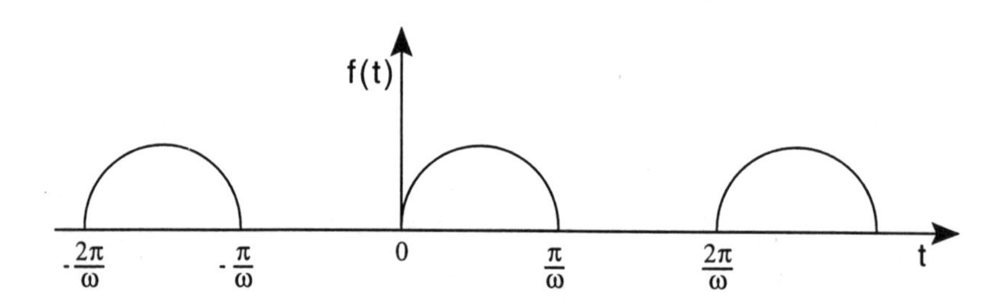

Figure 7.1: Half-wave rectifier

Example 2

Find the steady-state current produced in the circuit shown in Fig. 7.2 with the periodic voltage:

$$E(t) = \begin{cases} E_0 & 0 < t < 0.005 \\ -E_0 & 0.005 < t < 0.01. \end{cases}$$

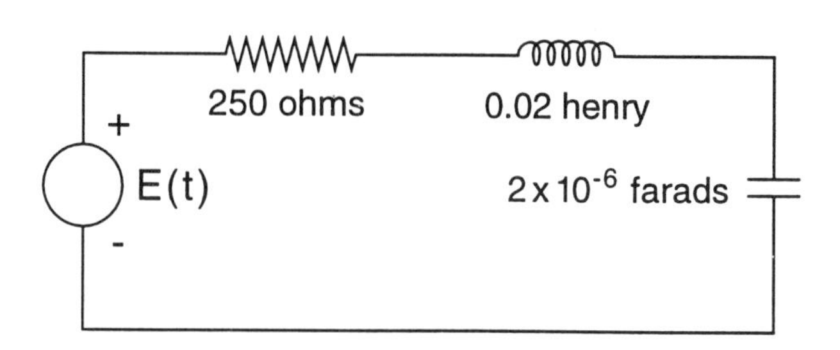

Figure 7.2: LRC circuit

Solution

Here we will use the method of complex impedance to find the steady-state current. The period is $2p = 0.01$. A complex Fourier series is given by

$$E(t) = \sum_{n=-\infty}^{\infty} c_n e^{\frac{in\pi t}{p}} = \sum_{n=-\infty}^{\infty} c_n e^{\frac{in\pi t}{0.005}}$$

where

$$\begin{aligned} c_n &= \frac{1}{0.01}\int_0^{0.01} E(t) e^{-\frac{in\pi t}{0.005}}\,dt \\ &= \frac{1-(-1)^n}{n\pi i}E_0 \\ &= \begin{cases} 0 & n \text{ even} \\ -\frac{2E_0 i}{n\pi} & n \text{ odd} \quad n \neq 0. \end{cases} \\ c_0 &= 0 \end{aligned}$$

Therefore, the Fourier series is given by

$$E(t) = \sum_{n=-\infty}^{\infty} (\frac{-2E_0 i}{n\pi}) e^{200n\pi i t} \qquad n \text{ odd}.$$

We know the steady-state current produced by the voltage of the form $Ae^{i\omega t}$ can be found by dividing the voltage by the complex impedance

$$Z(\omega) = R + i(\omega L - \frac{1}{\omega C}).$$

In this problem, $R = 250$ ohms, $L = 0.02$ henry, $C = 2 \times 10^{-6}$ farads and $\omega = 200n\pi$.
Thus

$$Z(\omega) = Z_n = 250 + i(4n\pi - \frac{2500}{n\pi}) \quad n \text{ odd}.$$

Therefore, the steady-state current is given by

$$\begin{aligned} I_{ss} &= \sum_{n=-\infty}^{\infty} \frac{(-\frac{2E_0 i}{n\pi})e^{200n\pi it}}{250 + i(4n\pi - \frac{2500}{n\pi})} \qquad n \text{ odd} \\ &= \sum_{n=-\infty}^{\infty} \frac{-2E_0 i e^{200n\pi it}}{250n\pi + i(4n^2\pi^2 - 2500)}. \end{aligned}$$

The real trigonometric form of the steady-state current I_{ss} is given by

$$\begin{aligned} I_{ss} &= \sum_{n=-\infty}^{\infty} \frac{-2E_0 i(250n\pi - i(4n^2\pi^2 - 2500))e^{200n\pi it}}{(250n\pi)^2 + (4n^2\pi^2 - 2500)^2} \\ &= \sum_{n=1,3,\cdots}^{\infty} (a_n \cos 200n\pi t + b_n \sin 200n\pi t) \end{aligned}$$

where

$$\begin{aligned} a_n &= \frac{-4E_0(4n^2\pi^2 - 2500)}{(250n\pi)^2 + (4n^2\pi^2 - 2500)^2} \\ b_n &= \frac{1000E_0 n\pi \sin 200n\pi t}{(250n\pi)^2 + (4n^2\pi^2 - 2500)^2}. \end{aligned}$$

Example 3

If $f(t)$ is a periodic function whose definition in one period is

$$f(t) = \begin{cases} 1 & 0 < t < \pi \\ 0 & \pi < t < 2\pi \end{cases}$$

find the solution of the following differential equation which will satisfy the condition

$$y'' - 3y' + 2y = f(t), \qquad y_0 = y_0' = 0.$$

Solution

The Fourier series of $f(t)$ with period 2π is given by

$$f(t) = \frac{a_0}{2} + \sum_{n=1}^{\infty}(a_n \cos nt + b_n \sin nt)$$

where

$$\begin{aligned} a_n &= \frac{1}{\pi}\int_0^{2\pi} f(t)\cos nt dt = 0, \quad n \neq 0 \\ a_0 &= 1 \end{aligned}$$

and

$$\begin{aligned} b_n &= \frac{1}{\pi}\int_0^{\pi} \sin nt dt = \frac{2}{n\pi}, \quad n \text{ odd} \\ b_n &= 0 \qquad n \text{ even.} \end{aligned}$$

Thus,

$$f(t) = \frac{1}{2} + \frac{2}{\pi}\sum_{n=1}^{\infty}\frac{\sin(2n-1)t}{(2n-1)}.$$

The complementary solution of the given differential equation is

$$y_c = Ae^{2t} + Be^{t}$$

and, using the operator method, the particular integral is given by

$$\begin{aligned}
y_p &= \frac{1}{D^2-3D+2}\left[\frac{1}{2} + \frac{2}{\pi}\sum_{n=1}^{\infty}\frac{\sin(2n-1)t}{(2n-1)}\right] \\
&= \frac{1}{4} + \frac{2}{\pi}\sum_{n=1}^{\infty}\frac{1}{(2n-1)}\frac{1}{D^2-3D+2}\sin(2n-1)t \\
&= \frac{1}{4} + \frac{2}{\pi}\sum_{n=1}^{\infty}\frac{1}{(2n-1)}\frac{1}{(2-(2n-1)^2)-3D}\sin(2n-1)t \\
&= \frac{1}{4} + \frac{2}{\pi}\sum_{n=1}^{\infty}\frac{(2-(2n-1)^2)+3D}{(2-(2n-1)^2)-9D^2}\sin(2n-1)t \\
&= \frac{1}{4} + \frac{2}{\pi}\sum_{n=1}^{\infty}\frac{(2-(2n-1)^2)\sin(2n-1)t + 3(2n-1)\cos(2n-1)t}{(2n-1)\{(2-(2n-1)^2)+9(2n-1)^2\}}.
\end{aligned}$$

Thus,

$$y = y_c + y_p.$$

Using the initial conditions at $t = 0$, we obtain

$$\begin{aligned}
A + B + \frac{1}{4} + \frac{2}{\pi}\sum_{n=1}^{\infty}\frac{3}{\{(2-(2n-1)^2)^2+9(2n-1)^2\}} &= 0 \\
2A + B + \frac{2}{\pi}\sum_{n=1}^{\infty}\frac{(2-(2n-1)^2)}{\{(2-(2n-1)^2)^2+9(2n-1)^2\}} &= 0.
\end{aligned}$$

Solving for A and B, we obtain

$$A = \frac{1}{4} + \frac{2}{\pi}\sum_{n=1}^{\infty}\frac{1+(2n-1)^2}{(2-(2n-1)^2)^2+9(2n-1)^2}$$

and

$$B = -\frac{1}{2} - \frac{2}{\pi}\sum_{n=1}^{\infty}\frac{4}{(2-(2n-1)^2)^2+9(2n-1)^2}.$$

Hence the solution is $y = y_c + y_p$, provided A and B are given by the above.

Example 4

The forced oscillations of a body of mass m on a spring (see Fig. 7.3) are governed by the differential equation

$$my'' + cy' + ky = f(t)$$

where k is the spring constant and c is the damping constant.

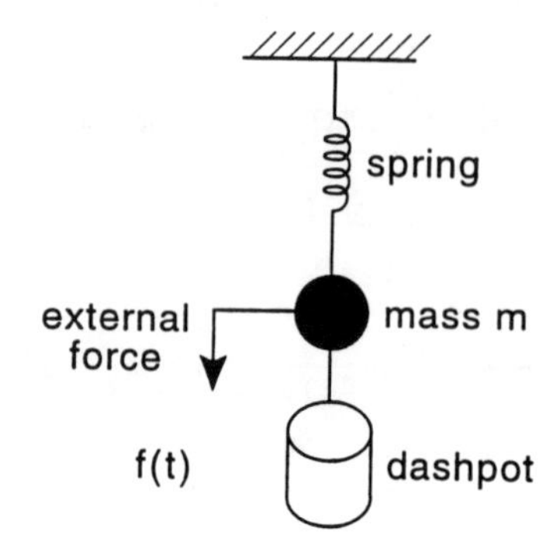

Figure 7.3: Forced oscillation of a mass

If the external force $f(t)$ is periodic and if the definition of one period is given by

$$f(t) = \frac{t}{12}(\pi^2 - t^2) \qquad \text{when} \qquad -\pi < t < \pi$$
$$\text{and} \qquad f(t+2) = f(t)$$

find the steady-state oscillation of the body.

Solution

The steady-state solution of the given problem can be obtained from the particular integral of the differential equation. We express $f(t)$ as the Fourier series expansion

$$f(t) = \frac{a_0}{2} + \sum_{n=1}^{\infty}(a_n \cos nt + b_n \sin nt)$$

where

$$a_n = \frac{1}{\pi}\int_{-\pi}^{\pi} f(t) \cos ntdt \quad , \quad n = 0, 1, 2, \cdots$$

and

$$b_n = \frac{1}{\pi}\int_{-\pi}^{\pi} f(t) \sin ntdt \quad , \quad n = 1, 2, \cdots$$

Thus,

$$a_n = \frac{1}{\pi}\int_{-\pi}^{\pi} \frac{t}{12}(\pi^2 - t^2) \cos ntdt = 0$$

because the integrand is an odd function and, because the integrand is an even function,

$$b_n = \frac{1}{\pi}\int_{-\pi}^{\pi}\frac{t}{12}(\pi^2 - t^2)\sin nt dt = -\frac{(-1)^n}{n^3} = \frac{(-1)^{n+1}}{n^3}.$$

Thus,

$$f(t) = \sum_{n=1}^{\infty} b_n \sin nt = \sum_{n=1}^{\infty}\frac{(-1)^{n+1}}{n^3}\sin nt.$$

From the differential equation we have

$$my'' + cy' + ky = \sum_{n=1}^{\infty}\frac{(-1)^{n+1}}{n^3}\sin nt.$$

The particular integral is given by

$$y_p = \sum_{n=1}^{\infty}\frac{(-1)^{n+1}}{n^3}\frac{1}{mD^2 + cD + k}\sin nt.$$

Here $D \equiv \frac{d}{dt}$, therefore

$$\begin{aligned} y_p &= \sum_{n=1}^{\infty}\frac{(-1)^{n+1}}{n^3}\frac{1}{-mn^2 + k + cD}\sin nt \\ &= \sum_{n=1}^{\infty}\frac{(-1)^{n+1}}{n^3}\frac{(k - mn^2) - cD}{(k - mn^2)^2 - c^2D^2}\sin nt \\ &= \sum_{n=1}^{\infty}\frac{(-1)^{n+1}}{n^3}\frac{(k - mn^2)\sin nt - cn\cos nt}{(k - mn^2)^2 + c^2n^2}. \end{aligned}$$

This is the required steady-state solution of the problem.

7.2 Applications of Fourier integrals

In the previous section, we demonstrated the application of Fourier series with some practical examples. This section deals with the application of Fourier integrals. While a Fourier series is usually employed when the forcing function is periodic, Fourier integrals can be used for non-periodic functions. In a Fourier series representation of a periodic function, there are discrete terms which are handled separately and then summed. However, in the integral representation of a non-periodic function, general infinitesimal segments of the continuous frequency representation are treated as if they were individual terms and then the total effect of their infinitesimal contributions is obtained by integration. In this section, we shall illustrate this process with some practical examples.

Example 5

Find the Fourier integral representation of the following rectangular pulse

$$f(t) = \begin{cases} 1 & t^2 < b^2 \\ 0 & t^2 > b^2. \end{cases}$$

Solution

The graphical representation of the given pulse is shown in the following figure:

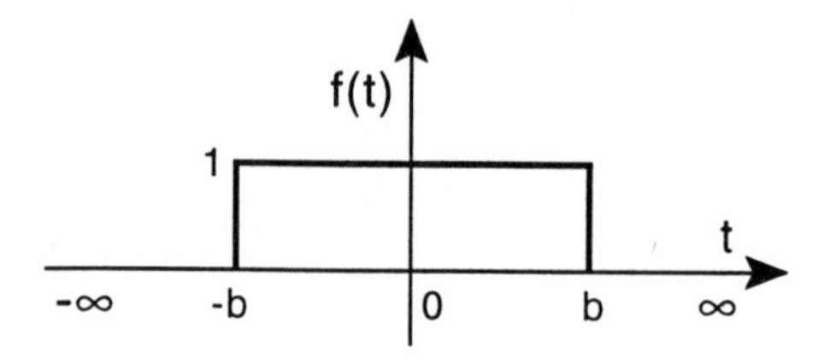

Figure 7.4(a): Graphical representation of $f(t)$

We shall first obtain the Fourier transform of the pulse and then take the inverse of the transform.

$$\mathcal{F}(f(t)) = g(\sigma) \quad = \quad \int_{-\infty}^{\infty} f(t)e^{-i\sigma t}dt = 2\frac{\sin b\sigma}{\sigma}.$$

By analogy with the theory of light, this $g(\sigma)$ is called the spectrum of the pulse $f(t)$, since it provides a measure of the intensity of $f(t)$ in the frequency domain.

The graphical representation of this spectrum is shown below.

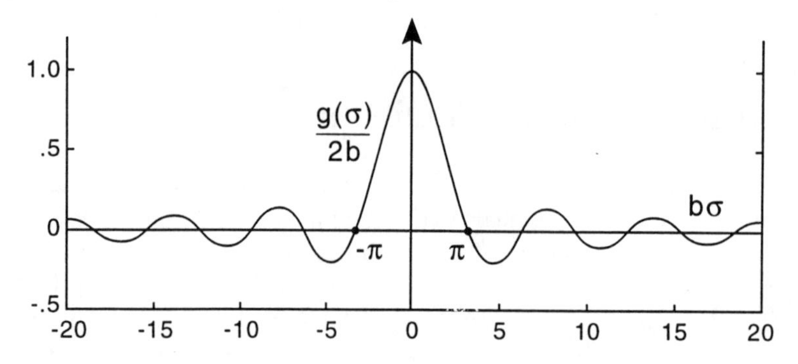

Figure 7.4(b): Spectrum of $f(t)$

From this graph, it is clear that most of the energy of the pulse lies within the central peak of width $\frac{2\pi}{b}$. Therefore, the longer the pulse, the more narrow the width into which its energy is concentrated. This width is usually known as the **spectral bandwidth**.

The Fourier integral representation is given by the inverse of this transform as follows:

$$f(t) = \frac{1}{2\pi}\int_{-\infty}^{\infty} g(\sigma)e^{i\sigma t}d\sigma = \frac{2}{\pi}\int_{0}^{\infty} \frac{\sin b\sigma \cos \sigma t}{\sigma}d\sigma.$$

This is the required representation.

Note

If we look for an approximate representation of the given pulse, then we can consider only the frequencies below σ_0 such that σ_0 is very large but finite. In that situation,

$$\begin{aligned} f(t) &\simeq \frac{2}{\pi}\int_0^{\sigma_0} \frac{\sin b\sigma \cos \sigma t}{\sigma}d\sigma \\ &= \frac{1}{\pi}\int_0^{\sigma_0} \frac{\sin(b+t)\sigma - \sin(t-b)\sigma}{\sigma}d\sigma \\ &= \frac{1}{\pi}\int_0^{\sigma_0} \frac{\sin(b+t)\sigma}{\sigma}d\sigma - \frac{1}{\pi}\int_0^{\sigma_0} \frac{\sin(t-b)\sigma}{\sigma}d\sigma. \end{aligned}$$

In the first integral, let $(b+t)\sigma = u$, and, in the second integral, let $(t-b)\sigma = u$. We have

$$f(t) \simeq \frac{1}{\pi}\int_0^{(b+t)\sigma_0} \frac{\sin u}{u}du - \frac{1}{\pi}\int_0^{(t-b)\sigma_0} \frac{\sin u}{u}du.$$

By using the definition of the sine-integral function of x

$$S_i(x) \equiv \int_0^x \frac{\sin u}{u}du$$

we obtain

$$f(t) \simeq \frac{1}{\pi}S_i[\sigma_0(t+b)] - \frac{1}{\pi}S_i[\sigma_0(t-b)].$$

Fig. 7.5 shows this approximation for $\sigma_0 = 4$, 8, and 16 radian per unit time for $b = 1$. Higher values of σ_0 give better approximations of $f(t)$. In connection with the Fourier series expansion of a periodic function, we have demonstrated the existence of Gibbs' phenomenon at both points of discontinuity; similar behaviour of the approximating curve can be observed in this case. The curve overshoots the right- and left-hand limits of the function by an amount which approaches

$$\frac{1}{2} - \frac{1}{\pi}S_i(\pi) \simeq 0.09$$

times the actual amount of the jump of the function. This behaviour, known as Gibbs' phenomenon, occurs even if σ_0 is infinite.

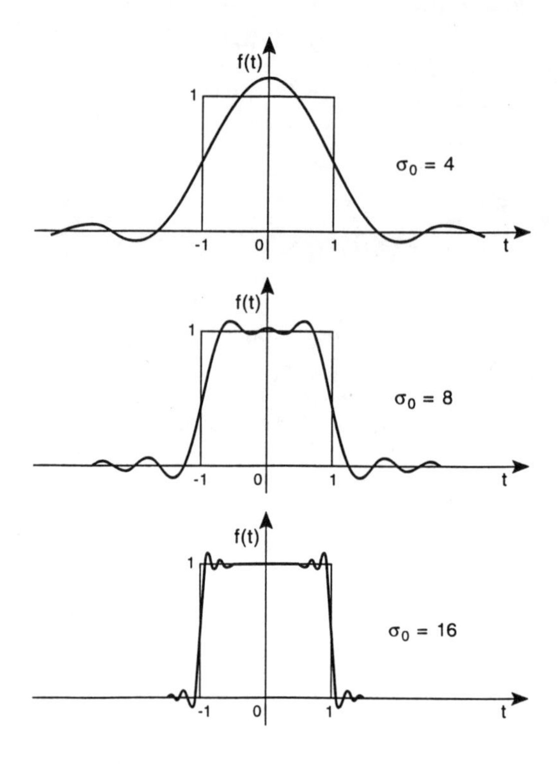

Figure 7.5: The approximate Fourier integral representation of a function $f(t)$ for frequencies $\sigma_0 = 4$, 8, and 16

Example 6

Find a particular integral of the equation

$$y'' - 5y' + 4y = f(t)$$

if $f(t)$ represents a rectangular pulse as given in Example 1.

Solution

The Fourier integral representation of $f(t)$ is studied in detail in Example 1 and is given by

$$f(t) = \frac{2}{\pi}\int_0^\infty \frac{\sin b\sigma \cos \sigma t}{\sigma} d\sigma.$$

The given differential equation can be written in the form

$$y'' - 5y' + 4y = \frac{2}{\pi}\int_0^\infty \frac{\sin b\sigma \cos \sigma t}{\sigma} d\sigma.$$

Then the particular integral can easily be obtained by using the operator method as follows:

$$\begin{aligned} y_p &= \frac{2}{\pi}\int_0^\infty \frac{1}{D^2 - 5D + 4}(\frac{\sin b\sigma}{\sigma}) \cos \sigma t d\sigma \\ &= \frac{2}{\pi}\int_0^\infty (\frac{\sin b\sigma}{\sigma}) \frac{1}{(4-\sigma^2) - 5D} \cos \sigma t d\sigma. \end{aligned}$$

Thus

$$\begin{aligned} y_p &= \frac{2}{\pi}\int_0^\infty (\frac{\sin \sigma b}{\sigma})(\frac{(4-\sigma^2) + 5D}{(4-\sigma^2)^2 - 25D^2}) \cos \sigma t d\sigma \\ &= \frac{2}{\pi}\int_0^\infty (\frac{\sin b\sigma}{\sigma})(\frac{(4-\sigma^2)\cos \sigma t - 5\sigma \sin \sigma t}{(4-\sigma^2)^2 + 25\sigma^2}) d\sigma. \end{aligned}$$

This is the required particular integral of the given differential equation.

This example illustrates that the Fourier integral representation may be used for evaluating a particular integral of a non-homogeneous differential equation.

7.3 Applications of Laplace transforms

We saw that the mathematical formulation of problems in mechanical, electrical, chemical, and industrial engineering often result in linear differential equations with constant coefficients. In this section, we will show how the Laplace transform is used to solve such problems. We shall illustrate the use of Laplace transforms in the following examples.

Example 7

A mass m at the end of a vertical vibrating spring of constant k undergoes vibrations about its equilibrium position according to the equation

$$m\frac{d^2y}{dt^2} + c\frac{dy}{dt} + ky = F_0 \cos \omega t$$

where c is the damping constant and y is the displacement of the mass from its equilibrium position at any time.

(a) Solve this equation subject to the initial conditions

$$y(0) = y'(0) = 0.$$

(b) What is the steady-state solution?

Solution

The given equation is

$$m\frac{d^2y}{dt^2} + c\frac{dy}{dt} + ky = F_0 \cos \omega t.$$

Taking the Laplace transforms of both sides of the above equation and using the initial conditions, we obtain

$$\begin{aligned}(ms^2 + cs + k)\mathcal{L}(y) &= F_0 \frac{s}{s^2+\omega^2} \\ \mathcal{L}(y) &= F_0 \frac{s}{(s^2+\omega^2)(ms^2+cs+k)}.\end{aligned}$$

Therefore,

$$\begin{aligned} y(t) &= F_0 \mathcal{L}^{-1}\{\frac{s}{(s^2+\omega^2)(ms^2+cs+k)}\} \\ &= \frac{F_0}{m}\mathcal{L}^{-1}\{\frac{s}{(s^2+\omega^2)((s+\frac{c}{2m})^2+(\frac{k}{m}-\frac{c^2}{4m^2}))}\}.\end{aligned}$$

We have to consider three cases in this problem:

Case (I): $4km > c^2$

In this case, let $\sqrt{\frac{k}{m} - \frac{c^2}{4m^2}} = b$ and $\frac{c}{2m} = a$, then

$$y(t) = \frac{F_0}{m}\mathcal{L}^{-1}\{\frac{s}{(s^2+\omega^2)((s+a)^2+b^2)}\}.$$

We use the Heaviside theorem of quadratic polynomials to obtain the inverse Laplace transform. The poles are given by the roots of $s = \pm i\omega$ and $s = -a \pm ib$.

The inverse corresponding to the poles $s = \pm i\omega$ is the following:

$$\begin{aligned}\phi(s) &= \frac{s}{(s+a)^2+b^2} \\ \phi(i\omega) &= \frac{i\omega}{(i\omega+a)^2+b^2} \\ \phi_r &= \frac{2a\omega^2}{(a^2+b^2-\omega^2)^2 - 4a^2\omega^2} \\ \phi_i &= \frac{\omega(a^2+b^2-\omega^2)}{(a^2+b^2-\omega^2)^2-4a^2\omega^2}.\end{aligned}$$

Therefore the inverse in this case is

$$\frac{1}{\omega}[\phi_i \cos\omega t + \phi_r \sin\omega t] = \frac{(a^2+b^2-\omega^2)\cos\omega t + 2a\omega \sin\omega t}{(a^2+b^2-\omega^2)^2 - 4a^2\omega^2}. \tag{7.1}$$

Similarly, the inverse corresponding to the poles $s = -a \pm ib$ is the following:

$$\begin{aligned}
\phi(s) &= \frac{s}{s^2 + \omega^2} \\
\phi(-a+ib) &= \frac{-a+ib}{(-a+ib)^2 + \omega^2} \\
\phi_r &= \frac{-2ab^2 - a(a^2 - b^2 + \omega^2)}{(a^2 - b^2 + \omega^2)^2 - 4a^2b^2} \\
\phi_i &= \frac{-2a^2b + b(a^2 - b^2 + \omega^2)}{(a^2 - b^2 + \omega^2)^2 - 4a^2b^2}.
\end{aligned}$$

The inverse in this case is therefore given by

$$\begin{aligned}
&\frac{1}{b} e^{-at}[\phi_i \cos bt + \phi_r \sin bt] \\
&= \frac{e^{-at}\{(-2a^2b + b(a^2 - b^2 + \omega^2)) \cos bt - (2ab^2 + a(a^2 - b^2 + \omega^2)) \sin bt\}}{(a^2 - b^2 + \omega^2)^2 - 4a^2b^2}.
\end{aligned} \tag{7.2}$$

The solution for this case is the sum of (7.1) and (7.2). Solution (7.1) is the oscillatory steady-state solution and (7.2) is the damped oscillatory solution.

Case (II): $4km = c^2$ **or** $b = 0$

and so

$$y(t) = \frac{F_0}{m}\mathcal{L}^{-1}\{\frac{s}{(s^2 + \omega^2)(s+a)^2}\}.$$

The poles are given by $s = \pm i\omega$ and $s = -a$. The latter is a double pole. The inverse corresponding to the poles $s = \pm i\omega$ is the following:

$$\begin{aligned}
\phi(s) &= \frac{s}{(s+a)^2} \\
\phi(i\omega) &= \frac{i\omega}{(a+i\omega)^2} = \frac{i\omega}{a^2 - \omega^2 + 2i\omega a}.
\end{aligned}$$

The real and imaginary parts of $\phi(i\omega)$ are respectively

$$\begin{aligned}
\phi_r &= \frac{+2a\omega^2}{(a^2 - \omega^2)^2 - 4a^2\omega^2} \\
\phi_i &= \frac{\omega(a^2 - \omega^2)}{(a^2 - \omega^2)^2 - 4a^2\omega^2}.
\end{aligned}$$

Therefore, the inverse in this case is

$$\begin{aligned}
&\frac{1}{\omega}\{\phi_i \cos \omega t + \phi_r \sin \omega t\} \\
&= \frac{(a^2 - \omega^2) \cos \omega t + 2a\omega \sin \omega t}{(a^2 - \omega^2) - 4a^2\omega^2}.
\end{aligned} \tag{7.3}$$

The inverse corresponding to the double pole at $s = -a$ is found by the residue theorem:

$$\begin{aligned}
\text{Inverse} &= \text{Residue at } s = -a \text{ a double pole} \\
&= \lim_{s \to -a} \frac{d}{ds}\{\frac{se^{st}}{(s^2+\omega^2)}\} \\
&= \lim_{s \to -a} [\frac{e^{st} + tse^{st}}{s^2+\omega^2} - \frac{2s^2e^{st}}{(s^2+\omega^2)^2}] \\
&= e^{-at}[\frac{1-at}{a^2+\omega^2} - \frac{2a^2}{(a^2+\omega^2)^2}]. \qquad (7.4)
\end{aligned}$$

The solution in this case is the sum of (7.3) and (7.4). Solution (7.3) is the steady-state solution and solution (7.4) the critically-damped solution.

Case (III): $4km < c^2$

In this case, we have

$$y(t) = \frac{F_0}{m}\mathcal{L}^{-1}\{\frac{s}{(s^2+\omega^2)(s+a+b)(s+a-b)}\}.$$

The poles are at $s = \pm i\omega$, $s = -(a+b) = -\alpha$ and $s = -(a-b) = -\beta$.

Using the theory of Residues:

$$\begin{aligned}
R_1 &= \text{Residue at } s = i\omega \\
&= \lim_{s \to i\omega} \frac{se^{st}}{(s+i\omega)(s+\alpha)(s+\beta)} \\
&= \frac{e^{i\omega t}}{2\{(\alpha\beta-\omega^2)+i\omega(\alpha+\beta)} \\
R_1 &= \frac{e^{i\omega t}[(\alpha\beta-\omega^2)-i\omega(\alpha+\beta)]}{2[(\alpha\beta-\omega^2)^2-\omega^2(\alpha+\beta)^2]}.
\end{aligned}$$

Similarly,

$$\begin{aligned}
R_2 &= \text{Residue at } s = -i\omega \\
&= \frac{e^{-i\omega t}[(\alpha\beta-\omega^2)+i\omega(\alpha+\beta)]}{2[(\alpha\beta-\omega^2)^2-\omega^2(\alpha+\beta)^2]}.
\end{aligned}$$

Thus,

$$\begin{aligned}
R_1 + R_2 &= \frac{2(\alpha\beta-\omega^2)\cos\omega t + 2\omega(\alpha+\beta)\sin\omega t}{2[(\alpha\beta-\omega^2)^2-\omega^2(\alpha+\beta)^2]} \\
&= \frac{(2\beta-\omega^2)\cos\omega t + \omega(\alpha+\beta)\sin\omega t}{(\alpha\beta-\omega^2)^2-\omega^2(\alpha+\beta)^2}. \qquad (7.5)
\end{aligned}$$

$$\begin{aligned} R_3 &= \text{Residue at } s = -\alpha \\ &= \lim_{s \to -\alpha} \frac{se^{st}}{(s+\beta)(s^2+\omega^2)} \\ &= \frac{-\alpha e^{-\alpha t}}{(-\alpha+\beta)(\alpha^2+\omega^2)} = \frac{\alpha e^{-\alpha t}}{(\alpha-\beta)(\alpha^2+\omega^2)}. \end{aligned}$$

$$\begin{aligned} R_4 &= \text{Residue at } s = -\beta \\ &= \lim_{s \to -\beta} \frac{se^{st}}{(s+\alpha)(s^2+\omega^2)} \\ &= \frac{-\beta e^{-\beta t}}{(\alpha-\beta)(\beta^2+\omega^2)}. \end{aligned}$$

$$R_3 + R_4 = \frac{\alpha e^{-\alpha t}}{(\alpha-\beta)(\alpha^2+\omega^2)} - \frac{\beta e^{-\beta t}}{(\alpha-\beta)(\beta^2+\omega^2)}. \tag{7.6}$$

The solution in this case is the sum of (7.5) and (7.6). Solution (7.5) is the steady-state solution and solution (7.6) is the damped solution.

Example 8

The vibrations of a mass m at the end of a vertical spring of constant k are given by

$$m\frac{d^2y}{dt^2} + ky = F(t) \tag{7.7}$$

where $F(t)$ is the applied external force at any time t and y is the displacement of m from its equilibrium position at any time. Suppose that the force is given by

$$F(t) = \begin{cases} F_0 & 0 < t < T \\ 0 & t > T. \end{cases} \tag{7.8}$$

Find the displacement at any time, assuming that the initial displacement and velocity are zero.

Solution

$F(t)$ in equation (7.8) can be expressed as a Heaviside unit step function

$$F(t) = F_0[u(t) - u(t-T)]. \tag{7.9}$$

Thus we can write (7.7) as

$$m\frac{d^2y}{dt^2} + ky = F_0[u(t) - u(t-T)]. \tag{7.10}$$

The initial conditions are:

$$y_0 = y_0' = 0. \tag{7.11}$$

Now, taking the Laplace transform of (7.10), and using the initial conditions, we obtain

$$\begin{aligned}(ms^2+k)\mathcal{L}(y) &= F_0(\frac{1}{s}-\frac{e^{-Ts}}{s}) \\ \mathcal{L}(y) &= F_0\frac{1-e^{Ts}}{(ms^2+k)s}.\end{aligned}$$

Therefore, the inverse Laplace transform yields

$$\begin{aligned}y(t) &= F_0\mathcal{L}^{-1}\{\frac{1}{s(ms^2+k)}-\frac{e^{-Ts}}{s(ms^2+k)}\} \\ &= \frac{F_0}{m}\mathcal{L}^{-1}\{(\frac{1}{s}-\frac{s}{s^2+k/m})(1-e^{-Ts})\frac{m}{k}\} \\ &= \frac{F_0}{k}[(1-\cos\sqrt{k/m}t)u(t)-(1-\cos\sqrt{k/m}(t-T))u(t-T)].\end{aligned}$$

Thus we have

$$y(t)=\begin{cases}\frac{F_0}{k}(1-\cos\sqrt{k/m}t) & t<T \\ \frac{F_0}{k}\{\cos\sqrt{k/m}(t-T)-\cos\sqrt{k/m}t\} & t>T.\end{cases}$$

Example 9

A beam which is hinged at its ends, $x=0$ and $x=a$, carries a uniform load W_0 per unit length. Find the deflection at any point.

Solution

From mechanics, we know that the differential equation and boundary conditions are

$$\frac{d^4y}{dx^4}=\frac{W_0}{EI} \qquad 0<x<a \tag{7.12}$$

at

$$\begin{aligned}x=0: \qquad y(0) &= y''(0)=0 \\ x=a: \qquad y(a) &= y''(a)=0 \end{aligned} \tag{7.13}$$

where EI is called the flexural rigidity of the beam and E is called Young's Modulus of Elasticity. These are both assumed to be constant.

This is a boundary value problem. We define the Laplace transform of $y(x)$ with respect to x as follows:

$$\mathcal{L}(y(x))=\int_0^\infty e^{-sx}y(x)dx \quad , \quad s>0.$$

Now, taking the Laplace transform of (7.12), we obtain

$$s^4\mathcal{L}(y)-s^3y_0-s^2y_0'-sy_0''-y_0'''=\frac{W_0}{EIs}.$$

Using the boundary conditions (7.13), and setting $y_0' = \alpha$ and $y_0''' = \beta$, we have

$$\mathcal{L}(y) = \frac{\alpha}{s^2} + \frac{\beta}{s^4} + \frac{W_0}{s^5 EI}. \tag{7.14}$$

Inverting, we obtain

$$y(x) = \alpha x + \beta \frac{x^3}{6} + \frac{W_0}{EI}\frac{x^4}{24}. \tag{7.15}$$

From the last two conditions in (7.13), we find

$$\begin{aligned} \alpha &= \frac{W_0 a^3}{24EI} \\ \beta &= \frac{W_0 a}{2EI}. \end{aligned}$$

Thus, the required deflection is given by

$$y(x) = \frac{W_0}{24EI} x(\ell - x)(\ell^2 + \ell x - x^2).$$

Example 10

An inductor L, capacitor C, and resistor R are connected in series. Initially, the charge on the capacitor is Q_0, while the current is zero. Show that the charge Q and current I will be oscillatory if $R < 2\sqrt{L/C}$ and then find the solutions for Q and I.

Solution

The differential equation and the initial conditions are given by

$$L\frac{d^2Q}{dt^2} + R\frac{dQ}{dt} + \frac{Q}{C} = 0 \tag{7.16}$$

$$t = 0: \quad Q = Q_0, \quad I = \frac{dQ}{dt} = 0. \tag{7.17}$$

Taking the Laplace transform of (7.16) and using the initial conditions, we find

$$\begin{aligned} \mathcal{L}(Q) &= \frac{LQ_0 s + RQ_0}{Ls^2 + Rs + \frac{1}{C}} \\ &= Q_0 \frac{s + R/L}{(s + \frac{R}{2L})^2 + (\frac{1}{LC} - \frac{R^2}{4L^2})}. \end{aligned}$$

We know that the solutions will be oscillatory if $\omega = \sqrt{\frac{1}{LC} - \frac{R^2}{4L^2}} > 0$ so that $R < 2\sqrt{\frac{L}{C}}$ and hence $\mathcal{L}(Q) = Q_0 \frac{s + \frac{R}{2L} + \frac{R}{2L}}{(s + \frac{R}{2L})^2 + \omega^2}$.

Therefore,

$$\begin{aligned} Q(t) &= Q_0\mathcal{L}^{-1}\{\frac{s+\frac{R}{2L}}{(s+\frac{R}{2L})^2+\omega^2}\}+\mathcal{L}^{-1}\{\frac{\frac{R}{2L}}{(s+\frac{R}{2L})^2+\omega^2}\} \\ &= e^{-\frac{Rt}{2L}}Q_0\{\cos\omega t+\frac{R}{2L\omega}\sin\omega t\} \\ &= \frac{Q_0}{2L\omega}e^{-\frac{Rt}{2L}}\sqrt{R^2+4L^2\omega^2}\sin(\omega t+\delta) \end{aligned}$$

and

$$\begin{aligned} I(t) &= \frac{dQ}{dt} \\ &= -\frac{Q_0(R^2+4\omega^2L^2)}{4\omega L^2}e^{-\frac{Rt}{2L}}\sin\omega t \end{aligned}$$

where $\omega=\sqrt{\frac{1}{LC}-\frac{R^2}{4L^2}}$ and $\delta=\tan^{-1}(\frac{2\omega L}{R})$.

Example 11

A resistor of R ohms and a capacitor of C farads are connected in series with a generator supplying E volts. Initially the charge on the capacitor is zero. Find the charge and current at any time $t>0$ if

(a) $E=E_0$, a constant;

(b) $E=E_0, e^{-\alpha t}$ $\alpha>0$.

Solution

The differential equation and the initial conditions are

$$R\frac{dQ}{dt}+\frac{Q}{C}=E(t) \tag{7.18}$$

at $t=0$: $Q=0$ and $I=\frac{dQ}{dt}$.

(a) When $E=E_0$, a constant, (7.18) can be written as

$$\frac{dQ}{dt}+\frac{Q}{RC}=\frac{E_0}{R}. \tag{7.19}$$

Taking the Laplace transform of (7.19) and using the initial condition, we obtain

$$\begin{aligned} s\mathcal{L}(Q)+\frac{1}{RC}\mathcal{L}(Q) &= \frac{E_0}{Rs} \\ (s+\frac{1}{RC})\mathcal{L}(Q) &= \frac{E_0}{Rs}. \end{aligned}$$

Therefore,

$$\begin{aligned}\mathcal{L}(Q) &= \frac{E_0}{R}\frac{1}{s(s+\frac{1}{RC})}\\ &= \frac{E_0}{R}[\frac{1}{s}-\frac{1}{s+\frac{1}{RC}}]RC = CE_0(\frac{1}{s}-\frac{1}{s+\frac{1}{RC}}) = CE_0(\frac{1}{s}-\frac{1}{s+\frac{1}{RC}}).\end{aligned}$$

Taking the inverse, we obtain

$$Q(t) = CE_0(1-e^{-t/RC})$$

and

$$I(t) = \frac{dQ}{dt} = \frac{E_0}{R}e^{-t/RC}.$$

(b) When $E = E_0e^{-\alpha t}$, the differential equation is

$$\frac{dQ}{dt}+\frac{Q}{RC} = \frac{E_0}{R}e^{-\alpha t}. \qquad (7.20)$$

Taking the Laplace transform of (7.20) and using the initial conditions, we obtain

$$\begin{aligned}\mathcal{L}(Q) &= \frac{E_0}{R}\frac{1}{(s+\alpha)(s+\frac{1}{RC})}\\ &= \frac{E_0C}{\alpha RC-1}[\frac{1}{s+\frac{1}{RC}}-\frac{1}{s+\alpha}].\end{aligned}$$

Thus

$$Q(t) = \frac{CE_0}{\alpha RC-1}[e^{-t/RC}-e^{-\alpha t}]$$

and

$$I(t) = \frac{dQ}{dt} = \frac{CE_0}{\alpha RC-1}[\alpha e^{-\alpha t}-\frac{e^{-t/RC}}{RC}] \quad \alpha \neq \frac{1}{RC}.$$

Example 12

A radioactive substance M_1 having a decay rate k_1 disintegrates into a second radioactive substance M_2 which has a decay rate k_2. Substance M_2 disintegrates into M_3, which is stable. If $M_1(t)$, $M_2(t)$, and $M_3(t)$ are the masses of the respective substances at time t, the governing equations are

$$\begin{aligned}\frac{dM_1}{dt} &= -k_1M_1\\ \frac{dM_2}{dt} &= -k_2M_2+k_1M_1\\ \frac{dM_3}{dt} &= k_2M_2.\end{aligned} \qquad (7.21)$$

Find the solution of the above system under the conditions

$$\begin{aligned}M_1(0) &= M_0\\ M_2(0) &= 0\\ M_3(0) &= 0.\end{aligned} \qquad (7.22)$$

Solution

Taking the Laplace transform of the above system of equations, we obtain

$$\begin{aligned}\mathcal{L}\{\dot{M}_1\} &= -k_1\mathcal{L}(M_1)\\ \mathcal{L}\{\dot{M}_2\} &= -k_2\mathcal{L}(M_2)+k_1\mathcal{L}(M_1)\\ \mathcal{L}\{\dot{M}_3\} &= -k_2\mathcal{L}(M_2)\end{aligned}\tag{7.23}$$

or

$$\begin{aligned}s\mathcal{L}(M_1)-M_1(0) &= -k_1\mathcal{L}(M_1)\\ s\mathcal{L}(M_2)-M_2(0) &= -k_2\mathcal{L}(M_2)+k_1\mathcal{L}(M_1)\\ s\mathcal{L}(M_3)-M_3(0) &= k_2\mathcal{L}(M_2).\end{aligned}\tag{7.24}$$

Using initial conditions (7.22), we obtain

$$\begin{aligned}(s+k_1)\mathcal{L}(M_1) &= M_0\\ (s+k_2)\mathcal{L}(M_2) &= k_1\mathcal{L}(M_1)\\ s\mathcal{L}(M_3) &= k_2\mathcal{L}(M_2).\end{aligned}\tag{7.25}$$

Solving, we have

$$\begin{aligned}\mathcal{L}(M_1) &= \frac{M_0}{s+k_1}\\ \mathcal{L}(M_2) &= \frac{k_1}{s+k_2}\mathcal{L}(M_1)=\frac{M_0k_1}{(s+k_1)(s+k_2)}\\ \mathcal{L}(M_3) &= \frac{k_2}{s}\mathcal{L}(M_2)=\frac{M_0k_1k_2}{s(s+k_1)(s+k_2)}.\end{aligned}$$

Taking the Laplace inverse:

$$\begin{aligned}M_1(t) &= M_0e^{-k_1t}\\ M_2(t) &= M_0k_1[\frac{e^{-k_1t}}{k_2-k_1}+\frac{e^{-k_2t}}{k_1-k_2}]\\ &= \frac{M_0k_1}{k_1-k_2}[e^{-k_2t}-e^{-k_1t}]\\ M_3(t) &= M_0k_1k_2[\frac{1}{k_1k_2}+\frac{e^{-k_1t}}{k_1(k_1-k_2)}+\frac{e^{-k_2t}}{k_2(k_2-k_1)}]\\ &= M_0[1+\frac{k_2e^{-k_1t}}{k_1-k_2}+\frac{k_1e^{-k_2t}}{k_2-k_1}]\\ &= M_0[1+\frac{1}{k_1-k_2}(k_2e^{-k_1t}-k_1e^{-k_2t})].\end{aligned}$$

7.4 Applications with PDE

Many problems in Science and Engineering can be formulated mathematically with the partial differential equations. In chapter 5 we examined the mathematical development that governs a physical system, and their corresponding solution techniques. The present chapter will examine applications of partial differential equations with problems arising in transmission lines, heat transfer, chemical diffusion, vibration of beams and strings, surface and tidal waves, etc. We shall derive these equations where necessary from physical principles and solve the problems using initial and boundary conditions, which we call **initial-boundary value problems**.

7.5 Transmission lines

The mathematical equations defining the transmission line problem are developed in Chapter 1. We will demonstrate here the solution to the transmission line problem.

Example 13

A semi-infinite transmission line which is initially dead has negligible resistance and ground conductance per unit length. A voltage $E_0(t)$ is applied at the sending end $x = 0$. Find the voltage and current at any position $x > 0$.

Solution

The telephone equation is given by

$$\frac{\partial^2 E}{\partial x^2} = LC\frac{\partial^2 E}{\partial t^2} + (RC + LG)\frac{\partial E}{\partial t} + RGE. \tag{7.26}$$

It is given that $R = 0$ or $G = 0$. Hence eqn (7.26) reduces to

$$\frac{\partial^2 E}{\partial x^2} = LC\frac{\partial^2 E}{\partial t^2}. \tag{7.27}$$

The boundary conditions are given by

$$x = 0 \quad E(0,t) = E_0(t) \tag{7.28}$$

$$x \to \infty \quad |E(x,t)| < M \tag{7.29}$$

which means that the solution must be bounded as $x \to \infty$. The initial condition is given by

$$t = 0: \quad E(x,0) = 0 \tag{7.30}$$

$$\frac{\partial E}{\partial t}(x,0) = 0. \tag{7.31}$$

Taking the Laplace transform of (7.27) and using the initial conditions (7.30) and (7.31), we obtain

$$\frac{d^2}{dx^2}\mathcal{L}(E) - LC\, s^2\mathcal{L}(E) = 0. \tag{7.32}$$

The solution of which can be written as

$$\mathcal{L}(E) = A\, e^{\sqrt{LC}sx} + B\, e^{-\sqrt{LC}sx}. \tag{7.33}$$

This solution must be bounded when $x \to \infty$. Therefore, $A = 0$ and thus (7.33) becomes

$$\mathcal{L}(E) = B\, e^{-\sqrt{LC}sx}. \tag{7.34}$$

The Laplace transform of boundary condition (7.28) gives

$$x = 0: \qquad \mathcal{L}E(0,t) = \mathcal{L}E_0(t). \tag{7.35}$$

Thus the solution of (7.33) can be written as

$$\mathcal{L}(E) = \mathcal{L}E_0(t)e^{-x\sqrt{LC}s}. \tag{7.36}$$

The inverse Laplace transform is then

$$E(x,t) = E_0(t - x\sqrt{LC})u(t - x\sqrt{LC}). \tag{7.37}$$

Now, to find the current, we use the following equation

$$\frac{\partial I}{\partial t} = -\frac{1}{L}\frac{\partial E}{\partial x}.$$

Performing this operation and then integrating with respect to time, we obtain

$$I(x,t) = \sqrt{\frac{C}{L}}E_0(t - x\sqrt{LC})u(t - x\sqrt{LC})$$

which is the required current. Here u is a unit step function.

7.6 The heat conduction problem

We saw in Chapter 1 the mathematical development describing the heat conduction problem. The governing equation turns out to be a partial differential equation. In the following, we will examine heat conduction in a semi-infinite solid in which radiation is present at the surface when it is at zero degrees.

Example 13

A semi infinite solid $x > 0$ is initially at temperature T_0. If the left-hand end radiates freely into a medium at zero degrees celsius, find the temperature of the solid at any subsequent time.

Solution

The governing differential equation can be written as

$$\frac{\partial T}{\partial t} = k\frac{\partial^2 T}{\partial x^2}. \tag{7.38}$$

The boundary conditions are

$$x = 0 \quad : \quad -\frac{\partial T}{\partial t} + hT = 0 \tag{7.39}$$

$$x \to \infty \quad : \quad |T(x,t)| < M. \tag{7.40}$$

The initial condition is

$$T = 0 \quad : \quad T(x,0) = T_0. \tag{7.41}$$

Let us use the following transformation

$$\theta = T - \frac{1}{h}\frac{\partial T}{\partial x}. \tag{7.42}$$

Then, equations (7.38) to (7.41) may be written as

$$\frac{\partial \theta}{\partial t} = k\frac{\partial^2 \theta}{\partial x^2} \tag{7.43}$$

$$x = 0 \quad : \quad \theta = 0 \tag{7.44}$$

$$x \to \infty \quad : \quad |\theta(x,t)| < M \tag{7.45}$$

$$t = 0 \quad : \quad \theta = T_0. \tag{7.46}$$

The solution of the boundary value problem (7.43) to (7.46) can be obtained by the Laplace transform method.

Taking the Laplace transform of (7.43), (7.44) and (7.45) and using the initial condition (7.46), we obtain

$$\frac{d^2}{dx^2}\mathcal{L}(\theta) - \frac{s}{k}\mathcal{L}(\theta) = -\frac{T_0}{k} \tag{7.47}$$

$$x = 0 \quad : \quad \mathcal{L}(\theta) = 0 \tag{7.48}$$

$$x \to \infty \quad : \quad |\mathcal{L}(\theta)| < M. \tag{7.49}$$

The solution to this boundary value problem is given by

$$\mathcal{L}(\theta) = \frac{T_0}{s}\{1 - e^{-\sqrt{\frac{s}{k}}x}\}. \tag{7.50}$$

The inverse Laplace transform is

$$\theta(x,t) = T_0 erf(\frac{x}{2\sqrt{kt}}) \tag{7.51}$$

where

$$erf(\frac{x}{2\sqrt{kt}}) = \frac{2}{\sqrt{\pi}}\int_0^{\frac{x}{2\sqrt{kt}}} e^{-\eta^2}\,d\eta.$$

and note that when $x \to \infty$, $\theta(x, y)$ has the limit T_0. Now, to determine T we have the equation as

$$\frac{\partial T}{\partial x} - hT = -h\,\theta(x,t) \tag{7.52}$$

which is a first order differential equation. The solution can be written as

$$T(x,t) = A\,e^{hx} - h\,e^{hx}\int_{\infty}^{x}\theta(\eta,t)e^{-h\eta}d\eta \tag{7.53}$$

where A is an integration constant.

Substituting $\eta = x + \xi$, (7.53) can be reduced to

$$T(x,t) = A\,e^{hx} + h\int_0^{\infty}\theta(x+\xi,t)e^{-h\xi}d\xi. \tag{7.54}$$

But, as $x \to \infty$, $\theta(x,t)$ must be bounded and have the limit T_0, and, as T must be finite, it follows that A must be zero. Thus, the solution of our problem reduces to

$$\begin{aligned} T(x,t) &= h\int_0^{\infty}\theta(x+\xi,t)e^{-h\xi}d\xi \\ &= hT_0\int_0^{\infty}e^{-h\xi}[erf(\frac{x+\xi}{2\sqrt{kt}})]d\xi \\ &= \frac{2hT_0}{\sqrt{\pi}}\int_0^{\infty}e^{-h\xi}[\int_0^{\frac{x+\xi}{2\sqrt{kt}}}e^{-\eta^2}d\eta]d\xi. \end{aligned} \tag{7.55}$$

We shall evaluate this integral by integrating by parts:

$$\begin{aligned} T(x,t) &= \frac{2T_0}{\sqrt{\pi}}[-e^{-h\xi}\int_0^{\frac{x+\xi}{2\sqrt{kt}}}e^{-\eta^2}d\eta]_0^{\infty} \\ &\quad +\frac{2T_0}{\sqrt{\pi}}\int_0^{\infty}e^{-h\xi}\frac{d}{d\xi}(\int_0^{\frac{x+\xi}{2\sqrt{kt}}}e^{-\eta^2}d\eta)d\xi. \end{aligned}$$

Thus we have

$$T(x,t) = \frac{2T_0}{\sqrt{\pi}}\int_0^{\frac{x}{2\sqrt{kt}}}e^{-\eta^2}d\eta + \frac{T_0}{\sqrt{\pi kt}}\int_0^{\infty}e^{-h\xi-\frac{(x+\xi)^2}{4kt}}d\xi$$

on differentiating this in the usual way.

Now,

$$\frac{4kt\,h\xi + (x+\xi)^2}{4kt} = \frac{(x+\xi+2hkt)^2}{4kt} - hx - h^2kt.$$

Substituting $\frac{x+\xi+2hkt}{2\sqrt{kt}} = \eta$ in the second integral we obtain

$$\begin{aligned} T(x,t) &= T_0\,erf(\frac{x}{2\sqrt{kt}}) \\ &\quad +\frac{2T_0}{\sqrt{\pi}}e^{hx+h^2kt}\int_{\frac{x+2hkt}{2\sqrt{kt}}}^{\infty}e^{-\eta^2}d\eta \\ &= T_0\,erf(\frac{x}{2\sqrt{kt}}) + T_0e^{hx+h^2kt}erfc(\frac{x}{2\sqrt{kt}} + h\sqrt{kt}) \end{aligned}$$

which can be written as

$$\frac{T}{T_0} = erf(\frac{x}{2\sqrt{kt}}) + e^{hx+h^2kt} erfc(\frac{x}{2\sqrt{kt}} + h\sqrt{kt}) \tag{7.56}$$

where $erfc(z) = \int_z^\infty e^{-\eta^2} d\eta = 1 - erf(z)$. The surface temperature of the solid, T_s, can be obtained by putting $x = 0$ in (7.56) and is given by

$$\frac{T_s}{T_0} = e^{h^2kt} erfc(h\sqrt{kt}). \tag{7.57}$$

When $t \to 0$, the value $\frac{T_s}{T_0} \to 1$ which is the initial condition. Now, for large values of time, i.e., when $t \to \infty$

$$\frac{T_s}{T_0} \simeq \frac{1}{h\sqrt{\pi xt}}\{1 - \frac{1}{2h^2kt} + \frac{3}{4h^4k^2t^2} - \cdots\}$$

and so

$$\frac{T_s}{T_0} \simeq \frac{1}{h\sqrt{\pi kt}}. \tag{7.58}$$

Thus, when cooling has been going on for a very long time, the surface temperature may be taken to be $T_s = \frac{T_0}{h\sqrt{\pi kt}}$, with an error less than $\frac{T_0}{2h^3\sqrt{\pi k^3 t^3}}$.

7.7 The chemical diffusion problem

The mathematical model of chemical diffusion is the same as that of heat conduction. In this section, we will look at an example using partial differential equations.

Example 14

A plane sheet of thickness 2ℓ is initially at zero concentration of the diffusing substance. Then, suddenly the surfaces of the sheet are maintained at constant concentrations c_0. Assuming the sheet occupies the region $-\ell \le x \le \ell$, so that there is symmetry about $x = 0$, find the concentration at any point x of the sheet at any time $t > 0$.

Solution

The partial differential equation which governs the physical system is given by

$$\frac{\partial c}{\partial t} = D\frac{\partial^2 c}{\partial x^2} \tag{7.59}$$

where c is the concentration of the diffusing substance and D is the diffusion coefficient which is assumed constant.

The boundary conditions may be written

$$c = c_0, \qquad x = \ell, \qquad t > 0 \tag{7.60}$$

$$\frac{\partial c}{\partial x} = 0, \qquad x = 0, \qquad t > 0. \tag{7.61}$$

The initial condition is

$$c = 0 \quad , \quad t = 0. \tag{7.62}$$

Equation (7.61) expresses the fact that there is no diffusion across the central plane of the sheet. Because of symmetry, it is often convenient to use the condition (7.61) and to consider only the half sheet $0 \leq x \leq \ell$, instead of using the condition $c = c_0$ at $x = -\ell$.

Taking the Laplace transform of (7.59), (7.60) and (7.61) with respect to time t and using initial condition (7.62), we obtain

$$\frac{d^2}{dx^2}\mathcal{L}(c) - (\frac{s}{D})\mathcal{L}(c) = 0 \tag{7.63}$$

with

$$\frac{d}{dx}\mathcal{L}(c) = 0, \qquad x = 0 \tag{7.64}$$

$$\mathcal{L}(c) = \frac{c_0}{s}, \qquad x = \ell. \tag{7.65}$$

The general solution of (7.63) is

$$\mathcal{L}(c) = A\cosh\sqrt{\frac{s}{D}}x + B\sinh\sqrt{\frac{s}{D}}x$$

and using the conditions (7.64) and (7.65), yields

$$\mathcal{L}(c) = \frac{c_0\cosh\sqrt{\frac{s}{D}}x}{s\cosh\sqrt{\frac{s}{D}}\ell}. \tag{7.66}$$

We will derive two important types of solutions, one useful for small values of time and the other one useful for large values of time.

(I) Solution with small t

To find the solution with small values of time, we express the hyperbolic functions in (7.66) in terms of negative exponentials and then expand in a series by the binomial theorem for large values of s.

Therefore, we have

$$\begin{aligned}
\mathcal{L}(c) &= \frac{c_0(e^{\sqrt{\frac{s}{D}}x} + e^{-\sqrt{\frac{s}{D}}x})}{s\,e^{\sqrt{\frac{s}{D}}\ell}(1 + e^{-2\sqrt{\frac{s}{D}}\ell})} \\
&= \frac{c_0}{s}\{e^{-\sqrt{\frac{s}{D}}(\ell-x)} + e^{-\sqrt{\frac{s}{D}}(x+\ell)}\}\sum_{n=0}^{\infty}(-1)^n e^{-2n\sqrt{\frac{s}{D}}\ell} \\
&= \frac{c_0}{s}\sum_{n=0}^{\infty}(-1)^n e^{-\sqrt{\frac{s}{D}}\{(2n+1)\ell-x\}} + \frac{c_0}{s}\sum_{n=1}^{\infty}(-1)^n e^{-\sqrt{\frac{s}{D}}\{(2n+1)\ell+x\}}.
\end{aligned} \tag{7.67}$$

Thus, the inverse Laplace transform yields

$$c = c_0 \sum_{n=0}^{\infty}(-1)^n\{erfc\frac{(2n+1)\ell - x}{2\sqrt{Dt}} + erfc\frac{(2n+1)\ell + x}{2\sqrt{Dt}}\}. \tag{7.68}$$

The series converges very rapidly for small values of $\frac{Dt}{\ell^2}$. As for example, the concentration at the centre of the sheet ($x = 0$) when $\frac{Dt}{\ell^2} = 1$ is given $c/c_0 = 0.8920$ and when $\frac{Dt}{\ell^2} = 0.25$ is $c/c_0 = 0.3145$.

(II) Solution with large t

This solution may be simply obtained by the method of residues. It can be easily shown that the integrand in the complex inversion formula

$$\frac{1}{2\pi i}\int_{a-i\infty}^{a+i\infty} \frac{\cosh\sqrt{\frac{s}{D}}x}{s\cosh\sqrt{\frac{s}{D}}\ell} e^{st} ds$$

contains all simple poles situated at

$$s = 0, \qquad \sqrt{\frac{s}{D}}\ell = (n + \frac{1}{2})\pi i, \qquad n = 0, \pm 1, \pm 2, \cdots$$

or

$$s = 0, \qquad -\frac{(2n+1)^2\pi^2 D}{4\ell^2}, \qquad n = 0, 1, 2, \cdots$$

Now,

$$\begin{aligned}
R_0 &= \text{Residue at } s = 0 \\
&= \lim_{s\to 0}(s)(\frac{e^{st}\cosh\sqrt{\frac{s}{D}}x}{s\cosh\sqrt{\frac{s}{D}}\ell}) = 1 \\
R_n &= \text{Residue at } s = -\frac{(2n+1)^2\pi^2 D}{4\ell^2} = s_n \\
&= \lim_{s\to s_n}(s - s_n)(\frac{e^{st}\cosh\sqrt{\frac{s}{D}}x}{s\cosh\sqrt{\frac{s}{D}}\ell}) \\
&= \{\lim_{s\to s_n}\frac{s - s_n}{\cosh\sqrt{\frac{s}{D}}\ell}\}\{\lim_{s\to s_n}\frac{e^{st}\cosh\sqrt{\frac{s}{D}}x}{s}\} \\
&= \{\lim_{s\to s_n}\frac{1}{\frac{\ell}{2\sqrt{Ds}}\sinh\sqrt{\frac{s}{D}}\ell}\}\{\lim_{s\to s_n}\frac{e^{st}\cosh\sqrt{\frac{s}{D}}x}{s}\} \\
&= \frac{\{4(-1)^n\}}{(2n+1)\pi} e^{\frac{-(2n+1)^2\pi^2 Dt}{4\ell^2}} \cos\frac{(2n+1)\pi x}{2\ell}
\end{aligned}$$

which is obtained by L'Hospital theorem. Thus, we obtain

$$c = c_0 - \frac{4c_0}{\pi}\sum_{n=0}^{\infty}\frac{(-1)^n}{2n+1}e^{-\frac{(2n+1)^2\pi^2 Dt}{4\ell^2}}\cos\frac{(2n+1)\pi x}{2\ell} \tag{7.69}$$

The series converges rapidly for large values of time. Thus, for the concentration at the centre of the sheet ($x = 0$), when $\frac{Dt}{\ell^2} = 1$ is $c/c_0 = 0.8920$ and when $\frac{Dt}{\ell^2} = 0.25$, $c/c_0 = 0.3145$.

7.8 Vibration of beams

We saw the deduction of the governing differential equations for the vibrating strings. In this section, we are concerned with another vibration problem of considerable interest which is the transverse vibration of a beam. The basic difference between a beam and a string is that a beam is stiff and resists bending. To obtain the partial differential equation describing these vibrations, consider the motion of a beam initially located on the x-axis as shown in Fig. 7.6 which is vibrating transversely, i.e., perpendicular to the x-axis.

We assume a general cross section of the beam at x to be $A(x)$ and the moment of inertia of that cross section about the x-axis to be $I(x)$. Let the weight of the beam per unit volume be ρ and the modulus of elasticity E.

In addition to the internal forces, the beam may be acted upon by a distributed load of known intensity $f(x, y, \dot{y}, t)$ per unit length of the beam in which the gravitational forces and frictional forces are also included. Consider an elementary length Δx having the beam cross sections at x and $x + \Delta x$, and therefore total external forces are given by

$$-f(x, y, \dot{y}, t)\Delta x.$$

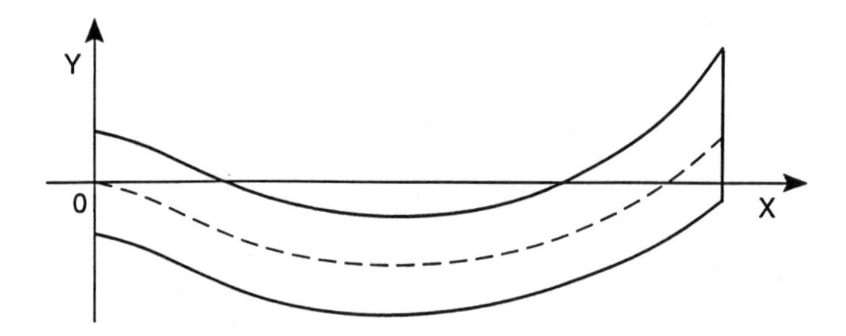

Figure 7.6 Transverse vibration of a beam

The negative sign is due to the fact that the load is considered downward while the forces are resolved upward. The two shearing forces acting on the interior faces of the segment can be written as

$$-s(x + \Delta x, t) + s(x, t)$$

where s is the shearing force. The mass of the segment Δx is

$$\frac{\rho A(x)\Delta x}{g}$$

where g is the acceleration due to gravity. Then, by Newton's second law,

$$\begin{aligned}
(\text{mass}) \times (\text{acceleration}) &= \text{external forces} \\
\frac{\rho A(x)\Delta x}{g}\frac{\partial^2 y}{\partial t^2} &= -[s(x \pm \Delta x, t) - s(x,t)] - f(x,y,\dot{y},t)\Delta x.
\end{aligned}$$

Dividing by Δx, and taking the limit as $\Delta x \to 0$, we obtain

$$\frac{\rho A(x)}{g}\frac{\partial^2 y}{\partial t^2} = -\frac{\partial s}{\partial x} - f(x,y,\dot{y},t) \tag{7.70}$$

but, the shearing force s is given by

$$s = \frac{\partial M}{\partial x}$$

where M is the bending moment.
Also, we know that

$$M = EI(x)\frac{\partial^2 y}{\partial x^2}.$$

Thus,

$$\frac{\partial s}{\partial x} = \frac{\partial^2 M}{\partial x^2} = \frac{\partial^2}{\partial x^2}(EI(x)\frac{\partial^2 y}{\partial x^2}).$$

Substituting this expression into (7.70) and rearranging the term, we obtain

$$\frac{\partial^2}{\partial x^2}(EI(x)\frac{\partial^2 y}{\partial x^2}) = -\frac{\rho A(x)}{g}\frac{\partial^2 y}{\partial t^2} - f(x,y,\dot{y},t). \tag{7.71}$$

In many important practical problems, the cross section $A(x)$ and moment of inertia $I(x)$ are considered constants and $f(x,y,\dot{y},t)$ is zero. Under these conditions, eqn (7.71) reduces to the following simple form:

$$\frac{\partial^2 y}{\partial t^2} + c^2\frac{\partial^4 y}{\partial x^4} = 0 \tag{7.72}$$

where $c^2 = \frac{EIg}{A\rho}$ is considered a parameter of the problem.

This one-dimensional analysis can be extended to a two-dimensional vibration problem of the rectangular elastic plate with the most simplified form of differential equation, the bi-harmonic equation

$$\frac{\partial^2 w}{\partial t^2} + c^2(\frac{\partial^4 w}{\partial x^4} + 2\frac{\partial^4 w}{\partial x^2 \partial y^2} + \frac{\partial^4 w}{\partial y^4}) = f(x,y,w,\dot{w},t) \tag{7.73}$$

where $f(x,y,w.\dot{w},t)$ is the intensity of the normal load and $w(x,y,t)$ is the deflection of the plate. For steady deflection of a rectangular plate, eqn (7.73) reduces to

$$\frac{\partial^4 w}{\partial x^4} + 2\frac{\partial^4 w}{\partial x^2 \partial y^2} + \frac{\partial^4 w}{\partial y^4} = f(x,y). \tag{7.74}$$

Example 15

A semi-infinite beam which is initially at rest on the x-axis is given a transverse displacement y_0 at its end $x = 0$. Determine the transverse displacement $y(x,t)$ of the beam at any position $x > 0$ and at any time $t > 0$.

Solution

The governing differential equation is

$$\frac{\partial^2 y}{\partial t^2} + c^2 \frac{\partial^4 y}{\partial x^4} = 0. \tag{7.75}$$

The initial conditions are

$$\begin{aligned} y(x,0) &= 0 \\ y_t(x,0) &= 0. \end{aligned} \tag{7.76}$$

The boundary conditions are

$$\begin{aligned} & y(0,t) = y_0 \\ & y_{xx}(0,t) = 0 \\ x \to \infty : \quad & |y(x,t)| < M. \end{aligned} \tag{7.77}$$

Taking the Laplace transform of (7.75) and using the initial conditions (7.76), we obtain

$$\frac{d^4}{dx^4}\mathcal{L}(y) + \frac{s^2}{c^2}\mathcal{L}(y) = 0. \tag{7.78}$$

The transformed boundary conditions are then

$$\begin{aligned} x = 0 : \quad & \mathcal{L}(y) = y_0/s \\ & \tfrac{d^2}{dx^2}\mathcal{L}(y) = 0 \\ x \to \infty : \quad & \mathcal{L}(y) \text{ is bounded.} \end{aligned} \tag{7.79}$$

The auxiliary equation of (7.78) is given by

$$m^4 + (\frac{s}{c})^2 = 0. \tag{7.80}$$

The four roots of this equation are

$$m = \sqrt{\frac{s}{c}} e^{(2n+1)\frac{\pi i}{4}} \quad , \quad n = 0, 1, 2, 3.$$

Thus we have

$$\begin{aligned} n = 0 \ &: \ m_1 = \sqrt{\tfrac{s}{c}} e^{\frac{\pi i}{4}} = \sqrt{\tfrac{s}{c}}(\tfrac{1}{\sqrt{2}} + i\tfrac{1}{\sqrt{2}}) \\ n = 1 \ &: \ m_2 = \sqrt{\tfrac{s}{c}} e^{\frac{3\pi i}{4}} = \sqrt{\tfrac{s}{c}}(-\tfrac{1}{\sqrt{2}} + i\tfrac{1}{\sqrt{2}}) \\ n = 2 \ &: \ m_3 = \sqrt{\tfrac{s}{c}} e^{\frac{5\pi i}{4}} = \sqrt{\tfrac{s}{c}}(-\tfrac{1}{\sqrt{2}} - i\tfrac{1}{\sqrt{2}}) \\ n = 3 \ &: \ m_4 = \sqrt{\tfrac{s}{c}} e^{\frac{7\pi i}{4}} = \sqrt{\tfrac{s}{c}}(\tfrac{1}{\sqrt{2}} - i\tfrac{1}{\sqrt{2}}). \end{aligned}$$

Here, m_1 and m_4 are complex conjugate and m_2 and m_3 are complex conjugate numbers. Therefore, the general solution is

$$\begin{aligned}\mathcal{L}(y) &= e^{\sqrt{\frac{s}{2c}}x}(A\cos\sqrt{\frac{s}{2c}}x + B\sin\sqrt{\frac{s}{2c}}x) \\ &\quad + e^{-\sqrt{\frac{s}{2c}}x}(C\cos\sqrt{\frac{s}{2c}}x + D\sin\sqrt{\frac{s}{2c}}x).\end{aligned}$$

From the boundedness condition in (7.79), we require that $A = B = 0$ so that

$$\mathcal{L}(y) = e^{-\sqrt{\frac{s}{2c}}x}(C\cos\sqrt{\frac{s}{2c}}x + D\sin\sqrt{\frac{s}{2c}}x).$$

Now, using the first and second conditions in (7.79), we obtain $C = y_0/s$ and $D = 0$, so the solution is given by

$$\mathcal{L}(y) = \frac{y_0}{s}e^{-\sqrt{\frac{s}{2c}}x}\cos\sqrt{\frac{s}{2c}}x. \tag{7.81}$$

The inverse Laplace transform of (7.81) can be obtained by the complex inversion formula

$$y(x,t) = \frac{1}{2\pi i}\int_{a-i\infty}^{a+i\infty}\frac{y_0 e^{st-\sqrt{\frac{s}{2c}}x}\cos\sqrt{\frac{s}{2c}}x ds}{s}.$$

We have a branch point at $s = 0$ of the integrand. To evaluate this integral, we use the following contour as shown in Fig. 7.7. This contour is usually known as the Bromwich contour.

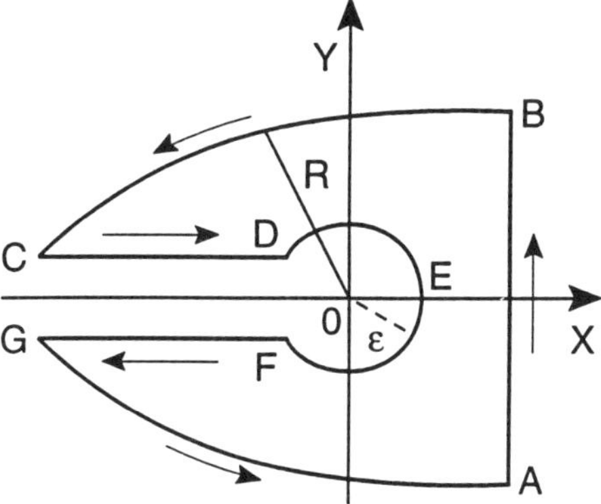

Figure 7.7: A Bromwich contour

Using this contour, if the integrand is represented by $F(s)$ then

$$\begin{aligned}\int_{AB} F(s)ds &+ \int_{BC} F(s)ds + \int_{CD} F(s)ds \\ &+ \int_{DEF} F(s)ds + \int_{FG} F(s)ds + \int_{GA} F(s)ds = 0.\end{aligned}$$

But, $\int_{BC} F(s)ds$ and $\int_{GA} F(s)ds$ can be shown to be zero, and hence

$$\int_{AB} F(s)ds = -\{\int_{CD} F(s)ds + \int_{DEF} F(s)ds + \int_{FG} F(s)ds\}$$

which yields then

$$\begin{aligned} y(x,t) &= \frac{1}{2\pi i}\int_{a-i\infty}^{a+i\infty} \frac{y_0 e^{st-\sqrt{\frac{s}{2c}}x}\cos\sqrt{\frac{s}{2c}}x}{s}ds \\ &= -\lim_{\substack{R\to\infty \\ \epsilon\to 0}} \frac{1}{2\pi i}\{\int_{CD} F(s)ds + \int_{DEF} F(s)ds + \int_{FG} F(s)ds\}. \end{aligned} \quad (7.82)$$

Now, along CD,

$$\begin{aligned} s &= X\,e^{\pi i} = -X \\ \sqrt{s} &= i\sqrt{X} \quad , \quad ds = -dX \end{aligned}$$

and so we find

$$\int_{CD} F(s)ds = \int_{R}^{\epsilon} \frac{y_0 e^{-Xt-i\sqrt{\frac{X}{2c}}x}\cosh\sqrt{\frac{X}{2c}}x}{X}dX.$$

Along FG,

$$\begin{aligned} s &= X\,e^{-\pi i} = -X \\ \sqrt{s} &= -i\sqrt{X} \quad , \quad d_s - dX \\ \int_{FG} F(s)ds &= \int_{\epsilon}^{R} \frac{y_0 e^{-Xt+i\sqrt{\frac{X}{2c}}x}\cosh\sqrt{\frac{X}{2c}}x}{X}dX. \end{aligned}$$

Thus

$$\int_{CD} F(s)ds + \int_{FG} F(s)ds = y_0\int_0^{\infty} \frac{e^{-Xt}(2i\sin\sqrt{\frac{X}{2c}}x)\cosh\sqrt{\frac{X}{2c}}x}{X}dX.$$

Also, along DEF,

$$\begin{aligned} s &= \epsilon\,e^{i\theta} \quad , \quad ds = \epsilon i\,e^{i\theta}d\theta \\ \sqrt{s} &= \sqrt{\epsilon}e^{i\theta/2} \end{aligned}$$

and so we find

$$\begin{aligned} \int_{DEF} F(s)ds &= \int_{\pi}^{-\pi} i\,y_0 e^{\epsilon\,e^{i\theta}t-\sqrt{\frac{\epsilon\,e^{i\theta}}{2c}}x}\cos(\sqrt{\frac{\epsilon\,e^{i\theta}}{2c}}x)d\theta \quad \text{when } \epsilon\to 0 \\ &= \int_{+\pi}^{-\pi} i\,y_0 d\theta \\ &= -2\pi i\,y_0. \end{aligned}$$

Thus we have

$$y(x,t) = y_0[1 - \frac{1}{\pi}\int_0^{\infty} \frac{e^{-Xt}\sin(\sqrt{\frac{X}{2c}}x)\cosh(\sqrt{\frac{X}{2c}}x)dX}{X}].$$

Substitute $\sqrt{\frac{X}{2c}} = \lambda^2$ then

$$y(x,t) = y_0[1 - \frac{2}{\pi}\int_0^{\infty} \frac{e^{-2c\lambda^2 t}\sin\lambda x\cosh\lambda x d\lambda}{x}] \quad (7.83)$$

which is the required result in integral form.

Example 16: The solution of the biharmonic equation arising in plate deflection theory

The differential equation for bending a rectangular elastic plate is given as follows:

$$w_{xxxx} + 2w_{xxyy} + w_{yyyy} = f(x, y), \tag{7.84}$$

where $w(x, y)$ is the normal deflection of the plate at the point x and y, and $f(x, y)$ is the intensity of the normal load. The simple boundary conditions when the edge of the plate is rigidly fixed are

$$\left.\begin{array}{lll} w(x,y) = 0 & , & \text{at } x = 0, a \\ & & \quad y = 0, b \\ w_{xx} = 0 & , & \text{at } x = 0, a \\ w_{yy} = 0 & , & \text{at } y = 0, b \end{array}\right\}. \tag{7.85}$$

Solution

In order to solve this boundary value problem (7.84) and (7.85), it is convenient to introduce the finite Fourier sine transform with respect to y coordinate defined by

$$W(x, \lambda) = \int_0^b w(x, y) \sin \lambda y dy, \tag{7.86}$$

$$F(x, \lambda) = \int_0^b f(x, y) \sin \lambda y dy, \tag{7.87}$$

where $\lambda = n\pi/b$, $n = 1, 2, 3, \cdots$
The inversions of these transforms are then given by

$$w(x, y) = (2/b) \sum_{m=1}^{\infty} W(x, \lambda) \sin \lambda y, \tag{7.88}$$

$$f(x, y) = (2/b) \sum_{n=1}^{\infty} F(x, \lambda) \sin \lambda y. \tag{7.89}$$

Application of (7.88) and (7.89) to (7.84) along with the boundary conditions (7.87) reduces the equation to

$$W^{iv} - 2\lambda^2 W'' + \lambda^4 W = F(x, \lambda), \tag{7.90}$$

$$W = W'' = 0 \quad , \quad \text{at } x = 0, a. \tag{7.91}$$

Here a prime represents the differentiation with respect to x. The solution of this set exists in the following form

$$W(x, \lambda) = W_c(x, \lambda) + W_p(x, \lambda), \tag{7.92}$$

where,

$$W_c(x, \lambda) = (A + Bx)e^{(\lambda x)} + (C + Dx)e^{(-\lambda x)}, \tag{7.93}$$

$$\begin{aligned} W_p(x, \lambda) &= -(1/2\lambda^3) \int_0^x \sinh(\lambda(x - t)) F(t, \lambda) dt \\ &\quad + (1/2\lambda^2) \int_0^x \int_0^\eta \cosh(\lambda(x - t)) F(t, \lambda) dt d\eta. \end{aligned} \tag{7.94}$$

Here, A, B, C and D are arbitrary constants and are determined by using conditions (7.85). Application of boundary conditions (7.85) shows that the four constants A, B, C and D exist as follows:

$$-C\Delta = A\Delta = \begin{vmatrix} -W_p(a,\lambda) & 2a\cosh(\lambda a) \\ -W_p''(a,\lambda) & 2\lambda^2 a\cosh(\lambda a) + 4\lambda\sinh(\lambda a) \end{vmatrix}, \tag{7.95}$$

$$D\Delta = B\Delta = \begin{vmatrix} 2\sinh(\lambda a) & -W_p(a,\lambda) \\ 2\lambda^2\sinh(\lambda a) & -W_p''(a,\lambda) \end{vmatrix}, \tag{7.96}$$

where,

$$\Delta = 8\lambda(\sinh\lambda a)^2. \tag{7.97}$$

Thus the inverse function of the transformation (7.86) is given by

$$w(x,y) = (2/b)\sum_{n=1}^{\infty} W_c(x,\lambda)\sin\lambda y + (2/b)\sum_{n=1}^{\infty} W_p(x,\lambda)\sin\lambda y. \tag{7.98}$$

This normal deflection of the plate can be explicitly determined once we assume the mathematical form of $f(x,y)$.

Particular Case

Assume,

$$f(x,y) = \sum_{m=1}^{\infty}\sum_{n=1}^{\infty} a_{mn}\sin\frac{m\pi x}{a}\frac{n\pi y}{b} \tag{7.99}$$

where,

$$a_{mn} = (4/ab)\int_0^a\int_0^b f(x,y)\sin\frac{m\pi x}{a}\sin\frac{n\pi y}{b}dxdy. \tag{7.100}$$

Thus we have

$$F(x,\lambda) = (b/2)\sum_{n=1}^{\infty} a_{mn}\sin\frac{m\pi x}{a}. \tag{7.101}$$

It is obvious that the particular integral $W_p(x,\lambda)$ exists in the following form:

$$W_p(x,\lambda) = (b/2)\sum_{m=1}^{\infty}\frac{a_{mn}}{\pi^4[\frac{m^2}{a^2}+\frac{n^2}{b^2}]^2}\sin\frac{m\pi x}{a}. \tag{7.102}$$

Application of the boundary conditions (7.85) shows that the constants are given by $A = B = C = D = 0$. Thus the transformed solution is,

$$W(x,\lambda) = (b/2)\sum_{m=1}^{\infty}\frac{a_{mn}}{\pi^4[\frac{m^2}{a^2}+\frac{n^2}{b^2}]^2}\sin\frac{m\pi x}{a}. \tag{7.103}$$

The inverse function is then given by

$$w(x,y) = \sum_{m=1}^{\infty}\sum_{n=1}^{\infty}\frac{a_{mn}}{\pi^4[\frac{m^2}{a^2}+\frac{n^2}{b^2}]^2}\sin\frac{m\pi x}{a}\sin\frac{n\pi x}{b}. \tag{7.104}$$

Extremum principle of functional

The differential equation (7.84), subject to the boundary conditions (7.85), is equivalent to the variational problem of minimizing the functional,

$$V(w) = \int_0^a \int_0^b \{w_{xx}^2 + 2w_{xx}w_{yy} + w_{yy}^2 - 2wf\}dx\,dy. \tag{7.105}$$

Let us assume that $f(x, y)$ is a continuous function in the domain defined by $0 \leq x \leq a$, $0 \leq y \leq b$. We seek a function $w(x, y)$ continuous in this domain along with its first and second derivatives, which minimizes the integral (7.105). In the interior of the domain, let us write

$$f(x, y) = \sum_{m=1}^{\infty}\sum_{n=1}^{\infty} a_{mn} \sin\frac{m\pi x}{a} \sin\frac{n\pi y}{b}, \tag{7.106}$$

and

$$w(x, y) = \sum_{m=1}^{\infty}\sum_{n=1}^{\infty} c_{mn} \sin\frac{m\pi x}{a} \sin\frac{n\pi y}{b},$$

where parameters c_{mn} are unknown and a_{mn} are the Fourier coefficients given by (7.100). Introducing the functional relation (7.106) into (7.105), it is found that

$$(4/ab)V(w) = \pi^4 \sum_{m=1}^{\infty}\sum_{n=1}^{\infty} c_{mn}^2 [\frac{m^2}{a^2} + \frac{n^2}{b^2}]^2 - 2\sum_{m=1}^{\infty}\sum_{n=1}^{\infty} c_{mn}a_{mn}. \tag{7.107}$$

Because of the completeness relation for the trigonometric functions, the variational problem becomes the problem of determining the coefficients c_{mn} such that (7.107) is as small as possible. We see immediately that the minimum is described by

$$c_{mn} = \frac{a_{mn}}{\pi^4[\frac{m^2}{a^2} + \frac{n^2}{b^2}]^2}. \tag{7.108}$$

Thus,

$$w(x, y) = \sum_{m=1}^{\infty}\sum_{n=1}^{\infty} \frac{a_{mn}}{\pi^4[\frac{m^2}{a^2} + \frac{n^2}{b^2}]^2} \sin\frac{m\pi x}{a} \sin\frac{n\pi y}{b} \tag{7.109}$$

is the normal deflection of the plate. Comparing (7.104) and (7.109), it is observed that these two expressions are identical.

7.9 The hydrodynamics of waves and tides

In this section, we use partial differential equations with problems arising in water waves. In the first part, we discuss the mathematical model of surface waves and their solutions, while in the second part, we deal with tidal waves. These problems can be regarded as boundary value problems.

7.9.1 Surface waves

The complete boundary value problem of surface waves can be mathematically stated as Laplace's equation (see Lamb (1932)), illustrated in Fig. 7.10.

$$\frac{\partial^2 \phi}{\partial x^2} + \frac{\partial^2 \phi}{\partial z^2} = 0. \tag{7.110}$$

There are two free surface boundary conditions:
Dynamic boundary condition:

$$\eta = \frac{1}{g}(\frac{\partial \phi}{\partial t}) \quad \text{on } z = 0 \tag{7.111}$$

and the Kinematic boundary condition:

$$\frac{\partial \eta}{\partial t} = -\frac{\partial \phi}{\partial z} \quad \text{on } z = 0. \tag{7.112}$$

The third boundary condition is the bottom boundary condition:

$$\frac{\partial \phi}{\partial z} = 0 \quad \text{on } z = -h. \tag{7.113}$$

Solution

The method of separation of variables is powerful for this solution. This method usually leads to the standing wave solution.

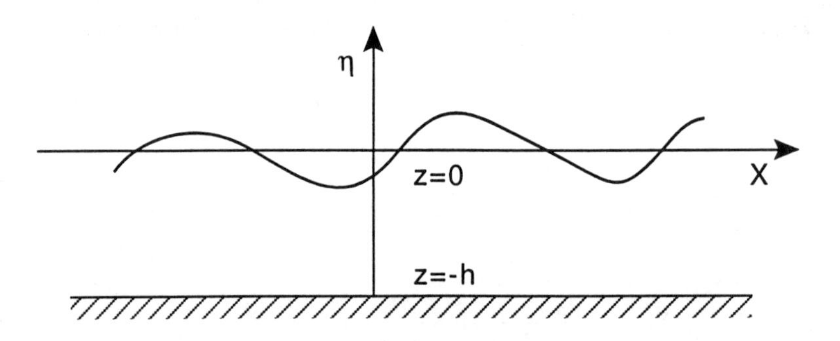

Figure 7.8: A progressive wave

Combining eqns (7.111) and (7.112), we obtain an equation in terms of ϕ

$$\frac{\partial^2 \phi}{\partial t^2} + g\frac{\partial \phi}{\partial z} = 0 \quad \text{on } z = 0. \tag{7.114}$$

Assuming the progressive wave solution, i.e. $\phi \sim e^{i(kx-\sigma t)}$ we can obtain a solution of the two dimensional Laplace's equation satisfied by the velocity potential ϕ.

The real form of the solution can be written as

$$\phi = Z(z)\cos(kx - \sigma t). \tag{7.115}$$

The Z-solution can be obtained as

$$Z(z) = A\cosh kz + B\sinh kz \tag{7.116}$$

and therefore the ϕ-solution is

$$\phi = (A\cosh kz + B\sinh kz)\cos(kx - \sigma t). \tag{7.117}$$

Using the bottom boundary condition (7.113) we obtain

$$\phi = A\frac{\cosh k(z+h)}{\cosh kh}\cos(kx - \sigma t). \tag{7.118}$$

Then using the free surface condition (7.111) at $z = 0$ and assuming that

$$\eta = a\sin(kx - \sigma t) \tag{7.119}$$

where a is the amplitude of the incident wave, we obtain $A = \frac{ga}{\sigma}$ and consequently:

$$\phi = \frac{ga}{\sigma}\frac{\cosh k(z+h)}{\cosh kh}\cos(kx - \sigma t). \tag{7.120}$$

If the progressive wave travels from $-\infty$ to $+\infty$ making an angle δ with the x-axis, then the forms of ϕ and η must be modified to yield:

$$\phi = \frac{ga}{\sigma}\frac{\cosh k(z+h)}{\cosh kh}\cos(k(x\cos\delta + y\sin\delta) - \sigma t) \tag{7.121}$$

$$\eta = a\sin(kx\cos\delta + ky\sin\delta - \sigma t). \tag{7.122}$$

Then, substituting the values of $\frac{\partial^2\phi}{\partial t^2}$ and $g\frac{\partial\phi}{\partial z}$ into (7.114) on $z = 0$, yields

$$\sigma^2 = gk\tanh(kh). \tag{7.123}$$

which is known as the dispersion relation of the surface waves from which we can find the wave length. The phase velocity C can be obtained as follows:

Consider $kx - \sigma t =$ constant. Then

$$C = \frac{dx}{dt} = \frac{\sigma}{k} \tag{7.124}$$

where σ is the frequency and k is the wave number, and so the dispersion relation becomes

$$C^2 = \frac{g}{k}\tanh kh. \tag{7.125}$$

In terms of wavelength, this relation can be written as

$$L = \frac{gT^2}{2\pi}\tanh(\frac{2\pi h}{L}) \tag{7.126}$$

where $L = \frac{2\pi}{k}$ in which L is the wavelength.

7.9.2 Tidal waves

Consider a tidal wave travelling along a straight canal, with a horizontal bed, and parallel vertical sides. Let us suppose that the x axis is parallel to the length of the canal and the z axis is perpendicular and the motion takes place entirely in these two dimensions x and z.

Then, from Bernoulli's equation, the pressure at any point (x, z) is equal to the static pressure due to the depth below the free surface

$$p - p_0 = \rho g(h + \eta - z) \tag{7.127}$$

where

p_0 is the uniform external pressure
h is the uniform depth of the canal
η is the water elevation.

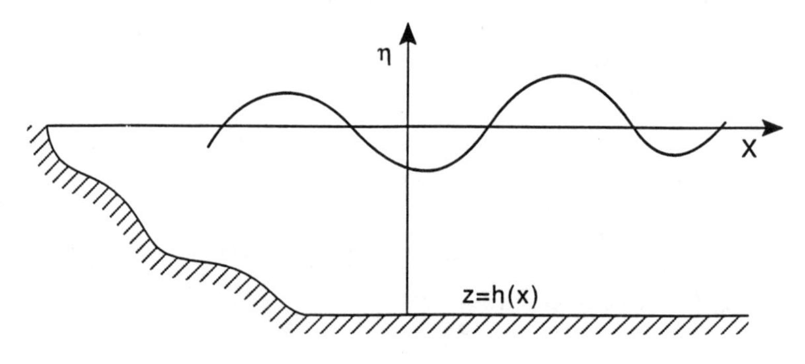

Figure 7.9: Flow configuration

Hence,

$$\frac{\partial p}{\partial x} = \rho g \frac{\partial \eta}{\partial x}. \tag{7.128}$$

The one-dimensional form of the equation of horizontal motion may be stated as (see Curle and Davies, 1968)

$$\frac{\partial u}{\partial t} + u\frac{\partial u}{\partial x} = -\frac{1}{\rho}\frac{\partial p}{\partial x}. \tag{7.129}$$

For infinitely small motions we can assume that the term $u\frac{\partial u}{\partial x}$ is very small compared to the other terms in (7.129). Then this equation can be written in the following form

$$\frac{\partial u}{\partial t} = -g\frac{\partial \eta}{\partial x}. \tag{7.130}$$

The equation of continuity in two-dimensional form may be written as

$$\frac{\partial u}{\partial x} + \frac{\partial w}{\partial z} = 0. \tag{7.131}$$

When equation (7.131) is integrated over a column of water extending from the bottom of the basin $z = -h$ to the free surface $z = \eta$ with respect to z, we obtain

$$\int_{-h}^{\eta} \frac{\partial u}{\partial x} dz + \int_{-h}^{\eta} \frac{\partial w}{\partial z} dz = 0. \tag{7.132}$$

Now performing the integrations, we have

$$\frac{\partial}{\partial x} \int_{-h}^{\eta} u dz = \int_{-h}^{\eta} \frac{\partial u}{\partial x} dz + u(\eta) \frac{\partial \eta}{\partial x} + u(-h) \frac{\partial h}{\partial x}, \qquad \int_{-h}^{\eta} \frac{\partial w}{\partial z} dz = w(\eta) - w(-h).$$

Also we have,

$$\frac{d\eta}{dt} = \frac{\partial \eta}{\partial t} + u(\eta) \frac{\partial \eta}{\partial x} = w(\eta).$$

Therefore,

$$\begin{aligned} w(\eta) &= \frac{\partial \eta}{\partial t} + u(\eta) \frac{\partial \eta}{\partial x} \\ w(-h) &= -u(-h) \frac{\partial h}{\partial x}. \end{aligned}$$

Using these relations, (7.132) can be written as

$$\frac{\partial}{\partial x} \int_{-h}^{\eta} u dz - u(\eta) \frac{\partial \eta}{\partial x} - u(-h) \frac{\partial h}{\partial x} + w(\eta) - w(-h) = 0.$$

Simplifying we obtain,

$$\frac{\partial}{\partial x} \int_{-h}^{\eta} u dz + \frac{\partial \eta}{\partial t} = 0. \tag{7.133}$$

Assuming that u is not a function of z, we obtain from eqn (7.133)

$$\frac{\partial}{\partial x} u(\eta + h) + \frac{\partial \eta}{\partial t} = 0. \tag{7.134}$$

For a small amplitude wave with constant depth, (7.134) can be written as

$$h \frac{\partial u}{\partial x} + \frac{\partial \eta}{\partial t} = 0. \tag{7.135}$$

Eliminating η between (7.130) and (7.135), we obtain

$$\frac{\partial^2 u}{\partial t^2} = gh \frac{\partial^2 u}{\partial x^2}. \tag{7.136}$$

The elimination of u gives an equation of the same form, i.e.

$$\frac{\partial^2 \eta}{\partial t^2} = gh \frac{\partial^2 \eta}{\partial x^2}. \tag{7.137}$$

(a) Deduction of the continuity equation

The continuity equation can also be obtained as follows: Consider the elementary volume of fluid as shown in Fig. 7.10.

Suppose,

$$\begin{aligned}\text{Mass present in the elementary volume at time } t &= \rho A \Delta x \\ \text{Mass present in the elementary volume at } t+\Delta t &= \rho A \Delta x + \frac{\partial}{\partial t}(\rho A)\Delta t \Delta x.\end{aligned}$$

Net increase of mass in t within the element

$$= \frac{\partial}{\partial t}(\rho A)\Delta t \Delta x. \tag{7.138}$$

The inflow of mass through the face $s_1 = \rho u A \Delta t$.

The inflow of mass through the face

$$s_2 = \rho u A \Delta t + \frac{\partial}{\partial x}(\rho u A)\Delta x \Delta t.$$

Net inflow in the element

$$= -\frac{\partial}{\partial x}(\rho u A)\Delta x \Delta t. \tag{7.139}$$

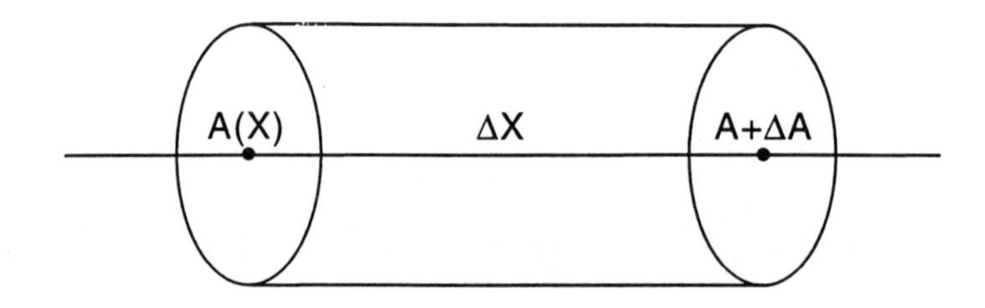

Figure 7.10: Elementary volume of tidal flow

Therefore, equating (7.138) and (7.139), we obtain

$$\frac{\partial \rho}{\partial t} + \frac{1}{A}\frac{\partial}{\partial x}(\rho u A) = 0 \tag{7.140}$$

where $A =$ area, $\rho =$ density of the fluid and $u =$ horizontal fluid velocity.

(b) Alternative form of the continuity equation

Another way of writing the continuity eqn (7.140) is the following. Referring to Fig. 7.11, the mass balance can be written,

$$\begin{aligned}&\text{Mass present at time } t = \rho\eta b dx\\&\text{Mass present at } t+\Delta t = \rho\eta b dx + \tfrac{\partial}{\partial t}(\rho\eta b)\Delta t\Delta x\\&\text{Net increase of mass in the elementary volume}\\&= \tfrac{\partial}{\partial t}(\rho\eta b)\Delta t\Delta x.\end{aligned} \tag{7.141}$$

where, η = water elevation and b = width of estuary.

Then, eqns (7.141) with (7.139), we obtain

$$\frac{\partial}{\partial t}(\rho\eta b) + \frac{\partial}{\partial x}(\rho u A) = 0.$$

For an incompressible fluid, we assume ρ = constant and if $b = b(x)$, then

$$\frac{\partial \eta}{\partial t} + \frac{1}{b}\frac{\partial}{\partial x}(uA) = 0. \tag{7.142}$$

If we assume $A = bh$, where h is the depth of the estuary and b is the breadth, then (7.142) can be written as

$$\frac{\partial \eta}{\partial t} + \frac{1}{b}\frac{\partial}{\partial x}(bhu) = 0. \tag{7.143}$$

We shall consider the motion and the continuity as given in (7.130) and (7.143) respectively.

Motion along the x-axis is

$$\frac{\partial u}{\partial t} + g\frac{\partial \eta}{\partial x} = 0. \tag{7.144}$$

Continuity:

$$\frac{\partial \eta}{\partial t} + \frac{1}{b}\frac{\partial}{\partial x}(bhu) = 0 \tag{7.145}$$

where h is the depth of the estuary, and in real estuaries these are functions of x only.

Suppose that b = constant and h = constant, then eqn (7.145) reduces to

$$\frac{\partial \eta}{\partial t} + h\frac{\partial u}{\partial x} = 0. \tag{7.146}$$

Eliminating u from (7.144) and (7.146), we obtain,

$$\frac{\partial^2 \eta}{\partial t^2} = gh\frac{\partial^2 \eta}{\partial x^2}.$$

We define $C_0 = \sqrt{gh}$ = wave speed, which is known as celerity. Therefore,

$$\frac{\partial^2 \eta}{\partial t^2} = C_0^2\frac{\partial^2 \eta}{\partial x^2}. \tag{7.147}$$

Similarly, we obtain

$$\frac{\partial^2 u}{\partial t^2} = C_0^2 \frac{\partial^2 u}{\partial x^2}. \tag{7.148}$$

Equations (7.147) and (7.148) are the well-known wave equations. These equations are satisfied for the specific case of periodic long waves entering a canal of uniform section and of infinite length by the harmonic function

$$\eta = a\cos(\sigma t - kx) \tag{7.149}$$

where at $x = 0$: $\eta = a\cos\sigma t$ (a wave communicating at the mouth). Now, substituting this expression into (7.147) we obtain the frequency as $\sigma = C_0 k = \frac{2\pi}{T}$ where $k = \frac{2\pi}{L}$ is the wave number, T the time period and L the wavelength.

Now, from (7.144), we have

$$\frac{\partial u}{\partial t} = (-g)(ak)\sin(\sigma t - kx).$$

Integrating with respect to t, we obtain

$$u = \frac{aC_0}{h}\cos(\sigma t - kx). \tag{7.150}$$

Therefore, the surface elevation and the velocity are given by

$$\begin{aligned} \eta &= a\cos(\sigma t - kx) \\ u &= \frac{aC_0}{h}\cos(\sigma t - kx). \end{aligned} \tag{7.151}$$

Note that velocities and surface elevations are in phase and are related by

$$u = (\frac{\eta}{h})C_0. \tag{7.152}$$

The velocity is of the same sign as the amplitude, hence the wave is a **progressive wave**.

$$u_{max} = (\frac{a}{h})C_0. \tag{7.153}$$

Chapter 8

Green's function

8.1 One-dimensional Green's function

Green's functions serve as mathematical characterizations of important physical concepts. These functions are of fundamental importance in many practical problems, and in understanding the theory of differential equations. In the following we illustrate a simple practical problem from mechanics whose solution can be composed of a Green's function, named for the English mathematical physicist George Green (1793–1841).

8.1.1 A distributed load of the string

We consider a perfectly flexible elastic string stretched to a length l under tension T. Let the string bear a distributed load per unit length $\omega(x)$ including the weight of the string. Also assume that the static deflections produced by this load are all perpendicular to the original, undeflected position of the strings.

Consider an elementary length Δx of the string as shown in Fig. 8.1. Since the

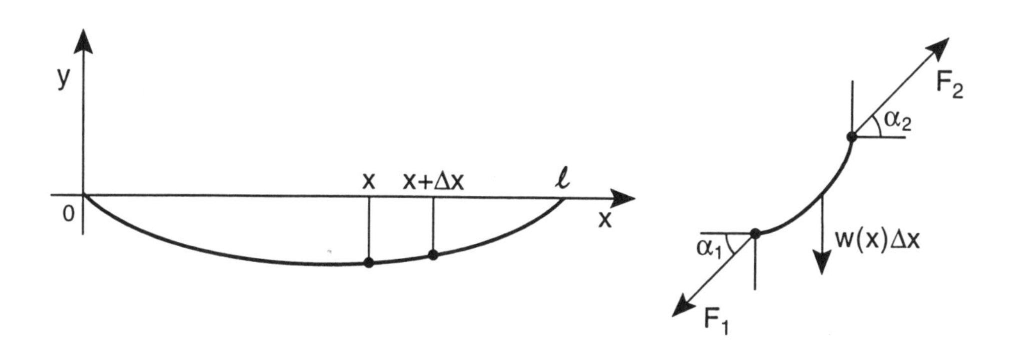

Figure 8.1: A stretched string deflected by a distributed load

deflected string is in equilibrium, the net horizontal and vertical forces must both be zero. Thus

$$F_1 \cos\alpha_1 = F_2 \cos\alpha_2 = T \tag{8.1}$$

$$F_2 \sin\alpha_2 = F_1 \sin\alpha_1 - \omega(x)\Delta x. \tag{8.2}$$

Here the positive direction of $\omega(x)$ has been chosen upward. We consider the deflection to be so small that the forms F_1, F_2 will not differ much from the tension of string T i.e., $T = F_1 = F_2$. So we obtain approximately

$$T \sin\alpha_2 = T \sin\alpha_1 - \omega(x)\Delta x$$

which is

$$\tan\alpha_2 = \tan\alpha_1 - \frac{\omega(x)\Delta x}{T}. \tag{8.3}$$

(Using $\sin\alpha \simeq \tan\alpha$ if α is small.) But we know

$$\tan\alpha_2 = \frac{dy}{dx}|_{x+\Delta x}$$

$$\tan\alpha_1 = \frac{dy}{dx}|_{x}.$$

Hence rewriting (8.3), and letting $\Delta x \to 0$, we obtain

$$\lim_{\Delta x \to 0} \frac{\frac{dy}{dx}|_{x+\Delta x} - \frac{dy}{dx}|_{x}}{\Delta x} = -\frac{\omega(x)}{T}$$

$$or, \quad T\frac{d^2y}{dx^2} = -\omega(x) \tag{8.4}$$

which is the differential equation satisfied by the deflection curve of the string. This equation is obtained using the distributed load of the string.

8.1.2 A concentrated load of the strings

We now consider the deflection of the string under the influence of a concentrated load rather than a distributed load. A concentrated load is, of course, a mathematical fiction which cannot be realized physically. Any nonzero load concentrated at a single point implies an infinite pressure which may break the string. The use of concentrated load in the investigation of physical systems is both common and fruitful.

It can be easily noticed from eqn (8.4) that if there is no distributed load ($\omega(x) = 0$) then $y'' = 0$ at all points of the string. This implies that y is a linear function which follows that the deflection curve of the string under the influence of a single concentrated load R consists of two linear segments as shown in Fig. 8.2.

Resolving the forces acting on the string can be given as

$$F_2 \cos\alpha_2 = F_1 \cos\alpha_1$$

$$F_1 \sin\alpha_1 + F_2 \sin\alpha_2 = -R. \tag{8.5}$$

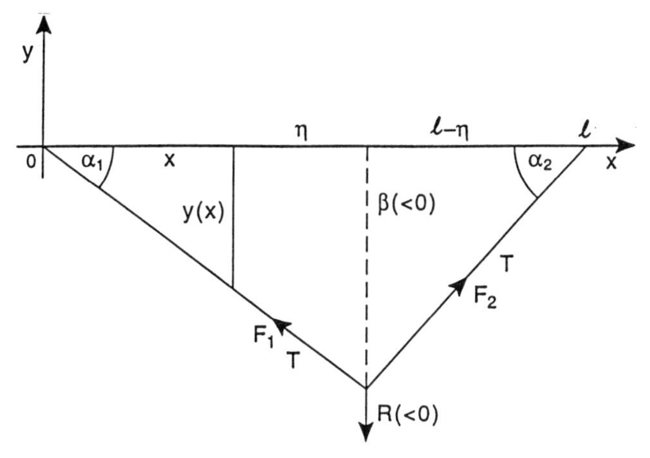

Figure 8.2: A stretched string deflected by a concentrated load

When the deflection is small we can assume that $F_2 \cos\alpha_2 = F_1 \cos\alpha_1 = T$, or simply $F_1 = F_2 = T$ and

$$\begin{aligned} \sin\alpha_1 &\simeq \tan\alpha_1 \\ \sin\alpha_2 &\simeq \tan\alpha_2. \end{aligned}$$

Then eqn (8.5) becomes

$$\begin{aligned} \tan\alpha_1 + \tan\alpha_2 &= -\frac{R}{T} \\ \frac{-\beta}{\eta} + \frac{-\beta}{l-\eta} &= -\frac{R}{T} \end{aligned}$$

and

$$\beta = \frac{R(l-\eta)\eta}{Tl}$$

where β is the transverse deflection of the string at $x = \eta$.

With the deflection β known it is a simple matter to use the similar triangles to find the deflection of the string at any point x. The results are

$$\frac{y(x)}{x} = \frac{\beta}{\eta} \qquad 0 \le x \le \eta$$

so,

$$y(x) = \frac{R(l-\eta)x}{Tl} \qquad 0 \le x \le \eta$$

also

$$\frac{y(x)}{l-x} = \frac{\beta}{l-\eta}$$

so,

$$y(x) = \frac{R(l-x)\eta}{Tl} \qquad \eta \le x \le l.$$

Thus the results are

$$y(x,\eta) = \begin{cases} \frac{R(l-\eta)x}{Tl} & 0 \le x \le \eta \\ \frac{R(l-x)\eta}{Tl} & \eta \le x \le l. \end{cases} \tag{8.6}$$

It is worth mentioning that $y(x,\eta)$ is used rather than $y(x)$ to indicate that the deflection of y depends on the point η where the concentrated load is applied and the point x where the deflection is observed. These two points are respectively called the 'source point' and the 'field point'.

It can be easily observed from (8.6) that the deflection of a string at a point x due to a concentrated load R applied at a point η is the same as the deflection produced at the point η by an equal load applied at the point x. When R is a unit load it is customary to use the notation $G(x,\eta)$ known as the Green's function corresponding to the function $y(x,\eta)$. Many authors call Green's function an **influence function**. It is observed that this Green's function is symmetric in the two variables x and η such that

$$G(x,\eta) = G(\eta, x). \tag{8.7}$$

Thus for this problem, in terms of unit positive load, Green's function is given by

$$G(x,\eta) = \begin{cases} \frac{(l-\eta)x}{Tl}, 0 \le x \le \eta \\ \frac{(l-x)\eta}{Tl}, \eta \le x \le l. \end{cases} \tag{8.8}$$

The symmetry of $G(x,\eta)$ is an important illustration discovered by Maxwell and Rayleigh known as Maxwell-Rayleigh reciprocity law which holds in many physical systems including mechanical and electrical. James Clerk Maxwell (1831–1879) was a Scottish mathematical physicist. Lord Rayleigh (1842–1919) was an English mathematical physicist.

Remark

It is interesting and important to note that with Green's function $G(x,\eta)$ an expression of deflection of a string under an arbitrary distributed load can be found without solving equation (8.4). To see this point clearly, we divide the interval $0 \le x \le l$ into n subintervals by the points $\eta_0 = 0, \eta_1, \eta_2, \ldots, \eta_n = l$ such that $\Delta\eta_i = \eta_i - \eta_{i-1}$. Let ξ_i be an arbitrary point in the subinterval $\Delta\eta_i$. Let us also consider that the position of the distributed load acting on the subinterval $\Delta\eta_i$, namely $\omega(\xi_i)\Delta\eta_i$, is concentrated at the point $\eta = \xi_i$. The deflection produced at the point x by this load is the product of the load and the deflection produced at x by a unit load at the point $\eta = \xi_i$, which is

$$(\omega(\xi_i)\Delta\eta_i)G(x,\xi_i).$$

Thus if we add up all the deflections produced at the point x by the various concentrated forces which together approximate the actual distributed load, we obtain the sum

$$\sum_{i=1}^{n} \omega(\xi_i)G(x,\xi_i)\Delta\eta_i.$$

This sum becomes an integral when in the limit $\Delta\eta_i \to 0$, and the deflection $y(x)$ at an arbitrary point x is given by

$$y(x) = \int_0^l \omega(\eta)G(x,\eta)d\eta. \tag{8.9}$$

Hence, once the function $G(x,\eta)$ is known, the deflection of the string under any piecewise continuous distributed load can be determined at once by the integral (8.9). Mathematically it is clear that (8.9) is a solution of the ordinary differential equation

$$Ty'' = -\omega(x)$$

because

$$\begin{aligned} y'' &= \frac{\partial^2}{\partial x^2}\int_0^l \omega(\eta)G(x,\eta)d\eta \\ &= \int_0^l \omega(\eta)\frac{\partial^2 G}{\partial x^2}(x,\eta)d\eta \end{aligned}$$

and so

$$\int_0^l \omega(\eta)(T\frac{\partial^2 G}{\partial x^2}(x,\eta))dx = -\omega(x).$$

This is only true provided

$$T\frac{\partial^2 G}{\partial x^2}(x,\eta) = -\delta(x-\eta)$$

such that

$$\int_0^l \omega(\eta)(-\delta(x-\eta))d\eta = -\omega(x)$$

where $\delta(x-\eta)$ is a Dirac delta function.

8.1.3 Properties of Green's function

This generalized function $\delta(x-\eta)$ has the following important properties (see Lighthill's Fourier Analysis and Generalized Function, 1963, and Rahman, 1991):

I. $$\delta(x-\eta) = \begin{cases} 0 & x \neq \eta \\ \infty & x = \eta. \end{cases}$$

II. $\int_{-\infty}^{\infty} \delta(x-\eta)dx = 1.$

III. If $f(x)$ is a piecewise continuous function in $-\infty < x < \infty$, then

$$\int_{-\infty}^{\infty} f(x)\delta(x-\eta)dx = f(\eta).$$

Thus integrating $TG_{xx} = -\delta(x-\eta)$ between $x = \eta + 0$ and $x = \eta - 0$ we obtain,

$$T \int_{\eta-0}^{\eta+0} G_{xx}dx = -\int_{\eta-0}^{\eta+0} \delta(x-\eta)dx$$

$$T \; [\frac{\partial G}{\partial x}|_{\eta+0} - \frac{\partial G}{\partial x}|_{\eta-0}] = -1.$$

Hence the jump at $x = \eta$ for this problem is

$$\frac{\partial G}{\partial x}|_{\eta+0} - \frac{\partial G}{\partial x}|_{\eta-0} = -(\frac{1}{T})$$

which is the downward jump. Thus when the tension $T = 1$, this downward jump is -1 at $x = \eta$.

Definition 1

A function $f(x)$ has a jump λ at $x = \eta$ if, and only if, the respective right- and left-hand limits $f(\eta + 0)$ and $f(\eta - 0)$ of $f(x)$ exists as x tends to η, and

$$f(\eta + 0) - f(\eta - 0) = \lambda.$$

This jump λ is said to be upward or downward depending on whether λ is positive or negative. At a point $x = \eta$ when $f(x)$ is continuous, the jump λ is, of course, zero.

With this definition of jump, it is an easy matter to show that $\frac{\partial G}{\partial x}$ has a downward jump of $-\frac{1}{T}$ at $x = \eta$ because we observe from eqn (8.8) that

$$\lim_{x \to \eta+0} \frac{\partial G}{\partial x} = \lim_{x \to \eta+0} \frac{-\eta}{lT} = -\frac{\eta}{lT}$$

$$\lim_{x \to \eta-0} \frac{\partial G}{\partial x} = \lim_{x \to \eta-0} \frac{l-\eta}{lT} = \frac{l-\eta}{lT}.$$

It is obvious that these limiting values are not equal and their difference is

$$-\frac{\eta}{lT} - \frac{l-\eta}{lT} = -\frac{1}{T}$$

which is a downward jump as asserted. As we have seen $G(x, \eta)$ consists of two linear expressions, it satisfies the linear homogeneous differential equation $Ty'' = 0$ at all points of the interval $0 \le x \le l$ except at $x = \eta$. In fact the second derivative $\frac{\partial^2 G}{\partial x^2}(x, \eta)$ does not exist because $\frac{\partial G}{\partial x}(x, \eta)$ is discontinuous at that point.

The properties of the function $G(x, \eta)$ which we have just observed are not accidental characteristics for just a particular problem. Instead, they are an important class of functions associated with linear differential equations with constants as well as variable coefficients. We define this class of functions with its properties as follows:

Definition 2

Consider the second-order homogeneous differential equation

$$y'' + P(x)y' + Q(x)y = 0 \tag{8.10}$$

with the homogeneous boundary conditions

$$\left.\begin{array}{ll} x = a: & \alpha_1 y(a) + \alpha_2 y'(a) = 0 \\ x = b: & \beta_1 y(b) + \beta_2 y'(b) = 0 \end{array}\right\} \tag{8.11}$$

where α_1 and α_2 are not both zero, and β_1 and β_2 are not both zero.

Consider a function $G(x,\eta)$ which satisfies the differential eqn (8.10) such that

$$G_{xx} + P(x)G_x + Q(x)G = -\delta(x-\eta) \tag{8.12}$$

with the following property

$$x = a: \quad \alpha_1 G(a,\eta) + \alpha_2 G_x(a,\eta) = 0 \tag{8.13}$$

$$x = b: \quad \beta_1 G(b,\eta) + \beta_2 G_x(b,\eta) = 0 \tag{8.14}$$

$$x = \eta \quad G(x,\eta) \text{ is continuous in } a \le x \le b \tag{8.15}$$

and

$$\lim_{x\to\eta+0} \frac{\partial G}{\partial x} - \lim_{x\to\eta-0} \frac{\partial G}{\partial x} = -1 \tag{8.16}$$

has a jump -1 at $x = \eta$.

Then this function $G(x,\eta)$ is called the Green's function of the problem defined by the given differential equation and its boundary conditions.

In the following we shall demonstrate how to construct a Green's function involving a few given boundary value problems.

Example 1

Construct a Green's function for the equation $y'' + y = 0$ with the boundary conditions $y(0) = y(\frac{\pi}{2}) = 0$.

Solution

Since $G(x,\eta)$ satisfies the given differential equation such that

$$G_{xx} + G = -\delta(x-\eta)$$

therefore a solution exists in the following form

$$G(x,\eta) = \begin{cases} A\cos x + B\sin x & 0 \le x \le \eta \\ C\cos x + D\sin x & \eta \le x \le \frac{\pi}{2}. \end{cases}$$

The boundary conditions at $x = 0$ and $x = \frac{\pi}{2}$ must be satisfied which yields $G(0,\eta) = A = 0$ and $G(\frac{\pi}{2},\eta) = D = 0$. Hence we have

$$G(x,\eta) = \begin{cases} B\sin x & 0 \le x \le \eta \\ C\cos x & \eta \le x \le \frac{\pi}{2}. \end{cases}$$

From the continuity condition at $x = \eta$, we have $B\sin\eta = C\cos\eta$ such that $\frac{B}{\cos\eta} = \frac{C}{\sin\eta} = \alpha$. Thus $B = \alpha\cos\eta$ and $C = \alpha\sin\eta$ where α is an arbitrary constant. The Green's function reduces to

$$G(x,\eta) = \begin{cases} \alpha\cos\eta\sin x & 0 \le x \le \eta \\ \alpha\sin\eta\cos x & \eta \le x \le \frac{\pi}{2}. \end{cases}$$

Finally to determine α we use the jump condition at $x = \eta$ which gives

$$\frac{\partial G}{\partial x}|_{\eta+0} - \frac{\partial G}{\partial x}|_{\eta-0} = -1$$

$$\alpha[\sin^2\eta + \cos^2\eta] = 1$$

and therefore $\alpha = 1$. With α known the Green's function is completely determined, and we have

$$G(x,\eta) = \begin{cases} \cos\eta\sin x & 0 \le x \le \eta \\ \cos x\sin\eta & \eta \le x \le \frac{\pi}{2}. \end{cases}$$

Example 2

Construct the Green's function for the equation $y'' + \nu^2 y = 0$ with the boundary conditions $y(0) = y(l) = 0$.

Solution

Since $G(x,\eta)$ satisfies the given differential equation such that

$$G_{xx} + \nu^2 G = -\delta(x-\eta)$$

therefore a solution can be formed as

$$G(x,\eta) = \begin{cases} A\cos\nu x + B\sin\nu x & 0 \le x \le \eta \\ C\cos\nu x + D\sin\nu x & \eta \le x \le 1. \end{cases}$$

The boundary conditions at $x = 0$ and at $x = 1$ must be satisfied which yields $G(0,\eta) = A = 0$ and $G(1,\eta) = C\cos\nu + D\sin\nu = 0$ such that $\frac{C}{\sin\nu} = \frac{-D}{\cos\nu} = \alpha$ which gives $C = \alpha\sin\nu$ and $D = -\alpha\cos\nu$. Substituting these values into our expression, $G(x,\eta)$ is obtained as:

$$G(x,\eta) = \begin{cases} B\sin\nu x & 0 \le x \le \eta \\ \alpha\sin\nu(1-x) & \eta \le x \le 1 \end{cases}$$

where α is an arbitrary constant. To determine the values of B and α we use the continuity condition at $x = \eta$. From the continuity at $x = \eta$, we have

$$B \sin \nu\eta = \alpha \sin \nu(1-\eta)$$

such that

$$\frac{B}{\sin \nu(1-\eta)} = \frac{\alpha}{\sin \nu\eta} = \gamma$$

which yields

$$\begin{aligned} \alpha &= \gamma \sin \nu\eta \\ \beta &= \gamma \sin \nu(1-\eta) \end{aligned}$$

where γ is an arbitrary constant. Thus the Green's function reduces to

$$G(x,\eta) = \begin{cases} \gamma \sin \nu(1-\eta) \sin \nu x & 0 \le x \le \eta \\ \gamma \sin \nu\eta \sin \nu(1-x) & \eta \le x \le 1. \end{cases}$$

Finally to determine γ, we use the jump condition at $x = \eta$ which is

$$\frac{\partial G}{\partial x}|_{\eta+0} - \frac{\partial G}{\partial x}|_{\eta-0} = -1.$$

From this condition we can obtain

$$\gamma = \frac{1}{\nu \sin \nu}.$$

With γ known, the Green's function is completely determined, and we have

$$G(x,\eta) = \begin{cases} \frac{\sin \nu x \sin \nu(1-\eta)}{\nu \sin \nu} & 0 \le x \le \eta \\ \frac{\sin \nu\eta \sin \nu(1-x)}{\nu \sin \nu} & \eta \le x \le 1. \end{cases}$$

This is true provided $\nu \ne n\pi, n = 0, 1, 2, \ldots$. It can be easily seen that $G(x,\eta) = G(\eta, x)$ which is known as the symmetry property of any Green's function. This property is inherent with this function. However, it should be noted that for some physical problems this Green's function may not be symmetric.

Example 3

Find the Green's function for the following boundary value problem:

$$y'' + \nu^2 y = 0; \quad y(0) = y(1); \quad y'(0) = y'(1).$$

Solution

As before the Green's function is given by

$$G(x,\eta) = \begin{cases} A\cos\nu x + B\sin\nu x & 0 \le x \le \eta \\ C\cos\nu x + D\sin\nu x & \eta \le x \le 1. \end{cases}$$

The boundary conditions at $x = 0$ and $x = 1$ must be satisfied which yields respectively

$$\begin{aligned} A &= C\cos\nu + D\sin\nu \\ \nu B &= -C\nu\sin\nu + D\nu\cos\nu \qquad \nu \neq 0. \end{aligned}$$

Solving for C and D in terms of A and B, we obtain

$$\begin{aligned} C &= \frac{\begin{vmatrix} A & \sin\nu \\ B & \cos\nu \end{vmatrix}}{\begin{vmatrix} \cos\nu & \sin\nu \\ -\sin\nu & \cos\nu \end{vmatrix}} = A\cos\nu - B\sin\nu \\ D &= \frac{\begin{vmatrix} \cos\nu & A \\ -\sin\nu & B \end{vmatrix}}{\begin{vmatrix} \cos\nu & \sin\nu \\ -\sin\nu & \cos\nu \end{vmatrix}} = A\sin\nu + B\cos\nu. \end{aligned}$$

After a little reduction, the Green's function can be written as

$$G(x,\eta) = \begin{cases} A\cos\nu x - B\sin\nu x & 0 \le x \le \eta \\ A\cos\nu(1-x) - B\sin\nu(1-x) & \eta \le x \le 1. \end{cases}$$

Now using the continuity condition at $x = \eta$, the constants A and B are determined as

$$\begin{aligned} A &= \alpha(\sin\nu\eta + \sin\nu(1-\eta)) \\ B &= \alpha(-\cos\nu\eta + \cos\nu(1-\eta)) \end{aligned}$$

where α is an arbitrary constant.

Hence the Green's function is given by

$$G(x,\eta) = \begin{cases} \alpha[(\sin\nu\eta + \sin\nu(1-\eta))\cos\nu x + \\ (\cos\nu(1-\eta) - \cos\nu\eta)\sin\nu x], & 0 \le x \le \eta \\ \alpha[(\sin\nu\eta + \sin\nu(1-\eta))\cos\nu(1-x) \\ -(\cos\nu(1-\eta) - \cos\nu\eta)\sin\nu(1-x)], & \eta \le x \le 1. \end{cases}$$

Now to determine the value of α we use the jump condition at $x = \eta$

$$\frac{\partial G}{\partial x}\Big|_{\eta+0} - \frac{\partial G}{\partial x}\Big|_{\eta-0} = -1.$$

And after considerable reduction, we obtain $2\alpha\nu(1-\cos\nu) = -1$ such that $\alpha = \frac{-1}{2\nu(1-\cos\nu)}$.

With α known, the Green's function is completely determined and we have

$$G(x,\eta) = \begin{cases} \alpha\{\sin\nu(\eta - x) + \sin\nu(1-\eta+x)\} & 0 \le x \le \eta \\ \alpha\{\sin\nu(x-\eta) + \sin\nu(1+\eta-x)\} & \eta \le x \le 1 \end{cases}$$

where $\alpha = \frac{-1}{2\nu(1-\cos\nu)}$.

8.2 Green's function using variation of parameters

In this section we shall explore the results of the method of variation of parameters to obtain the Green's function. Let us consider the linear nonhomogeneous second order differential equation:

$$y'' + P(x)y' + Q(x)y = f(x) \qquad a \le x \le b \tag{8.17}$$

with the homogeneous boundary conditions

$$x = a: \quad \alpha_1 y(a) + \alpha_2 y'(a) = 0 \tag{8.18}$$

$$x = b: \quad \beta_1 y(b) + \beta_2 y'(b) = 0 \tag{8.19}$$

where constants α_1 and α_2, and likewise β_1 and β_2, are not both zero. We shall assume that $f(x)$, $P(x)$ and $Q(x)$ are continuous in $a \le x \le b$. The homogeneous part of eqn (8.17) is

$$y'' + P(x)y' + Q(x)y = 0. \tag{8.20}$$

Let $y_1(x)$ and $y_2(x)$ be two linearly independent solutions of (8.20). Then the complementary function is given by

$$y_c = Ay_1(x) + By_2(x) \tag{8.21}$$

where A and B are two arbitrary constants. To obtain the complete general solution we vary the parameters A and B such that

$$y = A(x)y_1(x) + B(x)y_2(x) \tag{8.22}$$

which is now a complete solution of (8.17). Using the method of variation of parameters (see Rahman, 1991), we have the following two equations:

$$\left.\begin{aligned} A'y_1 + B'y_2 &= 0 \\ A'y_1' + B'y_2' &= f(x) \end{aligned}\right\} \tag{8.23}$$

solving for A' and B', we have

$$A' = \frac{-y_2 f}{W} \quad \text{and} \quad B' = \frac{y_1 f}{W} \tag{8.24}$$

where

$$W = \begin{vmatrix} y_1 & y_2 \\ y_1' & y_2' \end{vmatrix} = \text{Wronskian.}$$

Let us now solve for A and B by integrating their derivatives between a and x and x and b, respectively.

The integration of these results yields, using η as dummy variable

$$\begin{aligned} A &= -\int_a^x \frac{y_2(\eta)f(\eta)}{W(\eta)}d\eta \\ B &= \int_b^x \frac{y_1(\eta)f(\eta)}{W(\eta)}d\eta \\ &= -\int_x^b \frac{y_1(\eta)f(\eta)}{W(\eta)}d\eta. \end{aligned} \tag{8.25}$$

Thus a particular solution can be obtained as

$$y = -y_1(x)\int_a^x \frac{y_2(\eta)f(\eta)}{W(\eta)}d\eta - y_2(x)\int_x^b \frac{y_1(\eta)f(\eta)}{W(\eta)}d\eta. \tag{8.26}$$

Now moving $y_1(x)$ and $y_2(x)$ into the respective integrals, we have

$$y = -[\int_a^x \frac{y_1(x)y_2(\eta)}{W(\eta)}f(\eta)d\eta + \int_x^b \frac{y_2(x)y_1(\eta)}{W(\eta)}f(\eta)d\eta] \tag{8.27}$$

which is of the form

$$y = -\int_a^b G(x,\eta)f(\eta)d\eta \tag{8.28}$$

where

$$G(x,\eta) = \begin{cases} \frac{y_1(x)y_2(\eta)}{W(\eta)} & a \le \eta \le x \quad \text{i.e.} \quad \eta \le x \le b \\ \frac{y_2(x)y_1(\eta)}{W(\eta)} & x \le \eta \le b \quad \text{i.e.} \quad a \le x \le \eta. \end{cases} \tag{8.29}$$

From the way in which $y_1(x)$ and $y_2(x)$ were selected, it is clear that $G(x,\eta)$ satisfies the boundary conditions of the problem. It is also evident that $G(x,\eta)$ is a continuous function and at $x = \eta : G(\eta_{+0},\eta) = G(\eta_{-0},\eta)$. Furthermore except at $x = \eta, G(x,\eta)$ satisfies the homogeneous form of the given differential equation since $y_1(x)$ and $y_2(x)$ are solutions of this equation. This can be seen as follows: Equation (8.28) must satisfy the eqn (8.17) which means

$$-\int_a^b \{G_{xx} + P(x)G_x + Q(x)G\}f(\eta)d\eta = f(x).$$

The quantity under the bracket must be zero except at $x = \eta$ and thus it follows that

$$G_{xx} + P(x)G_x + Q(x)G = -\delta(\eta - x) \tag{8.30}$$

where $\delta(\eta - x)$ is a Dirac delta function.
Finally, for $\frac{\partial G}{\partial x}(x, \eta)$ we have

$$G_x(x,\eta) = \begin{cases} \frac{y_1'(x)y_2(\eta)}{W(\eta)} & \eta \le x \le b \\ \frac{y_2'(x)y_1(\eta)}{W(\eta)} & a \le x \le \eta. \end{cases}$$

Hence,

$$\frac{\partial G}{\partial x}|_{\eta+0} - \frac{\partial G}{\partial x}|_{\eta-0} = \frac{y_1'(\eta)y_2(\eta) - y_2'(\eta)y_1(\eta)}{W(\eta)} = -1$$

which shows that $\frac{\partial G}{\partial x}(x, \eta)$ has a jump of the amount -1 at $x = \eta$. This result can be shown to be true from the relation (8.30) by integrating from η_{+0} to η_{-0},

$$\begin{aligned} \int_{\eta-0}^{\eta+0} G_{xx}dx + \int_{\eta-0}^{\eta+0}(P(x)G_x + Q(x)G)dx &= -\int_{\eta-0}^{\eta+0}\delta(x-\eta)dx \\ \frac{\partial G}{\partial x}|_{\eta+0} - \frac{\partial G}{\partial x}|_{\eta-0} &= -1. \end{aligned}$$

Because

$$\begin{aligned} \int_{\eta-0}^{\eta+0} P(x)\frac{\partial G}{\partial x}dx &= P(x)\int_{\eta-0}^{\eta+0}\frac{\partial G}{\partial x}dx = P(x)[G(\eta^+,\eta) - G(\eta^-,\eta)] = 0 \\ \text{and} \int_{\eta-0}^{\eta+0} Q(x)Gdx &= Q(\eta)\int_{\eta-0}^{\eta+0} Gdx = 0, \end{aligned}$$

since $G(x, \eta)$ is a continuous function in $\eta_{-0} \le x \le \eta_{+0}$.

Definition 3

Consider the second-order non-homogeneous differential equation.

$$y'' + p(x)y' + Q'(x)y = f(x) \tag{8.31}$$

with the homogeneous boundary conditions

$$\left.\begin{aligned} x = a: \quad & \alpha_1 y(a) + \alpha_2 y'(a) = 0 \\ x = b: \quad & \beta_1 y(b) + \beta_2 y'(b) = 0 \end{aligned}\right\} \tag{8.32}$$

where α_1 and α_2 are not both zero, and β_1 and β_2 are not both zero.

Consider a particular solution which satisfies the differential equation (8.31) in the following form

$$y = -\int_a^b G(x,\eta)f(\eta)d\eta \tag{8.33}$$

where $G(x, \eta)$ satisfies the differential equation (8.31) such that

$$G_{xx} + P(x)G_x + Q(x)G = -\delta(\eta - x) \tag{8.34}$$

with the following properties:

(I) Boundary conditions:

$$\left.\begin{array}{ll} x = a: & \alpha_1 G(a,\eta) + \alpha_2 G_x(a,\eta) = 0 \\ x = b: & \beta_1 G(b,\eta) + \beta_2 G_x(b,\eta) = 0 \end{array}\right\}. \tag{8.35}$$

(II) Continuity condition:

$$x = \eta, \qquad G(\eta_{+0},\eta) = G(\eta_{-0},\eta) \tag{8.36}$$

i.e., $G(x,\eta)$ is continuous at $x = \eta$ on $a \le x \le b$.

(III) Jump discontinuity of the gradient of $G(x,\eta)$ at $x = \eta$ that means,

$$\frac{\partial G}{\partial x}|_{x=\eta+0} - \frac{\partial G}{\partial x}|_{x=\eta-0} = -1.$$

Then a particular solution of the given boundary value problem (8.31) and (8.32) can be obtained as

$$y = -\int_a^b G(x,\eta) f(\eta) d\eta$$

where $G(x,\eta)$ is the Green's function for the boundary value problem.

Remark

It is worth mentioning here that the Green's function obtained through this procedure may or may not be symmetric, which means there is no guarantee that $G(x,\eta) = G(\eta,x)$.

Example 4

Find the Green's function of the homogeneous boundary value problem $y'' + y = 0$, $y(0) = 0$ and $y'(\pi) = 0$. Then solve the non-homogeneous system

$$\begin{aligned} y'' + y &= -3\sin 2x \\ y(0) &= 0 \\ y'(\pi) &= 0. \end{aligned}$$

Solution

The Green's function is given by

$$G(x,\eta) = \begin{cases} A\cos x + B\sin x & 0 \le x \le \eta \\ C\cos x + D\sin x & \eta \le x \le \pi. \end{cases}$$

Using the boundary conditions, we have

$$\begin{aligned} x = 0: &\quad 0 = A \\ x = \pi: &\quad 0 = D. \end{aligned}$$

Thus we obtain

$$G(x,\eta) = \begin{cases} B\sin x & 0 \le x \le \eta \\ C\cos x & \eta \le x \le \pi. \end{cases}$$

$G(x,\eta)$ is a continuous function at $x = \eta$. Therefore, $B\sin\eta = C\cos\eta$ from which we have $B = \alpha\cos\eta$ and $C = \alpha\sin\eta$, where α is an arbitrary constant.

The Green's function is then given by

$$G(x,\eta) = \begin{cases} \alpha\cos\eta\sin x & 0 \le x \le \eta \\ \alpha\sin\eta\cos x & \eta \le x \le \pi. \end{cases}$$

This arbitrary constant α can be evaluated by the jump condition

$$\begin{aligned} \frac{\partial G}{\partial x}|_{\eta+0} - \frac{\partial G}{\partial x}|_{\eta-0} &= -1 \\ \text{or,} \quad \alpha[-\sin^2\eta - \cos^2\eta] &= -1 \end{aligned}$$

and hence $\alpha = 1$. Therefore the required Green's function is

$$G(x,\eta) = \begin{cases} \cos\eta\sin x & 0 \le x \le \eta \\ \sin\eta\cos x & \eta \le x \le \pi. \end{cases}$$

Hence the solution of the non-homogeneous boundary value problem is

$$\begin{aligned} y &= -\int_0^\pi G(x,\eta)f(x)d\eta \\ &= -[\int_0^x G(x,\eta)f(x)d\eta + \int_x^\pi G(x,\eta)f(x)d\eta] \\ &= -[\int_0^x \sin\eta\cos x(-3\sin 2\eta)d\eta + \int_x^\pi \sin x\cos\eta(-3\sin 2\eta)d\eta] \\ &= 3[\cos x\int_0^x \sin\eta\sin 2\eta d\eta + \sin x\int_x^\pi \cos\eta\sin 2\eta d\eta]. \end{aligned}$$

Performing the indicated integration, we obtain

$$\begin{aligned} y &= 3[\frac{2}{3}][\cos x\sin^3 x + \sin x(1+\cos^3 x)] \\ &= 2\sin x + \sin 2x. \end{aligned}$$

Using the elementary operator method, the solution of the non-homogeneous ODE is given by

$$y = A\cos x + B\sin x + \sin 2x.$$

With the given boundary conditions, we obtain $A = 0$ and $B = 0$, and hence the solution is

$$y = 2\sin x + \sin 2x$$

which is identical to the Green's function method.

Example 5

Find a particular solution of the following boundary value problem by using the Green's function method:

$$y'' = -x; \quad y(0) = 0, y(1) = 0.$$

Solution

The Green's function is obtained from the homogeneous part of the differential equation and is given by,

$$G(x,\eta) = \begin{cases} Ax + B & 0 \le x \le \eta \\ \alpha x + \beta & \eta \le x \le 1. \end{cases}$$

Using the boundary conditions, we obtain
$0 = B$ and $\alpha = -\beta$ and hence

$$G(x,\eta) = \begin{cases} Ax & 0 \le x \le \eta \\ \beta(1-x) & \eta \le x \le 1. \end{cases}$$

Now continuity condition yields $A\eta = \beta(1-\eta)$ which can be written as $\frac{A}{1-\eta} = \frac{\beta}{\eta} = \gamma$ such that $A = \gamma(1-\eta)$, where γ is the arbitrary constant. Hence we have

$$G(x,\eta) = \begin{cases} \gamma(1-\eta)x & 0 \le x \le \eta \\ \gamma(1-x)\eta & \eta \le x \le 1. \end{cases}$$

From the jump condition at $x = \eta$, we obtain

$$\begin{aligned} \frac{\partial G}{\partial x}|_{\eta+0} - \frac{\partial G}{\partial x}|_{\eta-0} &= -1 \\ \text{or,} \quad \gamma[-\eta - (1-\eta)] &= -1 \end{aligned}$$

such that $\gamma = 1$.
Hence the Green's function is determined as

$$G(x,\eta) = \begin{cases} (1-\eta)x & 0 \le x \le \eta \\ (1-x)\eta & \eta \le x \le 1. \end{cases}$$

Changing the roles of x and η, we have

$$G(x,\eta) = G(\eta,x) = \begin{cases} (1-x)\eta & 0 \le \eta \le x \\ (1-\eta)x & x \le \eta \le 1. \end{cases}$$

Thus a particular solution of this boundary problem is obtained as:

$$\begin{aligned} y(x) &= -\int_0^1 G(x,\eta)f(\eta) \\ &= -[\int_0^x G(x,\eta)f(\eta)d\eta + \int_x^1 G(x,\eta)f(\eta)d\eta] \end{aligned}$$

$$
\begin{aligned}
&= -[\int_0^x (1-x)\eta(-\eta)d\eta + \int_x^1 (1-\eta)x(-\eta)d\eta] \\
&= [(1-x)|\frac{\eta^3}{3}|_0^x + x|\frac{\eta^2}{2} - \frac{\eta^3}{3}|_x^1] \\
&= \frac{x^3}{3} + \frac{x}{6} - \frac{x^3}{2} = \frac{x}{6}(1-x^2).
\end{aligned}
$$

8.3 Developments of Green's function in 2D

We already saw the application of Green's function in one-dimension to boundary value problems in ordinary differential equations in the previous section. In this section, we illustrate the use of Green's function in two-dimensions to the boundary value problems in partial differential equations arising in a wide class of problems in engineering and mathematical physics.

As Green's function is always associated with the Dirac delta function, it is, therefore, useful to formally define this function. We define the Dirac delta function $\delta(x-\xi, y-\eta, z-\zeta)$ in two-dimensions by

$$
\text{I.} \quad \delta(x-\xi, y-\eta, z-\zeta) = \begin{cases} \infty, & x=\xi, y=\eta \\ 0 & \text{otherwise} \end{cases} \tag{8.37}
$$

$$
\text{II.} \quad \iint_{R_\varepsilon} \delta(x-\xi, y-\eta)dxdy = 1, R_\varepsilon : (x-\xi)^2 + (y-\eta)^2 < \varepsilon^2 \tag{8.38}
$$

$$
\text{III.} \quad \iint_R f(x,y)\delta(x-\xi, y-\eta)dxdy = f(\xi,\eta) \tag{8.39}
$$

for arbitrary continuous function $f(x,y)$ in the region R.

Remark

The Dirac delta function is not a regular function but it is a generalized function as defined by Lighthill (1958) in his book *Fourier Analysis and Generalized Function*, Cambridge University Press, 1958. This function is also called an impulse function which is defined as the limit of a sequence of regular well-behaved functions having the required property that the area remains a constant (unity) as the width is reduced. The limit of this sequence is taken to define the impulse function as the width is reduced toward zero. For more information and an elegant treatment of the delta function as a generalized function, interested readers are referred to *Theory of Distribution* by L. Schwartz (1843–1921).

If $\delta(x-\xi)$ and $\delta(y-\eta)$ are one-dimensional delta functions, then we have

$$
\iint_R f(x,y)\delta(x-\xi)\delta(y-\eta)dxdy = f(\xi,\eta). \tag{8.40}
$$

Since equations (8.39) and (8.40) hold for an arbitrary continuous function f, we conclude that

$$
\delta(x-\xi, y-\eta) = \delta(x-\xi)\delta(y-\eta) \tag{8.41}
$$

which simply implies that the two dimensional delta function is the product of one-dimensional delta functions. Higher dimensional delta functions can be defined in a similar manner.

8.3.1 Two-dimensional Green's function

The application of Green's function in 2D can best be described by considering the solution of the Dirichlet problem.

$$\nabla^2 u = h(x,y) \qquad \text{in the two dimensional region R} \tag{8.42}$$

$$u = f(x,y) \qquad \text{on the boundary C} \tag{8.43}$$

where $\nabla^2 = \frac{\partial^2}{\partial x^2} + \frac{\partial^2}{\partial y^2}$. Before attempting to solve this boundary value problem heuristically, we first define the Green's function for the Laplace operator. Then, the Green's function for the Dirichlet problem involving the Helmholtz operator may be defined in a similar manner.

The Green's function denoted by $G(x,y;\xi,\eta)$ for the Dirichlet problem involving the Laplace operator is defined as the function which satisfies the following properties:

I.
$$\nabla^2 G = \delta(x-\xi, y-\eta) \quad \text{in } R \tag{8.44}$$
$$G = 0 \quad \text{on } C \tag{8.45}$$

II. G is symmetric, that is,
$$G(x,y;\xi,\eta) = G(\xi,\eta;x,y) \tag{8.46}$$

III. G is continuous in x, y, ξ, η but $\frac{\partial G}{\partial n}$ the normal derivative has a discontinuity at the point (ξ,η) which is specified by the equation

$$\lim_{\varepsilon\to 0}\int_{C_\varepsilon} \frac{\partial G}{\partial n} ds = 1 \tag{8.47}$$

where n is the outward normal to the circle

$$C_\varepsilon : \quad (x-\xi)^2 + (y-\eta)^2 = \varepsilon^2.$$

Remark

The Green's function G may be interpreted as the response of the system at a field point (x,y) due to a delta function $\delta(x,y)$ input at the source point (ξ,η). G is continuous everywhere in the region R, and its first and second derivatives are continuous in R except at (ξ,η). Thus the property (i) essentially states that $\nabla^2 G = 0$ everywhere except at the source point (ξ,η).

Properties (ii) and (iii) pertaining to the Green's function G can be established directly by using the Green's second identity of vector calculus (Wylie & Barrett, 1982). This formula states that if ϕ and ψ are two functions having continuous second partial derivatives in a region R bounded by a curve C then

$$\int\int_R (\phi\nabla^2\psi - \psi\nabla^2\phi)dS = \int_C (\phi\frac{\partial\psi}{\partial n} - \psi\frac{\partial\phi}{\partial n})ds \tag{8.48}$$

where dS is the elementary area and ds the elementary length.

Now let us consider that $\phi = G(x, y; \xi; \eta)$ and $\psi = G(x, y; \xi^*\eta^*)$ then from (8.48)

$$\int\int_R \{G(x, y; \xi, \eta)\nabla^2 G(x, y; \xi^*\eta^*) - G(x, y; \xi^*, \eta^*)\nabla^2 G(x, y; \xi, \eta)\}dS$$
$$= \int_C \{G(x, y; \xi, \eta)\frac{\partial G}{\partial n}(x, y; \xi^*, \eta^*) - G(x, y; \xi^*, \eta^*)\frac{\partial G}{\partial n}(x, y; \xi, \eta)\}ds. \quad (8.49)$$

$$\begin{aligned} \text{Since} \quad G(x, y : \xi, \eta) &= 0 \quad \text{on } C \\ \text{and} \quad G(x, y; \xi^*, \eta^*) &= 0 \quad \text{on } C \\ \text{also} \quad \nabla^2 G(x, y; \xi^*, \eta^*) &= \delta(x - \xi, y - \eta) \quad \text{in } R \\ \text{and} \quad \nabla^2 G(x, y; \xi^*, \eta^*) &= \delta(x - \xi^*, y - \eta^*) \quad \text{in } R. \end{aligned}$$

Hence we obtain from (8.49)

$$\int\int_R G(x, y; \xi, \eta)\delta(x - \xi^*, y - \eta^*)dxdy = \int\int_R G(x, y; \xi^*, \eta^*)\delta(x, \xi; y - \eta)dxdy$$

which reduces to

$$G(\xi^*, \eta^*; \xi, \eta) = G(\xi, \eta; \xi^*, \eta^*).$$

Thus the Green's function is symmetric.

To prove property (iii) of Green's function we simply integrate both sides of (8.44) which yields

$$\int\int_{R_\varepsilon} \nabla^2 G dS = \int\int_{R_\varepsilon} \delta(x - \xi, y - \eta)dxdy = 1$$

where R_ε is the region bounded by the circle C_ε.
Thus it follows that

$$\lim_{\varepsilon \to 0} \int\int_{R_\varepsilon} \nabla^2 G dS = 1$$

and using the Divergence theorem (Wylie and Barrett, 1982),

$$\lim_{\varepsilon \to 0} \int_{C_\varepsilon} \frac{\partial G}{\partial n} ds = 1.$$

Theorem 1

Prove that the solution of the Dirichlet problem

$$\nabla^2 u = h(x, y) \qquad \text{in } R \quad (8.50)$$

subject to the boundary conditions

$$u = f(x, y) \qquad \text{on } C \quad (8.51)$$

is given by

$$u(x,y) = \int\int_R G(x,y;\xi,\eta)h(\xi,\eta)d\xi d\eta + \int_C f\frac{\partial G}{\partial n}ds \tag{8.52}$$

where G is the Green's function and n denotes the outward normal to the boundary C of the region R. It can be easily seen that the solution of the problem is determined if the Green's function is known.

Proof

In equation (8.48), let us consider that

$$\phi(\xi,\eta) = G(\xi,\eta;x,y) \quad \text{and } \psi(\xi,\eta) = u(\xi,\eta),$$

and we obtain

$$\begin{aligned}&\int\int_R [G(\xi,\eta;x,y)\nabla^2 u - u(\xi,\eta)\nabla^2 G]d\xi d\eta \\ &= \int_C [G(\xi,\eta;x,y)\frac{\partial u}{\partial n} - u(\xi,\eta)\frac{\partial G}{\partial n}]ds.\end{aligned} \tag{8.53}$$

Here $\nabla^2 u = h(\xi,\eta)$ and $\nabla^2 G = \delta(\xi - x, \eta - y)$ in the region R.
Thus we obtain from (8.53)

$$\begin{aligned}&\int\int_R \{G(\xi,\eta;x,y)h(\xi,\eta) - u(\xi,\eta)\delta(\xi - x,\eta - y)\}d\xi d\eta \\ &= \int_C \{G(\xi,\eta;x,y)\frac{\partial u}{\partial n} - u(\xi,\eta)\frac{\partial G}{\partial n}\}ds.\end{aligned} \tag{8.54}$$

Since $G = 0$ and $u = f$ on C, and noting that G is symmetric, it follows that

$$u(x,y) = \int\int_R G(x,y;\xi,\eta)h(\xi,\eta)d\xi d\eta + \int_C f\frac{\partial G}{\partial n}ds$$

which is the required proof as asserted.

8.3.2 Method of Green's function

It is often convenient to seek G as the sum of a particular integral of the non-homogeneous differential equation and the solution of the associated homogeneous differential equation. That is, G may assume the form

$$G(\xi,\eta;x,y) = g_h(\xi,\eta;x,y) + g_p(\xi,\eta;x,y) \tag{8.55}$$

where g_h, known as the free-space Green's function, satisfies

$$\nabla^2 g_h = 0 \quad \text{in } R \tag{8.56}$$

and g_p satisfies

$$\nabla^2 g_p = \delta(\xi - x, \eta - y) \qquad \text{in } R \tag{8.57}$$

so that by superposition $G = g_h + g_p$ satisfies

$$\nabla^2 G = \delta(\xi - x, \eta - y) \tag{8.58}$$

Note that (x, y) will denote the source point and (ξ, η) denotes the field point. Also $G = 0$ on the boundary requires that

$$g_h = -g_p \tag{8.59}$$

and that g_p need not satisfy the boundary condition.

In the following we will demonstrate the method to find g_p for the Laplace and Helmholtz operators.

8.3.3 The Laplace operator

In this case g_p must satisfy

$$\nabla^2 g_p = \delta(\xi - x, \eta - y) \qquad \text{in } R.$$

Then for $r = \sqrt{(\xi - x)^2 + (\eta - y)^2} > 0$, that is, for $\xi \neq x$, and $\eta \neq y$, we have by taking (x, y) as the centre (assume that g_p is independent of θ.)

$$\nabla^2 g_p = 0, \qquad \text{or,} \qquad \frac{1}{r}\frac{\partial}{\partial r}(r\frac{\partial g_p}{\partial r}) = 0.$$

The solution of which is given by

$$g_p = \alpha + \beta \ln r.$$

Now apply the condition

$$\lim_{\varepsilon \to 0} \int_{C_\varepsilon} \frac{\partial g_p}{\partial n} ds = \lim_{\varepsilon \to 0} \int_0^{2\pi} (\frac{\beta}{r}) r d\theta = 1.$$

Thus $\beta = \frac{1}{2\pi}$ and α is arbitrary. For simplicity we choose $\alpha = 0$. Then g_p takes the form

$$g_p = \frac{1}{2\pi} \ln r. \tag{8.60}$$

This is known as the fundamental solution of the Laplace's equation in two dimensions.

8.3.4 The Helmholtz operator

Here g_p satisfies the following equation

$$(\nabla^2 + \lambda^2)g_p = \delta(\xi - x, \eta - y). \tag{8.61}$$

Again for $r > 0$, we find

$$\begin{aligned} \frac{1}{r}\frac{\partial}{\partial r}(r\frac{\partial g_p}{\partial r}) + \lambda^2 g_p &= 0 \\ \text{or,} \quad r^2(g_p)_{rr} + r(g_p)_r + \lambda^2 r^2 g_p &= 0. \end{aligned} \tag{8.62}$$

This is the Bessel equation of order zero, the solution of which is

$$g_p = \alpha J_0(\lambda r) + \beta Y_0(\lambda r). \tag{8.63}$$

Since the behaviour of J_0 at $r = 0$ is not singular, we set $\alpha = 0$. Thus we have

$$g_p = \beta Y_0(\lambda r).$$

But for very small r, $Y_0 \approx \frac{2}{\pi}\ln r$.
Applying the condition $\lim_{\varepsilon\to 0}\int_{C_\varepsilon}\frac{\partial g_p}{\partial n}ds = 1$, we obtain

$$\begin{aligned} \lim_{\varepsilon\to 0}\int_{C_\varepsilon}\beta\frac{\partial Y_0}{\partial r}ds &= 1 \\ \lim_{\varepsilon\to 0}\int_0^{2\pi}\beta(\frac{2}{\pi})(\frac{1}{r})rd\theta &= 1 \\ \frac{2\beta}{\pi}(2\pi) &= 1 \\ \beta &= \frac{1}{4}. \end{aligned}$$

Thus

$$g_p = \frac{1}{4}Y_0(\lambda r). \tag{8.64}$$

Since $(\nabla^2 + \lambda^2) \to \nabla^2$ as $\lambda \to 0$, it follows that $\frac{1}{4}Y_0(\lambda r) \to \frac{1}{2\pi}\ln r$ as $\lambda \to 0$.

Theorem 2: Solution of Dirichlet problem using the Laplace operator

Show that the method of Green's function can be used to obtain the solution of the Dirichlet problem described by the Laplace operator:

$$\nabla^2 u = h \quad \text{in } R \tag{8.65}$$

$$u = f \quad \text{on } C \tag{8.66}$$

and the solution is given by

$$u(x,y) = \int\int_R G(x,y;\xi,\eta)h(\xi,\eta)d\xi d\eta + \int_C f\frac{\partial G}{\partial n}ds \tag{8.67}$$

where R is a circular region with boundary C.

Proof

Applying Green's second formula

$$\int\int_R (\phi\nabla^2\psi - \psi\nabla^2\phi)dS = \int_C (\phi\frac{\partial\psi}{\partial n} - \psi\frac{\partial\phi}{\partial n})ds \tag{8.68}$$

to the functions $\phi(\xi, \eta) = G(\xi, \eta; x, y)$ and $\psi(\xi, \eta) = u(\xi, \eta)$ we obtain

$$\int\int_R (G\nabla^2 u - u\nabla^2 G)dS = \int_C (G\frac{\partial u}{\partial n} - u\frac{\partial G}{\partial n})ds.$$

$$\begin{aligned} \text{But} \qquad \nabla^2 u &= h(\xi, \eta) \\ \text{and} \qquad \nabla^2 G &= \delta(\xi - x, \eta - y) \qquad \text{in } R. \end{aligned}$$

Thus we have

$$\begin{aligned} &\int\int_R \{G(\xi, \eta; x, y)h(\xi, \eta) - u(\xi, \eta)\delta(\xi - x, \eta - y)\}dS \\ &= \int_C \{G(\xi, \eta; x, y)\frac{\partial u}{\partial n} - u(\xi, \eta)\frac{\partial G}{\partial n}\}ds. \end{aligned} \tag{8.69}$$

Because $G = 0$ and $u = f$ on the boundary C, and G is symmetric, it follows that

$$u(x, y) = \int\int_R G(x, y; \xi, \eta)h(\xi, \eta)d\xi d\eta + \int_C f\frac{\partial G}{\partial n}ds.$$

which is the required proof as asserted.

Example 6

Consider the Dirichlet problem for the unit circle given by

$$\nabla^2 u = 0 \qquad \text{in } R \tag{8.70}$$

$$u = f(\theta) \qquad \text{on } C. \tag{8.71}$$

Find the solution by using the Green's function method.

Solution

We introduce the polar coordinates by means of the equations

$$\begin{aligned} x &= \rho\cos\theta \qquad & \xi &= \sigma\cos\beta \\ y &= \rho\sin\theta \qquad & \eta &= \sigma\sin\beta \end{aligned}$$

so that

$$r^2 = (x - \xi)^2 + (y - \eta)^2 = \sigma^2 + \rho^2 - 2\rho\sigma\cos(\beta - \theta).$$

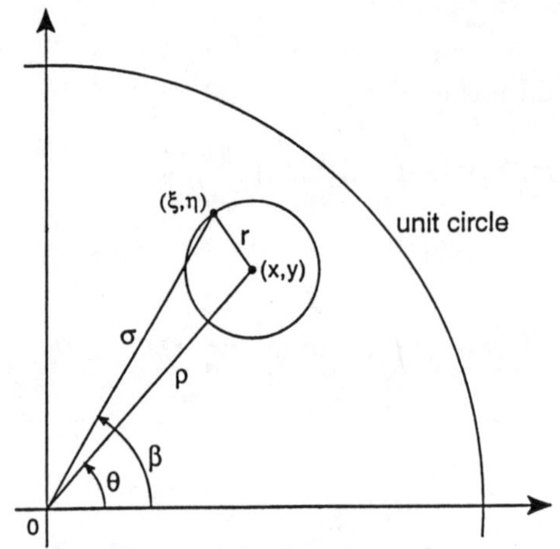

Figure 8.3: A unit circle

To find the Green's function let us consider $G(x,y)$ is the solution of the sum of two solutions, one regular and the other one singular. We know G satisfies

$$\begin{aligned}
\nabla^2 G &= \delta(\xi - x, \eta - y) \\
\text{and if } G &= g_h + g_p, \\
\text{then } \nabla^2 g_p &= \delta(\xi - x, \eta - y) \\
\text{and } \nabla^2 g_h &= 0.
\end{aligned}$$

These equations in polar coordinates (see Fig. 8.3) can be written as

$$\frac{1}{\rho}\frac{\partial}{\partial \rho}(\rho \frac{\partial g_p}{\partial \rho}) + \frac{1}{\rho^2}\frac{\partial^2 g_p}{\partial \theta^2} = \delta(\sigma - \rho, \beta - \theta) \tag{8.72}$$

$$\text{and} \quad \frac{1}{\sigma}\frac{\partial}{\partial \sigma}(\sigma \frac{\partial g_h}{\partial \sigma}) + \frac{1}{\sigma^2}\frac{\partial^2 g_h}{\partial \beta^2} = 0. \tag{8.73}$$

By the method of separation of variables, the solution of (8.73) can be written as (see Rahman, 1994)

$$g_h = \frac{a_0}{2} + \sum_{n=1}^{\infty} \sigma^n (a_n \cos n\beta + b_n \sin n\beta). \tag{8.74}$$

A singular solution for (8.72) is given by

$$g_p = \frac{1}{4\pi} \ln r^2 = \frac{1}{4\pi} \ln[\sigma^2 + \rho^2 - 2\rho\sigma \cos(\beta - \theta)].$$

Thus when $\sigma = 1$ on the boundary C,

$$g_h = -g_p = -\frac{1}{4\pi} \ln[1 + \rho^2 - 2\rho \cos(\beta - \theta)].$$

The relation

$$\ln[1+\rho^2-2\rho\cos(\beta-\theta)] = -2\sum_{n=1}^{\infty}\frac{\rho^n\cos n(\beta-\theta)}{n} \tag{8.75}$$

can be established as follows:

$$\begin{aligned}
\ln[1+\rho^2-\rho(e^{i(\beta-\theta)}+e^{-i(\beta-\theta)})] &= \ln\{(1-\rho e^{i(\beta-\theta)})(1-\rho e^{-i(\beta-\theta)})\} \\
&= \ln(1-\rho e^{i(\beta-\theta)})+\ln(1-\rho e^{-i(\beta-\theta)}) \\
&= -[\rho e^{i(\beta-\theta)}+\frac{\rho^2}{2}e^{2i(\beta-\theta)}+\ldots] \\
&- [\rho e^{-i(\beta-\theta)}-\frac{\rho^2}{2}e^{-2i(\beta-\theta)}+\ldots] \\
&= -2\sum_{n=1}^{\infty}\frac{\rho^n\cos n(\beta-\theta)}{n}.
\end{aligned}$$

When $\sigma = 1$ at the circumference of the unit circle, we obtain

$$\frac{1}{2\pi}\sum_{n=1}^{\infty}\frac{\rho^n\cos n(\beta-\theta)}{n} = \sum_{n=1}^{\infty} a_n\cos n\beta + b_n\sin n\beta.$$

Now equating the coefficients of $\cos n\beta$ and $\sin n\beta$ to determine a_n and b_n, we find

$$\begin{aligned}
a_n &= \frac{\rho^n}{2\pi n}\cos n\theta \\
b_n &= \frac{\rho^n}{2\pi n}\sin n\theta.
\end{aligned}$$

It therefore follows that (8.74) becomes

$$\begin{aligned}
g_h(\rho,\theta;\sigma,\beta) &= \frac{1}{2\pi}\sum_{n=1}^{\infty}\frac{(\rho\sigma)^n}{n}\cos n(\beta-\theta) \\
&= -\frac{1}{4\pi}\ln[1+(\rho a)^2-2(\rho\sigma)\cos(\beta-\theta)].
\end{aligned}$$

Hence the Green's function for this problem is

$$\begin{aligned}
G(\rho,\theta;\sigma,\beta) &= g_p+g_h \\
&= \frac{1}{4\pi}\ln[\sigma^2+\rho^2-2\sigma\rho\cos(\beta-\theta)] \\
&\quad -\frac{1}{4\pi}\ln[1+(\rho\sigma)^2-2(\rho\sigma)\cos(\beta-\theta)]
\end{aligned} \tag{8.76}$$

from which we find

$$\frac{\partial G}{\partial n}|_{\text{on } C} = (\frac{\partial G}{\partial \sigma})_{\sigma=1} = \frac{1}{2\pi}[\frac{1-\rho^2}{1+\rho^2-2\rho\cos(\beta-\theta)}].$$

If $h = 0$, the solution of the problem reduces to

$$u(\rho, \theta) = \frac{1}{2\pi} \int_0^{2\pi} \frac{1 - \rho^2}{1 + \rho^2 - 2\rho\cos(\beta - \theta)} f(\beta) d\beta$$

which is the Poisson's Integral Formula.

Theorem 3: Solution of Dirichlet problem using the Helmholtz operator

Show that the Green's function method can be used to solve the Dirichlet problem for the Helmholtz operator:

$$\nabla^2 u + \lambda^2 u = h \quad \text{in } R \tag{8.77}$$

$$u = f \quad \text{on } C \tag{8.78}$$

where R is a circular region of unit radius with boundary C.

Proof

The Green's function must satisfy the Helmholtz equation in the following form

$$\begin{aligned} \nabla^2 G + \lambda^2 G &= \delta(\xi - x, \eta - y) \quad \text{in } R \\ G &= 0 \quad \text{on } R. \end{aligned} \tag{8.79}$$

We seek the solution in the following form

$$G(\xi, \eta; x, y) = g_h(\xi, \eta; x, y) + g_p(\xi, \eta; x, y)$$

$$\text{such that} \quad (\nabla^2 + \lambda^2) g_h = 0 \tag{8.80}$$

$$\text{and} \quad (\nabla^2 + \lambda^2) g_p = \delta(\xi - x, \eta - y). \tag{8.81}$$

The solution of (8.81) yields (eqn (8.63))

$$g_p = \frac{1}{4} Y_0(\lambda r) \tag{8.82}$$

where

$$r = [(\xi - x)^2 + (\eta - y)^2]^{\frac{1}{2}}.$$

The solution for equation (8.80) can be determined by the method of separation of variables. Thus, the solution in polar coordinates as given below

$$x = \rho\cos\theta \qquad \xi = \sigma\cos\beta$$

$$y = \rho\sin\theta \qquad \eta = \sigma\sin\beta$$

may be written in the form

$$g_h(\rho, \theta; \sigma, \beta) = \sum_{n=0}^{\infty} J_n(\lambda\sigma)[a_n \cos n\beta + b_n \sin n\beta]. \tag{8.83}$$

But on the boundary C,

$$g_h + g_p = 0.$$

Therefore

$$g_h = -g_p = -\frac{1}{4}Y_0(\lambda r)$$

$$\begin{aligned} \text{where} \quad r &= [\rho^2 + \sigma^2 - 2\rho\sigma\cos(\beta - \theta)]^{\frac{1}{2}} \\ \text{and at } \sigma = 1: \quad r &= [1 + \rho^2 - 2\rho\cos(\beta - \theta)]^{\frac{1}{2}}. \end{aligned}$$

Thus on the boundary ($\sigma = 1$), these two solutions yield

$$-\frac{1}{4}Y_0(\lambda r) = \sum_{n=0}^{\infty} J_n(\lambda)[a_n \cos n\beta + b_n \sin n\beta]$$

which is a Fourier expansion. The Fourier coefficients are obtained as

$$\begin{aligned} a_0 &= -\frac{1}{8\pi J_0(\lambda)}\int_{-\pi}^{\pi} Y_0[\lambda\sqrt{1 + \rho^2 - 2\rho\cos(\beta - \theta)}]d\beta \\ a_n &= -\frac{1}{4\pi J_0(\lambda)}\int_{-\pi}^{\pi} Y_0[\lambda\sqrt{1 + \rho^2 - 2\rho\cos(\beta - \theta)}]\cos n\beta d\beta \\ b_n &= -\frac{1}{4\pi J_0(\lambda)}\int_{-\pi}^{\pi} Y_0[\lambda\sqrt{1 + \rho^2 - 2\rho\cos(\beta - \theta)}]\sin n\beta d\beta \\ & n = 1, 2, 3, \ldots \end{aligned}$$

From the Green's theorem, we have

$$\int\int_R \{G(\nabla^2 + \lambda^2)u - u(\nabla^2 + \lambda^2)G\}dS = \int_C \{G(\frac{\partial u}{\partial n}) - u(\frac{\partial G}{\partial n})\}ds.$$

But we know $(\nabla^2 + \lambda^2)G = \delta(\xi - x, \eta - y)$, and $G = 0$ on C, and $(\nabla^2 + \lambda^2)u = h$. Therefore

$$u(x, y) = \int\int_R h(\xi, \eta)G(\xi, \eta; x, y)d\xi d\eta + \int_C f(\xi, \eta)\frac{\partial G}{\partial n}ds$$

where G is given by

$$\begin{aligned} G(\xi, \eta; x, y) &= g_p + g_h \\ &= \frac{1}{4}Y_0(\lambda r) + \sum_{n=0}^{\infty} J_n(\lambda\sigma)\{a_n \cos n\beta + b_n \sin n\beta\}. \end{aligned}$$

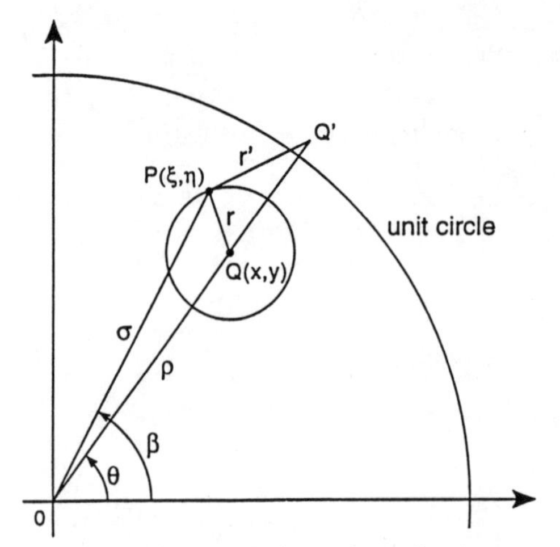

Figure 8.4: A unit circle

8.3.5 To obtain Green's function by the method of images

The Green's function can be obtained by using the method of images. This method is based essentially on the construction of Green's function for a finite domain. This method is restricted in the sense that it can be applied only to certain class of problems with simple boundary geometry. Let us consider the Dirichlet problem to illustrate this method.

Let $P(\xi, \eta)$ be the field point in the circle R, and let $Q(x, y)$ be the source point also in R. The distance between P and Q is $r = \sqrt{\sigma^2 + \rho^2 - 2\rho\sigma\cos(\beta - \theta)}$. Let Q' be the image which lies outside of R on the same ray from the origin opposite to the source point Q as shown in Fig. 8.4 such that

$$(OQ)(OQ') = \sigma^2$$

where σ is the radius of the circle through the point P centered at the origin.

Since the two triangles OPQ and OPQ' are similar by virtue of the hypothesis $(OQ)(OQ') = \sigma^2$ and by possessing the common angle at O, we have

$$\frac{r}{r'} = \frac{\sigma}{\rho} \tag{8.84}$$

where $r' = PQ'$ and $\rho = OQ$. If $\sigma = 1$, then (8.84) becomes

$$\frac{r}{r'}\frac{1}{\rho} = 1.$$

Then taking the logarithm of both sides with a multiple of $\frac{1}{2\pi}$, we obtain

$$\frac{1}{2\pi}\ln\left(\frac{r}{r'\rho}\right) = 0$$

$$\text{or} \quad \frac{1}{2\pi}\ln r - \frac{1}{2\pi}\ln r' + \frac{1}{2\pi}\ln\frac{1}{\rho} = 0. \tag{8.85}$$

This equation signifies that $\frac{1}{2\pi}\ln(\frac{r'}{r\rho})$ is harmonic in R except at Q and satisfies the Laplace's equation

$$\nabla^2 G = \delta(\xi - x, \eta - y). \tag{8.86}$$

Note that $\ln r'$ is harmonic everywhere except at Q', which is outside the domain R. This suggests that we can choose the Green's function as

$$G = \frac{1}{2\pi}\ln r - \frac{1}{2\pi}\ln r' + \frac{1}{2\pi}\ln\frac{1}{\rho}. \tag{8.87}$$

Given that Q' is at $(\frac{1}{\rho}, \theta)$, G in polar coordinates takes the form

$$\begin{aligned} G(\rho, \theta; \sigma, \beta) &= \frac{1}{4\pi}\ln(\sigma^2 + \rho^2 - 2\rho\sigma\cos(\beta - \theta)) \\ &\quad - \frac{1}{4\pi}\ln(\frac{1}{\sigma^2} + \rho^2 - \frac{2\rho}{\sigma}\cos(\beta - \theta)) + \frac{1}{2\pi}\ln\frac{1}{\sigma} \end{aligned} \tag{8.88}$$

which is the same as before.

Remark

Note that in the Green's function expression in (8.87) or (8.88), the first term represents the potential due to a unit line charge at the source point, whereas the second term represents the potential due to negative unit charge at the image point and the third term represents a uniform potential. The sum of these potentials makes up the potential field.

Example 7

Find the solution of the following boundary value problem by the method of images:

$$\begin{aligned} \nabla^2 u &= h \qquad \text{in } \eta > 0 \\ u &= f \qquad \text{on } \eta = 0. \end{aligned}$$

Solution

The image point should be obvious by inspection. Thus if we construct

$$G = \frac{1}{4\pi}\ln[(\xi - x)^2 + (\eta - y)^2] - \frac{1}{4\pi}\ln[(\xi - x)^2 + (\eta + y)^2] \tag{8.89}$$

the condition that $G = 0$ on $\eta = 0$ is clearly satisfied. Also G satisfies

$$\nabla^2 G = \delta(\xi - x, \eta - y)$$

and with $\frac{\partial G}{\partial n}|_C = [-\frac{\partial G}{\partial \eta}]_{\eta=0}$, the solution is given by

$$\begin{aligned} u(x, y) &= \int\int_R G(x, y; \xi, \eta)h(\xi, \eta)d\xi d\eta + \int_C f\frac{\partial G}{\partial n}ds \\ &= \frac{y}{\pi}\int_{-\infty}^{\infty}\frac{f(\xi)d\xi}{(\xi - x)^2 + y^2} + \frac{1}{4\pi}\int_0^{\infty}\int_{-\infty}^{\infty}\ln\{\frac{(\xi - x)^2 + (\eta - y)^2}{(\xi - x)^2 + (\eta + y)^2}\}h(\xi, \eta)d\xi d\eta. \end{aligned}$$

8.3.6 Method of eigenfunctions

Green's function can also be obtained by applying the method of eigenfunctions. We consider the boundary value problem

$$\begin{aligned} \nabla^2 u &= h \quad \text{in } R \\ u &= f \quad \text{on } C. \end{aligned} \tag{8.90}$$

The Green's function must satisfy

$$\begin{aligned} \nabla^2 G &= \delta(\xi - x, \eta - y) \quad \text{in } R \\ G &= 0 \quad \text{on } C \end{aligned} \tag{8.91}$$

and hence the associated eigenvalue problem is

$$\begin{aligned} \nabla^2 \phi + \lambda \phi &= 0 \quad \text{in } R \\ \phi &= 0 \quad \text{on } C. \end{aligned} \tag{8.92}$$

Let ϕ_{mn} be the eigenfunctions and λ_{mn} be the corresponding eigenvalues. We then expand G and δ in terms of the eigenfunctions ϕ_{mn}

$$G(\xi, \eta; x, y) = \sum_m \sum_n a_{mn}(x, y)\phi(\xi, \eta) \tag{8.93}$$

$$\delta(\xi, \eta; x, y) = \sum_m \sum_n b_{mn}(x, y)\phi_{mn}(\xi, \eta) \tag{8.94}$$

where

$$\begin{aligned} b_{mn} &= \frac{1}{||\phi_{mn}||^2} \int\int_R \delta(\xi, \eta; x, y)\phi_{mn}(\xi, \eta) d\xi d\eta \\ &= \frac{\phi_{mn}(x, y)}{||\phi_{mn}||^2} \end{aligned} \tag{8.95}$$

in which $||\phi_{mn}||^2 = \int\int_R \phi_{mn}^2 d\xi d\eta$. Now substituting (8.93) and (8.94) into (8.91) and using the relation from (8.92),

$$\nabla^2 \phi_{mn} + \lambda_{mn}\phi_{mn} = 0$$

we obtain

$$-\sum_m \sum_n \lambda_{mn} a_{mn}(x, y)\phi_{mn}(\xi, \eta) = \sum_m \sum_n \frac{\phi_{mn}(x, y)\phi_{mn}(\xi, \eta)}{||\phi_{mn}||^2}.$$

Therefore, we have

$$a_{mn}(x, y) = \frac{-\phi_{mn}(x, y)}{\lambda_{mn}||\phi_{mn}||^2}. \tag{8.96}$$

Thus the Green's function is given by

$$G(\xi, \eta; x, y) = -\sum_m \sum_n \frac{\phi_{mn}(x, y)\phi_{mn}(\xi, \eta)}{\lambda_{mn}||\phi_{mn}||^2}. \tag{8.97}$$

We shall demonstrate this method by the following example.

Example 8

Find the Green's function for the following boundary value problem

$$\begin{aligned} \nabla^2 u &= h \quad \text{in } R \\ u &= 0 \quad \text{on } C. \end{aligned}$$

Solution

The eigenfunction can be obtained explicitly by the method of separation of variables.
We assume a solution in the form

$$\phi(\xi, \eta) = X(\xi)Y(\eta).$$

Substitution of this into $\nabla^2\phi + \lambda\phi = 0$ in R and $\phi = 0$ on C, yields

$$\begin{aligned} X'' + \alpha^2 X &= 0 \\ Y'' + (\lambda - \alpha^2)Y &= 0 \end{aligned}$$

where α^2 is a separation constant. With the homogeneous boundary conditions $X(0) = X(a) = 0$ and $Y(0) = Y(b) = 0, X$ and Y are found to be

$$X_m(\xi) = A_m \sin\frac{m\pi\xi}{a} \quad \text{and} \quad Y_n(\eta) = B_n \sin\frac{n\pi\eta}{b}.$$

We then have

$$\lambda_{mn} = \pi^2\left(\frac{m^2}{a^2} + \frac{n^2}{b^2}\right) \quad \text{with} \quad \alpha = \frac{m\pi}{a}.$$

Thus the eigenfunctions are given by

$$\phi_{mn}(\xi, \eta) = \sin\frac{m\pi\xi}{a}\sin\frac{n\pi\eta}{b}.$$

Hence

$$\begin{aligned} ||\phi_{mn}||^2 &= \int_0^a\int_0^b \sin^2\frac{m\pi\xi}{a}\sin^2\frac{n\pi\eta}{b}d\xi d\eta \\ &= \frac{ab}{4} \end{aligned}$$

so that the Green's function can be obtained as

$$G(\xi, \eta; x, y) = -\frac{4ab}{\pi^2}\sum_{m=1}^{\infty}\sum_{n=1}^{\infty}\frac{\sin\frac{m\pi x}{a}\sin\frac{n\pi y}{b}\sin\frac{m\pi\xi}{a}\sin\frac{n\pi\eta}{b}}{(m^2b^2 + a^2n^2)}.$$

8.4 Development of Green's function in 3D

Since most of the problems encountered in the physical sciences are in three dimension, we can extend the Green's function to three or more dimensions. Let us extend our definition of Green's function in three dimensions.

Consider the Dirichlet problem involving Laplace operator. The Green's function satisfies the following:

(I)

$$\left.\begin{aligned} \nabla^2 G &= \delta(x-\xi, y-\eta, z-\zeta) \quad \text{in } R \\ G &= 0 \quad \text{on } S. \end{aligned}\right\} \tag{8.98}$$

(II)

$$G(x,y,z;\xi,\eta,\zeta) = G(\xi,\eta,\zeta;x,y,z) \tag{8.99}$$

(III)

$$\lim_{\varepsilon\to 0}\iint_{S_\varepsilon} \frac{\partial g}{\partial n} ds = 1 \tag{8.100}$$

where n is the outward normal to the surface
$S_\varepsilon : (x-\xi)^2 + (y-\eta)^2 + (z-\zeta)^2 = \varepsilon^2$.

Proceeding as in the two-dimensional case, the solution of Dirichlet problem

$$\left.\begin{aligned} \nabla^2 \phi &= h \quad \text{in } R \\ \phi &= f \quad \text{on } S \end{aligned}\right\} \tag{8.101}$$

is
$$u(x,y,z) = \iiint_R GhdR + \iint_S fG_n dS. \tag{8.102}$$

To obtain the Green's function, we let the Green's function have two parts as follows

$$G(\xi,\eta,\zeta;x,y,z) = g_h(\xi,\eta,\zeta;x,y,z) + g_p(\xi,\eta,\zeta;x,y,z)$$

$$\begin{aligned} \text{where} \quad \nabla^2 g_h &= \delta(x-\xi, y-\eta, z-\zeta) \quad \text{in } R \\ \text{and} \quad \nabla^2 g_p &= 0 \quad \text{on } S \\ G &= 0 \quad \text{i.e.,} \quad g_p = -g_h \quad \text{on } S. \end{aligned}$$

Example 9

Obtain the Green's function for the Laplace's equation in the spherical domain.

Solution

Within the spherical domain with radius a, we consider

$$\nabla^2 G = \delta(\xi - x, \eta - y, \zeta - z)$$

which means

$$\nabla^2 g_h = \delta(\xi - x, \eta - y, \zeta - z)$$

and

$$\nabla^2 g_p = 0.$$

For

$$r = [(\xi - x)^2 + (\eta - y)^2 + (\zeta - z)^2]^{\frac{1}{2}} > 0$$

with (x, y, z) as the origin, we have

$$\nabla^2 g_h = \frac{1}{r^2}\frac{d}{dr}(r^2 \frac{dg_h}{dr}) = 0.$$

Integration then yields

$$g_h = A + \frac{B}{r} \quad \text{for } r > 0.$$

Applying the condition (III), we obtain

$$\lim_{\varepsilon \to 0} \int\int_{S_\varepsilon} G_n dS = \lim_{\varepsilon \to 0} \int\int_{S_\varepsilon} (g_h)_r dS = 1$$

from which we obtain $B = -\frac{1}{4\pi}$ and A is arbitrary. If we set $A = 0$ for convenience, (this is the boundedness condition at infinity), we have

$$g_h = -\frac{1}{4\pi r}.$$

To obtain the complete Green's functions, we need to find the solution for g_p. If we draw a three-dimensional diagram analogous to the two-dimensional as depicted in the last section, we will have a similar relation

$$r' = \frac{ar}{\rho} \tag{8.103}$$

where r' and ρ are measured in three-dimensional space.

Thus we have

$$g_p = \frac{(\frac{a}{\rho})}{4\pi r'}$$

and hence

$$G = \frac{-1}{4\pi r} + \frac{\frac{a}{\rho}}{4\pi r'} \tag{8.104}$$

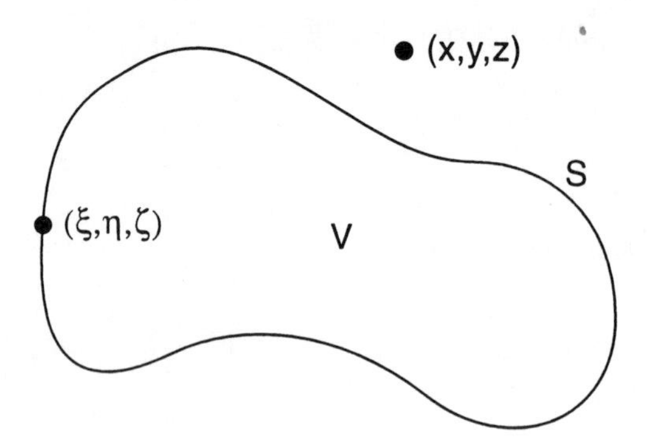

Figure 8.5: An arbitrary body surface

which is harmonic everywhere in r except at the source point, and is zero on the surface S. In terms of spherical coordinates:

$$\xi = \tau \cos\psi \sin\alpha \qquad x = \rho\cos\phi\sin\theta$$

$$\eta = \tau \sin\psi \sin\alpha \qquad y = \rho\sin\phi\sin\theta$$

$$\zeta = \tau\cos\alpha \qquad z = \rho\cos\theta.$$

G can be written in the form

$$G = \frac{-1}{4\pi(\tau^2+\rho^2-2\rho\tau\cos\gamma)^{\frac{1}{2}}} + \frac{1}{4\pi(\frac{\tau^2\rho^2}{a^2}+a^2-2\tau\rho\cos\gamma)^{\frac{1}{2}}} \tag{8.105}$$

where γ is the angle between r and r'.
Now differentiating G, we have

$$[\frac{\partial G}{\partial\tau}]_{\tau=a} = \frac{a^2+\rho^2}{4\pi a(a^2+\rho^2-2a\rho\cos\gamma)^{\frac{1}{2}}}.$$

Thus the Dirichlet problem for $h=0$ is

$$u(x,y,z) = \frac{\alpha(a^2-\rho^2)}{4\pi}\int_{\psi=0}^{2\pi}\int_{\alpha=0}^{\pi}\frac{f(\alpha,\psi)\sin\alpha d\alpha d\psi}{(a^2+\rho^2-2a\rho\cos\gamma)^{\frac{1}{2}}} \tag{8.106}$$

where $\cos\gamma = \cos\alpha\cos\theta+\sin\alpha\sin\theta\cos(\psi-\phi)$. This integral is called the three dimensional Poisson integral formula.

8.4.1 Development of Green's function in 3D for physical problems

It is known concerning the distribution of singularities in a flow field, that it is possible for certain bodies to be represented by combinations of sources and sinks and doublets. The combination of these singularities on the surrounding fluid may be used to represent complex body structures. We may think of a particular body as being composed of a continuous distribution of singularities: such singularities on the body surface will enable us to evaluate the velocity potential ϕ on that surface.

To deduce the appropriate formula, we consider two solutions of Laplace's equation in a volume V of fluid bounded by a closed surface S. Consider two potentials ϕ and ψ such that ϕ and ψ satisfy Laplace's equation in the following manner:

$$\nabla^2\phi = 0 \tag{8.107}$$

and

$$\nabla^2\psi = \delta(\vec{r} - \vec{r_0}) = \delta(x - \xi, y - \eta, z - \zeta), \tag{8.108}$$

where $\phi(x, y, z)$ is a regular potential of the problem and $\psi(x, y, z; \xi, \eta, \zeta)$ is the source potential which has a singularity at $\vec{r} = \vec{r_0}$. Here δ is a Dirac delta function defined as

$$\delta(\vec{r} - \vec{r_0}) = \begin{cases} 0 & \text{when } \vec{r} \neq \vec{r_0} \\ \infty & \text{when } \vec{r} = \vec{r_0}. \end{cases}$$

The source point $\vec{r_0} \equiv (\xi, \eta, \zeta)$ is situated at the body surface or inside the body. The point $\vec{r} = (x, y, z)$ can be regarded as the field point. There are three cases to investigate.

Case (I)

The field point (x, y, z) lies outside the body of surface area S and volume V (see Fig. 8.5).
By applying Green's theorem,

$$\begin{aligned} \int\int_S [\phi\frac{\partial\psi}{\partial n} - \psi\frac{\partial\phi}{\partial n}]dS &= \int\int\int_V \vec{\nabla}(\phi\vec{\nabla}\psi) - \psi\vec{\nabla}\phi)dV \\ &= \int\int\int_V [\phi\nabla^2\psi + (\vec{\nabla}\phi)(\vec{\nabla}\psi) - \psi\nabla^2\phi - (\vec{\nabla}\psi)(\vec{\nabla}\phi)]dV \\ &= 0. \end{aligned} \tag{8.109}$$

Case (II)

The field point (x,y,z) lies outside the body of surface area S and volume V (see Fig. 8.5).

$$\begin{aligned} \int\int_S [\phi\frac{\partial\psi}{\partial n} - \psi\frac{\partial\eta}{\partial n}]dS &= \int\int\int_V [\phi\nabla^2\psi - \psi\nabla^2\phi]dV \\ &= \int\int\int_V (\phi\nabla^2\psi)dV \\ &= \int\int\int_V (\phi(x, y, z)\delta(x - \xi, y - \eta, z - \zeta))dV \\ &= \phi(\xi, \eta, \zeta). \end{aligned} \tag{8.110}$$

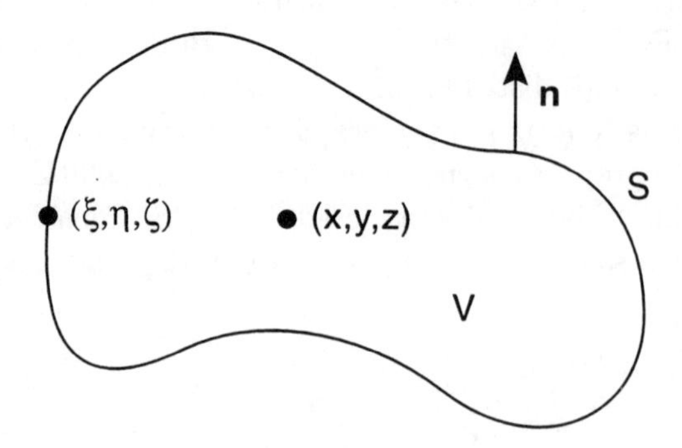

Figure 8.6: Surface of integration for Green's theorem

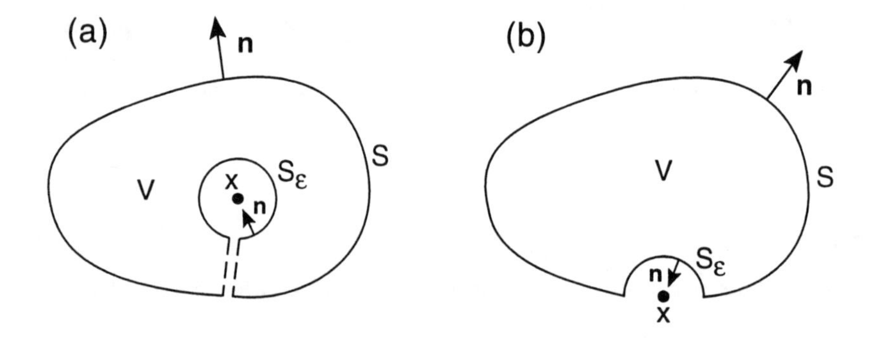

Figure 8.7: Surfaces of integration for Green's theorem

Now, changing the roles of (x, y, z) and (ξ, η, ζ), we obtain from eqn (8.100)

$$\begin{aligned}\phi(x,y,z) &= \int\int_S [\phi(\xi,\eta,\zeta)\frac{\partial\psi}{\partial n}(x,y,z;\xi,\eta,\zeta) \\ &\quad -\psi(x,y,z;\xi,\eta,\zeta)\frac{\partial\phi}{\partial n}(\xi,\eta,\zeta)]ds. \qquad (8.111)\end{aligned}$$

Referring to Fig. 8.7 note that in (a) the point is interior to S, surrounded by a small spherical surface S_ε; (b) the field point is on the boundary surface S and S_ε is a hemisphere.

Case (III)

The field point (x, y, z) lies on the body surface S within the volume V.

Referring to the work of Newman and Mei (1978) on this subject, we may write:

$$\int\int_S [\phi\frac{\partial\psi}{\partial n} - \psi\frac{\partial\phi}{\partial n}]dS = \int\int\int_V \phi(x,y,z)\delta(x-\xi, y-\eta, z-\zeta)dV = \frac{1}{2}\phi(\xi,\eta,\zeta).$$

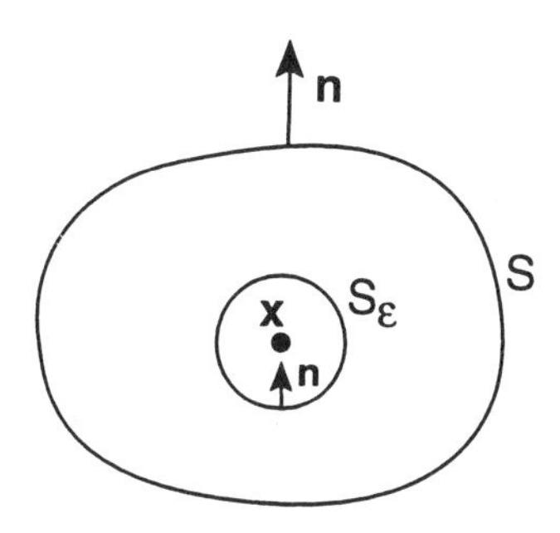

Figure 8.8: Surface of integration

Changing the roles of (x, y, z) and (ξ, η, ζ), we obtain

$$\begin{aligned}\phi(x,y,z) &= 2\iint_S [\phi(\xi,\eta,\zeta)\frac{\partial\psi}{\partial n}(x,y,z;\xi,\beta,\zeta) \\ &\quad - \psi(x,y,z;\xi,\eta,\zeta)\frac{\partial\phi}{\partial n}(\xi,\eta,\zeta)dS. \end{aligned} \tag{8.112}$$

Alternative method of deducing (II) and (III)

Consider the source potential

$$\psi = \frac{1}{4\pi r} = \frac{1}{4\pi}[(x-\xi)^2 + (y-\eta)^2 + (z-\zeta)^2]^{-\frac{1}{2}}$$

where the field point (x, y, z) is well inside the body S. Then with reference to Fig. 8.7, Green's divergence theorem yields

$$\iint_{S+S_\varepsilon} (\phi\frac{\partial\psi}{\partial n} - \psi\frac{\partial\phi}{\partial n})dS = 0,$$

where S_ε is a sphere of radius r. It follows that

$$\iint_S (\phi\frac{\partial\psi}{\partial n} - \psi\frac{\partial\phi}{\partial n})dS = -\int_{S_\varepsilon} (\phi\frac{\partial\psi}{\partial n} - \psi\frac{\partial\phi}{\partial n})dS$$

$$\frac{1}{4\pi}\iint_S (\phi\frac{\partial}{\partial n}\frac{1}{r} - \frac{1}{r}\frac{\partial\phi}{\partial n})dS = -\frac{1}{4\pi}\iint_{S_\varepsilon} [\phi\frac{\partial}{\partial n}\frac{1}{r} - \frac{1}{r}\frac{\partial\phi}{\partial n}]dS.$$

Now using the concept of an outward normal, n, from the fluid,

$$\frac{\partial}{\partial n}\frac{1}{r} = -\frac{\partial}{\partial r}\frac{1}{r} = \frac{1}{r^2}.$$

The right hand side integral becomes

$$\begin{aligned}
\frac{1}{4\pi}\iint_{S_\varepsilon}\{\phi\frac{\partial}{\partial n}(\frac{1}{r})-\frac{1}{r}\frac{\partial\phi}{\partial n}\}dS &= \frac{1}{4\pi}\iint_{S_\varepsilon}\frac{\phi}{r^2}dS-\frac{1}{4\pi}\iint_{S_\varepsilon}\frac{1}{r}(\frac{\partial\phi}{\partial n})dS \\
&= \frac{1}{4\pi}.\frac{\phi(x,y,z)}{r^2}.\iint_{S_\varepsilon}dS-\frac{1}{4\pi}.\frac{1}{r}(\frac{\partial\phi}{\partial n})_{S_\varepsilon}\iint_{S_\varepsilon}dS \\
&= \frac{1}{4\pi}\frac{\phi(x,y,z)}{r^2}(4\pi r^2)-\frac{1}{4\pi}\frac{1}{r}(\frac{\partial\phi}{\partial n})_{S_\varepsilon}(4\pi r^2) \\
&= \phi(x,y,z)-(\frac{\partial\phi}{\partial n})_{S_\varepsilon}r.
\end{aligned}$$

Thus, when $r\to 0$, $\quad(\frac{\partial\phi}{\partial n})_{S_\varepsilon}r\to 0$.

Combining these results, we have

$$\phi(x,y,z)=-\frac{1}{4\pi}\iint_S[\phi\frac{\partial}{\partial n}\frac{1}{r}-\frac{1}{r}\frac{\partial\phi}{\partial n}\phi]dS.$$

This is valid when the field point (x,y,z) is inside S. The velocity potential $\phi(x,y,z)$ at a point inside a boundary surface S, can be written as

$$\phi(x,y,z)=\frac{1}{4\pi}\iint_S\frac{1}{r}(\frac{\partial\phi}{\partial n})dS-\frac{1}{4\pi}\iint_S\phi\frac{\partial}{\partial n}(\frac{1}{r})dS.$$

We know that the velocity potential due to a distribution of sources of strength m over a surface S is

$$\iint_S\frac{m}{r}dS,$$

and of a distribution of doublets of moments μ, the axes of which point inward along the normals to S, is

$$\iint_S\mu\frac{\partial}{\partial n}(\frac{1}{r})dS.$$

Thus, the velocity potential ϕ at a point (x,y,z) as given above, is the same as if the motion in the region bounded by the surface S due to a distribution over S of simple sources, of density $(\frac{1}{4\pi})(\frac{\partial\phi}{\partial n})$ per unit area, together with a distribution of doublets, with axes pointing inwards along the normals to the surfaces, of density $(\frac{\phi}{4\pi})$ per unit area. When the field point (x,y,z) is on S, as shown in Fig. 8.8, the surface S_ε is a hemisphere of surface area $2\pi r^2$.

When $\varepsilon\to 0$, $\quad\iint_{S_\varepsilon}\frac{\phi}{r^2}dS=2\pi\phi(x,y,z)$, then

$$\phi(x,y,z)=-\frac{1}{2\pi}\iint[\phi\frac{\partial}{\partial n}\frac{1}{r}-\frac{1}{r}\frac{\partial}{\partial n}\phi]ds.$$

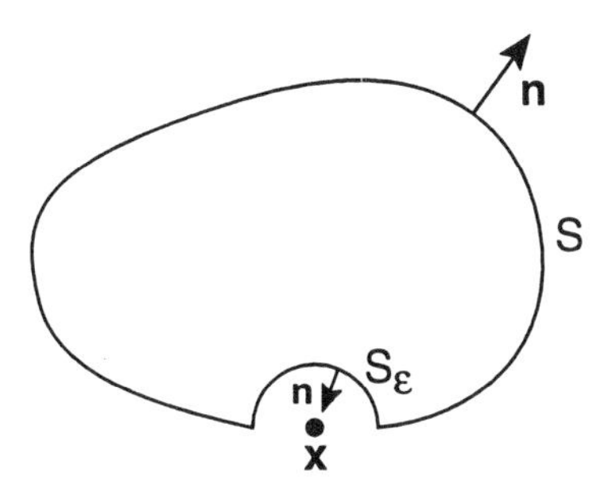

Figure 8.9: Surface of integration

Summary

Thus, summarizing all these results, we obtain

(I) When the point (x, y, z) lies outside S, then

$$\int\int_S (\phi \frac{\partial}{\partial n}\frac{1}{r} - \frac{1}{r}\frac{\partial \phi}{\partial n}) dS = 0.$$

(II) When the point (x, y, z) lies inside S, then

$$\phi(x, y, z) = -\frac{1}{4\pi}\int\int_S (\phi \frac{\partial}{\partial n}\frac{1}{r} - \frac{1}{r}\frac{\partial \phi}{\partial n}) dS.$$

(III) When the point (x, y, z) lies on the boundary S, then

$$\phi(x, y, z) = -\frac{1}{2\pi}\int\int_S (\phi \frac{\partial}{\partial n}\frac{1}{r} - \frac{1}{r}\frac{\partial \phi}{\partial n}) dS,$$

where ϕ is known as the velocity potential of the problem.

Note that the last equation is frequently used for obtaining the velocity potential due to the motion of a ship's hull. The normal velocity, $\frac{\partial \phi}{\partial n}$, is known on the body, so that the last equation is an integral equation for determining the unknown potential; this may be done by numerical integration.

In many practical problems, however, the body may move in a fluid bounded by other boundaries, such as the free surface, the fluid bottom, or possibly lateral boundaries such as canal walls.

In this context, we use Green's function

$$G(x, y, z; \xi, \eta, \zeta) = \frac{1}{r} + H(x, y, z; \xi, \eta, \zeta), \tag{8.113}$$

where,

$$\nabla^2 H = \delta(x - \xi, y - \eta, z - \zeta). \tag{8.114}$$

Green's function, defined above, can be stated as

$$\int\int_S (\phi\frac{\partial G}{\partial n} - G\frac{\partial \phi}{\partial n})dS = \begin{cases} 0 \\ -2\pi\phi(x,y,z) \\ -4\pi\phi(x,y,z) \end{cases} \tag{8.115}$$

for (x, y, z) outside, on or inside the closed surface S, respectively.

If a function H can be found with the property that $\frac{\partial\phi}{\partial n} = 0$ on the boundary surfaces of the fluid, then eqn (8.106) may be rewritten as

$$\int\int_S G(x,y,z;\xi,\eta,\zeta)\frac{\partial \phi}{\partial n}(\xi,\eta,\zeta)dS = \begin{cases} 0 \\ 2\pi\phi(x,y,z) \\ 4\pi\phi(x,y,z) \end{cases} \tag{8.116}$$

for (x, y, z) outside, on or inside the closed surface S, respectively. Here $\frac{\partial\phi}{\partial n}(\xi, \eta, \zeta) = Q(\xi, \eta, \zeta)$ is defined to be the unknown source density (strength) and has to be e-valuated by numerical methods from the above integral. Once the source density is known, the field potential $\phi(x, y, z)$ can be easily obtained.

8.4.2 Application: hydrodynamic pressure forces

One of the main purposes for studying the fluid motion past a body is to predict the wave forces and moments acting on the body due to the hydrodynamic pressure of the fluid.

The wave forces and moments can be calculated using the following formulae:

$$\vec{F} = \int\int_{S_B} (P\vec{n}dS) \tag{8.117}$$

$$\vec{M} = \int\int_{S_B} P(\vec{r}X\vec{n})dS, \tag{8.118}$$

where S_B is the body of the structure and $\vec{n}$ is the unit normal vector, which is positive when pointing out of the fluid volume.

From Bernoulli's equation, we know that

$$P = -\rho[\frac{\partial\phi}{\partial t} + \frac{1}{2}(\nabla\phi)^2 + gz], \tag{8.119}$$

where $\frac{\partial\phi}{\partial t}$ is the transient pressure, $\frac{1}{2}(\nabla\phi)^2$ is the dynamic pressure and gz is the static pressure. Then using eqn (8.110) with eqns (8.108) and (8.109), we obtain

$$\vec{F} = -\rho\int\int_{S_B}[\frac{\partial\phi}{\partial t} + \frac{1}{2}(\nabla\phi)^2 + gz]\vec{n}dS \tag{8.120}$$

$$\vec{M} = -\rho\int\int_{S_B}[\frac{\partial\phi}{\partial t} + \frac{1}{2}(\nabla\phi)^2 + gz](\vec{r}X\vec{n})dS. \tag{8.121}$$

In the following section, we shall deduce Green's function which is needed in the evaluation of the velocity potential from the integral equation (8.107). The solution is mainly due to Wehausen and Laitone (1960).

8.4.3 Derivation of Green's function

We will obtain Green's function solution (singular solution) for the cases of infinite depth and of finite depth. Mathematical solutions will be first obtained for the infinite depth case, and then extended to the case of finite depth.

Case I: Infinite depth

Consider Green's function of the form

$$G(x,y,z;\xi,\eta,\zeta,t) = Re\{\vec{g}(x,y,z;\xi,\eta,\zeta)e^{-i\sigma t}\}, \tag{8.122}$$

where Re stands for the real part of the complex function, and $\vec{g}$ is a complex function which can be defined as

$$\vec{g}(x,y,z;\xi,\eta,\zeta) = g_1(x,y,z;\xi,\eta,\zeta) + ig_2(x,y,z;\xi,\eta,\zeta). \tag{8.123}$$

Here g_1 and g_2 are real functions. Thus

$$G(x,y,z;\xi,\eta,\zeta,t) = g_1\cos\sigma t + g_2\sin\sigma t. \tag{8.124}$$

G satisfies Laplace's equation except at the point (ξ,η,ζ), where

$$\nabla^2 G = \frac{\partial^2 G}{\partial x^2} + \frac{\partial^2 G}{\partial y^2} + \frac{\partial^2 G}{\partial z^2} = \delta(x,y,z;\xi,\eta,\zeta). \tag{8.125}$$

From the above, the equations and conditions to be satisfied by g_1 and g_2 are as follows:

$$\nabla^2 g_1 = \delta(x,y,z;\xi,\eta,\zeta), \quad \nabla^2 g_2 = \delta(x,y,z;\xi,\eta,\zeta). \tag{8.126}$$

The linear surface boundary condition given by

$$\frac{\partial^2 G}{\partial t^2} + g\frac{\partial G}{\partial z} = 0, \qquad \text{for } z \leq 0, \tag{8.127}$$

yields to

$$\frac{\partial g_1}{\partial z} - \frac{\sigma^2}{g}g_1 = 0, \qquad \frac{\partial g_2}{\partial z} - \frac{\sigma^2}{g}g_2 = 0 \quad \text{at} \quad z = 0. \tag{8.128}$$

The bottom boundary conditions for the infinite depth case are

$$\lim_{z\to-\infty}\vec{\nabla}g_1 = 0, \qquad \lim_{z\to-\infty}\vec{\nabla}g_2 = 0. \tag{8.129}$$

The radiation condition can be stated as

$$\lim_{R\to\infty}\sqrt{R}(\frac{\partial\vec{g}}{\partial R} - ik\vec{g}) = 0, \tag{8.130}$$

which yields

$$\lim_{R\to\infty}\sqrt{R}(\frac{\partial g_1}{\partial R} + kg_2) = 0, \quad \lim_{R\to\infty}\sqrt{R}(\frac{\partial g_2}{\partial R} - kg_1) = 0. \tag{8.131}$$

Here k is the wavenumber and R is the radial distance in the xy-plane and is given by $R^2 = (x-\xi)^2 + (y-\eta)^2$.

We shall now assume a solution of G, as given in eqn (8.115) to be in the following form:

$$G = (\frac{1}{r} + g_0)\cos\sigma t + g_2 \sin\sigma t$$
$$= \frac{1}{r}\cos\sigma t + g_0\cos\sigma t + g_2\sin\sigma t, \quad (8.132)$$

where $\quad r^2 = (x-\xi)^2 + (y-\eta)^2 + (z-\zeta)^2$, and $g_1 = \frac{1}{r} + g - 0$.

Denoting the double Fourier transform in x and y of $\vec{g}$ by g^* as

$$\vec{g}(x,y,z;\xi,\eta,\zeta) = \frac{1}{2\pi}\int_0^\infty \int_{-\pi}^{\pi} g^*(k,\theta,z;\xi,\eta,\zeta) e^{ik(x\cos\theta + y\sin\theta)} d\theta dk, \quad (8.133)$$

we then can write

$$g_0(x,y,z;\xi,\eta,\zeta) = \frac{1}{2\pi}\int_0^\infty \int_{-\pi}^{\pi} g_0^*(k,\theta,z;\xi,\eta,\zeta) e^{ik(x\cos\theta + y\sin\theta)} d\theta dk, \quad (8.134)$$

and

$$g_2(x,y,z;\xi,\eta,\zeta) = \frac{1}{2\pi}\int_0^\infty \int_{-\pi}^{\pi} g_2^*(k,\theta,z;\xi,\eta,\zeta) e^{ik(x\cos\theta + y\sin\theta)} d\theta dk. \quad (8.135)$$

Note that functions g_0 and g_2 happen to be regular functions of x, y, z and satisfy Laplace's equation.

Applying the double Fourier transform in x and y of g_0 yields

$$\frac{1}{2\pi}\int_0^\infty \int_{-\pi}^{\pi} (\frac{\partial^2 g_0^*}{\partial z^2} - k^2 g_0^*) e^{ik(x\cos\theta + y\sin\theta)} d\theta dk = 0,$$

and consequently, we obtain

$$\frac{\partial^2 g_0^*}{\partial z^2} - k^2 g_0^* = 0. \quad (8.136)$$

Solving eqn (8.127) we get

$$g_0^* = A(k,\theta)e^{kz} + B(k,\theta)e^{-kz}. \quad (8.137)$$

Since g_0^* must be bounded as $z \to -\infty$, then $B = 0$, and the solution becomes

$$g_0^* = A(k,\theta)e^{kz}. \quad (8.138)$$

We know that

$$\frac{1}{\sqrt{x^2+y^2+z^2}} = \frac{1}{2\pi}\int_0^\infty \int_{-\pi}^{\pi} e^{-k|z|} e^{ik(x\cos\theta + y\sin\theta)} d\theta dk. \quad (8.139)$$

Extending this result we obtain

$$\begin{aligned}
\frac{1}{r} &= \frac{1}{\sqrt{(x-\xi)^2+(y-\eta)^2+(z-\zeta)^2}} \\
&= \frac{1}{2\pi}\int_0^\infty\int_{-\pi}^{\pi} e^{-k|z-\zeta|}e^{ik((x-\xi)\cos\theta+(y-\eta)\sin\theta)}d\theta dk \\
&= \frac{1}{2\pi}\int_0^\infty\int_{-\pi}^{\pi} e^{-k|z-\zeta|}e^{-ik(\xi\cos\phi+\eta\sin\theta)} \\
&\quad \times e^{ik(x\cos\theta+y\sin\theta)}d\theta dk \\
&= \frac{1}{2\pi}\int_0^\infty\int_{-\pi}^{\pi}(\frac{1}{r})^* e^{ik(x\cos\theta+y\sin\theta)}d\theta dk,
\end{aligned} \tag{8.140}$$

where

$$(\frac{1}{r})^* = e^{-k|z-\zeta|}e^{-ik(\xi\cos\theta+\eta\sin\theta)}. \tag{8.141}$$

Taking the double Fourier transform of the surface boundary condition (8.119) yields

$$\{\frac{\partial g_0^*}{\partial z} - k(\frac{1}{r})^*\} - \frac{\sigma^2}{g}(g_0^* + (\frac{1}{r})^*) = 0 \qquad \text{at} \quad z = 0. \tag{8.142}$$

Rearranging the terms yields

$$\frac{\partial g_0^*}{\partial z} - \frac{\sigma^2}{g}g_0^* = (k + \frac{\sigma^2}{g})(\frac{1}{r})^* \qquad \text{at} \quad z = 0. \tag{8.143}$$

From eqn (8.134) and the boundary condition (8.129), we have

$$A(k,\theta) = \frac{k+\sigma^2/g}{k-\sigma^2/g}e^{k\zeta}e^{-ik(\xi\cos\theta+\eta\sin\theta)}. \tag{8.144}$$

Therefore

$$g_0^* = \frac{k+\sigma^2/g}{k-\sigma^2/g}e^{k(z+\zeta)}e^{-ik(\xi\cos\theta+\eta\sin\theta)}. \tag{8.145}$$

Inverting this function gives

$$g_0 = \frac{1}{2\pi}\int_0^\infty\int_{-\pi}^{\pi}\frac{k+\sigma^2/g}{k-\sigma^2/g}e^{k(z+\zeta)}e^{ik((x-\xi)\cos\theta+(y-\eta)\sin\theta)}d\theta dk. \tag{8.146}$$

If we define $\frac{\sigma^2}{g} = \upsilon$, then the potential g_1 may be written as

$$\begin{aligned}
g_1(x,y,z) &= \frac{1}{r} + g_0(x,y,z) \\
&= \frac{1}{r} + \frac{1}{2\pi}PV\int_0^\infty\int_{-\pi}^{\pi}\frac{k+\upsilon}{k-\upsilon}e^{k(z+\zeta)} \\
&\quad \times e^{ik((x-\xi)\cos\theta+(y-\eta)\sin\theta)}d\theta dk.
\end{aligned} \tag{8.147}$$

This may be written as

$$g_1(x,y,z) = \frac{1}{r} + \frac{1}{r_1} + \frac{v}{\pi} PV \int_0^\infty \int_{-\pi}^{\pi} \frac{e^{k(z+\zeta)}}{k-v} e^{ik((x-\xi)\cos\theta + (y-\eta)\sin\theta)} d\theta dk. \quad (8.148)$$

where

$$r_1 = \sqrt{(x-\xi)^2 + (y-\eta)^2 + (z+\zeta)^2}.$$

Here PV stands for the Cauchy-Principal value. Note that g_1 satisfies all the given boundary conditions except the radiation condition. Now to satisfy the radiation condition, we need the asymptotic expansion of g_1. The solution (8.148) may be written in the form

$$g_1(x,y,z) = \frac{1}{r} + \frac{1}{r_1} + \frac{4v}{\pi} PV \int_0^\infty \int_0^{\frac{\pi}{2}} \frac{e^{k(z+\zeta)}}{k-v} \cos(kR\cos\theta) d\theta dk, \quad (8.149)$$

where $R = \sqrt{(x-\xi)^2 + (y-\eta)^2}$. But $\cos\theta = \lambda$ and $-\sin\theta d\theta = d\lambda$, which on substitution into eqn (8.149) yields

$$g_1(x,y,z) = \frac{1}{r} + \frac{1}{r_1} + \frac{4v}{\pi} PV \int_0^\infty \int_0^1 \frac{e^{k(z+\zeta)}}{k-v} \frac{\cos(kR\lambda)}{\sqrt{1-\lambda^2}} d\lambda dk. \quad (8.150)$$

We know that

$$\frac{2}{\pi} \int_0^1 \frac{\cos(kR\lambda)}{\sqrt{1-\lambda^2}} d\lambda = J_0(kR),$$

and hence

$$\begin{aligned} g_1(x,y,z) &= \frac{1}{r} + \frac{1}{r_1} + 2\lambda PV \int_0^\infty \frac{e^{k(z+\zeta)}}{k-v} J_0(kR) dk \\ &= \frac{1}{r} + PV \int_0^\infty \frac{k+v}{k-v} e^{k(z+\zeta)} J_0(kR) dk. \quad (8.151) \end{aligned}$$

To determine the asymptotic form of g_1, when R goes to infinity, we will use the following Fourier integrals:

$$\left. \begin{aligned} \int_a^\infty f(x) \tfrac{\sin R(x-x_0)}{x-x_0} dx &= \pi f(x_0) + O(\tfrac{1}{R}) \\ PV \int_a^\infty f(x) \tfrac{\cos R(x-x_0)}{x-x_0} dx &= O(\tfrac{1}{R}) \end{aligned} \right\} \quad (8.152)$$

for $a \le x_0 \le \infty$. We know that when $R \to \infty$

$$\frac{1}{r} = O(\frac{1}{R}), \quad \frac{1}{r_1} = O(\frac{1}{R})$$

and

$$g_1(x,y,z) = \frac{4v}{\pi} PV \int_0^\infty \int_0^1 \frac{1}{\sqrt{1-\lambda^2}} \cdot \frac{e^{k(z+\zeta)}}{k-v}.$$

$$[\cos(R\lambda\upsilon)\cos R\lambda(k-\upsilon) - \sin(R\lambda\upsilon)\sin R\lambda(k-\upsilon)]d\lambda dk + O(\frac{1}{R}).$$

Using the formulas (8.152), this equation can be reduced to

$$g_1(x,y,z) = -4\lambda e^{\upsilon(z+\zeta)} \int_0^1 \frac{\sin(r\lambda\upsilon)}{\sqrt{1-\lambda^2}} d\lambda + O(\frac{1}{R}),$$

which subsequently (see Erdelyi (1956)) can be reduced to the following:

$$g_1(x,y,z) = -2\pi\upsilon e^{\upsilon(z+\zeta)} \sqrt{\frac{2}{\pi R\upsilon}} \sin(R\upsilon - \frac{\pi}{4}) + O(\frac{1}{R}). \quad (8.153)$$

From the radiation conditions, we can at once predict the asymptotic form of $g_2(x,y,z)$, which is

$$g_2(x,y,z) = 2\pi\upsilon e^{\upsilon(z+\zeta)} \sqrt{\frac{2}{\pi R\upsilon}} \cos(R\upsilon - \frac{\pi}{4}) + O(\frac{1}{R}). \quad (8.154)$$

Thus, the asymptotic form of $G = g_1 \cos\sigma t + g_2 \sin\sigma t$ is

$$-2\pi\upsilon e^{\upsilon(z+\zeta)} \sqrt{\frac{2}{\pi R\upsilon}} \sin(R\upsilon - \sigma t - \frac{\pi}{4})| + O(\frac{1}{R}). \quad (8.155)$$

It can be easily verified the g_1 has the same asymptotic behavior as

$$-2\pi\upsilon e^{(z+\zeta)} Y_0(R\upsilon),$$

and therefore the function $g_2(x,y,z)$ will be

$$g_2 = 2\pi\upsilon e^{\upsilon(z+\zeta)} J_0(R\upsilon) \quad (8.156)$$

which satisfies all the required conditions. Here, $J_0(R\upsilon)$ and $Y_0(R\upsilon)$ are the Bessel functions of the first and second kind, respectively.

Combining all these results, the final real solution form of G is

$$\begin{aligned} G(x,y,z;\xi,\eta,\zeta,t) &= [\frac{1}{r} + PV \int_0^\infty \frac{k+\upsilon}{k-\upsilon} e^{k(z+\zeta)} J_0(kR) dk] \cos\sigma t \\ & \quad +2\pi\upsilon e^{\upsilon(z+\zeta)} J_0(\upsilon R) \sin\sigma t. \end{aligned} \quad (8.157)$$

The complex form is

$$\begin{aligned} \vec{g} &= g_1 + ig_2 \\ &= [\frac{1}{r} + PV \int_0^\infty \frac{k+\upsilon}{k-\upsilon} e^{k(z+\zeta)} J_0(kR) dk] \\ &+ i2\pi\upsilon e^{\upsilon(z+\zeta)} J_0(\upsilon r). \end{aligned} \quad (8.158)$$

Case II: Finite depth case

Consider Green's function of the form

$$G = Re\{\vec{g}e^{-i\sigma t}\} = g_1 \cos \sigma t + g_2 \sin \sigma t, \tag{8.159}$$

where

$$g_1 = \frac{1}{r} + \frac{1}{r_2} + g_0(x, y, z)$$

and

$$r_2^2 = (x - \xi)^2 + (y - \eta)^2 + (z + 2h + \zeta)^2.$$

The function g_1 satisfies Laplace's equation

$$\nabla^2 g_1 = 0,$$

and therefore

$$\nabla^2 (\frac{1}{r} + \frac{1}{r_2} + g_0) = 0.$$

We can easily verify that

$$\nabla^2 (\frac{1}{r}) = 0 \qquad \text{and,} \quad \nabla^2 (\frac{1}{r_2}) = 0.$$

Therefore

$$\nabla^2 g_0 = 0. \tag{8.160}$$

Applying the double Fourier transform

$$g_0(x, y, z) = \frac{1}{2\pi} \int_0^\infty \int_{-\pi}^{\pi} g_0^*(k, \theta, z) e^{ik(x \cos \theta + y \sin \theta)} d\theta dk$$

to Laplace's equation (8.160), we obtain

$$\frac{\partial^2 g_0^*}{\partial z^2} - k g_0^* = 0, \tag{8.161}$$

the solution of which can be written as

$$g_0^* = Ae^{kz} + Be^{-kz}. \tag{8.162}$$

The constants A and B must be evaluated using the bottom boundary conditions and the free surface boundary condition.

The bottom boundary condition is given by

$$\frac{\partial g_1}{\partial z}(x, y, z = -h) = 0. \tag{8.163}$$

Thus

$$\frac{\partial}{\partial z}(\frac{1}{r}+\frac{1}{r_2})+(\frac{\partial g_0}{\partial z})=0 \qquad \text{at } z=-h.$$

It can be easily verified that

$$\frac{\partial}{\partial z}(\frac{1}{r}+\frac{1}{r_2})=0 \qquad \text{at } z=-h,$$

if we choose

$$r_2^2=(x-\xi)^2+(y-\eta)^2+(z+2h+\zeta)^2.$$

Thus the bottom boundary condition to be satisfied by g_0 is

$$\frac{\partial g_0}{\partial z}=0 \qquad \text{at } z=-h. \tag{8.164}$$

The double Fourier transform of eqn (8.164) yields

$$\frac{\partial g_0^*}{\partial z}=0 \qquad \text{at } z=-h. \tag{8.165}$$

Using this condition in the solution (8.162), we obtain $B=Ae^{-2kh}$ and hence

$$g_0^*=Ae^{-kh}[e^{k(z+h)}+e^{-k(z+h)}]=C\cosh k(z+h) \tag{8.166}$$

where $C=2Ae^{-kh}$ a redefined constant. To evaluate C, we have to satisfy the free surface condition

$$\frac{\partial g_1}{\partial z}-v g_1=0 \qquad \text{at } z=0,$$

that is

$$\frac{\partial}{\partial z}(\frac{1}{r}+\frac{1}{r_2}+g_0)-v(\frac{1}{r}+\frac{1}{r_2}+g_0)=0 \qquad \text{at } z=0. \tag{8.167}$$

Taking the double Fourier transform of eqn (8.167), we obtain

$$\frac{\partial g_0^*}{\partial z}-v g_0^*=(k+v)(\frac{1}{r}+\frac{1}{r_2})^*, \qquad \text{at } z=0. \tag{8.168}$$

Using this condition in eqn (8.166), we obtain

$$C=\frac{k+v}{k\sinh kh-v\cosh kh}(\frac{1}{r}+\frac{1}{r_2})^* \qquad z=0. \tag{8.169}$$

We know that

$$(\frac{1}{r})^*=e^{-k|z-\zeta|}e^{-ik(\xi\cos\theta+\eta\sin\theta)}$$

and

$$(\frac{1}{r_2})^*=e^{-k|z+2h+\zeta|}e^{-ik(\xi\cos\theta+\eta\sin\theta)}.$$

Hence, at $z = 0$

$$(\frac{1}{r})^* = e^{k\xi}e^{-ik(\xi\cos\theta+\eta\sin\theta)}$$

and

$$(\frac{1}{r_2})^* = e^{-k\zeta}e^{-2kh}e^{-ik(\xi\cos\theta\eta\sin\theta)}.$$

Therefore

$$\begin{aligned}(\frac{1}{r})^* + (\frac{1}{r_2}^*) &= e^{-kh}(e^{k(h+\zeta)} + e^{-k(h+\zeta)})e^{-ik(\xi\cos\theta+\eta\sin\theta)} \\ &= 2e^{-kh}\cosh k(h+\zeta)e^{-ik(\xi\cos\theta+\eta\sin\theta)}.\end{aligned}$$

Substituting this into eqn (8.166) yields

$$C = \frac{2e^{-kh}(k+v)\cosh k(h+\zeta)}{k\sinh kh - v\cosh kh}e^{-ik(\xi\cos\theta+\eta\sin\theta)}, \tag{8.170}$$

and consequently

$$g_0^* = \frac{2(k+v)e^{-kh}\cosh k(h+\zeta)\cosh k(z+h)}{k\sin hkh - v\cosh kh}e^{-ik(\xi\cos\theta+\eta\sin\theta)}. \tag{8.171}$$

Inverting this expression, we can write the g_1 solution as

$$\begin{aligned}g_1 &= \frac{1}{r} + \frac{1}{r_2} + \frac{1}{2\pi}\int_0^\infty \int_{-\pi}^{\pi} \frac{2(k+v)e^{-kh}\cosh k(h+\zeta)}{k\sinh kh - v\cosh kh} \\ &\quad \times \cosh k(z+h)e^{-ik((x-\xi)\cos\theta+(y-\eta)\sin\theta)}d\theta dk \end{aligned} \tag{8.172}$$

which can subsequently be written as

$$g_1 = \frac{1}{r} + \frac{1}{r_2} + \int_0^\infty \frac{2(k+v)e^{-kh}\cosh k(h+\zeta)\cosh k(z+\zeta)}{k\sinh kh - v\cosh kh}J_0(kR)dk.$$

To satisfy the radiation condition we must first obtain the asymptotic expansion $(R \to \infty)$ for g_1.

Consider the integral

$$\begin{aligned}g_0 &= \frac{1}{2\pi}\int_0^\infty \int_{-\pi}^{\pi} \frac{2(k+v)e^{-kh}\cosh k(h+\zeta)\cosh k(z+h)}{k\sinh kh - v\cosh kh} \\ &\quad \times e^{ik((x-\xi)\cos\theta+(y-\eta)\sin\theta)}d\theta dk.\end{aligned}$$

Since $x - \xi = R\cos\delta$ and $y - \eta = R\sin\delta$, then

$$\begin{aligned}R &= (x-\xi)\cos\delta + (y-\eta)\sin\delta \\ &= \sqrt{(x-\xi)^2 + (y-\eta)^2}.\end{aligned}$$

Also

$$
\begin{aligned}
e^{ik(x-\xi)\cos\theta+(y-\eta)\sin\theta} &= e^{ikR\cos(\theta-\delta)} \\
\text{and hence} \quad \int_{-\pi}^{\pi} e^{ikR\cos(\theta-\delta)}d\theta &= 4\int_0^{\frac{\pi}{2}} \cos(kR\cos\theta)d\theta \\
&= 4\int_0^1 \frac{\cos(kr\lambda)}{\sqrt{1-\lambda^2}}d\lambda \\
&= 2\pi J_0(kR),
\end{aligned}
$$

where

$$
J_0(kR) = \frac{2}{\pi}\int_0^1 \frac{\cos(kR\lambda)}{\sqrt{1-\lambda^2}}d\lambda.
$$

Thus

$$
\begin{aligned}
g_0 &= \frac{2}{\pi}\int_0^1 [\int_0^\infty \frac{2(k+\upsilon)e^{-kh}\cosh k(z+\zeta)\cosh k(h+\zeta)}{\frac{(k\sinh kh-\upsilon\cosh kh)}{k-m_0}} \\
&\quad \times \frac{1}{\sqrt{1-\lambda^2}}(\frac{\cosh(kR\lambda)}{k-m_0})dk]d\lambda,
\end{aligned}
$$

where

$$
\cos(kR\lambda) = \cos(\lambda Rm_0)\cos\lambda R(k-m_0) - \sin(\lambda Rm_0)\sin\lambda R(k-m_0).
$$

Using the Fourier integrals in the form (8.152) gives

$$
\begin{aligned}
g_0 &= \frac{-4(m_0+\upsilon)e^{-m_0h}\cosh m_0(z+\zeta)\cosh m_0(h+\zeta)}{\lim_{R\to m_0}\frac{(k\sinh kh-\upsilon\cosh kh)}{k-m_0}} \\
&\quad \times \int_0^1 \frac{\sin(R\lambda m_0)}{\sqrt{1-\lambda^2}}d\lambda + O(\frac{1}{R}),
\end{aligned}
$$

where m_0 is a root of $m_0h\tanh m_0h = \upsilon h$. Now

$$
\begin{aligned}
\text{limit} &= \lim_{k\to m_0}\frac{k\sinh kh-\upsilon\cosh kh}{k-m_0} \\
&= \lim_{k\to m_0}[kh\cosh kh + \sinh kh - \upsilon h\sinh kh] \\
&= m_0h\cosh m_0h + \sinh m_0h - \upsilon h\sinh m_0h \\
&= \frac{\upsilon h\cosh^2 m_0h + \sinh^2 m_0h - \upsilon h\sinh^2 m_0h}{\sinh m_0h} \\
&= \frac{\upsilon h + \sinh^2 m_0h}{\sinh m_0h}.
\end{aligned}
$$

Therefore the asymptotic behaviour of $g_0(x, y, z)$ as $R \to \infty$ is

$$g_0 = -\frac{4(m_0 + v)\cosh m_0(z+\zeta)\cosh m_0(h+\zeta)\sinh m_0 h}{vh + \sinh^2 m_0 h} \times \int_0^1 \frac{\sin(R\lambda m_0)}{\sqrt{1-\lambda^2}} d\lambda + O(\frac{1}{R}).$$

However, the asymptotic behaviour of the integral term is given by

$$\int_0^1 \frac{\sinh(\lambda R m_0)}{\sqrt{1-\lambda^2}} d\lambda = \frac{\pi}{2}\sqrt{\frac{2}{\pi R m_0}} \sin(R m_0 - \frac{\pi}{4}) + O(\frac{1}{R}). \tag{8.173}$$

Thus

$$g_0 = -\frac{2\pi(m_0 + v)e^{-m_0 h}\sinh m_0 h \cosh m_0(z+\zeta)\cosh m_0(h+\zeta)}{vh + \sinh^2 m_0 h} \times \sqrt{\frac{2}{\pi m_0 R}} \sin(R m_0 - \frac{\pi}{4}) + O(\frac{1}{R}). \tag{8.174}$$

Now, using the radiation condition

$$\sqrt{R}[\frac{\partial g_0}{\partial R} + m_0 g_2] = 0 \ , \quad \text{as } R \to \infty,$$

we obtain

$$g_2 = \frac{2\pi(m_0 + v)e^{-m_0 h}\sinh m_0 h \cosh m_0(z+\zeta)\cosh m_0(h+\zeta)}{vh + \sinh^2 m_0 h} \times \sqrt{\frac{1}{\pi m_0 R}} \cos(R m_0 - \frac{\pi}{4}) + O(\frac{1}{R}).$$

It can be easily verified that the function

$$g_2 = \frac{2\pi(m_0 h)e^{-m_0 h}\sinh m_0 h \cosh m_0(z+\zeta)\cosh m_0(h+\zeta)}{vh + \sinh^2 m_0 h} J_0(m_0 R) \tag{8.175}$$

will satisfy all the required boundary conditions including Laplace's equation.

Thus, combining these results, the final form of the velocity potential G may be expressed as

$$\begin{aligned} G(x, y, z; \xi, \eta, \zeta, t) &= [\frac{1}{r} + \frac{1}{r_2} + PV \int_0^\infty \frac{2(k+v)e^{-kh}\cosh k(h+\zeta)}{k \sinh kh - v \cosh kh} \\ &\quad \times \cosh k(z+h) J_0(kR) dk] \cos \sigma t \\ &\quad + \frac{2\pi(m_0 + v)e^{-m_0 h}\sinh m_0 h \cosh m_0(h+\zeta)}{vh + \sinh^2 m_0 h} \\ &\quad \times \cosh m_0(z+h) J_0(m_0 R) \sin \sigma t, \end{aligned} \tag{8.176}$$

where

$$m_0 \tanh m_0 h = \upsilon = \frac{\sigma^2}{g}.$$

The above results are due to Wehausen and Laitone (1960). John (1950) has derived the following Green's function in terms of the infinite series:

$$\begin{aligned} G(x, y, z; \xi, \eta, \zeta, t) &= 2\pi \frac{\upsilon^2 - m_0^2}{hm_0^2 - h\upsilon^2 + \upsilon} \cosh m_0(z+h) \\ & \times \cosh m_0(h+\zeta)[Y_0(m_0R)\cos\sigma t - J_0(m_0R)\sin\sigma t] \\ & +4\sum_{k=1}^{\infty} \frac{m_k^2 + \upsilon^2}{hm_k^2 + h\upsilon^2 - \upsilon} \cos m_k(z+h) \cosh m_k(h+\zeta) \\ & \times K_0(m_kR)\cos\sigma t, \end{aligned} \tag{8.177}$$

where $m_k, k > 0$ are the positive real roots of $m \tan mh + \upsilon = 0$, and $K_0(m_kR)$ is the modified Bessel function of second kind of zeroth order.

8.5 Numerical formulation

Green's function provides an elegant mathematical tool to obtain the wave loadings on arbitrarily shaped structures. We know that the total complex velocity potential $\phi(x, y, z, t)$, in diffraction theory, may be expressed as the sum of an incident wave potential ϕ_I and a scattered potential ϕ_S in the following manner:

$$\Phi(x, y, z, t) = Re\{\phi(x, y, z)e^{-i\sigma t}\}, \tag{8.178}$$

where Φ is the real velocity potential, $\phi(x, y, z)$ is the complex potential and σ is the oscillatory frequency of the harmonic motion of the structure. Thus

$$\phi = \phi_I + \phi_S. \tag{8.179}$$

The complex potentials should satisfy the following conditions:

$$\left.\begin{array}{lc} \nabla^2\phi = 0 & \text{in the fluid interior} \\ \frac{\partial\phi}{\partial z} - \upsilon\phi = 0 & \text{on the free surface} \\ \frac{\partial\phi_I}{\partial n} = -\frac{\partial\phi_S}{\partial n} & \text{on the fixed structures} \\ \frac{\partial\phi_I}{\partial n} = \frac{\partial\phi_S}{\partial n} = 0 & \text{on the sea bottom} \\ \lim_{R\to\infty}\sqrt{R}(\frac{\partial\phi_S}{\partial R} - ik\phi_S) = 0 & \text{at far field} \end{array}\right\} \tag{8.180}$$

where R is the horizontal polar radius.

Hydrodynamic pressure from the linearized Bernoulli equation is

$$P = -\rho(\frac{\partial \phi}{\partial t}) = \rho Re[i\sigma\phi] = \rho Re[i\sigma(\phi_I + \phi_S)]. \tag{8.181}$$

The force and moment acting on the body may be calculated from the formulae

$$\vec{F} = -\int\int_S P\vec{n}dS \tag{8.182}$$

$$\vec{M} = -\int\int_S P(\vec{r}X\vec{n})dS. \tag{8.183}$$

In evaluating the wave loading on the submerged structures, we first need to find the scattered potentials, ϕ_S, which may be represented as being due to a continuous distribution of point wave sources over the immersed body surface. As indicated earlier, a solution for ϕ_S at the point (x, y, z) may be expressed as:

$$\phi_S(x, y, z) = \frac{1}{4\pi}\int\int_S Q(\xi, \eta, \zeta)G(x, y, z; \xi, \eta, \zeta)dS. \tag{8.184}$$

The integral is over all points (ξ, η, ζ) lying on the surface of the structure, $Q(\xi, \eta, \zeta)$ is the source strength distribution function, and dS is the differential area on the immersed body surface. Here $G(x, y, z; \xi, \eta, \zeta)$ is a complex Green's function of a wave source of unit strength located at the point (ξ, η, ζ).

Such a Green's function, originally developed by John (1950) and subsequently extended by Wehausen and Laitone (1960), was illustrated in the previous section.

The complex form of G may be expressed either in integral form or in infinite series form.

The integral form of G is as follows:

$$\begin{aligned} G(x, y, z; \xi, \eta, \zeta) &= \frac{1}{r} + \frac{1}{r_2} \\ &+ PV\int_0^\infty \frac{2(k+\upsilon)e^{-kh}\cosh(k(\zeta+h))\cosh(k(z+h))}{k\sinh kh - \upsilon\cosh kh}J_0(kR)dk - 2\pi i \\ &\times \frac{(m_0+\upsilon)e^{-m_0h}\sinh m_0h\cosh m_0(h+\zeta)\cosh m_0(z+h)}{\upsilon h + \sinh^2 m_0h}J_0(m_0R). \end{aligned} \tag{8.185}$$

Since

$$\frac{e^{-m_0h}\sinh m_0h}{\upsilon h + \sinh^2 m_0h} = \frac{m_0 - \upsilon}{(m_0^2 - \upsilon^2)h + \upsilon}, \tag{8.186}$$

then eqn (8.186) may be expressed as

$$\begin{aligned} G(x, y, z; \xi, \eta, \zeta) &= \frac{1}{r} + \frac{1}{r_2} \\ &+ PV\int_0^\infty \frac{2(k+\upsilon)e^{-kh}\cosh(k(\zeta+h))\cosh(k(z+h))}{k\sinh kh - \upsilon\cosh kh}J_0(kR)dk \\ &- 2\pi i\frac{m_0^2 - \upsilon^2}{(m_0^2 - \upsilon^2)h + \upsilon}\cosh m_0(\zeta+h)\cosh m_0(z+h)J_0(m_0R). \end{aligned} \tag{8.187}$$

The second form of Green's function involves a series representation

$$G(x,y,z;\xi,\eta,\zeta) = 2\pi\frac{v^2 - m_0^2}{(m_0^2 - v^2)h + v}\cosh m_0(z+h)$$

$$\times \cosh m_0(h+\zeta)[Y_0(m_0R) - iJ_0(m_0R)] + 4\sum_{k=1}^{\infty}\frac{(m_k^2 + v^2)}{(m_k^2 + v^2)h - v}$$

$$\times \cos m_k(z+h)\cos m_k(h+\zeta)K_0(m_kR), \tag{8.188}$$

for which $m_k, k > 0$ are the positive real roots of

$$m\tan mh + v = 0. \tag{8.189}$$

We must find $Q(\xi,\eta,\zeta)$, the source density distribution function. This function must satisfy the third condition of (8.181), which applies at the body surface. The component of fluid velocity normal to the body surface must be equal to that of the structure itself, and may be expressed in terms of ϕ_S. Thus, following the treatment illustrated by Kellog (1953), the normal derivative of the potential ϕ_S in eqn (8.184) assumes the following form:

$$\begin{aligned} -2\pi Q(x,y,z) \quad &+ \quad \iint_S Q(\xi,\eta,\zeta)\frac{\partial G}{\partial n}(x,y,z;\xi,\eta,\zeta)dS \\ &= \quad -\frac{\partial \phi_I}{\partial n}(x,y,z) \times 4\pi. \end{aligned} \tag{8.190}$$

This equation is known as Fredholm's integral equation of the second kind, which applies to the body surface S and must be solved for $Q(\xi,\eta,\zeta)$.

A suitable way of solving eqn (8.190) for unknown source density Q is the matrix method. In this method, the surface of the body is discretized into a large number of panels, as shown in Fig. 8.10. For each panel the source density is assumed to be constant. This replaces eqn (8.190) by a set of linear algebraic equations with Q on each being an unknown.

These equations may be written as

$$\sum_{j=1}^{N} A_{ij}Q_j = a_i \qquad \text{for } i = 1,2,\ldots,N \tag{8.191}$$

where N is the number of panels. The coefficients a_i and A_{ij} are given, respectively, by

$$a_i = -2\frac{\partial \phi_I}{\partial n}(x_i,y_i,z_i) \tag{8.192}$$

$$A_{ij} = -\delta_{ij} + \frac{1}{2\pi}\iint_{\Delta S_j}\frac{\partial G}{\partial n}(x_i,y_i,z_i;\xi_j,\eta_j,\zeta_j)dS \tag{8.193}$$

where δ_{ij} is the Kronecker delta, ($\delta_{ij} = 0$ for $i \neq j, \delta_{ii} = 1$), the point (x_i,y_i,z_i) is the centroid of the i-th panel of area ΔS_i, and n is measured normal to the surface

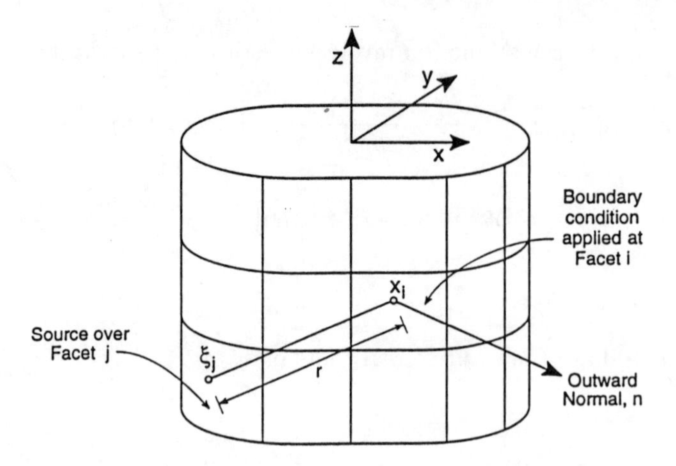

Figure 8.10: Boundary condition at panel i due to source density over panel j

at that point. Assuming the value of $\frac{\partial G}{\partial n}$ is constant over the panel and equal to the value at the centroid, the expression for A_{ij} can be approximated as

$$\begin{aligned} A_{ij} &= -\delta_{ij} + \frac{\Delta S_j}{2\pi}\frac{\partial G}{\partial n}(x_i, y_i, z_i; \xi_j, \eta_j, \zeta_j) \\ &= -\delta_{ij} + \frac{\Delta S_j}{2\pi}[\frac{\partial G}{\partial x}n_x + \frac{\partial G}{\partial y}n_y + \frac{\partial G}{\partial z}n_z], \end{aligned} \tag{8.194}$$

in which n_x, n_y and n_z are the unit normal vectors defining the panel orientation.

Note that when $i = j, \delta_{ii} = 1$ and the last term in eqn (8.194) is equal to zero, and therefore may be omitted.

The column vector a_i in eqn (8.192) may be evaluated as

$$a_i = -2[\frac{\partial \phi_I}{\partial x}n_x + \frac{\partial \phi_I}{\partial z}n_z], \tag{8.195}$$

where

$$\phi_I = \frac{gA}{\sigma}\cosh k(z+h)\cosh khe^{ikx}.$$

Thus

$$a_i = -2(\frac{gAk}{\sigma})e^{ikx}[i\frac{\cosh k(z+h)}{\cosh kh}n_x + \frac{\sinh k(z+h)}{\cosh kh}n_z]. \tag{8.196}$$

Once A_{ij} and a_i are known, the source distribution Q_j may be obtained by a complex matrix inversion procedure.

We then obtain the potential ϕ_S around the body surface by using a discrete version of eqn (8.184) which is

$$\phi_S(x_i, y_i, z_i) = \sum_{j=1}^{\infty} B_{ij}Q_j, \tag{8.197}$$

where

$$B_{ij} = \frac{1}{4\pi}\int\int_{\Delta S_j} G(x_i, y_i, z_i; \xi_j, \eta_j, \zeta_j)dS$$
$$= \frac{\Delta S_j}{4\pi} G(x_i, y_i, z_i; \xi_j, \eta_j, \zeta_j). \tag{8.198}$$

But when $i = j$, there exists a singularity in Green's function; however, we can still evaluate B_{ij} if we retain the dominant term in Green's Function as follows:

$$B_{ij} = \frac{1}{4\pi}\int\int_{\Delta S_i} \frac{dS}{r}. \tag{8.199}$$

This may be integrated for a given panel.

Once the potentials, ϕ_S and ϕ_I, have been determined, we can obtain the pressure force P, surface elevation, η, forces, $\vec{F}$, and moments, $\vec{M}$, in sequence.

The matrix equation (8.197) may be solved using a standard complex matrix inversion subroutine, such as that available in the IMSL library. The numerical prediction can then be compared with available experimental measurements. Hogben *et al.* (1977) have published a number of comparisons between computer predictions and experimental results. Fig. 8.11 shows the comparisons of computed and measured forces and moments for both circular and square cylinders reported by Hogben and Standing (1975). More comprehensive published results dealing with offshore structures of arbitrary shapes can be found in the works of Boreel (1974), Van Oortmerssen (1972), Faltinsen and Michelsen (1974), Garrison and Rao (1971), Garrison and Chau (1972) and Mogridge and Jamieson (1975).

Fenton (1978) has applied Green's function to obtain the scattered wave field and forces on a body which is axisymmetric about a vertical axis. Starting with John's (1950) equations, he expressed them as a Fourier series in terms of an azimuthal angle about the vertical axis of symmetry. He obtained a one-dimensional integral equation which was integrated analytically. Fenton (1978) then applied his theory to a right circular cylinder fixed to the ocean bed, in water of depth h; the cylinder was $0.7h$ and had a diameter of $0.4h$. He compared his theoretical computations with the experimental results of Hogben and Standing (1975), and his results are reproduced in Fig. 8.11. Fenton recorded that convergence of the Fourier series for Green's function was extremely rapid, and he was able to achieve an accuracy of 0.0001.

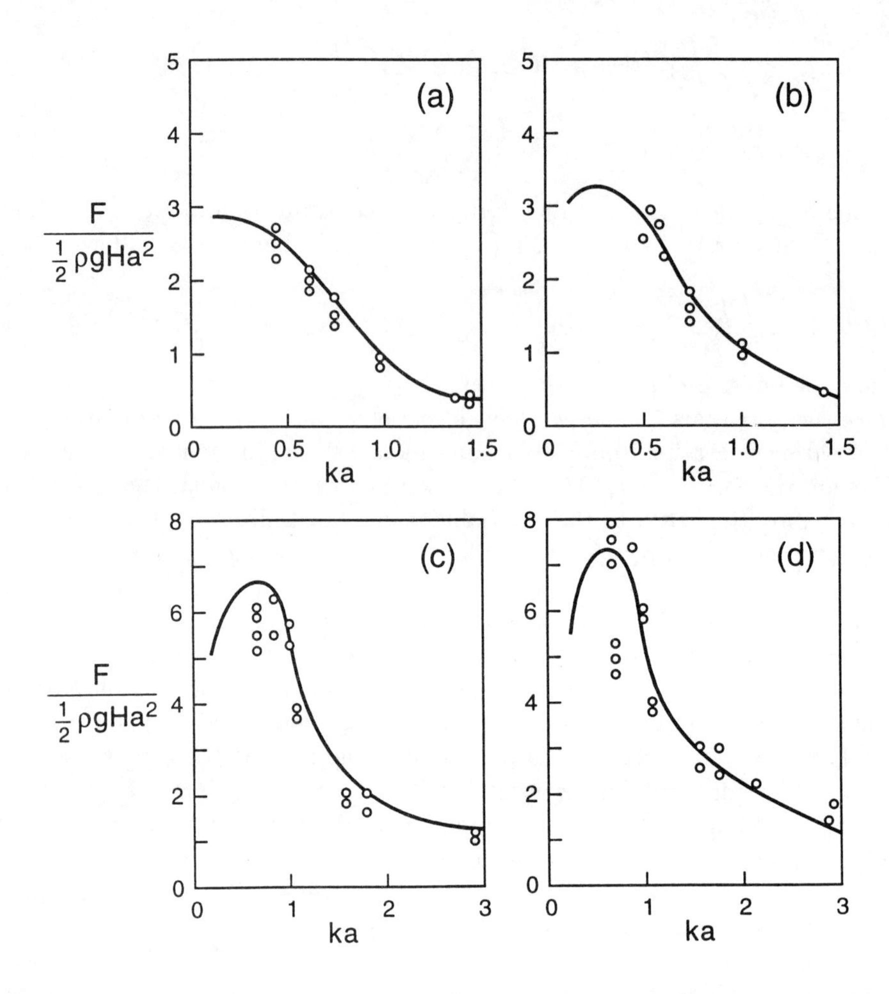

Figure 8.11: Comparison of computer predictions with experimental data of Hogben and Standing (1975) for horizontal wave forces on circular and square columns. (a) circular, $h'/h = 0.7$, (b) square, $h'/h = 0.7$, (c) circular, surface-piercing, (d) square, surface-piercing. Here, h' is the height of the cylinder

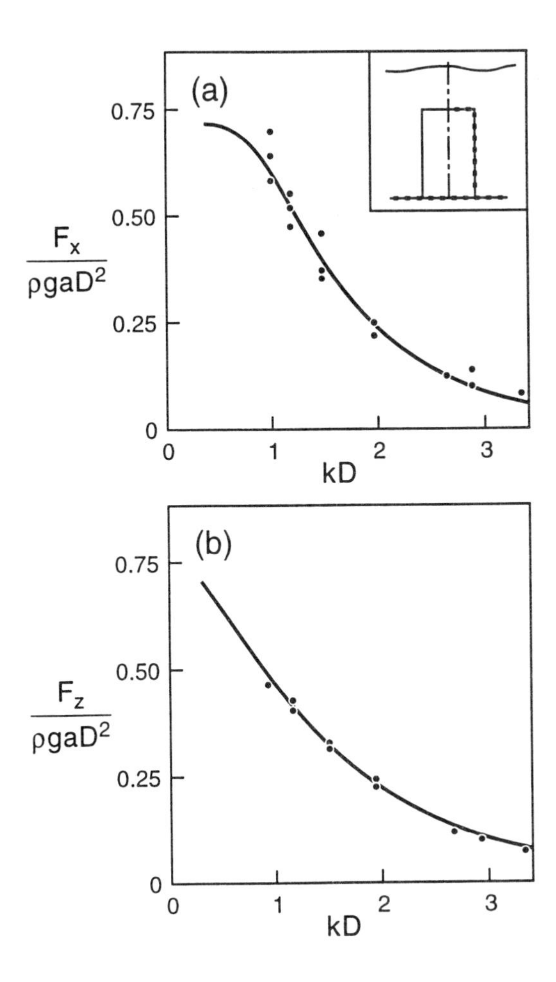

Figure 8.12: Variation of (a) dimensionless drag force and (b) dimensionless vertical force with wavenumber for a truncated circular cylinder of height $h' = 0.7h$ and diameter $D = 0.4h$, where h is the water depth. Experimental results from Hogben and Standing (1975) .... Fenton's (1978) computer predictions ——

Chapter 9

Integral transforms

9.1 Introduction

The classical methods of solution of the boundary value problems of mathematical physics, namely the integral transforms, originated from the pioneering work of Fourier and Heaviside during the eighteenth century. This chapter is mainly concerned giving an outline of the use of integral transforms to obtain solutions of problems governed by the partial differential equations with assigned boundary and initial conditions. We have already seen the use of Fourier and Laplace transforms to the applied problems of mathematical physics. Therefore these methods will not be repeated here, but we will instead introduce the Hankel, Mellin and Z-transforms with some applications.

About 1890, Heaviside was originally interested in the solution of the ordinary differential equations with constant coefficients occurring in the theory of electric circuits. Later he extended his theory to the partial differential equations occurring in the problems of electromagnetism and heat conduction. Later investigations by Bromwich, Carson and van der Pol placed the Heaviside calculus on a sound footing.

The use of an integral transform will often reduce a partial differential equation in n independent variables to one in $(n-1)$ variables, thus reducing the difficulty of the problem. Successive operations of this type sometimes reduce the problem to the solution of an ordinary differential equation. Successive operations could, in fact, reduce the problem to the solution of an ordinary algebraic equation.

9.2 The Hankel transform

Suppose $f(x,y)$ is a function of two variables x and y defined in $-\infty \leq (x,y) \leq \infty$. Then the double Fourier transform of $f(x,y)$ and its inverse may be written as

$$\mathcal{F}\{f(x,y)\} = \int_{-\infty}^{\infty}\int_{-\infty}^{\infty} f(x,y)e^{-i(\alpha x+\beta y)}dxdy = F(\alpha,\beta) \tag{9.1}$$

$$f(x,y) = \frac{1}{2\pi}\int_{-\infty}^{\infty}\int_{-\infty}^{\infty} F(\alpha,\beta)e^{i(\alpha x+\beta y)}d\alpha d\beta, \tag{9.2}$$

provided these integrals exist.

In these integrals we introduce the polar coordinates by writing

$$\begin{aligned}
x &= r\cos\theta \\
y &= r\sin\theta \\
\alpha &= \lambda\sin\phi \\
\beta &= \lambda\cos\phi \\
\text{and} \quad dxdy &= rd\theta dr, \qquad d\alpha d\beta = (\lambda d\phi)d\lambda
\end{aligned} \tag{9.3}$$

so that

$$F(\alpha,\beta) = \int_{r=0}^{\infty}\int_{\theta=-\pi}^{\pi} f(r,\theta)e^{-i\lambda r\sin(\theta+\phi)}(rd\theta)dr \tag{9.4}$$

$$f(r,\theta) = \frac{1}{2\pi}\int_{\lambda=0}^{\infty}\int_{\phi=-\pi}^{\pi} F(\lambda,\phi)e^{i\lambda r\sin(\theta+\phi)}(\lambda d\phi)d\lambda. \tag{9.5}$$

We next write that the functions f and F contain the eternal functions as follows

$$\left.\begin{aligned}
f(r,\theta) &= e^{in\theta}f(r) \\
F(\lambda,\phi) &= e^{-in\phi}F(\lambda)
\end{aligned}\right\} \tag{9.6}$$

where n is an integer or zero so that the above result becomes

$$e^{-in\phi}F(\lambda) = \int_{r=0}^{\infty} rf(r)\{\frac{1}{2\pi}\int_{\theta=-\pi}^{\pi} e^{in\theta-i\lambda r\sin(\theta+\phi)}d\theta\}dr \tag{9.7}$$

$$e^{in\theta}f(r) = \int_{\lambda=0}^{\infty} \lambda F(\lambda)\{\frac{1}{2\pi}\int_{\phi=-\pi}^{\pi} e^{-in\phi+i\lambda r\sin(\theta+\phi)}d\phi\}d\lambda. \tag{9.8}$$

Now using the standard integral representation of the Bessel function $J_n(\lambda r)$ in the form

$$e^{-in\phi}J_n(\lambda r) = \frac{1}{2\pi}\int_{-\pi}^{\pi} e^{-in\theta-i\lambda r\sin(\theta+\phi)}d\theta. \tag{9.9}$$

The above results assume the form

$$F(\lambda) = H_n\{f(r)\} = \int_0^{\infty} rf(r)J_n(\lambda r)dr \tag{9.10}$$

$$f(r) = H_n^{-1}\{F(\lambda)\} = \int_0^{\infty} \lambda F(\lambda)J_n(\lambda r)d\lambda \tag{9.11}$$

where (9.11) is obtained by the complex conjugate of (9.9).

Now the function $F(\lambda)$ defined by (9.10) is called the Hankel transform of $f(r)$ of order n. Integral (9.10) exists for certain classes of functions of practical interest. The inverse Hankel transformation is defined by (9.11).

In particular, the Hankel transforms of the functions of the zero order ($n = 0$) and of the first order ($n = 1$) are widely used for solving problems involving Laplace's equation in cylindrical geometry.

Example 1

Find the zero order Hankel transforms of the following functions

$$
\begin{aligned}
(a) \quad f(r) &= \frac{1}{r}e^{-ar}, \quad a > 0 \\
(b) \quad f(r) &= \frac{1}{r}\delta(r).
\end{aligned}
$$

Solution

$$
\begin{aligned}
(a) \quad F(\lambda) = H_0\{f(r)\} &= \int_0^\infty \{rf(r)\}J_0(\lambda r)dr \\
&= \int_0^\infty e^{-ar}J_0(\lambda r)dr = \frac{1}{\sqrt{\lambda^2 + a^2}}
\end{aligned}
$$

$$
\begin{aligned}
(b) \quad F(\lambda) = H_0\{(f(r)\} &= \int_0^\infty \{rf(r)\}J_0(\lambda r)dr \\
&= \int_0^\infty \delta(r)J_0(\lambda r)dr \\
&= J_0(0) = 1.
\end{aligned}
$$

Example 2

Find the first order Hankel transforms of the function $f(r) = e^{-ar}, a > 0$.

Solution

$$
\begin{aligned}
F(\lambda) = H_1\{f(r)\} &= \int_0^\infty \{rf(r)\}J_1(\lambda r)dr \\
&= \int_0^\infty re^{-ar}J_1(\lambda r)dr \\
&= \frac{\lambda}{(a^2 + \lambda^2)^{\frac{3}{2}}}.
\end{aligned}
$$

Remark

$\int_0^\infty e^{-ar}J_0(\lambda r)dr$ has been evaluated by term by term integration of the series expansion of $J_0(\lambda r) = \sum_{m=0}^\infty \frac{(-1)^m(\lambda r)^{2m}}{2^{2m}(m!)^2}$. Note that the integral is nothing but the Laplace transform of $J_0(\lambda r)$. To evaluate the integral $\int_0^\infty re^{-ar}J_1(\lambda r)dr$, we have used the recurrence relations of Bessel's functions such as

$$
\frac{d}{dx}\{x^{-\nu}J_\nu(x)\} = -x^{-\nu}J_{\nu+1}(x).
$$

In our case $\nu = 0$ and so

$$\frac{d}{dx}\{J_0(x)\} = -J_1(x).$$

Thus for example, differentiating the results in Example 1(a) both sides with respect to λ we obtain,

$$\frac{\partial}{\partial \lambda}\int_0^\infty e^{-ar}J_0(\lambda r)dr = \frac{\partial}{\partial \lambda}[\frac{1}{\sqrt{\lambda^2+a^2}}]$$

which is the same as

$$\int_0^\infty re^{-ar}J_1(\lambda r)dr = \frac{\lambda}{(\lambda^2+a^2)^{\frac{3}{2}}}.$$

Similarly differentiating the results in Example 1(a) both sides with respect to a, we obtain

$$\frac{\partial}{\partial a}\int_0^\infty e^{-ar}J_0(\lambda r)dr = \frac{\partial}{\partial a}[\frac{1}{\sqrt{\lambda^2+a^2}}]$$

which yields

$$\int_0^\infty re^{-ar}J_0(\lambda r)dr = \frac{a}{(\lambda^2+a^2)^{\frac{3}{2}}}.$$

Properties of Hankel transforms

We state the following properties of the Hankel transform:

(I) The Hankel transform is a linear integral operator based on the properties of integration and therefore superposition applies. Thus for any arbitrary constants a and b,

$$H_n\{af(r)+bg(r)\} = aH_n\{f(r)\}+bH_n\{g(r)\}. \tag{9.12}$$

This follows from the integral definition of the Hankel transform.

(II) The Hankel transform satisfies the Parseval relation,

$$\int_0^\infty rf(r)g(r)dr = \int_0^\infty \lambda F(\lambda)g(\lambda)d\lambda \tag{9.13}$$

where $F(\lambda)$ and $G(\lambda)$ are the Hankel transforms of $f(r)$ and $g(r)$ respectively.

Proof

Let us consider the right hand side

$$\int_0^\infty \lambda F(\lambda)g(\lambda)d\lambda = \int_{\lambda=0}^\infty \lambda F(\lambda)(\int_{r=0}^\infty rg(r)J_n(\lambda r)dr)d\lambda.$$

Interchanging the order of integration and rearranging the terms

$$\int_0^\infty \lambda F(\lambda)g(\lambda)d\lambda = \int_{r=0}^\infty rg(r)(\int_{\lambda=0}^\infty \lambda F(\lambda)J_n(\lambda r)d\lambda)dr.$$

which can be written as

$$\int_{\lambda=0}^\infty \lambda F(\lambda)G(\lambda)d\lambda = \int_{r=0}^\infty rg(r)f(r)dr$$

which is the required proof.

Now if $f(r) = g(r)$, we obtain the Parseval's identity as

$$\int_0^\infty r|f(r)|^2dr = \int_0^\infty \lambda|F(\lambda)|^2d\lambda.$$

(III) The Hankel transform of a derivative of a function $f(r)$,

$$H_n\{f'(r)\} = \frac{\lambda}{2n}[(n-1)F_{n+1}(\lambda) - (n+1)F_{n-1}(\lambda)]$$

provided $rf(r)$ vanishes as $r \to 0$ and $r \to \infty$.

In the following the symbol for Hankel transform $F(\lambda)$ has been used as $F_n(\lambda)$ to specify the order.

Proof

$$\begin{aligned} H_n\{f'(r)\} &= \int_0^\infty \{f'(r)\}J_n(\lambda r)dr \\ &= [rf(r)J_n(\lambda r)]_0^\infty - \int_0^\infty f(r)\frac{d}{dr}(rJ_n(\lambda r))dr. \end{aligned}$$

Since $rf(r) = 0$ as $r \to 0$ and $r \to \infty$,
We have

$$\begin{aligned} H_n\{f'(r)\} &= -\int_0^\infty f(r)\frac{d}{dr}(rJ_n(\lambda r))dr \\ &= -\int_0^\infty f(r)\{rJ_n'(\lambda r)\lambda + J_n(\lambda r)\}dr. \end{aligned}$$

But we know $J_n'(\lambda r) = \frac{1}{2}[J_{n-1}(\lambda r) - J_{n+1}(\lambda r)]$

and $J_n(\lambda r) = \frac{r}{2n}[J_{n+1}(\lambda r) + J_{n-1}(\lambda r)]$.

Using these Bessel identities, we obtain

$$H_n\{f'(r)\} = \frac{\lambda}{2n}[(n-1)F_{n+1}(\lambda) - (n+1)F_{n-1}(\lambda)].$$

(IV) The fourth property states that

$$H_n\{\frac{1}{r}\frac{d}{dr}(r\frac{df}{dr}) - \frac{n^2}{r^2}f(r)\} = -\lambda^2 F_n(\lambda),$$

provided that both $(rf'(r))$ and $rf(r))$ vanish as $r \to 0$ and $r \to \infty$.

Proof

We have, by definition,

$$\begin{aligned}
& H_n\{\frac{1}{r}\frac{d}{dr}(r\frac{df}{dr}) - \frac{n^2}{r^2}f(r)\} \\
&= \int_0^\infty \frac{d}{dr}(r\frac{df}{dr})J_n(\lambda r)dr - \int_0^\infty \frac{n^2}{r^2}(rf(r))J_n(\lambda r)dr \\
&= [r\frac{df}{dr}J_n(\lambda r)]_0^\infty - \int_0^\infty (r\frac{df}{dr})\lambda J_n'(\lambda r)dr - \int_0^\infty \frac{n^2}{r^2}(rf(r))J_n(\lambda r)dr \\
&\qquad \text{(by partial integration)} \\
&= -[f(r)\lambda r J_n'(\lambda r)]_0^\infty + \int_0^\infty \frac{d}{dr}\{\lambda r J_n'(\lambda r)\}f(r)dr - \int_0^\infty \frac{n^2}{r^2}(rf(r))J_n(\lambda r)dr \\
&\qquad \text{(by integration by parts).}
\end{aligned}$$

Thus we obtain,

$$H_n\{\frac{1}{r}\frac{d}{dr}(r\frac{df}{dr}) - \frac{n^2}{r^2}f(r)\} = \int_0^\infty \frac{d}{dr}\{\lambda r J_n'(\lambda r)\}f(r)dr - \int_0^\infty \frac{n^2}{r^2}(rf(r))J_n(\lambda r)dr.$$

But we know from Bessel's equation

$$\frac{1}{r}\frac{d}{dr}\{(\lambda r)J_n'(\lambda r)\} = -(\lambda^2 - \frac{n^2}{r^2})J_n(\lambda r).$$

Therefore

$$\begin{aligned}
& H_n\{\frac{1}{r}\frac{d}{dr}(r\frac{df}{dr}) - \frac{n^2}{r^2}f(r)\} \\
&= -\int_0^\infty (\lambda^2 - \frac{n^2}{r^2})rf(r)J_n(\lambda r)dr - \int_0^\infty \frac{n^2}{r^2}(rf(r))J_n(\lambda r)dr \\
&= -\lambda^2 \int_0^\infty rf(r)J_n(\lambda r)dr \\
&= -\lambda^2 H_n\{f(r)\} \\
&= -\lambda^2 F_n(\lambda).
\end{aligned}$$

This result is very useful in solving partial differential equations in axisymmetric cylindrical configurations. We demonstrate this point by considering the following practical examples.

Example 3

Find the solution of the free vibration of a large circular membrane governed by the following initial-value problem.

$$\begin{aligned} u_{rr} + \frac{1}{r}u_r &= \frac{1}{c^2}u_{tt} \quad 0 < r < \infty \ , \quad t > 0. \\ u(r,0) &= f(r), \quad u_t(r,0) = g(r), \quad 0 < r < \infty. \end{aligned}$$

where $c^2 = \frac{Tg}{\omega}$ = constant, T is the tension in the membrane, g the acceleration due to gravity and ω is the weight per unit area of the membrane.

Solution

Application of the Hankel problem of order zero

$$U(\lambda, t) = \int_0^\infty ru(r,t)J_0(\lambda r)dr$$

to the vibration problem yields

$$\frac{d^2U}{dt^2} + k^2c^2U = 0$$

because $H_0\{\frac{1}{r}\frac{d}{dr}(ru_r)\} = -\lambda^2 U.$

The initial conditions become

$$\begin{aligned} U(r,0) &= F(\lambda) \quad \text{and} \\ U_t(r,0) &= G(\lambda). \end{aligned}$$

The general solution of the transformed system is

$$U(\lambda,t) = F(\lambda)\cos(c\lambda t) + \frac{G(\lambda)}{c\lambda}\sin(c\lambda t).$$

The inverse Hankel transformation gives

$$u(r,t) = \int_0^\infty \lambda F(\lambda)\cos(c\lambda t)J_0(\lambda r)d\lambda + \frac{1}{c}\int_0^\infty G(\lambda)\sin(c\lambda t)J_0(\lambda r)d\lambda. \tag{9.14}$$

This is the required solution.

In particular, we consider the following initial conditions

$$\begin{aligned} u(r,0) &= f(r) = \frac{A}{(1+\frac{r^2}{a^2})^{\frac{1}{2}}} \\ u_t(r,0) &= g(r) = 0 \end{aligned}$$

so that

$$\begin{aligned} G(\lambda) &= 0 \quad \text{and} \\ F(\lambda) &= Aa \int_0^\infty \frac{rJ_0(\lambda r)dr}{\sqrt{a^2+r^2}} = \frac{Aa}{\lambda} e^{-a\lambda} \end{aligned}$$

by means of Ex. 1(a).
Thus the solution (10.14) becomes

$$\begin{aligned} u(r,t) &= Aa \int_0^\infty e^{-a\lambda} \cos(c\lambda t) J_0(\lambda r) d\lambda \\ &= (Aa) Re \int_0^\infty e^{-\lambda(a-ict)} J_0(\lambda r) d\lambda \\ &= (Aa) Re\{ \frac{1}{\sqrt{r^2+(a-ict)^2}} \}. \end{aligned}$$

Note

$$\begin{aligned} F(\lambda) &= \int_0^\infty rf(r) J_0(\lambda r) dr \\ f(r) &= \int_0^\infty \lambda F(\lambda) J_0(\lambda r) d\lambda. \end{aligned}$$

Example 4

Find the steady-state solution of the axisymmetric acoustic radiation problem governed by the wave equation in cylindrical polar coordinates (r, θ, z):

$$\begin{aligned} u_{tt} &= c^2 \nabla^2 u, \qquad 0 < r < \infty, \ z > 0, \ t > 0 \\ u_z &= f(r,t) \quad \text{on } z = 0. \end{aligned}$$

where $f(r,t)$ is a given function and c is a constant. We also assume that the solution is bounded and behaves as outgoing spherical waves. This is referred to as the Sommerfield radiation.

Solution

We seek a steady-state solution of the radiation potential $u = e^{i\omega t}\phi(r,z)$ so that ϕ satisfies the Helmholtz equation

$$\phi_{rr} + \frac{1}{r}\phi_r + \phi_{zz} + \frac{\omega^2}{c^2}\phi = 0, \qquad 0 < r < \infty, \ z > 0,$$

with the boundary condition representing the normal velocity prescribed on the $z = 0$ plane:

$$\phi_z = f(r) \quad \text{on } z = 0$$

where $f(r)$ is a known function of r.
We solve the problem by the zero order Hankel transformation

$$\Phi(\lambda, z) = \int_0^\infty r J_0(\lambda r)\phi(r, z)dr$$

so that the given differential equation becomes

$$\begin{aligned} \Phi_{zz} &= k^2\Phi \qquad z > 0 \\ \Phi_z &= F(\lambda) \qquad \text{on } z = 0 \end{aligned}$$

where $k = (\lambda^2 - \frac{w^2}{c^2})^{\frac{1}{2}}$. The solution of this system is

$$\Phi(\lambda, z) = -\frac{1}{k}F(\lambda)e^{-kz}$$

where k is real and positive for $k > \frac{\omega}{c}$, and purely imaginary for $k < \frac{\omega}{c}$.
The inverse transformation yields the solution

$$\phi(r, z) = -\int_0^\infty F(\lambda)e^{-kz}(\frac{\lambda}{k})J_0(\lambda r)d\lambda.$$

Since the exact evaluation of this integral is difficult, we choose a simple form of $f(r)$ as

$$f(r) = \begin{cases} \alpha & 0 < r < a \\ 0 & r > a \end{cases}$$

where α is a given constant so that

$$\begin{aligned} F(\lambda) &= \int_0^\infty r f(r) J_0(\lambda r)dr \\ &= \alpha\int_0^a r J_0(\lambda r)dr \\ &= \frac{\alpha a}{\lambda}J_1(a\lambda). \end{aligned}$$

This result is obtained from Bessel's identity

$$\frac{d}{dx}(xJ_1(x)) = xJ_0(x).$$

Hence the solution for this special case is given by

$$\phi(r, z) = -\alpha a\int_0^\infty \frac{1}{k}e^{-kz}J_0(\lambda r)J_1(a\lambda)d\lambda.$$

This result can be simplified further if we seek the asymptotic evaluation of the integral. To do this, we express it in terms of the spherical polar coordinates (R, θ, ψ) where

$$\begin{aligned} x &= R\sin\theta\cos\psi \\ y &= R\sin\theta\sin\psi \\ z &= R\cos\theta \end{aligned}$$

combined with the asymptotic result

$$J_0(\lambda r) \sim (\frac{2}{\pi\lambda r})^{\frac{1}{2}} \cos(\lambda r - \frac{\pi}{4}) \quad \text{as } r \to \infty$$

so that the acoustic potential $u = e^{i\omega t}\phi$ is

$$u \sim -\frac{\alpha a\sqrt{2}e^{i\omega t}}{\sqrt{\pi R\sin\theta}} \int_0^\infty \frac{k^{-1}}{\sqrt{\lambda}} J_1(\lambda a)\cos(\lambda R\sin\theta - \frac{\pi}{4})e^{-kz}d\lambda.$$

This integral can be evaluated asymptotically for $R \to \infty$ by using the stationary phase approximation formula to obtain (see Rahman (1994))

$$u \sim \frac{\alpha \dot{a} c}{\omega R\sin\theta} J_1(k_1 a)e^{i(\omega t - \omega\frac{R}{c})}$$

where $k_1 = \frac{\omega}{(c\sin\theta)}$ is the stationary point. This solution represents the outgoing spherical waves with constant velocity c and decaying amplitude as $R \to \infty$.

9.3 The Mellin transform

Consider the Fourier transform pair in complex form:

$$\begin{aligned} F(\omega) &= \mathcal{F}\{f(t)\} = \int_{-\infty}^{\infty} f(t)e^{-i\omega t}dt \\ \text{and} \quad f(t) &= \frac{1}{2\pi}\int_{-\infty}^{\infty} F(\omega)e^{i\omega t}d\omega. \end{aligned}$$

It is possible to take the Fourier transform of the function $e^{-at}f(t), a > 0$ so that

$$\begin{aligned} \mathcal{F}\{f(t)e^{-at}\} &= \int_{-\infty}^{\infty}\{e^{-at}f(t)\}e^{-i\omega t}dt \\ &= \int_{-\infty}^{\infty} f(t)e^{-(a+i\omega)t}dt \\ &= F(a+i\omega) \end{aligned} \tag{9.15}$$

such that the inverse becomes

$$e^{-at}f(t) = \frac{1}{2\pi}\int_{-\infty}^{\infty} F(a+i\omega)e^{i\omega t}d\omega.$$

Therefore

$$f(t) = \frac{1}{2\pi}\int_{-\infty}^{\infty} F(a+i\omega)e^{(a+i\omega)t}d\omega. \tag{9.16}$$

Now put $a + i\omega = p \implies d\omega = \frac{dp}{i}$. Then (9.15) and (9.16) become

$$F(p) = \int_{-\infty}^{\infty} f(t)e^{-pt}dt \tag{9.17}$$

$$\text{and} \quad f(t) = \frac{1}{2\pi i}\int_{a-i\infty}^{a+i\infty} F(p)e^{pt}dp. \tag{9.18}$$

By substituting $e^{-t} = x \implies t = \ln(\frac{1}{x})$

$$\begin{aligned} -e^{-t}dt &= dx \\ dt &= -\frac{1}{x}dx \\ \text{and} \quad f(t) &= f(\ln\frac{1}{x}) = g(x) \end{aligned}$$

(9.17) and (9.18) become

$$F(p) = \int_0^\infty g(x)x^{p-1}dx = G(p) = \mathcal{M}\{g(x)\} \tag{9.19}$$

$$f(t) = g(x) = \frac{1}{2\pi i}\int_{a-i\infty}^{a+i\infty} G(p)x^{-p} = \mathcal{M}^{-1}\{G(p)\}. \tag{9.20}$$

The function $G(p)$ is called the Mellin transform of $g(x)$. The inverse Mellin transform is given by (9.20).

Properties of the Mellin transforms

We state the following operational properties of the Mellin transforms:
(I) The Mellin transform is a linear operator such that

$$\mathcal{M}\{af(x) + bg(x)\} = a\mathcal{M}\{f(x)\} + b\mathcal{M}\{g(x)\}$$

where a and b are arbitrary constants.
(II) Scaling property:

$$\mathcal{M}\{f(ax)\} = a^{-p}\mathcal{M}\{f(x)\} = a^{-p}F(p).$$

(III) Shifting property:

$$\mathcal{M}\{x^a f(x)\} = F(p+a).$$

Proof

$$\begin{aligned} \mathcal{M}\{x^a f(x)\} &= \int_0^\infty (x^a f(x))x^{p-1}dx \\ &= \int_0^\infty f(x)x^{(p+a)-1}dx \\ &= F(p+a) \qquad \text{as asserted.} \end{aligned}$$

(IV) Differentiation property:

$$\mathcal{M}\{f'(x)\} = -(p-1)F(p-1) \quad \text{provided} \quad [f(x)x^{p-1}]_0^\infty = 0.$$

Proof

$$\begin{aligned}\mathcal{M}\{f'(x)\} &= \int_0^\infty f'(x)x^{p-1}dx \\ &= [f(x)x^{p-1}]_0^\infty - (p-1)\int_0^\infty f(x)x^{p-2}dx \\ &= 0-(p-1)F(p-1) \\ &= -(p-1)F(p-1) \quad \text{as asserted.}\end{aligned}$$

Similarly we obtain

$$\begin{aligned}\mathcal{M}\{f''(x)\} &= (p-1)(p-2)F(p-2) \\ \mathcal{M}\{f^{(n)}(x)\} &= \frac{(-1)^n\Gamma(p)}{\Gamma(p-n)}F(p-n)\end{aligned}$$

provided

$$\lim_{x\to 0} x^{p-r-1}f^{(r)}(x) = 0, \quad r = 0,1,2,\ldots(n-1).$$

(V) Some more differentiation properties:

$$\begin{aligned}\mathcal{M}\{xf'(x)\} &= -pM\{f(x)\} = -pF(p) \quad \text{provided } [x^pf(x)]_0^\infty = 0 \\ \mathcal{M}\{x^2f''(x) &= (p+p^2)F(p) \\ \mathcal{M}\{(x\frac{d}{dx})^n f(x)\} &= (-1)^n p^n F(p), \quad n = 1,2,\ldots\end{aligned}$$

(VI) Convolution property:

$$\begin{aligned}\mathcal{M}[\int_0^\infty f(x\eta)g(\eta)d\eta] &= F(p)G(1-p) \\ \mathcal{M}[\int_0^\infty f(\frac{x}{\eta})g(\eta)\frac{d\eta}{\eta} &= F(p)G(p).\end{aligned}$$

Proof

$$\mathcal{M}[\int_0^\infty f(x\eta)g(\eta)d\eta] = \int_0^\infty x^{p-1}(\int_0^\infty f(x\eta)g(\eta)d\eta)dx.$$

Interchanging the order of integration, we obtain

$$\begin{aligned}\mathcal{M}[\int_0^\infty f(x\eta)g(\eta)d\eta] &= \int_{\eta=0}^\infty g(\eta)\{\int_{x=0}^\infty f(x\eta)x^{p-1}dx\}d\eta \\ &\qquad \text{put} \quad x\eta = y, \quad x = \tfrac{y}{\eta}, \quad \eta dx = dy \\ &= \int_0^\infty g(\eta)\eta^{-p}(\int_{y=0}^\infty f(y)y^{p-1}dy)d\eta \\ &= G(1-p)F(p) \\ &= F(p)G(1-p) \quad \text{as asserted.}\end{aligned}$$

(VII) Some more convolution properties:

If $F(p) = M\{f(x)\}$ and $G(p) = \mathcal{M}\{g(x)\}$ then the following convolution result holds:

$$\mathcal{M}\{f(x)g(x)\} = \frac{1}{2\pi i}\int_{a-i\infty}^{a+i\infty} F(S)G(p-S)dS.$$

In particular when $p = 1$, we obtain the Parseval's formula:

$$\int_0^\infty f(x)g(x)dx = \frac{1}{2\pi i}\int_{a-i\infty}^{a+i\infty} F(S)G(1-S)dS.$$

Proof

We know,

$$\mathcal{M}\{f(x)\} = \int_0^\infty f(x)x^{p-1}dx = F(p) \quad (9.21)$$

$$f(x) = \frac{1}{2\pi i}\int_{a-i\pi}^{a+i\pi} F(p)x^{-p}dp. \quad (9.22)$$

Multiplying (9.22) by $g(x)$ and taking the Mellin transform, we have now

$$\begin{aligned}\int_0^\infty f(x)g(x)x^{p-1}dx &= (\frac{1}{2\pi i})\int_{a-i\pi}^{a+i\pi} F(S)\{\int_0^\infty g(x)x^{p-S-1}dx\}ds \\ &= \frac{1}{2\pi i}\int_{a-i\pi}^{a+i\pi} F(S)G(p-S)dS \qquad \text{as asserted.}\end{aligned}$$

Example 5

Show that the Mellin transform of $(1+x)^{-1}$ is $\pi cosec \pi p, 0 < \text{Re } p < 1$.

Solution

$$\mathcal{M}\{(1+x)^{-1}\} = \int_0^\infty \frac{x^{p-1}}{1+x}dx.$$

Put $t = \frac{x}{1+x}$ such that

$$\begin{aligned}1-t &= 1-\frac{x}{1+x} = \frac{1}{1+x} \\ 1+x &= \frac{1}{1-t} \\ x &= \frac{1}{1-t}-1 = \frac{t}{1-t} \\ dx &= \frac{1}{(1-t)^2}dt.\end{aligned}$$

Thus we obtain

$$\begin{aligned}\mathcal{M}\{(1+x)^{-1}\} &= \int_0^1 (1-t)^{-1}t^{p-1}(1-t)^{-p+1}\frac{dt}{(1-t)^2}\\ &= \int_0^1 (1-t)^{-p}t^{p-1}dt.\end{aligned}$$

Consider now the standard Beta function

$$B(m,n) = \int_0^1 x^{m-1}(1-x)^{n-1}dx = \frac{\Gamma(m)\Gamma(n)}{\Gamma(m+n)}.$$

and hence,

$$\begin{aligned}\mathcal{M}\{(1+x)^{-1}\} &= B(1-p,p)\\ &= \frac{\Gamma(1-p)\Gamma(p)}{\Gamma(1)}\\ &= \Gamma(1-p)\Gamma(p)\\ &= \pi cosec\, \pi p, 0 < \text{Re } p < 1\end{aligned}$$

which is a standard result.

Example 6

Obtain the solution of the boundary value problem

$$\begin{aligned}x^2u_{xx} + xu_x + u_{yy} &= 0 \quad 0 \le x < \infty, 0 < y < 1.\\ u(x,0) &= 0 \quad \text{and}\\ u(x,1) &= \begin{cases}\alpha, & 0 \le x \le 1.\\ 0, & x > 1.\end{cases}\end{aligned}$$

where α is a constant.

Solution

We apply the Mellin transform

$$U(p,y) = \int_0^\infty u(x,y)x^{p-1}dx$$

to reduce the system in the form

$$\begin{aligned}\mathcal{M}\{x^2u_{xx} + xu_x + u_{yy}\} &= 0\\ (p+p^2)U(p,y) + (-pU(p,y)) + \frac{d^2}{dy^2}U(p,y) &= 0.\end{aligned}$$

Simplifying we obtain,

$$\frac{d^2}{dy^2}U + p^2U = 0, \quad 0 < y < 1$$

$$\begin{aligned} U(p,0) = 0 \quad \text{and} \quad U(p,1) &= \int_0^\infty \alpha x^{p-1} dx \\ &= \alpha \int_0^1 x^{p-1} dx \\ &= (\frac{\alpha}{p}). \end{aligned}$$

The solution of this boundary value problem is

$$U(p,y) = \frac{\alpha}{p}(\frac{\sin py}{\sin p}), \qquad 0 < \text{Re } p < 1.$$

The inverse Mellin transform yields

$$u(x,y) = \frac{\alpha}{2\pi i}\int_{a-i\infty}^{a+i\infty} \frac{x^{-p}}{p}\frac{\sin py}{\sin p} dp$$

where $U(p,y)$ is analytic in the vertical strip $0 < \text{Re } p < \pi$ and hence $0 < a < \pi$. The integral has simple poles at $p = n\pi, n = 1, 2, 3, \ldots$ which lie inside a semicircular contour inside the right-half plane. Application of the theory of residues gives the solution for $x > 1$,

$$u(x,y) = \frac{\alpha}{\pi}\sum_{n=1}^{\infty}\frac{(-1)^n x^{-n\pi}}{n}\sin n\pi y.$$

9.4 The Z-transform

Definition 1

When $f(t)$, a continuous function of time, is sampled at regular intervals of period T the usual Laplace transform techniques are modified. The diagrammatic form of a simple sample together with its associated input-output wave forms is depicted in the following Fig. 9.1.

Let us consider the sampling frequency to be $f_S = \frac{1}{T}$ = cycles/sec = Hertz. Here $f(t)$ is the continuous function and $f_T(t)$ is the discretized version of $f(t)$.
Defining the set of impulse functions $\delta_T(t)$ by

$$\delta_T(t) \equiv \sum_{n=0}^{\infty}\delta(t - nT) \tag{9.23}$$

the input-output relationship of the samples becomes

$$f_T(t) = f(t).\delta_T(t) = \sum_{n=0}^{\infty} f(t)\delta(t - nT) \tag{9.24}$$

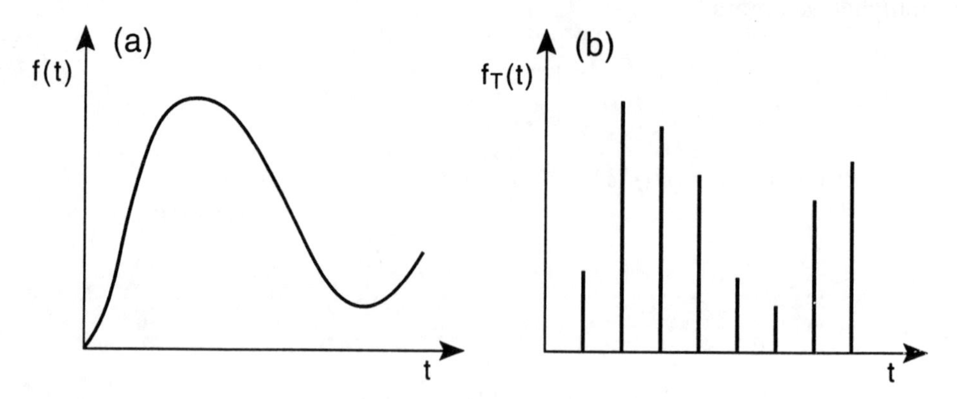

Figure 9.1: (a) Sketch of the continuous function $f(t)$ (b) Sketch of the discrete function $f_T(t)$

while for a given $f(t)$ and T, the $f_T(t)$ is unique, the converse is not true.

The Laplace transform can be used to define $f_T(t)$ as follows:

$$\begin{aligned} F_T(s) = \mathcal{L}\{f_T(t)\} &= \int_0^\infty f_T(t)e^{-st}dt \\ &= \int_0^\infty \sum_{n=0}^\infty f(t)\delta(t-nT)e^{-st}dt. \end{aligned}$$

Interchanging the integral and summation signs,

$$F_T(s) = \sum_{n=0}^\infty \int_0^\infty f(t)\delta(t-nT)e^{-st}dt = \sum_{n=0}^\infty f(nT)e^{-nTs}. \tag{9.25}$$

The variable z is introduced by means of the transformation

$$z = e^{Ts} \tag{9.26}$$

and since any function of s can now be replaced by a corresponding function of z, we have

$$F(z) = \sum_{n=0}^\infty f(nT)z^{-n} \tag{9.27}$$

$$\text{where} \quad F_T(s) = F(z) \tag{9.28}$$

$$\text{and} \quad s = \frac{1}{T}\ln z.$$

The Z operator can now be defined in terms of the Laplace operator by the relationship

$$\mathcal{Z}\{f(t)\} = \mathcal{L}\{f_T(t)\}. \tag{9.29}$$

Definition 2

An alternative definition (quoted without proof) is

$$\mathcal{Z}\{f(t)\} = \sum \text{residues of } [(\frac{1}{1 - z^{-1}e^{Ts}})F(s)].$$

The inverse Z transform is

$$f(t) = \mathcal{Z}^{-1}\{F(z)\} = \frac{1}{2\pi i}\oint_{r-i\infty}^{r-i\infty} F(z)z^{n-1}dz$$

where the contour of integration encloses all the singularities of the integrand.

Definition 3

We know the Laplace transformation of $f(t)$ is given by

$$\mathcal{L}\{f(t)\} = \int_0^\infty f(t)e^{-st}dt = F(s)$$

which is valid when the given function is continuous in the interval between 0 and ∞. Now if $f(t)$ is given to be discrete in that interval then

$$\int_0^\infty f(t)e^{-st}dt \approx \sum_{n=0}^{\infty} f(nT)e^{-snT}$$

where T = sampling interval, $t = nT$ and $e^{sT} = z$ and $\Delta t \to 1$
Then

$$\int_0^\infty f(t)e^{-st}dt \approx \sum_{n=0}^{\infty} f(nT)z^{-n}.$$

Note that when the sampling interval is very large then

$$\Delta t = \frac{\text{total length of time}}{\text{number of intervals}} \Longrightarrow 1.$$

Thus under these circumstances the

$$\mathcal{L}\{f(t)\} = \int_0^\infty f(t)e^{-st}dt \approx \sum_{n=0}^{\infty} f(nT)z^{-n} = \mathcal{Z}\{f(t)\}.$$

Hence the Z-transform of $f(t) = f(nT)$ is given by

$$\mathcal{Z}\{f(t)\} = \sum_{n=0}^{\infty} f(nT)z^{-n} = F(z),$$

where z is the complex number, T is the sampling interval and the transformation goes from t-domain to z-domain.

9.4.1 Deduction of some elementary Z-transforms

Example 7

Find the Z-transform of the function

$$f(t) = 1 = u(t) \quad t \geq 0.$$

Solution

$$\begin{aligned}
\mathcal{Z}\{f(t)\} &= \sum_{n=0}^{\infty} f(nT)z^{-n} \\
&= \sum_{n=0}^{\infty} (1)z^{-n} \\
&= 1 + \frac{1}{z} + \frac{1}{z^2} + \dots \\
&= (1 - \frac{1}{z})^{-1} \qquad |z| > 1 \text{ for convergence} \\
&= (\frac{z}{z-1}).
\end{aligned}$$

Note that

$$\mathcal{Z}\{f(t)\} = \mathcal{Z}(u(t-1)) = \sum_{n=1}^{\infty} z^{-n} = -1 + \frac{z}{z-1} = \frac{1}{z-1}.$$

Example 8

Find the Z-transform of the function

$$f(t) = t \qquad t \geq 0.$$

Solution

$$\begin{aligned}
\mathcal{Z}\{f(t)\} &= \sum_{n=0}^{\infty} (nT)z^{-n} \\
&= T\sum_{n=0}^{\infty} nz^{-n} \\
&= T[z^{-1} + 2z^{-2} + 3z^{-3} + \dots] \\
&= \frac{T}{z}[1 + 2z^{-1} + 3z^{-2} + \dots], \quad |z| > 1
\end{aligned}$$

$$= \frac{T}{z}(1 - z^{-1})^{-2}$$
$$= \frac{Tz}{(z-1)^2}.$$

Note: A convergent series can be differentiated term by term.

Example 9

Find the Z-transform of the function

$$f(t) = e^{2t} \qquad t \geq 0.$$

Solution

$$\begin{aligned}
\mathcal{Z}\{f(t)\} &= \sum_{n=0}^{\infty} e^{2nT} z^{-n} \\
&= \sum_{n=0}^{\infty} (e^{2T} z^{-1})^n \\
&= 1 + e^{2T} z^{-1} + e^{4T} z^{-2} + \ldots, \qquad |z| > e^{2T} \\
&= (1 - \frac{e^{2T}}{z})^{-1} \\
&= \frac{z}{z - e^{2T}}.
\end{aligned}$$

Theorem 1

Find the Z-transform of the function $f(t - kT)$ if $f(t) = 0$ for $t < 0$ and $\mathcal{Z}\{f(t)\} = F(z)$.

Solution

$$\begin{aligned}
\mathcal{Z}\{f(t - kT)\} &= \sum_{n=0}^{\infty} f(nT - kT) z^{-n} \\
&= \sum_{n=0}^{\infty} f\{(n-k)T\} z^{-n} \\
&= \sum_{n=k}^{\infty} f\{(n-k)T\} z^{-n}.
\end{aligned}$$

Changing n to $n+k$, we obtain

$$\begin{aligned}\mathcal{Z}\{f(t-kT)\} &= \sum_{n=0}^{\infty} f(nT)z^{-n-k} \\ &= z^{-k}\sum_{n=0}^{\infty} f(nT)z^{-n} \\ &= z^{-n}\mathcal{Z}\{f(t)\} \\ &= z^{-k}F(z).\end{aligned}$$

Theorem 2

Given that $\mathcal{Z}\{f(t)\} = F(z)$ and $f(t) = 0$ for $t < 0$, then find $\mathcal{Z}\{f(t+kT)\}$.

Solution

$$\begin{aligned}\mathcal{Z}\{f(t+kT)\} &= \sum_{n=0}^{\infty} f(nT+kT)z^{-n} \\ &= \sum_{n=0}^{\infty} f((n+k)T)z^{-n}.\end{aligned}$$

Consider for $k = 1$,

$$\begin{aligned}\mathcal{Z}\{f(t+kT)\} &= \sum_{n=0}^{\infty} f((n+1)T)z^{-n} \\ &= -zf(0) + \sum_{n=-1}^{\infty} f((n+1)T)z^{-n}.\end{aligned}$$

Changing n to $n-1$;

$$\begin{aligned}&= -zf(0) + \sum_{n=0}^{\infty} f(nT)z^{-n+1} \\ &= -zf(0) + z\sum_{n=0}^{\infty} f(nT)z^{-n} \\ &= -zf(0) + z\mathcal{Z}\{f(t)\}.\end{aligned}$$

Thus

$$\mathcal{Z}\{f((n+1)T)\} = zF(z) - zf(0).$$

In a similar manner, we can develop for $n = 2$ which is

$$\mathcal{Z}\{f((n+2)T)\} = z\mathcal{Z}\{f((n+1)T)\} - zf(1)$$

$$
\begin{aligned}
&= z[z\mathcal{Z}\{f(nT)\} - zf(0)] - zf(1) \\
&= z^2\mathcal{Z}\{f(nT)\} - z^2 f(0) - zf(T).
\end{aligned}
$$

Hence extending this idea to k factor,

$$\mathcal{Z}\{f((n+k)T)\} = z^k\mathcal{Z}\{f(nT)\} - [z^k f(0) + z^{k-1}f(T) + \ldots + zf((k-1)T)].$$

Note

Thus we have

$$\mathcal{Z}\{f(t-kT)\} = z^{-k}F(z)$$

and

$$\mathcal{Z}\{f(t+kT)\} = z^k F(z) - [z^k f(0) + z^{k-1}f(T) + \ldots + zf((k-1)T)].$$

Formula 1

$$\mathcal{Z}\{\sin\omega t\} = \frac{z\sin\omega T}{z^2 - 2z\cos\omega T + 1}.$$

Formula 2

$$\mathcal{Z}\{\cos\omega t\} = \frac{1}{2}\,\frac{z(z-\cos\omega T)}{z^2 - 2z\cos\omega T + 1}.$$

Proof of formulas (1) and (2)

Consider the function $f(t) = e^{i\omega t}$ such that $Re\{f(t)\} = \cos\omega t$ and $Im\{f(t)\} = \sin\omega t$.

Then the Z-transform is

$$
\begin{aligned}
\mathcal{Z}\{f(t)\} &= Z(e^{i\omega t}) \\
&= \sum_{n=0}^{\infty} e^{i\omega nT} z^{-n} \\
&= \sum_{n=0}^{\infty} (\frac{e^{i\omega T}}{z})^n \\
&= [1 - \frac{e^{i\omega T}}{z}]^{-1} \\
&= \frac{z}{z - e^{i\omega T}} \\
&= \frac{z}{(z-\cos\omega T) - i\sin\omega T} \\
&= \frac{z(z-\cos\omega T + i\sin\omega T)}{(z-\cos\omega T)^2 + \sin^2\omega t} \\
&= \frac{z(z-\cos\omega T) + iz\sin\omega T}{z^2 - 2z\cos\omega T + 1}.
\end{aligned}
$$

Thus

$$\begin{aligned}\mathcal{Z}(\cos\omega t) &= \frac{z(z\cos\omega T)}{z^2 - 2z\cos\omega T + 1}\\ \mathcal{Z}(\sin\omega t) &= \frac{z\sin\omega T}{z^2 - 2z\cos\omega T + 1}.\end{aligned}$$

Formula 3

$$\mathcal{Z}\{\cosh\omega T\} = \frac{z(z - \cosh\omega T)}{z^2 - 2z\cosh\omega T + 1}.$$

Formula 4

$$\mathcal{Z}\{\sinh\omega T\} = \frac{z\sinh\omega T}{z^2 - 2z\cosh\omega T + 1}.$$

Proof of formulas (3) and (4)

Consider the function

$$f(t) = e^{\omega t} = \cosh\omega t + \sinh\omega t.$$

Then Z-transform is

$$\begin{aligned}\mathcal{Z}\{f(t)\} &= \mathcal{Z}\{e^{\omega T}\}\\ &= \sum_{n=0}^{\infty}(e^{\omega nT})z^{-n}\\ &= \sum_{n=0}^{\infty}(\frac{e^{\omega T}}{z})^n\\ &= (1 - \frac{e^{\omega T}}{z})^{-1}\\ &= \frac{z}{z - e^{\omega T}}\\ &= \frac{z}{(z - \cosh\omega T) - \sinh\omega T}\\ &= \frac{z(z - \cosh\omega T) + z\sinh\omega t}{z^2 - 2z\cosh\omega T + 1}\end{aligned}$$

$$\begin{aligned}\mathcal{Z}(\cosh\omega t) &= \frac{z(z - \cosh\omega T)}{z^2 - 2z\cosh\omega T + 1}\\ \mathcal{Z}(\sinh\omega T) &= \frac{z\sinh\omega T}{z^2 - 2z\cosh\omega T + 1}.\end{aligned}$$

Example 10

Find $\mathcal{Z}\{\delta(t)\}$ where $\delta(t) = \lim_{T\to 0} \frac{u(t)-u(t-T)}{T}$.

Solution

$$\begin{aligned}
\mathcal{Z}\{\delta(t)\} &= z\{\delta(nT)\} \\
&= \sum_{n=0}^{\infty} \delta(nT) z^{-n} \\
&= \sum_{n=0}^{\infty} \frac{1}{T}[u(nT) - u((n-1)T)] z^{-n} \\
&= \frac{1}{T} \sum_{n=0}^{\infty} \{u(nT) - u((n-1)T)\} z^{-n} \\
&= \frac{1}{T}[\sum_{n=0}^{\infty} z^{-n} - \sum_{n=1}^{\infty} z^{-n}] \\
&= \frac{1}{T}.
\end{aligned}$$

Example 11

Find $\mathcal{Z}\{\delta(t - mT)\}$.

Solution

$$\begin{aligned}
\mathcal{Z}\{\delta(t - mT)\} &= \sum_{n=0}^{\infty} \delta(nT - mT) z^{-n} \\
&= \sum_{n=0}^{\infty} \delta((n-m)T) z^{-n} \\
&= \sum_{n=0}^{\infty} \frac{1}{T}[u((n-m)T) - u((n-m-1)T)] z^{-n} \\
&= \frac{1}{T}\{\sum_{n=m}^{\infty} z^{-n} - \sum_{n=m+1}^{\infty} z^{-n}\} \\
&= \frac{1}{T} z^{-m}.
\end{aligned}$$

Formula 5

Prove that $\mathcal{Z}\{a(u(t) - u(t-T))\} = a$.

Proof Taking Z-transform

$$\begin{aligned}
\mathcal{Z}\{a(u(t) - u(t-T))\} &= a\sum_{n=0}^{\infty}[u(nT) - u((n-1)T)]z^{-n} \\
&= a[\sum_{n=0}^{\infty} z^{-n} - \sum_{n=1}^{\infty} z^{-n}] \\
&= a.
\end{aligned}$$

Formula 6

If $f(t) = a^{\frac{(t-T)}{T}}u(t-T)$ prove that

$$\mathcal{Z}\{f(t)\} = (\frac{1}{z-a}).$$

Proof

$$\begin{aligned}
\mathcal{Z}\{f(t)\} &= \mathcal{Z}\{a^{\frac{(t-T)}{T}}u(t-T)\}, \qquad t = nT \\
&= \sum_{n=0}^{\infty} a^{n-1}u((n-1)T)z^{-n} \\
&= \sum_{n=1}^{\infty} a^{n-1}z^{-n}.
\end{aligned}$$

Changing n to $n+1$:

$$\begin{aligned}
&= \sum_{n=0}^{\infty} a^{n}z^{-n}z^{-1} \\
&= z^{-1}\sum_{n=0}^{\infty}(\frac{a}{z})^{n} \\
&= (z^{-1})(1-\frac{a}{z})^{-1} \\
&= \frac{1}{z-a}.
\end{aligned}$$

9.4.2 Difference equation

From the differential equation we can form the difference equation as follows: The forward difference is

$$y'(t) = \frac{y(t+T) - y(t)}{T}$$

$$
\begin{aligned}
y''(t) &= \frac{y'(t+T) - y'(t)}{T} \\
&= \frac{y(t+2T) - 2y(t+T) = y(t)}{T^2}.
\end{aligned}
$$

$$
\begin{aligned}
\text{If} \quad T = 1 \qquad y'(t) &= y(t+2) - 2y(t+1) + y(t) \\
&= y_{n+2} - 2y_{n+1} + y_n.
\end{aligned}
$$

Example 12

Consider the equation

$$y_{n+2} - 6y_{n+1} + 9y_n = 0$$

with initial conditions $y_0 = 5, \quad y_1 = 12$.

Solution

Let $y_n = \lambda^n$ be a trial solution, where λ is a parameter.
The characteristic equation is

$$\lambda^2 - 6\lambda + 9 = 0.$$

Therefore $\lambda = 3$ is a double root.
The general solution is, therefore,

$$y_n = c_1 3^n + c_2 n 3^n = (c_1 + n c_2) 3^n.$$

Now using the initial conditions, we obtain

$$
\begin{aligned}
y_0 &= c_1 = 5, \quad y_1 = 12 = (5 + c_2)3 \Longrightarrow c_2 = 4 - 5 = -1 \\
y_n &= (5 - n)3^n.
\end{aligned}
$$

By the Z-transform method, we obtain

$$
\begin{aligned}
\mathcal{Z}(y_{n+2}) - 6\mathcal{Z}(y_{n+1}) + 9\mathcal{Z}(y_n) &= 0 \\
\{z^2 \mathcal{Z}(y_n) - z^2 y_0 - z y_1\} - 6\{z\mathcal{Z}(y_n) - z y_0\} + 9\mathcal{Z}(y_n) &= 0 \\
(z^2 - 6z + 9)\mathcal{Z}(y_n) - (z^2 - 6z)y_0 - y_1 z &= 0 \\
(z^2 - 6z + 9)\mathcal{Z}(y_n) &= 5(z^2 - 6z) + 12z \\
&= 5z^2 - 18z.
\end{aligned}
$$

Hence,

$$
\begin{aligned}
\mathcal{Z}\{y_n\} &= \frac{5z^2 - 18z}{(z^2 - 6z + 9)} \\
&= z\frac{(5z - 18)}{(z-3)^2} = z[\frac{5(z-3) - 3}{(z-3)^2}].
\end{aligned}
$$

Therefore,

$$\begin{aligned} y_n &= \mathcal{Z}^{-1}[\frac{5z}{(z-3)} - \frac{3z}{(z-3)^2}] \\ &= (5-t)3^t \end{aligned}$$

because $\mathcal{Z}\{a^t\} = \frac{z}{z-a}$ and $\mathcal{Z}\{t.a^t\} = \frac{az}{(z-a)^2}$.

9.4.3 Methods of evaluating inverse Z-transforms

The Z-transform pair is given by

$$\begin{aligned} \text{If } \ \mathcal{Z}\{f(t)\} &= f(z) = \sum_{n=0}^{\infty} f(nT)z^{-n} \\ \text{then, } \ f(t) &= \mathcal{Z}^{-1}\{F(z)\} = \frac{1}{2\pi i}\int_{r-i\infty}^{r+i\infty} F(z)z^{n-1}dz. \end{aligned}$$

(I) Cauchy's residue theorem

By using the Cauchy's residue theorem, we obtain for $t = nT$

$$f(t) = \sum_{\text{all } z_k} [\text{residues of } F(z)z^{n-1} \text{ at } z = z_k]$$

where the z_k defines all the poles of $F(z)z^{n-1}$ inside the closed contour known as a Bromwich Contour.

(II) Partial fraction

Expand $\{\frac{F(z)}{z}\}$ into partial fractions. The product of z with each of the partial fractions will then be recognizable from the standard forms in the table of Z-transforms. Note however that the continuous functions obtained are only valid at the sampling instants.

(III) Power series expansion

$F(z)$ is expanded into a power series in z^{-1} and the coefficient of the term in z^{-n} is the value of $F(nT)$, i.e., the value of $f(t)$ at the n-th sampling instant.

Example 13

In constructing a mathematical model of population, it is assumed that the probability P_n that a couple produced exactly n offsprings satisfies the difference equation

$$P_n = 0.7P_{n-1}.$$

Find P_n in terms of P_0 and determine P_0 from the fact that

$$P_0 + P_1 + P_2 + \ldots = 1.$$

Solution

The given difference equation is

$$P_n - 0.7P_{n-1} = 0. \tag{9.30}$$

Let us consider

$$P_n = C\lambda^n. \tag{9.31}$$

A solution where λ a parameter to be determined. Using (9.31) into (9.30) we obtain $\lambda^n - \frac{7}{10}\lambda^{n-1} = 0$ or $\lambda = \frac{7}{10}$. Thus a general solution is $P_n = C(\frac{7}{10})^n$. Thus for $n = 0$ the solution yields $P_0 = C$. Hence $P_n = P_0(\frac{7}{10})^n$. Given that $\sum_{n=0}^{\infty} P_n = 1$ and so $P_0 \sum_{n=0}^{\infty}(\frac{7}{10})^n = 1$ which reduces to $P_0 = \frac{3}{10}$. Thus the solution is $P_n = (\frac{3}{10})(\frac{7}{10})^n$.

Example 14

An alternative model to Example 8 is given by

$$P_n = \frac{1}{n}P_{n-1}.$$

Find P_n in terms of P_0 and prove that $P_0 = \frac{1}{e}$.

Solution

$$P_n = \frac{1}{n}P_{n-1}.$$

Thus we can obtain

$$\begin{aligned}
P_1 &= \frac{1}{1}P_0 \\
P_2 &= \frac{1}{2}P_1 = \frac{1}{2!}P_0 \\
P_3 &= \frac{1}{3}P_2 = \frac{1}{3!}P_0 \\
P_n &= \frac{1}{n!}P_0.
\end{aligned}$$

We know

$$\begin{aligned}
\sum_{n=0}^{\infty} P_n &= 1 \\
P_0 \sum_{n=0}^{\infty} \frac{1}{n!} &= 1 \quad (\text{since} \quad e^x = 1 + x + \frac{x^2}{2!} + \ldots) \\
P_n &= \frac{1}{n!}(\frac{1}{e}) \\
P_0 e^1 &= 1.
\end{aligned}$$

$$\text{Therefore} \qquad P_0 = \frac{1}{e}.$$

Example 15

Solve the difference equation

$$y(t+1) + 2y(t) = 1, \quad y(0) = 0.$$

Solution

Taking the Z-transform of the given difference equation, we obtain

$$\begin{aligned}
\mathcal{Z}\{y_{n+1}\} + 2\mathcal{Z}\{y_n\} &= \mathcal{Z}\{1\} \\
z\mathcal{Z}\{y_n\} - zy_0 + 2\mathcal{Z}\{y_n\} &= \frac{z}{z-1} \\
(z+2)\mathcal{Z}\{y_n\} &= \frac{z}{z-1} \\
\mathcal{Z}\{y_n\} &= \frac{z}{(z-1)(z+2)}.
\end{aligned}$$

By using the partial fraction method,

$$\mathcal{Z}\{y_n\} = \frac{2}{3}(\frac{1}{z+2}) + \frac{1}{3}(\frac{1}{z-1}).$$

Then the inverse yields

$$y_n = \frac{2}{3}\mathcal{Z}^{-1}(\frac{1}{z+2}) + \frac{1}{3}\mathcal{Z}^{-1}(\frac{1}{z-1}).$$

Using formula 6, we have

$$\begin{aligned}
y_n &= (\frac{2}{3})(-2)^{t-1}u(t-1) + \frac{1}{3}(1)^{t-1}u(t-1) \\
&= [\frac{2}{3}(-2)^{t-1} + \frac{1}{3}]u(t-1).
\end{aligned}$$

Thus

$$y(t) = \{\frac{2}{3}(-2)^{t-1} + \frac{1}{3}\}u(t-1).$$

Example 16

Solve the difference equation

$$y(t+1) + 2y(t) = 0; \quad y(0) = 4.$$

Solution

Taking the Z-transform of the given equation

$$\begin{aligned} \mathcal{Z}\{y(t+1)+2y(t)\} &= 0 \\ z\mathcal{Z}y(t)-zy(0)+2\mathcal{Z}y(t) &= 0 \\ (z+2)\mathcal{Z}\{y(t)\} &= 4z \\ \mathcal{Z}\{y(t)\} &= \frac{4z}{z+2} = 4 - \frac{8}{z+2}. \end{aligned}$$

Taking the inverse, we obtain

$$\begin{aligned} y(t) &= 4\mathcal{Z}^{-1}(1) - 8\mathcal{Z}^{-1}(\frac{1}{z+2}) \\ &= 4[u(t)-u(t-1)] - 8(-2)^{t-1}u(t-1) \\ &= 4u(t) - (4+8.(-2)^{t-1})u(t-1) \end{aligned}$$

which can be expressed as

$$y(t) = \begin{cases} 4 & t=0 \\ 4-(4+8(-2)^{t-1}), & t=1,2,3,\ldots \end{cases}$$

$$\text{or,} \quad y(t) = \begin{cases} 4 & t=0 \\ 4(-2)^t & t=1,2,3,\ldots \end{cases}$$

which is the required solution.

Example 17

In the study of infectious diseases a record is kept of the outbreaks of measles in a particular school. It is estimated that the probability of at least one new infection occurring in the n-th week after an outbreak is

$$P_n = P_{n-1} - \frac{1}{5}P_{n-2}.$$

(a) If $P_0 = 0$ and $P_1 = 1$, what is P_n?

(b) After how many weeks will the probability of occurrence of a new case of measles be less than 10%?

Solution

Let the trial solution be $P_n = \lambda^n$. Then the characteristic equation is

$$\lambda^2 - \lambda + \frac{1}{5} = 0$$

$$\text{or} \quad \lambda = \frac{1 \pm \sqrt{1-\frac{4}{5}}}{2} = \frac{1 \pm \sqrt{\frac{1}{5}}}{2}.$$

Hence the general solution is

$$P_n = C_1(\frac{1+\sqrt{\frac{1}{5}}}{2})^n + C_2(\frac{1-\sqrt{\frac{1}{5}}}{2})^n$$

$$n=0: \qquad P_0 = 0 = C_1 + C_2$$

$$n=1: \qquad P_1 = 1 = C_1\frac{1+\sqrt{\frac{1}{5}}}{2} + C_2\frac{1-\sqrt{\frac{1}{5}}}{2}.$$

Solving for C_1 and C_2 we obtain $C_1 = \sqrt{5}$ and $C_2 = -\sqrt{5}$.
Thus
(a) $P_n = \sqrt{5}\{(\frac{1+\sqrt{\frac{1}{5}}}{2})^n - (\frac{1-\sqrt{\frac{1}{5}}}{2})^n\}$

(b) when $P_n = 10\%$, then $\frac{1}{10} = \sqrt{5}\{(\frac{1+\sqrt{\frac{1}{5}}}{2})^n - (\frac{1-\sqrt{\frac{1}{5}}}{2})^n\}$.
Simplifying

$$\begin{aligned} \frac{2^n}{10\sqrt{5}} &= (1+\sqrt{\frac{1}{5}})^n - (1-\sqrt{\frac{1}{5}})^n \\ &\approx (1+n\sqrt{\frac{1}{5}} - 1 + n\sqrt{\frac{1}{5}}) \\ &= 2n\sqrt{\frac{1}{5}} \\ \frac{2^n}{2n} &= 10. \end{aligned}$$

Hence

$$2^n = 20n \qquad \text{which yields } n \approx 7.$$

Thus approximately 7 weeks after the outbreak, there is a probability of occurrence of a new case of measles.

Bibliography

[1] Ahlfors, L.V.(1966). *Complex Analysis*, 2nd ed., McGraw-Hill Book Company Inc., New York.

[2] Bromwell, A.(1953). *Advanced Mathematics in Physics and Engineering*, McGraw-Hill, New York, pp. 386-413.

[3] Chestnut, H. and Meyer, R.W.(1951). *Servo-mechanisms and Regulating System Design*, John Wiley, New York.

[4] Churchill, R.V.(1972). *Operational Mathematics*, 3rd ed., McGraw-Hill, New York.

[5] Churchill, R.V.(1963). *Fourier Series and Boundary Value Problems*, 2nd ed., McGraw-Hill, New York.

[6] Churchill, R.V. and Brown, J.W.(1990). *Complex Variables and Applications* McGraw-Hill, New York.

[7] Copson, E.T.(1957). *Theory of Functions of a Complex Variable*, Oxford University Press, London.

[8] Curle, N. and Davies, H.(1968). *Modern Fluid Dynamics*, Vol. 1. Van Nostrand, London.

[9] Dettman, J.W.(1965). *Applied Complex Variables*, Macmillan Company, New York.

[10] Dirac, P.M.A.(1935). *The Principles of Quantum Mechanics*, Oxford University Press, Oxford.

[11] Kellogg, O.D.(1973). *Foundations of Potential Theory*, Dover Publications, New York.

[12] Lamb, H.(1945). *Hydrodynamics*, Dover Publications, New York.

[13] Lighthill, J.(1986). *An Informal Introduction to Theoretical Fluid Mechanics*, Oxford University Press, Oxford.

[14] Love, A.E.H.(1944). *Elasticity*, Dover Publications, New York.

[15] McLachlan, N.W.(1963). *Complex Variable Theory and Transform Calculus*, Cambridge University Press, Cambridge.

[16] Milne-Thomson, L.M.(1955). *Theoretical Hydrodynamics*, Macmillan & Co. Ltd., London.

[17] Rahman, M.(1988). *The Hydrodynamics of Waves and Tides, with Applications: Topics in Engineering*, Vol. 4, Computational Mechanics Publications, Southampton, U.K.

[18] Rahman, M.(1994). *Water Waves: Relating Modern Theory to Advanced Engineering Applications*, Oxford University Press, Oxford.

[19] Rahman, M.(1997). *Complex Variables and Transform Calculus*, Computational Mechanics Publications, Southampton, U.K.

[20] Sneedon, I.N.(1951). *Fourier Transforms*, McGraw-Hill, New York.

[21] Thaler, G.J. and Brown, R.G.(1960). *Analysis and Design of Feedback Control Systems*, McGraw-Hill, New York.

[22] Titchmarsh, E.C.(1939). *Theory of Functions*, Oxford University Press, London.

[23] Whittaker, E.T. and Watson, G.N.(1950). *Modern Analysis*, Cambridge University Press, London.

[24] Wylie, C.R. and Barrett, L.C.(1982). *Advanced Engineering Mathematics*, McGraw-Hill, New York.

Index